Spatial Analysis

WILEY SERIES IN PROBABILITY AND STATISTICS

Established by *Walter A. Shewhart and Samuel S. Wilks*

The ***Wiley Series in Probability and Statistics*** is well established and authoritative. It covers many topics of current research interest in both pure and applied statistics and probability theory. Written by leading statisticians and institutions, the titles span both state-of-the-art developments in the field and classical methods.

Reflecting the wide range of current research in statistics, the series encompasses applied, methodological and theoretical statistics, ranging from applications and new techniques made possible by advances in computerized practice to rigorous treatment of theoretical approaches.

This series provides essential and invaluable reading for all statisticians, whether in academia, industry, government, or research.

Spatial Analysis

John T. Kent
University of Leeds, UK

Kanti V. Mardia
University of Leeds, UK
University of Oxford, UK

WILEY

This edition first published 2022

Registered Offices
John Wiley & Sons, Inc., 111 River Street, Hoboken, NJ 07030, USA
John Wiley & Sons Ltd, The Atrium, Southern Gate, Chichester, West Sussex, PO19 8SQ, UK

Editorial Office
9600 Garsington Road, Oxford, OX4 2DQ, UK

For details of our global editorial offices, customer services, and more information about Wiley products visit us at www.wiley.com.

Wiley also publishes its books in a variety of electronic formats and by print-on-demand. Some content that appears in standard print versions of this book may not be available in other formats.

Library of Congress Cataloging-in-Publication Data applied for

ISBN 9780471632054 (hardback); ISBN 9781118763568 (adobe pdf);
ISBN 9781118763575 (epub); 9781118763551 (obook)

Cover Design: Wiley
Cover Images: Courtesy of John Kent and Kanti Mardia

Set in 9.5/12.5pt STIXTwoText by Straive, Chennai, India
Printed and bound by CPI Group (UK) Ltd, Croydon, CR0 4YY

C9780471632054_250422

To my wife ***Sue*** *for all her patience during the many years it has taken to complete the book*

(John Kent)

To my son ***Hemant*** *and daughter-in-law* ***Preeti***
— with Jainness

(Kanti Mardia)

*"Whatever there is in all the three worlds, which are possessed of moving and non-moving beings, cannot exist as apart from the **'Ganita'** (mathematics/statistics)."*

Acharya Mahavira (Jain monk, AD 850)

Contents

List of Figures

List of Tables

Preface

Spatial statistics is concerned with data collected at various spatial locations or sites, typically in a Euclidean space $\mathbb{R}^d$, $d \geq 1$. The important cases in practice are $d = 1,2,3$, corresponding to the data on the line, in the plane, or in 3-space, respectively. A common property of spatial data is "spatial continuity," which means that measurements at nearby locations will tend to be more similar than measurements at distant locations. Spatial continuity can be modeled statistically using a covariance function of a stochastic process for which observations at nearby sites are more highly correlated than at distant sites. A stochastic process in space is also known as a random field.

One distinctive feature of spatial statistics, and related areas such as time series, is that there is typically just one realization of the stochastic process to analyze. Other branches of statistics often involve the analysis of independent replications of data.

The purpose of this book is to develop the statistical tools to analyze spatial data. The main emphasis in the book is on Gaussian processes. Here is a brief summary of the contents. A list of Notation and Terminology is given at the start for ease of reference. An introduction to the overall objectives of spatial analysis, together with some exploratory methods, is given in Chapter 1. Next is the specification of possible covariance functions (Chapter 2 for the stationary case and Chapter 3 for the intrinsic case). It is helpful to distinguish discretely indexed, or lattice, processes from continuously indexed processes. In particular, for lattice processes, it is possible to specify a covariance function through an autoregressive model (the SAR and CAR models of Chapter 4), with specialized estimation procedures (Chapter 6). Model fitting through maximum likelihood and related ideas for continuously indexed processes is covered in Chapter 5. An important use of spatial models is kriging, i.e. the prediction of the process at a collection of new sites, given the values of the process at a collection of training sites (Chapter 7), and in particular the links to machine learning are explained. Some additional topics, for which there was not space for in the book, are summarized in Chapter 8. The technical mathematical tools have been collected in Appendix A for ease of reference. Appendix B contains a short historical review of the spatial linear model.

The development of statistical methodology for spatial data arose somewhat separately in several academic disciplines over the past century.

(a) *Agricultural field trials.* An area of land is divided into long, thin plots, and different crop is grown on each plot. Spatial correlation in the soil fertility can cause spatial correlation in the crop yields (Webster and Oliver, 2001).
(b) *Geostatistics.* In mining applications, the concentration of a mineral of interest will often show spatial continuity in a body of ore. Two giants in the field of spatial analysis came out of this field. Krige (1951) set out the methodology for spatial prediction (now known as kriging) and Matheron (1963) developed a comprehensive theory for stationary and intrinsic random fields; see Appendix B.
(c) *Social and medical science.* Spatial continuity is an important property when describing characteristics that vary across a region of space. One application is in geography and environmetrics and key names include Cliff and Ord (1981), Anselin (1988), Upton and Fingleton (1985, 1989), Wilson (2000), Lawson and Denison (2002), Kanevski and Maignan (2004), and Schabenberger and Gotway (2005). Another application is in public health and epidemiology, see, e.g., Diggle and Giorgi (2019).
(d) *Splines.* A very different approach to spatial continuity has been pursued in the field of nonparametric statistics. Spatial continuity of an underlying smooth function is ensured by imposing a *roughness penalty* when fitting the function to data by least squares. It turns out that fitted spline is identical to the kriging predictor under suitable assumptions on the underlying covariance function. Key names here include Wahba (1990) and Watson (1984). A modern treatment is given in Berlinet and Thomas-Agnan (2004).
(e) *Mainstream statistics.* From at least the 1950s, mainstream statisticians have been closely involved in the development of suitable spatial models and suitable fitting procedures. Highlights include the work by Whittle (1954), Matérn (1960, 1986), Besag (1974), Cressie (1993), and Diggle and Ribeiro (2007).
(f) *Probability theory and fractals.* For the most part, statisticians interested in asymptotics have focused on "outfill" asymptotics – the data sites cover an increasing domain as the sample size increases. The other extreme is "infill asymptotics" in which the interest is on the local smoothness of realizations from the spatial process. This infill topic has long been of interest to probabilists (e.g. Adler, 1981). The smoothness properties of spatial processes underlie much of the theory of fractals (Mandelbrot, 1982).
(g) *Machine learning.* Gaussian processes and splines have become a fundamental tool in machine learning. Key texts include Rasmussen and Williams (2006) and Hastie et al. (2009).

(h) *Morphometrics.* Starting with Bookstein (1989), a pair of thin-plate splines have been used for the construction of deformations of two-dimensional images. The thin-plate spline is just a special case of kriging.

(i) *Image analysis.* Stationary random fields form a fundamental model for randomness in images, though typically the interest is in more substantive structures. Some books include Grenander and Miller (2007), Sonka et al. (2013), and Dryden and Mardia (2016). The two edited volumes Mardia and Kanji (1993) and Mardia (1994) are still relevant for the underlying statistical theory in image analysis; in particular, Mardia and Kanji (1993) contains a reproduction of some seminal papers in the area.

The book is designed to be used in teaching. The statistical models and methods are carefully explained, and there is an extensive set of exercises. At the same time the book is a research monograph, pulling together and unifying a wide variety of different ideas.

A key strength of the book is a careful description of the foundations of the subject for stationary and related random fields. Our view is that a clear understanding of the basics of the subject is needed before the methods can be used in more complicated situations. Subtleties are sometimes skimmed over in more applied texts (e.g. how to interpret the "covariance function" for an intrinsic process, especially of higher order, or a generalized process, and how to specify their spectral representations). The unity of the subject, ranging from continuously indexed to lattice processes, has been emphasized. The important special case of self-similar intrinsic covariance functions is carefully explained. There are now a wide variety of estimation methods, mainly variants and approximations to maximum likelihood, and these are explored in detail.

There is a careful treatment of kriging, especially for intrinsic covariance functions where the importance of drift terms is emphasized. The link to splines is explained in detail. Examples based on real data, especially from geostatistics, are used to illustrate the key ideas.

The book aims at a balance between theory and illustrative applications, while remaining accessible to a wide audience. Although there is now a wide variety of books available on the subject of spatial analysis, none of them has quite the same perspective. There have been many books published on spatial analysis, and here we just highlight a few. Ripley (1981) was one of the first monographs in the mainstream Statistics literature. Some key books that complement the material in this book, especially for applications, include Cressie (1993), Diggle and Ribeiro (2007), Diggle and Giorgi (2019), Gelfand et al. (2010), Chilés and Delfiner (2012), Banerjee et al. (2015), van Lieshout (2019), and Rasmussen and Williams (2006).

What background does a reader need? The book assumes a knowledge of the ideas covered by intermediate courses in mathematical statistics and linear algebra. In addition, some familiarity with multivariate statistics will be helpful. Otherwise, the book is largely self-contained. In particular, no prior knowledge of stochastic processes is assumed. All the necessary matrix algebra is included in Appendix A. Some knowledge of time series is not necessary, but will help to set some of the ideas into context.

There is now a wide selection of software packages to carry out spatial analysis, especially in R, and it is not the purpose in this book to compare them. We have largely used the package geoR (Ribeiro Jr and Diggle, 2001) and the program of Pardo-Igúzquiza et al. (2008), with additional routines written where necessary. The data sets are available from a public repository at https://github.com/jtkent1/spatial-analysis-datasets.

Several themes receive little or no coverage in the book. These include point processes, discretely valued processes (e.g. binary processes), and spatial–temporal processes. There is little emphasis on a full Bayesian analysis when the covariance parameters needed to be estimated. The main focus is on methods related to maximum likelihood.

The book has had a long gestation period. When we started writing the book the 1980s, the literature was much sparser. As the writing of the book progressed, the subject has evolved at an increasing rate, and more sections and chapters have been added. As a result the coverage of the subject feels more complete. At last, this first edition is finished (though the subject continues to advance).

A series of workshops at Leeds University (the Leeds Annual Statistics Research [LASR] workshops), starting from 1979, helped to develop the cross-disciplinary fertilization of ideas between Statistics and other disciplines. Some leading researchers who presented their work at these meetings include Julian Besag, Fred Bookstein, David Cox, Xavier Guyon, John Haslett, Chris Jennison, Hans Künsch, Alain Marechal, Richard Martin, Brian Ripley, and Tata Subba-Rao.

We are extremely grateful to Wiley for their patience and help during the writing of the book, especially Helen Ramsey, Sharon Clutton, Rob Calver, Richard Davies, Kathryn Sharples, Liz Wingett, Kelvin Matthews, Alison Oliver, Viktoria Hartl-Vida, Ashley Alliano, Kimberly Monroe-Hill, and Paul Sayer. Secretarial help at Leeds during the initial development was given by Margaret Richardson, Christine Rutherford, and Catherine Dobson.

We have had helpful discussions with many participants at the LASR workshops and with colleagues and students about the material in the book. These include Robert Adler, Francisco Alonso, Jose Angulo, Robert Aykroyd, Andrew Baczkowski, Noel Cressie, Sourish Das, Pierre Delfiner, Peter Diggle, Peter Dowd, Ian Dryden, Alan Gelfand, Christine Gill, Chris Glasbey, Arnaldo Goitía, Colin Goodall, Peter Green, Ulf Grenander, Luigi Ippoliti, Anil Jain, Giovanna

Jona Lasinio, André Journel, Freddie Kalaitzis, David Kendall, Danie Krige, Neil Lawrence, Toby Lewis, John Little, Roger Marshall, Georges Matheron, Lutz Mattner, Charles Meyer, Michael Miller, Mohsen Mohammadzadeh, Debashis Mondal, Richard Morris, Ali Mosammam, Nitis Mukhopadhyay, Keith Ord, E Pardo-Igúzquiza, Anna Persson, Sophia Rabe, Ed Redfern, Allen Royale, Sujit Sahu, Paul Sampson, Bernard Silverman, Nozer Singpurwalla, Paul Switzer, Charles Taylor, D. Vere-Jones, Alan Watkins, Geof Watson, Chris Wikle, Alan Wilson, and Jim Zidek.

John is grateful to his wife Sue for her support in the writing of this book, especially with the challenges of the Covid pandemic. Kanti would like to thank the Leverhulme Trust for an Emeritus Fellowship and Anna Grundy of the Trust for simplifying the administration process. Finally, he would like to express his sincere gratitude to his wife and his family for continuous love, support and compassion during his research writings such as this monograph.

We would be pleased to hear about any typographical or other errors in the text.

30 June 2021

John T. Kent
Kanti V. Mardia

List of Notation and Terminology

Here is a list of some of the key notations and terminology used in the book.

- $\mathbb{R}$ and $\mathbb{Z}$ denote the real numbers and integers.
- For a dimension $d \geq 1$, a *site* is a location $t \in \mathbb{R}^d$ or $t \in \mathbb{Z}^d$. The elements or components of a site t are written using square brackets

$$t = (t[1], \ldots, t[d]).$$

  Note t is not in bold face.
- A *random field* is synonymous with a *stochastic process*. A random field on $\mathbb{R}^d$ is written as $X(t) = X(t[1], \ldots, t[d]),\ t \in \mathbb{R}^d$, using function notation. A random field on the *lattice* $\mathbb{Z}^d$ is written as $X_t = X_{(t[1],\ldots,t[d])},\ t \in \mathbb{Z}^d$, using subscript notation. A random field is often assumed to be a Gaussian process (GP).
- The mean function and covariance function are written as $E\{X(t)\} = \mu(t)$ and covariance function $\text{cov}\{X(s), X(t)\} = \sigma(s,t)$. In the stationary case, $\mu(t) = \mu$ is constant and $\sigma(s,t) = \sigma(h)$ depends only on the *lag* $h = s - t$. In the lattice case, use subscripts, e.g. $\mu_t,\ \sigma_{s,t}$.
- A stationary covariance function $\sigma(h) = \sigma^2\rho(h)$ can be written as a product of a *marginal variance* σ^2 and an *autocorrelation function* $\rho(h)$.
- An *intrinsic* random field extends the idea of a stationary random field. Write $X_I(t)$ for an intrinsic random field of order $k \geq 0$ (IRF-k) with *intrinsic* covariance function $\sigma_I(h)$. For an intrinsic random field of order 0 (IRF-0), the *semivariogram* is given by $\gamma(h) = \sigma_I(0) - \sigma_I(h)$. A *registered* version of an intrinsic random field is denoted $X_R(t)$.
- For a stationary model, a *scheme* is a parameterized family of covariance functions. For an intrinsic model, a scheme is a parameterized family of intrinsic covariance functions (or equivalently for an IRF-0 model, a parameterized family of semivariograms).
- A *nugget effect* refers to observations from a random field subject to measurement error, with variance typically denoted τ^2.

- The vector of covariance parameters for a stationary or intrinsic model, possibly including a nugget effect, is denoted θ and can be partitioned as $\theta = (\sigma^2, \theta_c)$ in terms of an overall scale parameter and the remaining parameters.
- Spaces of polynomials in $\mathbb{R}^d$ (Section 3.4):
  - $\mathcal{H}_k$: Space of homogeneous polynomials of degree $k \geq 0$, with dimension denoted $p_H(k) = \dim(\mathcal{H}_k)$
  - $\mathcal{F}_k$: Space of all polynomials of degree $\leq k$ in $\mathbb{R}^d$, with dimension denoted $p_F(k) = \dim(\mathcal{F}_k)$.
- $\mathrm{IRF}_d(\alpha, k)$ denotes the isotropic *self-similar* intrinsic random field of index $\alpha > 0$ and with drift space $\mathcal{F}_k$ (Section 3.10). The intrinsic covariance function is denoted $\sigma_\alpha(h)$ and spectral density is denoted $f_\alpha(\omega)$.
- Most of the book is concerned with *ordinary* random fields. There are also *generalized* random fields indexed by functions rather than sites and written as $X_G(\cdot)$ with covariance functional $\sigma_G(\cdot, \cdot)$.
- The surface area of the unit sphere in $\mathbb{R}^d$ is denoted $\pi_d = 2\pi^{d/2}/\Gamma(d/2)$.
- D denotes a domain of sites in $\mathbb{R}^d$ or $\mathbb{Z}^d$. The notation encompasses several possibilities, including the following:
  - An open subset $D \subset \mathbb{R}^d$, e.g. $D = \mathbb{R}^d$
  - A finite collection of sites $D = \{t_1, \dots, t_n\}$ in $\mathbb{R}^d$ or $\mathbb{Z}^d$
  - The infinite lattice $D = \mathbb{Z}^d$
  - A finite rectangular lattice in $\mathbb{Z}^d$,

  $$D = \{t \in \mathbb{Z}^d : 1 \leq t[\ell] \leq n[\ell],\ \ell = 1, \dots, d\}$$

  with dimension vector $N = (n[1], \dots, n[d])$ and of size $|D| = |N| = n[1] \times \cdots \times n[d]$. In the lattice case, sites in D can be denoted using letters such as $t = (t[1], \dots, t[d])$ to emphasize the link to the continuous case, or using letters such as $j = (j[1], \dots, j[d])$ to emphasize the fact that the components are integers.

  For a finite domain, the notation $|D|$ stands for the number of sites in D.
- Frequencies in the Fourier domain are denoted $\omega = (\omega[1], \dots, \omega[d])$.
- Vectors indexing data are treated as column vectors and are written as $\boldsymbol{x} = [x_1, \dots, x_n]^T$ in bold lowercase letters, with the components indicated by subscripts. The transpose of $\boldsymbol{x}$ is denoted $\boldsymbol{x}^T$. This subscript convention is typical in multivariate analysis. Note the difference from the convention for sites t and frequencies ω.
- Random vectors, e.g., $\boldsymbol{x} = [x_1, \dots, x_n]^T$ or $\boldsymbol{X} = [X_1, \dots, X_n]^T$ are written in bold letters, with the components indicated by subscripts. In particular, upper case is used when the distinction between a random quantity and its possible values needs emphasis.
- Matrices are written using nonbold uppercase letters, e.g. A and Γ, with the elements of A written as a_{ij} or as $(A)_{ij}$. The two notations are synonymous. The columns of A are written using bracketed subscripts, $\boldsymbol{a}_{(j)}$. For a square matrix, the determinant is denoted by either $\det(A)$ or $|A|$; the notation $|A|$ should not be confused with $|D|$, the size of a domain D described above.

- If s and t are sites, then $s^T t = \sum_{\ell=1}^{d} s[\ell]t[\ell]$ is the *inner product* and $|t|^2 = t^T t$ is the *squared Euclidean norm*.
- Modulo notation mod (for numbers) and *Mod* (for vectors) (Section A.1)
- *Check* and *convolution* notation. If $\varphi(u)$ is a function of $u \in \mathbb{R}^d$, let $\check{\varphi}(u) = \varphi(-u)$. Then

$$(\varphi * \psi)(h) = \int \varphi(u)\psi(h-u)\, du, \quad (\varphi * \check{\varphi})(h) = \int \varphi(u)\varphi(u-h)\, du,$$

  and the latter is symmetric in h.
- The *Kronecker delta* and *Dirac delta* functions are denoted δ_h, $h \in \mathbb{Z}^d$ and $\delta(h)$, $h \in \mathbb{R}^d$, respectively.
- $\mathcal{N}$. A finite symmetric neighborhood of the origin in $\mathbb{Z}^2$. The augmented neighborhood $\mathcal{N}_0 = \mathcal{N} \cup \{0\}$ includes the origin. Half of the neighborhood $\mathcal{N}$ is denoted $\mathcal{N}^\dagger$ (Section 4.4).
- $\mathcal{H}$. A half-space in $\mathbb{Z}^d$, especially the lexicographic half-space $\mathcal{L}$ (Section 4.8). Related ideas are the weak past $\mathcal{B}$ and quadrant past $\mathcal{Q}$ (Section 4.8), and the partial past (Section 5.9).
- *Kriging* is essentially prediction for random fields. It comes in various forms including *simple kriging*, *ordinary kriging*, *universal kriging*, and *Bayesian kriging*. In each case, there is a *kriging predictor* at every site, which depends on the data through a *kriging vector*. Combining the kriging predictor for all sites yields a *kriging surface*. The *kriging variance* describes the accuracy of the predictor at each site. Tables 7.1 and 7.2 set out the notation for kriging and Table 7.3 provides a comparison with some related notation used in machine learning.
- The *transfer covariance matrix* and *transfer drift matrix* are used to construct the kriging predictor (Section 7.6).
- *Bordered covariance matrix*. This is an $(n+1) \times (n+1)$ matrix, Section 7.6.4, also used to construct the kriging predictor.
- Autoregression (AR) and related spatial models come in various forms in Chapter 4 including:
  - MA: Moving average (Section 4.3)
  - SAR: Simultaneous autoregression (Section 4.5)
  - CAR: Conditional autoregression (Section 4.6)
  - ICAR: Intrinsic CAR (Section 4.6.3)
  - QICAR: Quasi-intrinsic CAR (Section 4.6.3)
  - UAR: Unilateral autoregression (Section 4.8.2)
  - QAR: Quadrant unilateral autoregression (Section 4.8.3)
- Types of matrix
  - Tensor product matrices (Section A.3.9)
  - Toeplitz, circulant, folded circulant matrices (Sections A.3.8 and A.10).
  - All $n \times n$ circulant matrices in $d = 1$ dimension have the same eigenvectors. These can be represented in complex coordinates by the unitary matrix $G_n^{(\mathrm{DFT,com})}$ or in real coordinates by the orthogonal matrix $G_n^{(\mathrm{DFT,rea})}$ (Section A.7.2).

- Abbreviations and terminology for estimation and testing:
  - MLE: maximum likelihood estimation
  - AIC: Akaike information criterion
  - REML: restricted maximum likelihood
  - MINQUE: minimum quadratic unbiased estimation
  - GLS: generalized least squares
  - OLS: ordinary least squares
  - PMSE: prediction mean squared error for a kriging predictor
  - profile likelihood
  - likelihood ratio test
  - Vecchia approximation to the likelihood
  - moment estimation
  - Fisher information
  - composite likelihood
- Other abbreviations:
  - i.i.d.: independent and identically distributed
  - RF: random field
  - IRF: intrinsic random field
  - GP: Gaussian process = Gaussian random field
  - MRF: Markov random field
  - GMRF: Gaussian Markov random field
  - SLM: spatial linear model
  - FT: Fourier transform
  - IFT: inverse Fourier transform
  - DFT: discrete Fourier transform
  - DCT: discrete cosine transform
  - SPDE: stochastic partial differential equation
  - RKHS: reproducing kernel Hilbert space

1

Introduction

1.1 Spatial Analysis

Spatial analysis involves the analysis of data collected in a spatial region. A key aspect of such data is that observations at nearby sites tend to be highly correlated with one another. Any adequate statistical analysis should take these correlations into account.

The region in which the data lie is a subset of d-dimensional space, $\mathbb{R}^d$, for some $d \geq 1$. The important cases in practice are $d = 1,2,3$, corresponding to the data on the line, in the plane, or in 3-space, respectively.

The one-dimensional case, $d = 1$, is already well known from the analysis of time series. Therefore, it will come as no surprise that many of the techniques introduced in this book represent generalizations of standard methodology from time-series analysis. However, just as multivariate analysis contains techniques with no counterpart in univariate statistical analysis, spatial analysis includes techniques with no counterpart in time-series analysis.

Spatial data arise in many applications. In mining we may have measurements of ore grade at a set of boreholes. If all the observations along each borehole are averaged together, we obtain data in $d = 2$ dimensions, whereas if we retain the depth information at which each observation in the borehole is made, we obtain three-dimensional data. In agriculture, experiments are usually performed on experimental plots, which are regularly spaced in a field. For environmental monitoring, data are collected at an array of monitoring sites, possibly irregularly located. There may also be a temporal component to this monitoring application as data are collected through time.

Digital images can also be viewed as spatial data sets. Examples include Landsat satellite images of areas of the earth's surface, medical images of the interior of the human body, and fingerprint images.

Spatial Analysis, First Edition. John T. Kent and Kanti V. Mardia.

1.2 Presentation of the Data

The points in $\mathbb{R}^d$ at which the data are collected are known as *sites*. A data set consists of a collection of sites t_i and real-valued *observations* or *values* $x(t_i) \in \mathbb{R}^d, i = 1, \ldots, n$. Note that each $t_i = (t_i[1], \ldots, t_i[d])$ represents a vector in $\mathbb{R}^d$.

If the sites t_i are located arbitrarily in $\mathbb{R}^d$, the data are known as *irregularly spaced* data. However, if the components of the sites t_i are restricted to have integer values, $t_i \in \mathbb{Z}^d$ and the sites cover a rectangular region in $\mathbb{Z}^d$, then the data are known as *regular lattice* or *regularly spaced* data. (There is also the case of *irregular lattice* data for which the data do not fill a rectangular region.) For convenience, we shall often write lattice data using subscripts x_t rather than with parentheses $x(t)$ to emphasize the link with sequences of data in $d = 1$ dimension. For example, in $d = 2$ dimensions x_t stands for $x_{(t[1],t[2])}$, though we shall usually avoid the need to expand the suffix t in full.

There are two ways to represent spatial data.

Regularly spaced data can be represented as a two-way table of numbers. The other representation, which can be used both for regularly spaced and irregularly spaced data, is a list of spatial sites and data values.

Example 1.1 ***Illustrative data***

Table 1.1 gives a simple illustrative regularly spaced data set in $\mathbb{Z}^2$. In Panel (a), the data are presented as a two-way array of numbers. Panel (b) shows a *matrix coordinate system* in which the origin is at the upper left of the table, with $t[1]$ increasing down the rows of the table and $t[2]$ increasing across the rows. Although the matrix coordinate system is conventional for multivariate analysis, we do not use matrix coordinates in this book. Instead, we use *graphical coordinates*, as in Panel (c), for which the $t[1]$-axis increases horizontally to the right, and the $t[2]$-axis increases vertically upward. Finally, in Panel (d) the data are presented as a list of spatial sites and data values. □

A digital image can be regarded as a spatial data set on a large regular grid; typically, $d = 2$ and $n = 256 \times 256$ or 512×512. In this context, the sites are known as *pixels* (picture elements).

Example 1.2 ***Fingerprint data***

Figure 1.1 shows the gray level image of a fingerprint of R A Fisher. The sites t of the data lie on a rectangular grid 3003 pixels wide by 3339 pixels high. The values of x_t have been scaled to lie between 0 and 1. The marked rectangular section is investigated in more detail in Example 1.8. □

Table 1.1 Illustrative data x_t, $t = (t[1], t[2])$, on a 3×4 regular grid, represented in various ways, $n = 12$.

(a) Table of values on a 3×4 grid

6	7	3	10
13	2	4	3
22	9	2	5

(b) Matrix coordinates (not generally used in this book)

		t[2] 1	2	3	4
	1	6	7	3	10
t[1]	2	13	2	4	3
	3	22	9	2	5

(c) Graphical coordinates **(used for all spatial data sets in this book)**. The asterisks are explained in Example 1.7

	3	*6	*7	*3	10
t[2]	2	*13	*2	*4	3
	1	*22	*9	*2	5
		1	2	3	4
			t[1]		

(d) List of graphical coordinates and values

t[1]	t[2]	x_t	t[1]	t[2]	x_t
1	1	22	3	2	4
2	1	9	4	2	3
3	1	2	1	3	6
4	1	5	2	3	7
1	2	13	3	3	3
2	2	2	4	3	10

Example 1.3 ***Elevation data***

The topographic elevation data of Davis (1973) are given in Table 1.2 and consist of $n = 52$ irregularly spaced observations. The data contain geographic coordinates and elevations of control points for a surveying problem. The elevation is measured in feet above the sea level. The coordinates are expressed in 50-feet units measured

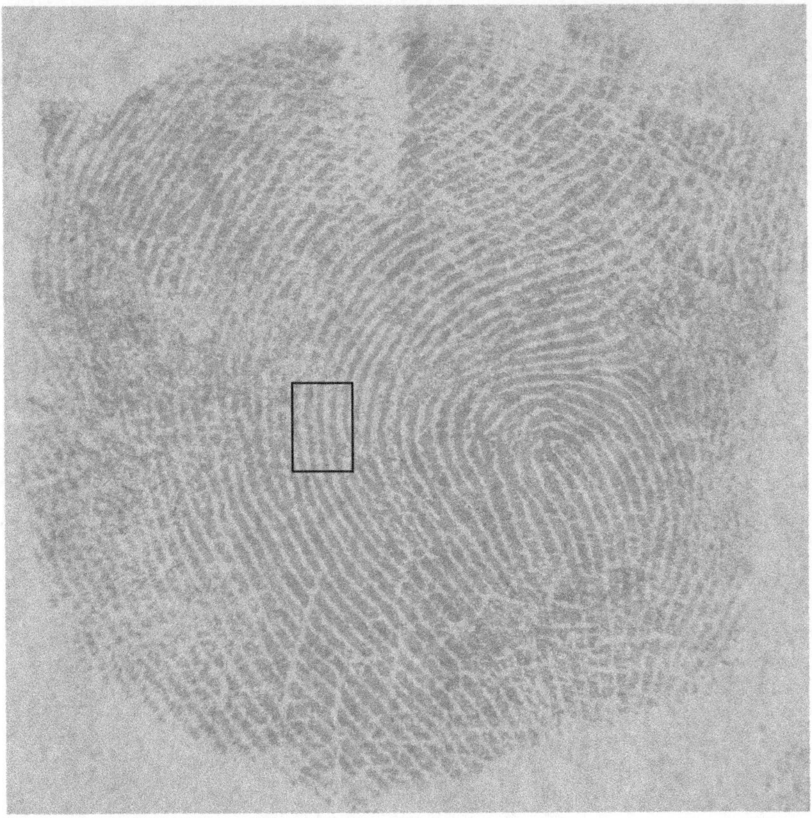

Figure 1.1 Fingerprint of R A Fisher, taken from Mardia's personal collection. A blowup of the marked rectangular section is given in Figure 1.9.

from an arbitrary origin located in the southwest corner; t_1 is the East–West coordinate and t_2 is the North–South coordinate. Figure 1.2 gives two plots of the data. The raw plot in Panel (a) shows the elevation values printed at each site. The bubble plot in Panel (b) shows a circle plotted at each site, where the size of the circle encodes graphically the elevation information; larger elevations are indicated by bigger circles. Patterns in the data are often easier to pick out using the bubble plot. Note that the elevations are high near the edges of the region with a basin in the middle. There are extra features associated with the data such as river locations, but we will limit ourselves here to just the elevation information for illustrative purposes.

One of the objectives for this sort of data is to predict the elevation throughout the region and to represent the result graphically. Using a statistical method called kriging (see Chapter 7 for details), the elevation was predicted or smoothed

Table 1.2 Elevation data: elevation $x(t)$ in feet above the sea level, where $t = (t[1], t[2])$, $n = 52$.

E-W $t[1]$	N-S $t[2]$	Elevation $x(t)$	E-W $t[1]$	N-S $t[2]$	Elevation $x(t)$
0.3	6.1	870	5.2	3.2	805
1.4	6.2	793	6.3	3.4	840
2.4	6.1	755	0.3	2.4	890
3.6	6.2	690	2.0	2.7	820
5.7	6.2	800	3.8	2.3	873
1.6	5.2	800	6.3	2.2	875
2.9	5.1	730	0.6	1.7	873
3.4	5.3	728	1.5	1.8	865
3.4	5.7	710	2.1	1.8	841
4.8	5.6	780	2.1	1.1	862
5.3	5.0	804	3.1	1.1	908
6.2	5.2	855	4.5	1.8	855
0.2	4.3	830	5.5	1.7	850
0.9	4.2	813	5.7	1.0	882
2.3	4.8	762	6.2	1.0	910
2.5	4.5	765	0.4	0.5	940
3.0	4.5	740	1.4	0.6	915
3.5	4.5	765	1.4	0.1	890
4.1	4.6	760	2.1	0.7	880
4.9	4.2	790	2.3	0.3	870
6.3	4.3	820	3.1	0.0	880
0.9	3.2	855	4.1	0.8	960
1.7	3.8	812	5.4	0.4	890
2.4	3.8	773	6.0	0.1	860
3.7	3.5	812	5.7	3.0	830
4.5	3.2	827	3.6	6.0	705

Coordinates are expressed in 50-feet units measured from an arbitrary origin located in the southwest corner, with $t[1]$ being the East–West coordinate and $t[2]$ being the South–North coordinate.
Source: Davis (1973).

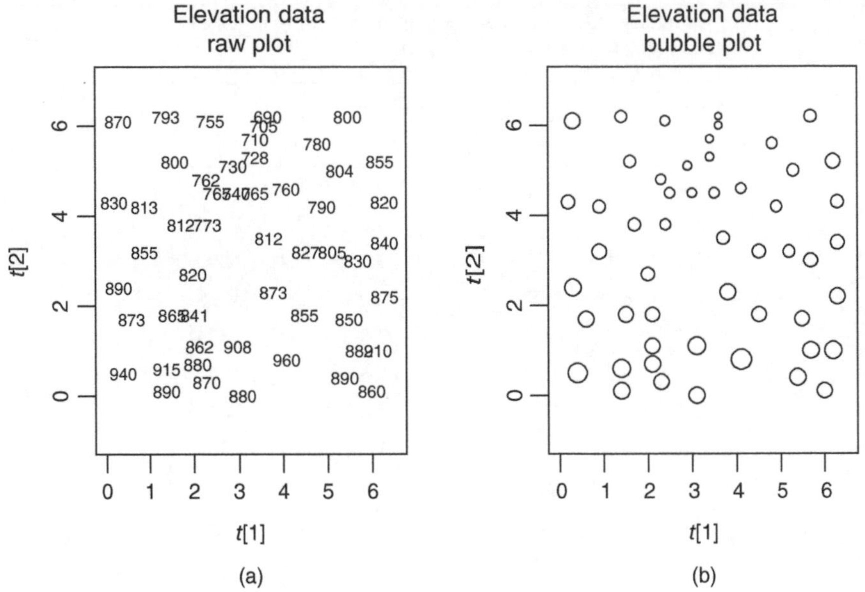

Figure 1.2 Elevation data: (a) raw plot giving the elevation at each site and (b) bubble plot where larger elevations are indicated by bigger circles.

on a fine grid of points throughout the region. The result for this data set can be summarized visually in different ways including:

- A contour map (Figure 1.3a)
- A perspective plot (Figure 1.3b), viewed from the top of the region
- A digital image using gray level (or color) to indicate ore grade, where white denotes the lower values and black denotes higher values (Figure 1.3c)

These images all show that the data have a valley in the top middle of the the image and a peak in the bottom middle.

In addition, the contour plot in Figure 1.3d shows the standard error of the predictor. Notice that the predictor is perfect with zero standard error at the data sites, and it has a larger standard error in places where the data sites are sparse. See Example 7.2 for more details. □

Example 1.4 ***Bauxite data***

Figure 1.4a shows the bauxite ore grade in percentages at $n = 33$ irregularly spaced sites in $\mathbb{R}^2$ (constructed from Marechal and Serra, 1970). Figure 1.4b shows the same information in a "bubble" plot for which larger data values are represented by larger circles. The data are also listed in Table 1.3.

Each representation makes clear the presence of hills on the left side and the bottom of the region and a valley in the middle. □

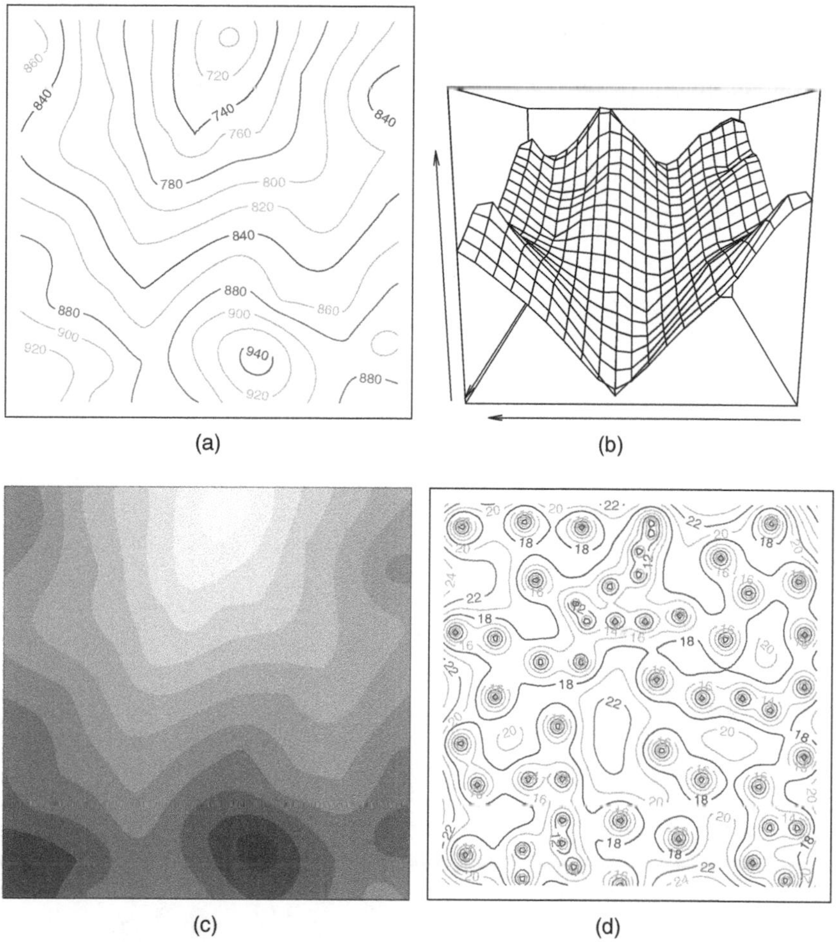

Figure 1.3 Panels (a), (b), and (c) show interpolated plots for the elevation data, as a contour map, a perspective plot (viewed from the top of the region), and an image plot, respectively. Panel (d) shows a contour map of the corresponding standard errors.

Example 1.5 ***Landsat data***
Band 2 (Landsat 7) is used to distinguish soil from vegetation and deciduous from coniferous vegetation. This data set contains values on a 200×200 grid of sites from a Landsat image of a rural field in western Canada. More details are available from Mardia and Pardo-Iguzquiza (2006). An image view of the data is shown in Figure 1.5. The distance between adjacent sites is 30 m. Except for a large-scale ridge running vertically through the middle of the image and some thin straight white lines representing farms, the small-scale structure of the data appears to be random alternating patches of light and dark. □

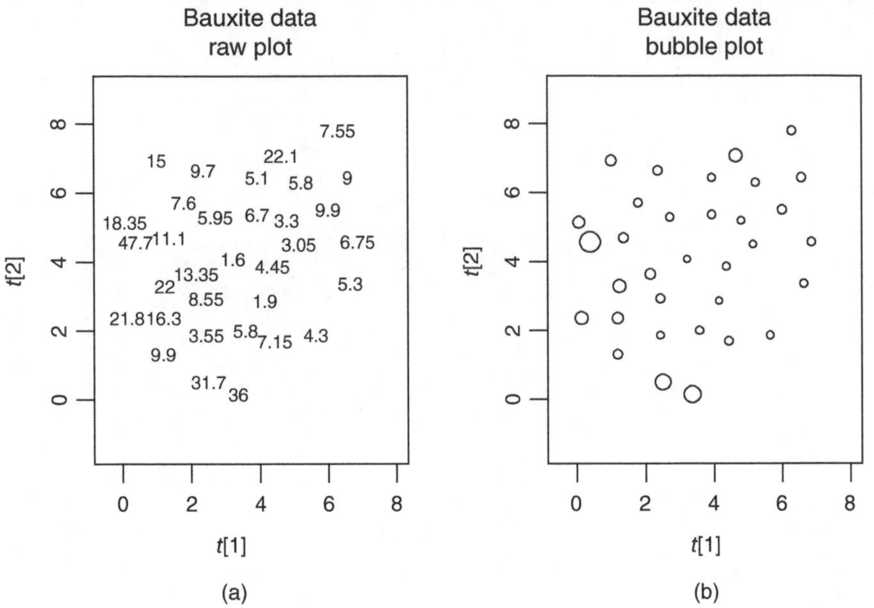

Figure 1.4 Bauxite data: (a) raw plot giving the ore grade at each site and (b) bubble plot where larger ore grades are indicated by bigger circles.

Table 1.3 Bauxite data: percentage ore grade for bauxite at $n = 33$ locations.

$t[1]$	$t[2]$	$x(t)$	$t[1]$	$t[2]$	$x(t)$	$t[1]$	$t[2]$	$x(t)$
0.07	5.14	18.35	2.43	1.86	3.55	4.43	1.69	7.15
0.14	2.36	21.80	2.43	2.93	8.55	4.64	7.07	22.10
0.39	4.57	47.70	2.50	0.50	31.70	4.79	5.19	3.30
1.00	6.93	15.00	2.71	5.29	5.95	5.14	4.50	3.05
1.19	1.31	9.90	3.21	4.07	1.60	5.21	6.29	5.80
1.19	2.36	16.30	3.36	0.14	36.00	5.64	1.86	4.30
1.24	3.29	22.00	3.57	2.00	5.80	6.64	3.36	5.30
1.36	4.69	11.10	3.93	5.36	6.70	6.00	5.50	9.90
1.79	5.71	7.60	3.93	6.43	5.10	6.29	7.79	7.55
2.14	3.64	13.35	4.14	2.86	1.90	6.57	6.43	9.00
2.36	6.64	9.70	4.36	3.86	4.45	6.86	4.57	6.75

Source: Based on Marechal and Serra (1970).

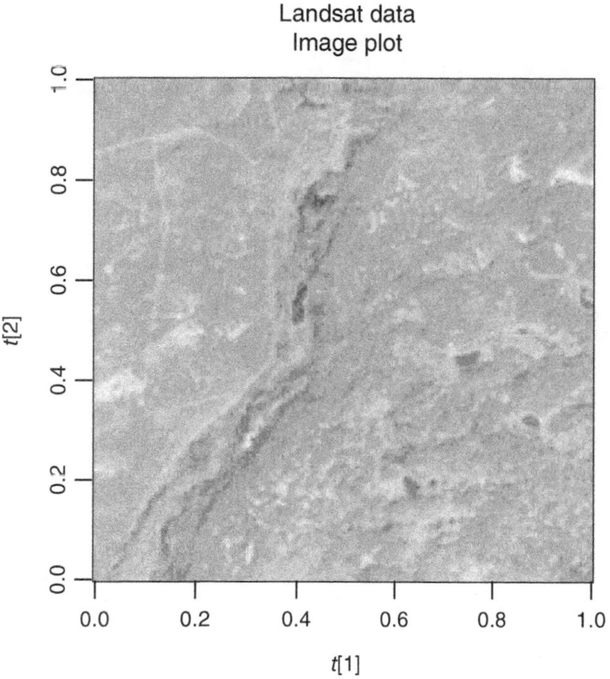

Figure 1.5 Landsat data (200 × 200 pixels): image plot.

Example 1.6 ***Synthetic Landsat data***
This is a simulated set of data designed to mimic the small-scale behavior of the Landsat data. However, the parameters of the underlying model are known in this case, so that it can be used to investigate different estimation procedures. The underlying model is described in Chapter 5. The data set is plotted as a 200×200 image in Figure 1.6. Notice the presence of multiple dark and light areas in the image extending over regions with a diameter of about 20 pixels. □

1.3 Objectives

There can be many possible objectives in a spatial analysis depending on the applications. It is convenient to set out these objectives in terms of increasing complexity. Our basic assumption is usually that some aspect of the data can be usefully modeled as a stationary random field described by its first two moments. The terms "random field" and "stochastic process" are synonymous; both refer to collections of random variables indexed by a collection of sites. However, the term "stochastic process" is often used when the sites lie in one dimension (especially

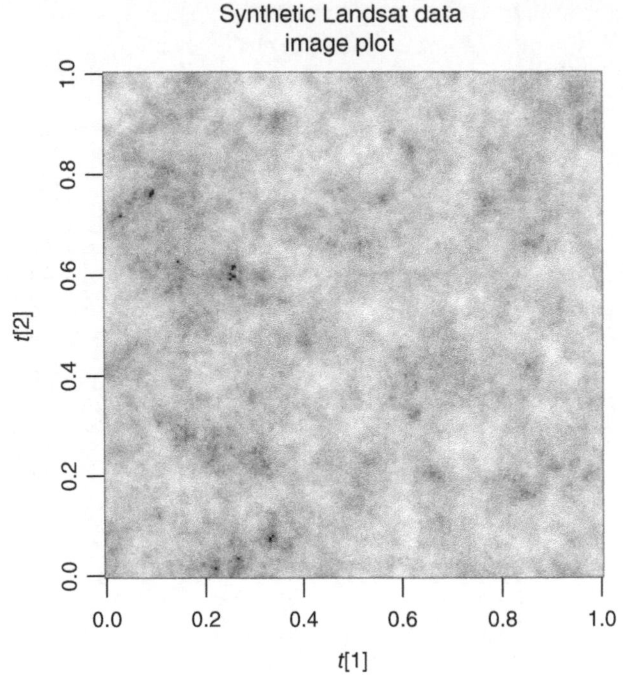

Figure 1.6 Synthetic Landsat data: image plot.

where the process is evolving through time), and the term "random field" when the sites lie in higher dimensions.

(a) *Spatial correlation.* A key aspect in most applications is that observations at nearby sites will tend to be correlated. Hence, in a data set that can be regarded as a stationary random field, one of the first objectives is to describe and quantify the extent of this spatial correlation.
(b) *Prediction.* Using a sample of observations one might want to predict the value of the random field at a new site. Using the spatial correlation structure will improve the accuracy of the predictor. A typical example arises in mining where measurements are made available at a set of boreholes, and one wants to predict the ore content at a new borehole or in a block of rock. In time-series analysis, "prediction" usually means predicting the future given the past. However, in spatial analysis, even for the one-dimensional case, there is generally no concept of "past" or "future." Instead, prediction involves either interpolation within or extrapolation beyond the set of data sites.
(c) *The spatial linear model.* A more realistic model than a stationary random field might include trend terms (such as linear or quadratic drift) or treatment effects (such as in a designed agricultural experiment). If the spatial correlation structure is known, then estimation of these parameters in the

spatial linear model is straightforward. If the spatial correlation structure also needs to be estimated, the problem is more challenging (Chapter 5). The simpler model of trend terms plus independent errors is known as "trend surface modeling."

(d) *Smoothing.* Extending the motivation behind the spatial linear model, we might imagine our data consist of "signal" plus "noise." One objective is then to estimate the signal by smoothing away the noise.

1.4 The Covariance Function and Semivariogram

1.4.1 General Properties

As the first step in the spatial data analysis, we often calculate the mean value and the covariance function. For this procedure to be useful, we imagine that the data came from a stationary random field.

Let $\{X(t) : t \in \mathbb{R}^d\}$ denote a d-dimensional real-valued stationary random field (see Chapter 2) with *mean value*

$$E\{X(t)\} = \mu \quad \text{(not depending on } t\text{)}$$

and *covariance function*

$$E\{[X(t) - \mu]\,[X(t+h) - \mu]\} = \sigma(h) \quad \text{(not depending on } t\text{)},$$

where $h = (h[1], \ldots, h[d]) \in \mathbb{R}^d$ and $|h| = (h[1]^2 + \cdots + h[d]^2)^{1/2}$. The *lag* h describes the separation between two sites t and $t + h$. Usually, we shall only consider models for which $\sigma(h) \to 0$ as $|h| \to \infty$ so that distant observations are nearly uncorrelated. The value of $\sigma(0) = \text{var}\{X(t)\}$ represents the marginal variance of the process and does not depend on t.

The second-order behavior of the process can also be described using the *semivariogram*, which is defined by

$$\begin{aligned}\gamma(h) &= \frac{1}{2}E\{X(t+h) - X(t)\}^2 = \frac{1}{2}\text{var}\{X(t+h) - X(t)\}\\ &= \sigma(0) - \sigma(h),\end{aligned} \tag{1.1}$$

the variance of an increment of lag $h \in \mathbb{R}^d$. Note that both $\sigma(h)$ and $\gamma(h)$ are *even* functions of h,

$$\sigma(h) = \sigma(-h), \quad \gamma(h) = \gamma(-h).$$

The semivariogram and the covariance function are essentially equivalent ways to describe the second-order behavior of a stationary random field. In particular, provided $\gamma(h)$ has a finite limiting value as $|h| \to \infty$, the covariance function can be recovered from the semivariogram by

$$\sigma(0) = \lim_{|h|\to\infty} \gamma(h), \quad \sigma(h) = \sigma(0) - \gamma(h).$$

The covariance function is well known from its use in time-series analysis. However, the semivariogram is valid in more settings in which $\gamma(h) \to \infty$. Since both constructions are popular, we shall work with both of them through the book.

An important simplification occurs when the covariance function and semivariogram depend only on the radial component $r = |h|$. In this case, the random field is said to be *isotropic* and the notation

$$\sigma^{\#}(r) = \sigma(h), \quad \gamma^{\#}(r) = \gamma(h)$$

is used to distinguish between a function with a scalar argument and a function with a vector argument. A more complete theoretical discussion of the covariance function and semivariogram is given in Chapters 2 and 3, respectively, and statistical methods of analysis are covered in Chapter 5. In this section, we limit ourselves to some simple initial observations.

Consider the sketch of an isotropic semivariogram given in Figure 1.7. The circles represent sample values of the semivariogram and the solid curve represents an underlying model fitted to the data. Several typical features of a semivariogram should be noted in the plot.

(a) *Monotonicity.* A semivariogram tends to be a monotone increasing function of the lag $|h|$.

(b) *Nugget effect.* The fitted semivariogram can be extrapolated toward lag $|h| = 0$, where it intersects the vertical axis at a nonzero value (0.2 in Figure 1.7). This nonzero value is known as a "nugget effect." That is, however close two observations are in space, there is still be some residual variability between them. A nugget effect can be explained in terms of either small-scale variation in the data or measurement error. The term nugget effect comes from mining, where the presence of small nuggets of mineral gives rise to very short-range autocorrelation effects. The topic is discussed in more detail in Chapter 5.

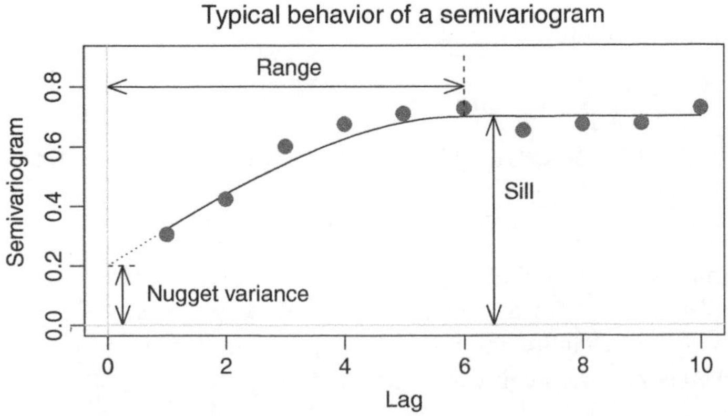

Figure 1.7 Typical semivariogram, showing the range, nugget variance, and sill.

(c) *Sill.* The fitted semivariogram can be extrapolated toward lag $|h| = \infty$. In Figure 1.7, the semivariogram increases to a finite sill with a value of 0.7. More generally, for stationary processes the sill is always finite. However, in the wider setting of intrinsic processes, the sill may be infinite; see Chapter 3.

(d) *Range.* In Figure 1.7, the sill is attained by the semivariogram when $|h| = 6$. This value of the lag is known as the "range." Observations at sites separated by a distance greater than the range are uncorrelated. In many examples, the range is not finite; the semivariogram approaches the sill only asymptotically as $|h| \to \infty$. But even in this situation, it is helpful to define an "approximate range" such that when the lag has reached this value, the semivariogram has nearly reached the sill. See Section 5.2 for further discussion.

1.4.2 Regularly Spaced Data

Next, we describe how to calculate the sample covariance function and the sample semivariogram from a set of data. Start with the case of regular lattice data $\{x_t : t \in D\}$ in d dimensions, where

$$D = \{t \in \mathbb{Z}^d : 1 \leq t[\ell] \leq n[\ell] \text{ for } \ell = 1, \ldots, d\}$$

denotes a rectangular region in $\mathbb{Z}^d$ with $n[1] \times \cdots \times n[d] = |D|$ sites. Let

$$D_h = \{t \in \mathbb{Z}^d : t \in D \text{ and } t + h \in D\} \tag{1.2}$$

denote those sites t for which t and $t+h$ lie in D. Provided $|h[\ell]| \leq n[\ell]$ for $\ell = 1, \ldots, d$, the set D_h contains $|D_h| = (n[1] - |h[1]|) \times \cdots \times (n[d] - |h[d]|)$ sites.

The sample mean is defined by

$$\bar{x} = \frac{1}{|D|} \sum_{t \in D} x_t.$$

The sample covariance function and sample semivariogram are defined by

$$s_h = \frac{1}{|D_h|} \sum_{t \in D_h} [x_{t+h} - \bar{x}][x_t - \bar{x}] \qquad g_h = \frac{1}{2} \frac{1}{|D_h|} \sum_{t \in D_h} [x_{t+h} - x_t]^2.$$

In each case, the sum is divided by the number of terms.

Note that the identity (1.1) does not hold in the sample case; in general

$$g_h \neq s_0 - s_h. \tag{1.3}$$

However, in practice the difference between g_h and $s_0 - s_h$ is usually negligible, provided $|D_h|$ is not too small.

If $h[\ell]$ is close to $n[\ell]$ for all $\ell = 1, \ldots, d$, then the number of terms in the summations for s_h and g_h becomes small. Thus, s_h and g_h become susceptible to large sampling fluctuations for such values of h.

Example 1.7 Consider the illustrative data of Table 1.1 where $n = 12$. To give a stronger feeling for the intuition behind the formulas of this section, we illustrate some of the calculations by hand.

The mean and variance are given by

$$\bar{x} = (22 + 9 + \cdots + 3 + 10)/12 = 7.17,$$
$$s_{(0,0)} = [(22 - 7.17)^2 + \cdots + (10 - 7.17)^2]/12 = 30.81.$$

For lag $h = (1,0)$ the subset of data $D_{(1,0)}$ in (1.2) has been marked with asterisks (*) in Table 1.1c; there are 9 such values. Thus, the calculations for the sample covariance function and sample semivariogram take the form

$$s_{(1,0)} = \{(9 - 7.17)(22 - 7.17) + \cdots + (10 - 7.17)(3 - 7.17)\} /9 = 1.94,$$
$$g_{(1,0)} = \left\{(9 - 22)^2 + (2 - 13)^2 + (7 - 6)^2 + (2 - 9)^2 + (4 - 2)^2 + (3 - 7)^2 + (5 - 2)^2 + (3 - 4)^2 + (10 - 3)^2\right\} /(2 \times 9) = 23.28.$$

Note that

$$23.28 = g_{(1,0)} \neq s_{(0,0)} - s_{(1,0)} = 28.87,$$

in accordance with (1.3). □

1.4.3 Irregularly Spaced Data

For irregularly spaced data at distinct sites $t_i\ :\ i = 1, \ldots, n$, it will be rare for two pairs of sites to have the same lag. Therefore, it is necessary to pool pairs of sites having approximately the same lag for the computation of the sample covariance function and semivariogram.

It is useful to distinguish between the isotropic and anisotropic cases. Start with the anisotropic case. Let $\delta > 0$ be a "smoothness parameter." For a lag $h \in \mathbb{R}^d$, let $M(h)$ define a set of site pairs lying in a window (depending on δ) of h. All the choices of window have the property that each site pair is included at most once; that is, if $t_i - t_j \in M(h)$, then $t_j - t_i \notin M(h)$. Then smoothed versions of the sample covariance function and sample semivariogram can be defined by

$$s(h) = \frac{1}{|M(h)|} \sum_{(i,j) \in M(h)} \{x(t_i) - \bar{x}\}\{x(t_j) - \bar{x}\}, \tag{1.4}$$

$$g(h) = \frac{1}{2|M(h)|} \sum_{(i,j) \in M(h)} \{x(t_i) - x(t_j)\}^2. \tag{1.5}$$

There are several choices for the window of h:

1. *Circular*. For $|h| \geq \delta$, set

   $$M(h) = \{(i,j)\ :\ i < j \text{ and } |t_i - t_j - h| < \delta\} \tag{1.6}$$

   denote a circular disk about h.

2. *Square*. For h such that $|h[\ell]| \geq \delta$ for at least one $\ell = 1, \ldots, d$, let
$$M(h) = \{(i,j) : i < j \text{ and } |t_i[\ell] - t_j[\ell] - h[\ell]| \leq \delta, \quad \ell = 1, \ldots, d\} \tag{1.7}$$
denote a square centered at h. Isaaks and Srivastava (1989) suggest letting h vary over a grid of vectors with components given by half-integers times δ, i.e. $h[\ell] = (\frac{1}{2} + k[\ell])\delta$, where the $k[\ell]$ are integers. Thus, the space of site differences becomes partitioned into nonoverlapping square blocks (unless any site differences lie exactly on the boundary between two blocks).
3. *Sector*. In $d = 2$ dimensions, it is natural to use switch to polar coordinates. In addition to the radial smoothness parameter $\delta > 0$, let $0 < \alpha_0 < \pi$ denote an angular smoothness parameter, e.g. $\alpha_0 = \pi/8$ (22.5°). Set
$$\begin{aligned} M(h) = \{(i,j) : \ & i < j, \ |\ |h| - |t_i - t_j|\ | \leq \delta \text{ and} \\ & h^T(t_i - t_j)/(|h|\ |t_i - t_j|)^{\frac{1}{2}} \geq \cos\alpha_0, \end{aligned} \tag{1.8}$$
where the notation $|\cdot|$ is used both for the absolute value of a real number and for the Euclidean norm of a vector in $\mathbb{R}^d$. Then $M(h)$ is an angular sector with radial width δ and angular semiwidth α_0. This is the window choice used in the computer package geoR (Ribeiro Jr and Diggle, 2001). It is common to fix $\theta = 0°, 45°, 90°, 135°$ to be one of the four principal directions and to plot the sample covariance function or the sample semivariogram vs. r for $r = \delta,\ 3\delta,\ 5\delta,\ \ldots$.

Note that $\bar{x}$ does not depend on the choice of window or on δ. If $M(h)$ is ever empty, the corresponding value of $s(h)$ or $g(h)$ is undefined. The validity of these window definitions requires that h and δ be chosen so that the origin does not lie in $M(h)$. This condition is also important in practice since the semivariogram will usually be nondifferentiable (and sometimes appear to be discontinuous) at the origin.

For isotropic data, it is natural to use an annulus for the set indexing site pairs. Thus, for $r = |h| > 0$, set
$$N(r) = \{(i,j) : i < j \text{ and } r - \delta < |t_i - t_j| \leq r + \delta\}, \tag{1.9}$$
and define the smoothed sample covariance function and the sample semivariogram by
$$s^{\#}(r) = \frac{1}{|N(r)|} \sum_{(i,j)\in N(r)} \{x(t_i) - \bar{x}\}\{x(t_j) - \bar{x}\}, \tag{1.10}$$
$$g^{\#}(r) = \frac{1}{|N(r)|} \sum_{(i,j)\in N(r)} \{x(t_i) - x(t_j)\}^2. \tag{1.11}$$
These functions can be plotted e.g. at the values $r = \delta,\ 3\delta,\ 5\delta,\ \ldots$, so that the annuli defining the $N(h)$ partition the space of site differences into nonoverlapping regions. Note that each site difference $t_i - t_j$ is listed just once in $N(h)$.

Some examples of these smoothing procedures are given in Section 1.5.

1.5 Behavior of the Sample Semivariogram

In Section 1.4.2, we described how to calculate the sample semivariogram g_h for a regular two-dimensional data set on an $n[1] \times n[2]$ grid, for $h = (h[1], h[2])$ satisfying $|h[1]| < n[1], |h[2]| < n[2]$. To understand this large array of numbers, it is helpful to use polar coordinates, $h = (r \sin\theta,\ r\cos\theta)$. Writing g_h as $g(r, \theta)$ when convenient, plot $g(r, \theta)$ vs. r for several choices of θ (typically the four main directions $\theta = 0°,\ 45°,\ 90°,\ 135°$).

Throughout this chapter, we have used and adapted the computer package `geoR` (Ribeiro Jr and Diggle, 2001) to plot the data and semivariograms. In particular, we have used the same convention as `geoR` to measure the angles. Recall that in graphical coordinates the $h[1]$-axis increases horizontally to the right, and the $h[2]$-axis increases vertically upward. Here, θ represents the direction of a line, measured clockwise from the vertical axis; see Figure 1.8. Further, since $g_h = g_{-h}$ for all h, the plots corresponding to θ and $\theta + 180°$ are identical. The following values of h are involved for each choice of θ:

$$
\begin{array}{lllll}
\theta = 0° & : & h = (0,0), & (0,1), & (0,2), & \text{etc.} \\
\theta = 45° & : & h = (0,0), & (1,1), & (2,2), & \text{etc.} \\
\theta = 90° & : & h = (0,0), & (1,0), & (2,0), & \text{etc.} \\
\theta = 135° & : & h = (0,0), & (-1,1), & (-2,2), & \text{etc.}
\end{array}
$$

Note that for $\theta = 0°$ and $90°, g_h$ is computed for $r = 0, 1, 2, \ldots$ whereas for $\theta = 45°$ and $135°$, g_h is computed for $r = 0, \sqrt{2}, 2\sqrt{2}, \ldots$.

If the data are believed to be isotropic, then values with the same $|h|$ can be combined together. For example, define

$$
g_1^\sharp = \frac{1}{2}(g_{(1,0)} + g_{(0,1)}), \quad g_2^\sharp = \frac{1}{2}(g_{(2,0)} + g_{(0,2)}),
$$

$$
g_{\sqrt{2}}^\sharp = \frac{1}{2}(g_{(1,1)} + g_{(1,-1)}), \quad g_{2\sqrt{2}}^\sharp = \frac{1}{2}(g_{(2,2)} + g_{(2,-2)}), \text{etc.}
$$

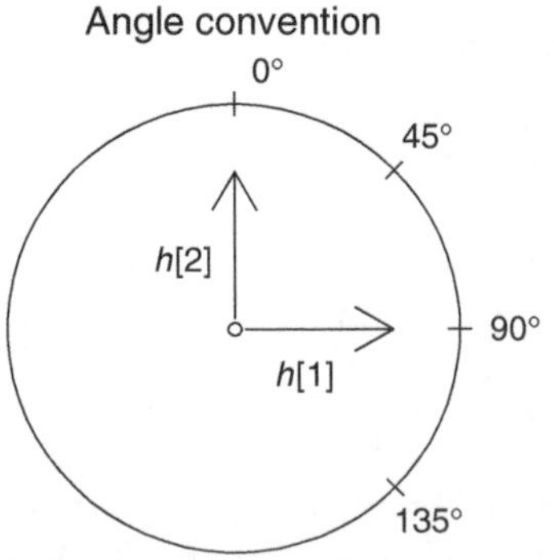

Figure 1.8 Angle convention for polar coordinates. Angles are measured clockwise from vertical.

In general, combining the data from the 0°, 45°, 90° and 135° axes suffices to provide a sufficient summary of the covariance structure of the data.

In the examples that follow, the exact semivariogram has been used for regular data, and the smoothed semivariogram based on angular sectors has been used for irregular data. In both cases, the directional semivariograms for different directions can be combined together to give the *omnidirectional* semivariogram.

Example 1.8 ***Fingerprint section – semivariograms***
A rectangular section, 218 pixels wide by 356 pixels high, was highlighted in the fingerprint in Figure 1.1 for Example 1.2. An image plot of this section is given in Figure 1.9a. This image contains several ridges and valleys parallel to the vertical axis. The semivariograms in the four principal directions are given in Figure 1.9b. The oscillations in the semivariograms in the directions 45°, 90°, and 135° are due to the oscillations in the data between the ridges and valleys. In particular, for the horizontal direction, $\theta = 90°$, there is a cycle of length about 50 pixels, which corresponds to the width between successive ridges. The semivariogram is much lower in the vertical direction, 0°, since it is limited to the variability along a single ridge or valley. An idealized version of a section of a fingerprint with this oscillation behavior is studied in Exercise 1.5. □

Example 1.9 ***Elevation data – semivariograms***
The elevation data was presented in Example 1.3. Semivariograms in the four principal directions are given in Figure 1.10a. They clearly indicate some anisotropy;

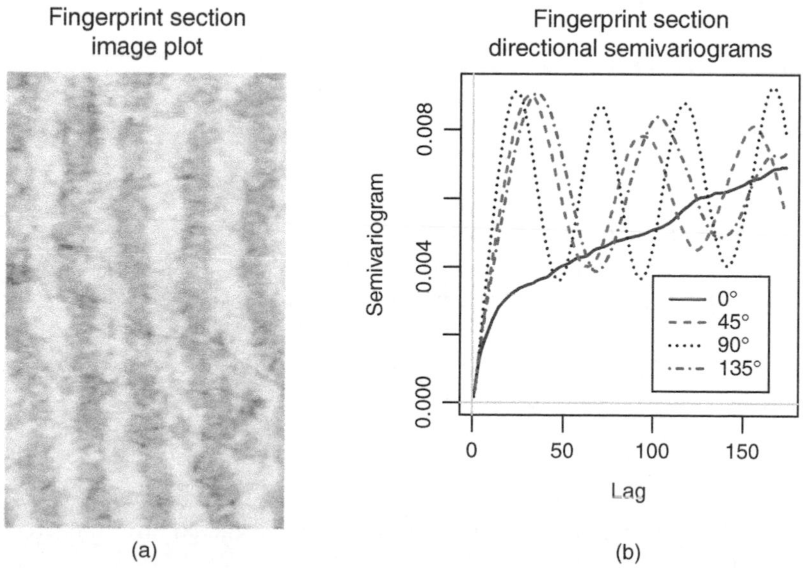

Figure 1.9 Fingerprint section data (218 pixels wide by 356 pixels high): (a) image plot and (b) directional semivariograms.

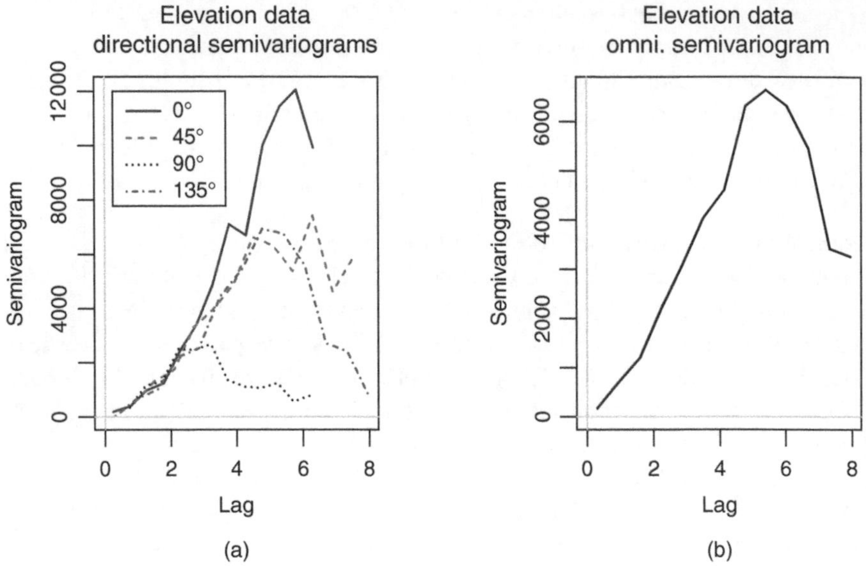

Figure 1.10 Elevation data: (a) directional semivariograms and (b) omnidirectional semivariogram.

the semivariogram in the 90° direction increases more slowly, whereas the semivariogram in the 0° direction increases more quickly. In Chapter 5, we introduce drift terms to describe the large-scale features in the data. Note there is no suggestion of a nugget effect in Figure 1.10b. □

Example 1.10 ***Bauxite data – semivariograms***
Consider the irregularly spaced bauxite data of Example 1.4. Since the data set is rather small, it is not meaningful to fit anything more complicated than an isotropic semivariogram using (1.11). For information, the directional semivariograms are plotted in Figure 1.11a, but they are very noisy. The omnidirectional semivariogram is plotted in Figure 1.11b using (1.9). Note that the omnidirectional semivariogram increases approximately linearly for small lags. It has a range of about $r = 5$ and a possible small nugget effect with a value of about 20. □

Example 1.11 ***Landsat data and synthetic Landsat data – semivariograms***
The Landsat data set was presented in Example 1.5 and the semivariograms are given here in Figure 1.12a. The data appear to be approximately isotropic, with a nugget effect of about 5 and a range of about 20. The synthetic Landsat data set of Example 1.6 was simulated to have a similar semivariogram, as shown in Figure 1.12b. Although the vertical units are different, the other characteristics are very similar. □

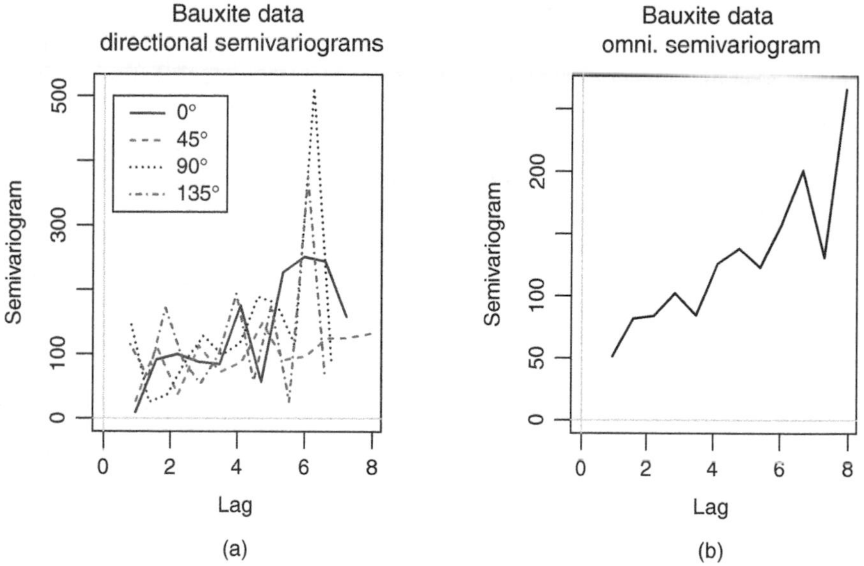

Figure 1.11 Bauxite data: (a) directional semivariograms and (b) omnidirectional semivariogram.

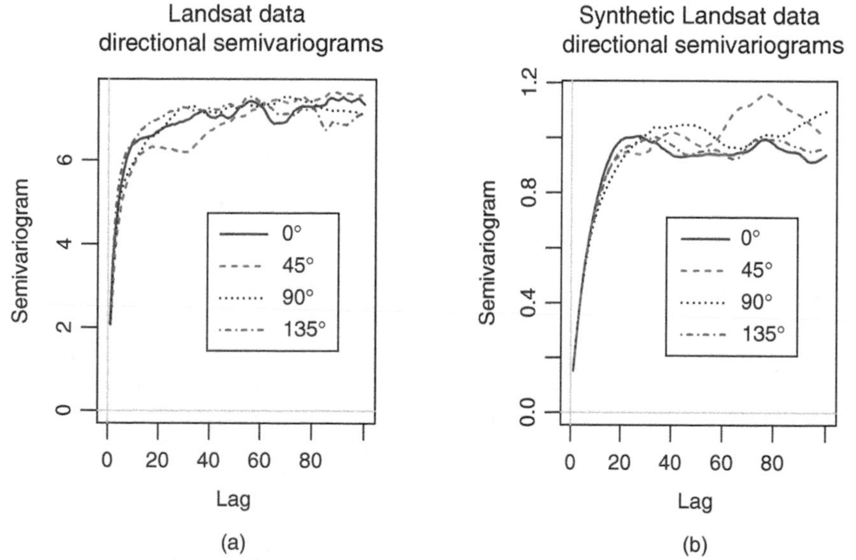

Figure 1.12 Directional semivariograms for (a) the Landsat data and (b) the synthetic Landsat data.

Example 1.12 ***Gravimetric data***
Table 1.4 gives a set of gravimetric data (Fraser, 1957) consisting of local gravity measurements over a regular 10×10 grid in Quebec, Canada. Small-scale disturbances in gravity are caused by many factors, including the presence of certain types of ore-bearing rock (good deposits increase the local gravitation). Hence, areas with unusually large gravimetric readings are worth further investigation using more specific, but much more expensive, methods such as borehole drilling. The values of the sample semivariogram in the four principal directions are listed in Table 1.5. A bubble plot and a plot of the semivariograms are given in Figure 1.13.

Note that for $\theta = 135°$, $g_h \propto |h|^2$, approximately. This feature strongly suggests the presence of a linear trend in the data along this direction. (See Section 5.2.2 for a more general discussion.) This behavior also carries over into the $\theta = 0°$ and $90°$ directions.

Table 1.4 Gravimetric data: local gravity measurements in Quebec, Canada.

13	25	14	13	16	2	−2	−49	−67	−73
39	26	14	18	12	7	−23	−57	−64	−87
13	10	5	20	9	−5	−35	−55	−74	−82
0	10	1	10	0	−26	−49	−75	−91	−94
−14	−6	0	−13	−25	−47	−50	−86	−97	−117
−15	−16	−15	−38	−55	−65	−88	−111	−120	−127
−26	−32	−53	−55	−80	−91	−130	−120	−138	−140
−40	−68	−96	−97	−94	−128	−135	−139	−165	−167
−62	−91	−119	−133	−133	−147	−154	−158	−171	−174
−89	−122	−144	−139	−155	−159	−181	−189	−199	−214

Source: Fraser (1957).

Table 1.5 Semivariograms in each direction for the gravimetric data.

$\|h\|$	$n(h)$	$\theta = 0°$	$\theta = 90°$	$\|h\|$	$n(h)$	$\theta = 45°$	$\theta = 135°$
1	90	304	435	1.4	81	183	1136
2	80	943	1506	2.8	64	460	4296
3	70	1923	3283	4.2	49	860	9554
4	60	3199	5619	5.7	36	1100	16 354
5	50	4742	8546	7.1	25	1224	24 024
6	40	6412	11 820	8.5	16	1138	31 189
7	30	8086	15 421	9.9	9	1077	37 486
8	20	10 302	18 929	11.3	4	550	46 925
9	10	12 158	22 419	12.7	1	256	51 529

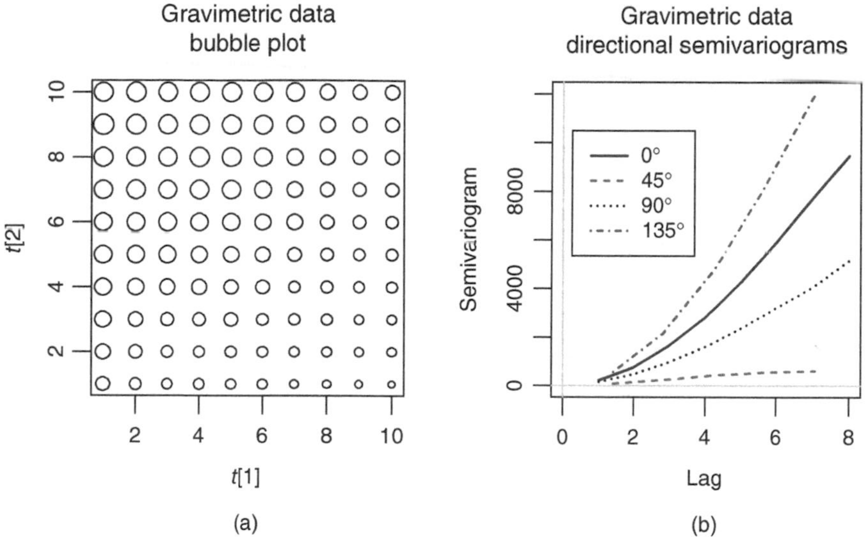

Figure 1.13 Gravimetric data: (a) bubble plot and (b) directional semivariograms.

The behavior of g_h for $\theta = 45°$ is more typical of a stationary random field. Though its behavior is difficult to see from the figure, $g(r, 45°)$ increases with $r = |h|$ up to approximately $r = 5$ and then is approximately constant, $g(r, 45°) \cong 0.055$, for $r \geq 5$. That is, in this direction, the semivariogram has an approximate range $r = 5$ and an approximate sill 0.055. □

Example 1.13 ***Soil data***
Table 1.6 gives the soil surface pH in $CaCl_2$ at 121 sites on an 11×11 square grid (Laslett et al., 1987). A bubble plot and a plot of the semivariogram in the four

Table 1.6 Soil data: surface pH in $CaCl_2$ on an 11×11 grid.

4.80	4.38	4.33	4.31	4.49	4.38	4.44	4.46	4.54	4.50	4.24
4.42	4.29	4.19	4.28	4.58	4.89	4.74	4.68	4.54	4.86	4.33
4.30	4.30	4.87	4.70	4.68	5.04	5.03	4.86	4.43	4.14	4.32
4.26	4.64	4.54	4.54	4.64	4.76	4.42	4.61	4.30	4.54	4.30
4.20	4.42	4.50	4.80	4.90	4.76	4.53	4.23	4.26	4.58	4.12
4.19	4.40	4.32	4.48	4.59	4.67	4.50	4.80	4.28	4.50	4.44
4.34	4.54	4.52	4.73	4.32	4.90	4.34	4.36	4.23	4.31	4.30
4.54	4.20	4.44	4.60	4.84	4.46	4.39	4.36	4.27	4.03	4.37
4.56	4.64	4.64	4.64	4.69	4.36	4.36	4.62	4.30	4.34	4.47
4.44	4.83	4.80	4.84	4.50	4.30	4.29	4.49	4.21	4.16	4.64
4.53	4.39	4.74	4.70	4.36	4.51	4.34	4.44	4.30	4.30	4.15

Source: Data from Laslett et al. (1987).

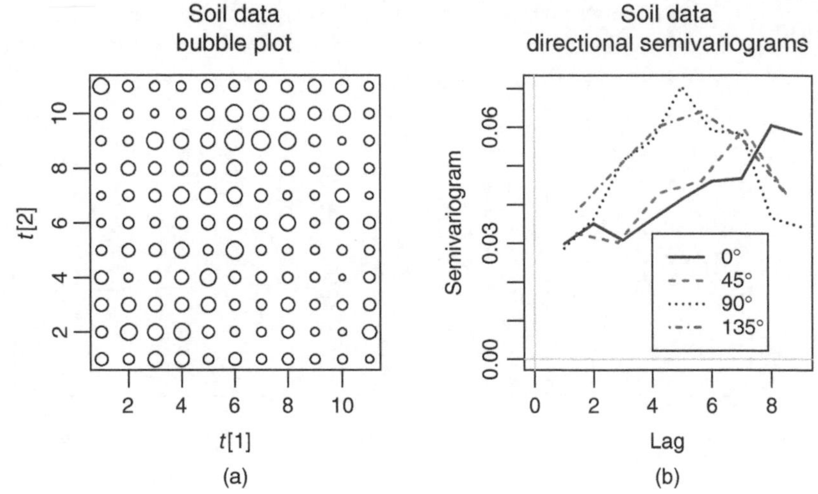

Figure 1.14 Soil data: (a) bubble plot and (b) directional semivariograms.

principal directions are given in Figure 1.14. The semivariogram displays some anisotropy. It increases more slowly in the 45° and 90° directions and more quickly in the 0° and 135° (= −45°) directions.

Note that the plotted semivariograms, when extrapolated to lag $r = 0$, appear to have a (common) nonzero intercept with a value of about 0.05. Since the sill is about 0.12, the nugget effect is very pronounced on this data set. □

1.6 Some Special Features of Spatial Analysis

(a) *Ordering*. There is a lack of ordering in spatial analysis unlike in time series. The ARIMA models in time series are built from a one-sided neighborhood at each time t. In contrast, natural spatial models, e.g. the conditional autoregression (CAR) models of Chapter 4, involve symmetric finite neighborhoods at each site. However, there are also some spatial models with an artificial ordering of space, e.g. unilateral models; see Chapter 4.

(b) *Edge effects*. Certain problems can arise in analysis of spatial data because some sites lie at the edge of the region rather than the interior. For spatial models defined in terms of neighborhoods, sites at the boundary do not have a full neighborhood. Other models use a torus approximation in which opposite edges are wrapped onto one another. In each case, the presence of edges can lead to artifacts in the statistical analysis if care is not taken; see Chapter 6. This problem of edge effects increases dramatically with dimension. In a time series of length n, there are only two end points. In a two-dimensional $n \times n$ array of data, there are $4n - 4$ edge and corner sites. In general, in d dimensions, there are $O(n^{d-1})$ edge sites in an n^d array of data. The reason that $g_h \neq s_0 - s_h$ for the sample semivariogram and the sample covariance function in Section 1.4 is due to these edge effects.

(c) *Asymptotics*. There are basically two types of limiting behavior for spatial data as more data are collected: infill and outfill asymptotics. Infill asymptotics imply collecting data at more finely spaced intervals on a region of fixed size, whereas in outfill asymptotics, the region of data expands with the spacing between the sites held fixed. In outfill asymptotics, widely spaced observations are usually modeled to be asymptotically independent, with most of the dependence occurring between nearby observations. Thus, outfill asymptotics represents a natural generalization of the classical situation of independent, identically distributed observations. In contrast, in infill asymptotics much heavier dependence is involved as the details of a smooth random function are gradually filled in. Infill asymptotics are often not relevant in time series because of a fixed sampling interval, e.g. in an economic series collected monthly, say. However, infill asymptotics can be more relevant in a spatial setting as increasing amounts of information are collected within a fixed area. See Section 5.14.

(d) *Anisotropy*. Another aspect of spatial analysis not present in time series (or one-dimensional spatial statistics) is lack of isotropy, i.e. a semivariogram may look different in different directions. See, for example, the fingerprint data in Example 1.8. The simplest model for anisotropy, called geometric anisotropy, is described in Chapter 5.

(e) *Role of increments*. In time-series analysis, the use of successive differences $x_t - x_{t-1}$ provides a powerful tool for the analysis of nonstationary data. However, in spatial analysis, there is no notion of successive/consecutive sites. Thus, it is necessary to look at *all* differences of the data simultaneously. An increment (of order 0) is defined as a linear combination of the data for which the coefficients sum to zero. The successive differences defined above are examples of increments in one dimension. Note that a constant term is filtered out or annihilated by such an increment, $(x_t + c) - (x_{t-1} + c) = x_t - x_{t-1}$. It is also possible to define higher order increments for which polynomial terms in t are annihilated. Increments play a key role in describing and working with intrinsic random fields. See Chapter 3.

(f) *Periodicity*. Periodic models are an important tool to describe seasonal behavior in time-series models. The analogue in two dimensions of a sine wave in one dimension looks like a corrugated iron sheet used for roofing. Such a wave is constant in the direction parallel to the wave front and oscillates in the direction normal to the wave front. In general, periodic models are less important

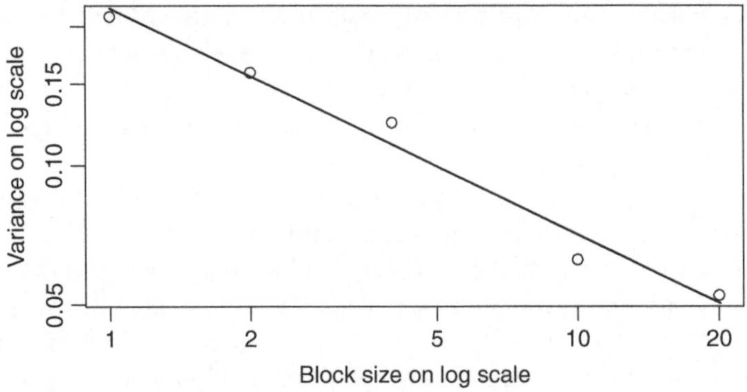

Figure 1.15 Mercer–Hall wheat data: log–log plot of variance vs. block size.

in higher dimensions than in one dimension, but for a good example where they are important, see Example 1.8 involving the fingerprint data.

(g) *Self-similarity.* A process is called self-similar if it looks essentially the same at all scales of measurement. The concept of self-similarity is closely related to the concept of fractals (Mandelbrot, 1982). Two aspects of self-similarity are important for spatial processes. One aspect is the long-range correlation and other is short-term regularity. These themes are discussed in Chapter 3; see also Example 1.14 below.

(h) *Aggregated data.* In spatial data, observations sometimes involve pooling of information over blocks of some size. Similar considerations arise in time series where the measurements might be aggregated over a month or year. Obviously, the covariance structure of the data will depend on the size of the blocks. See Section 2.13 for more details about the effect of aggregation on the covariance function.

Example 1.14 ***Mercer–Hall wheat data***

Mercer and Hall (1911) analyzed the results of a uniformity trial of wheat yields, measured on a 20×25 array of 500 plots (see Table 1.7). The plots were nearly square, of dimensions 10.82feet $\times$ 8.50feet. Since the mean yield is expected to be the same in each plot, all of the difference between plot yields can be modeled by autocorrelation. The main objective of the analysis is to investigate this autocorrelation by looking at how the variability of the data changes under different levels of aggregation. The plots are pooled together to form larger blocks of size b, say, as described in Table 1.8 and the observed yield per unit area is recorded for each block. Note that the sample variance over the blocks of the observed yield per unit area decreases as the block size b increases. A plot of log variance against $\log b$ in Figure 1.15 is very nearly a straight line, with the slope estimated by least squares to be −0.49. This result suggests a power law relationship,

$$\text{block variance} \propto \text{block area}^{-0.5}.$$

A theoretical explanation of this result in terms of long-range correlation is given in Section 2.13. □

Table 1.7 Mercer–Hall wheat yield (in lbs.) for 20 (rows) × 25 (columns) agricultural plots (Mercer and Hall, 1911), where top–bottom corresponds to West–East, and left–right corresponds to North–South.

(a) First 12 columns

3.63	4.15	4.06	5.13	3.04	4.48	4.75	4.04	4.14	4.00	4.37	4.02
4.07	4.21	4.15	4.64	4.03	3.74	4.56	4.27	4.03	4.50	3.97	4.19
4.51	4.29	4.40	4.69	3.77	4.46	4.76	3.76	3.30	3.67	3.94	4.07
3.90	4.64	4.05	4.04	3.49	3.91	4.52	4.52	3.05	4.59	4.01	3.34
3.63	4.27	4.92	4.64	3.76	4.10	4.40	4.17	3.67	5.07	3.83	3.63
3.16	3.55	4.08	4.73	3.61	3.66	4.39	3.84	4.26	4.36	3.79	4.09
3.18	3.50	4.23	4.39	3.28	3.56	4.94	4.06	4.32	4.86	3.96	3.74
3.42	3.35	4.07	4.66	3.72	3.84	4.44	3.40	4.07	4.93	3.93	3.04
3.97	3.61	4.67	4.49	3.75	4.11	4.64	2.99	4.37	5.02	3.56	3.59
3.40	3.71	4.27	4.42	4.13	4.20	4.66	3.61	3.99	4.44	3.86	3.99
3.39	3.64	3.84	4.51	4.01	4.21	4.77	3.95	4.17	4.39	4.17	4.17
4.43	3.70	3.82	4.45	3.59	4.37	4.45	4.08	3.72	4.56	4.10	3.07
4.52	3.79	4.41	4.57	3.94	4.47	4.42	3.92	3.86	4.77	4.99	3.91
4.46	4.09	4.39	4.31	4.29	4.47	4.37	3.44	3.82	4.63	4.36	3.79
3.46	4.42	4.29	4.08	3.96	3.96	3.89	4.11	3.73	4.03	4.09	3.82
5.13	3.89	4.26	4.32	3.78	3.54	4.27	4.12	4.13	4.47	3.41	3.55
4.23	3.87	4.23	4.58	3.19	3.49	3.91	4.41	4.21	4.61	4.27	4.06
4.38	4.12	4.39	3.92	4.84	3.94	4.38	4.24	3.96	4.29	4.52	4.19
3.85	4.28	4.69	5.16	4.46	4.41	4.68	4.37	4.15	4.91	4.68	5.13
3.61	4.22	4.42	5.09	3.66	4.22	4.06	3.97	3.89	4.46	4.44	4.52

Table 1.7 (Continued)

(b) Last 13 columns

4.58	3.92	3.64	3.66	3.57	3.51	4.27	3.72	3.36	3.17	2.97	4.23	4.53
4.05	3.97	3.61	3.82	3.44	3.92	4.26	4.36	3.69	3.53	3.14	4.09	3.94
3.73	4.58	3.64	4.07	3.44	3.53	4.20	4.31	4.33	3.66	3.59	3.97	4.38
4.06	3.19	3.75	4.54	3.97	3.77	4.30	4.10	3.81	3.89	3.32	3.46	3.64
3.74	4.14	3.70	3.92	3.79	4.29	4.22	3.74	3.55	3.67	3.57	3.96	4.31
3.72	3.76	3.37	4.01	3.87	4.35	4.24	3.58	4.20	3.94	4.24	3.75	4.29
4.33	3.77	3.71	4.59	3.97	4.38	3.81	4.06	3.42	3.05	3.44	2.78	3.44
3.72	3.93	3.71	4.76	3.83	3.71	3.54	3.66	3.95	3.84	3.76	3.47	4.24
4.05	3.96	3.75	4.73	4.24	4.21	3.85	4.41	4.21	3.63	4.17	3.44	4.55
3.37	3.47	3.09	4.20	4.09	4.07	4.09	3.95	4.08	4.03	3.97	2.84	3.91
4.09	3.29	3.37	3.74	3.41	3.86	4.36	4.54	4.24	4.08	3.89	3.47	3.29
3.99	3.14	4.86	4.36	3.51	3.47	3.94	4.47	4.11	3.97	4.07	3.56	3.83
4.09	3.05	3.39	3.60	4.13	3.89	3.67	4.54	4.11	4.58	4.02	3.93	4.33
3.56	3.29	3.64	3.60	3.19	3.80	3.72	3.91	3.35	4.11	4.39	3.47	3.93
3.57	3.43	3.73	3.39	3.08	3.48	3.05	3.65	3.71	3.25	3.69	3.43	3.38
3.16	3.47	3.30	3.39	2.92	3.23	3.25	3.86	3.22	3.69	3.80	3.79	3.63
3.75	3.91	3.51	3.45	3.05	3.68	3.52	3.91	3.87	3.87	4.21	3.68	4.06
4.49	3.82	3.60	3.14	2.73	3.09	3.66	3.77	3.48	3.76	3.69	3.84	3.67
4.19	4.41	3.54	3.01	2.85	3.36	3.85	4.15	3.93	3.91	4.33	4.21	4.19
3.70	4.28	3.24	3.29	3.48	3.49	3.68	3.36	3.71	3.54	3.59	3.76	3.36

Table 1.8 Aggregated Mercer–Hall wheat data for plots aggregated into blocks, giving the array layout for the blocks, the block dimensions, the block size b (number of plots in each block), the number of blocks n, and sample variance s^2.

Block layout	Block dimensions	b	n	s^2
20 × 25	1 × 1	1	500	0.2100
10 × 25	2 × 1	2	250	0.1587
5 × 25	4 × 1	4	125	0.1236
10 × 5	2 × 5	10	50	0.0625
5 × 5	4 × 5	20	25	0.0523

Exercises

1.1 Take the data used in Example 1.1. Using lags $h = (1,1)$ and $h = (1,-1)$, verify that $|D_h| = 6$ in both cases, and that

$$s_{(1,1)} = -8.22, \quad g_{(1,1)} = 41.58,$$
$$s_{(1,-1)} = 11.36, \quad g_{(1,-1)} = 3.50.$$

Also show that $s_{(0,0)} = 30.81$. Hence confirm Eq. (1.3) for this data set; namely $g_{(1,1)} \neq s_{(0,0)} - s_{(1,1)}$ and $g_{(1,-1)} \neq s_{(0,0)} - s_{(1,-1)}$. That is, that the population identity (1.1) does not hold in the sample case for this data set.

1.2 *Matheron (1962).* For a set of data giving bauxite ore concentration at $n = 300$ sites, the following semivariograms were calculated. (Note this is different from Example 1.4.)

h	$\theta = 0$		$\theta = 90°$	
(unit 50 m)	$n(h)$	$g(h)$	$n(h)$	$g(h)$
1	266	26.0	267	23.7
2	243	24.6	245	26.4
3	233	28.4	223	26.6
4	222	26.8	199	30.2
5	204	31.4	180	28.8
6	185	31.0	159	31.0
7	173	29.4	138	35.3
8	161	25.7	118	30.5
9	147	30.5	95	30.5
10	130	27.8	76	31.6

h	$\theta = 45°$		$\theta = 135°$	
(unit 50 m)	$n(h)$	$g(h)$	$n(h)$	$g(h)$
1.41	249	27.1	248	21.75
2.83	224	33.3	216	29.8
4.24	204	29.0	189	30.3
5.65	182	31.8	160	28.6
7.06	161	31.4	129	33.6
8.49	137	34.9	100	31.8
9.90	119	38.2	73	33.8
11.30	103	28.3	52	27.6

Draw the semivariograms and show that the data appear to be isotropic.

1.3 The following data provided by R. Webster give the thickness of topsoil in cm for a field, where points on a regular grid were inspected, 10 m apart in each direction.

				75.0	40.0	70.0			
		30.0	85.0	55.0	50.0	50.0	40.0		
	90.0	30.0	75.0	80.0	45.0	90.0	80.0	90.0	
45.0	35.0	85.0	80.0	90.0	90.0	90.0	90.0	90.0	
90.0	45.0	45.0	85.0	90.0	90.0	90.0	90.0	90.0	30.0
40.0	40.0	80.0	90.0	90.0	90.0	90.0	90.0	35.0	30.0
40.0	35.0	35.0	90.0	90.0	90.0	90.0	90.0	40.0	5.0
35.0	50.0	30.0	90.0	90.0	55.0	90.0	90.0	75.0	80.0
40.0	40.0	35.0	90.0	45.0	90.0	90.0	90.0	90.0	40.0
35.0	90.0	90.0	85.0	90.0	90.0	90.0	90.0	90.0	30.0

Plot the semivariograms for each of the four main directions. Show that there is a substantial nugget effect and that the semivariogram increases more quickly in the 90° direction than in the 0° direction.

1.4 *Krige* (1976). For the subsection 302 of the Hartbeesfontein Mine, the directional semivariograms for gold ore concentrations in the four main directions are as follows.

	$g(h)$			$g(h)$	
h	N/S	E/W	h	NW/SE	NE/SW
1	0.405	0.430	$\sqrt{2}$	0.455	0.470
2	0.463	0.543	$2\sqrt{2}$	0.542	0.577
3	0.493	0.588	$3\sqrt{2}$	0.605	0.590
4	0.522	0.620	$4\sqrt{2}$	0.623	0.618
5	0.544	0.632	$5\sqrt{2}$	0.658	0.621
6	0.553	0.660	$6\sqrt{2}$	0.675	0.642
7	0.570	0.657	$7\sqrt{2}$	0.690	0.652
8	0.589	0.670	$8\sqrt{2}$	0.680	0.655
9	0.577	0.682	$9\sqrt{2}$	0.695	0.653
10	0.581	0.688	$10\sqrt{2}$	0.685	0.644
12	0.620	0.676	$11\sqrt{2}$	0.640	0.660

h	g(h) N/S	g(h) E/W	h	g(h) NW/SE	g(h) NE/SW
14	0.618	0.692	$12\sqrt{2}$	0.650	0.660
16	0.637	0.674	$13\sqrt{2}$	0.635	0.652
18	0.640	0.660			
20	0.622	0.660			

Plot the semivariograms $g(h)$ vs. $\log h$ in the four directions N/S, E/W, NW/SE, and NE/SW. Show that the semivariograms in all four directions can be extrapolated to lag 0 to give a common intercept or nugget effect of about 0.3. Further show that the semivariogram in the N/S direction increases to a sill or a maximal value of about 0.6 by about lag 10, and that the semivariogram in the E/W direction increases to the same sill by about lag 15, with the semivariogram in the other two directions lying in-between.

1.5 Consider an idealized section of a fingerprint given as an $n[1] \times n[2]$ table of numbers for which the odd rows consist entirely of zeros, and the even rows consist entirely of ones. Show that the semivariogram $g(h)$ is identically 0 in the horizontal direction $\theta = 0°$, whereas it oscillates between 0 and 1 in the other three principal directions. What happens when the bands of zeros and ones are each repeated three times, $3 \ll \min\{n[1], n[2]\}$?

1.6 The Mercer–Hall wheat data are given in Table 1.7 as a 20×25 matrix. Suppose $b[1]$ and $b[2]$ are two integers dividing 20 and 25, respectively. Partition the original matrix into $b[1] \times b[2]$ blocks and average the data values over each block to get an aggregated matrix of size $n[1]/b[1] \times n[2]/b[2]$. Table 1.8 gives five choices for the dimensions $(b[1], b[2])$ with the size given by $b = b[1]b[2]$.

(a) Show that the upper-left corner entries in each of the aggregated matrices take the values 3.6300, 3.8500, 4.0275, 4.1110, 4.1445, respectively.

(b) Complete the calculation of the aggregated matrices, and show that the variances (over the $n[1]/b[1] \times n[2]/b[2]$ data values in each aggregated matrix) take the values in the final column of Table 1.8.

2

Stationary Random Fields

2.1 Introduction

A *random field* or *random process* is a collection of random variables $\{X(t) : t \in D\}$ where $D \subset \mathbb{R}^d$ is a subset of d-dimensional space. This chapter lays the foundations for the statistical analysis of random fields through the study of their first and second moment properties, that is, the means and variances of the random field at different sites together with the covariances $\text{cov}\{X(t), X(s)\} = \sigma(s,t)$, say, at pairs of sites. A covariance function is essential in spatial analysis because it specifies the extent to which the values of $X(t)$ and $X(s)$ are likely to be close to each other, especially when s is near t.

General properties of covariance functions are given in Section 2.2. The important special case of stationary covariance functions, for which $\sigma(s,t) = \sigma(h)$, say, depends only on $t - s = h$, say, is explored in Section 2.3. For these covariance functions, a spectral representation is derived, which gives both a frequency-domain interpretation and a useful mathematical tool for constructing valid covariance functions. A further specialization to *isotropic* covariance functions is given in Section 2.4 with some examples in Sections 2.5–2.7.

Starting with a given random field, it is possible to carry out various linear operations to derive new random fields. One example is differentiation for sufficiently smooth random fields, which is explored in Section 2.9. A converse operation is regularization of random fields, which involves integration, and is covered in Section 2.10.

Most of this chapter is focused on continuously indexed random fields, with $t \in \mathbb{R}^d$. However, there are also important examples indexed by the integer lattice, $t \in \mathbb{Z}^d$. These random fields are described in Section 2.11.

Typically, the covariance function $\sigma(h)$ dies away to 0 as $|h| \to \infty$. If $\sigma(h)$ dies away quickly enough, averages from the random field bchave in some sense like the averages of uncorrelated random variables. However, if $\sigma(h) \to 0$ slowly, then there is "long-range correlation" between distant observations from the random

Spatial Analysis, First Edition. John T. Kent and Kanti V. Mardia.

field, and the behavior of averages is qualitatively different. This situation is analyzed in Section 2.13.

The last section, Section 2.14, is devoted to simulation of Gaussian random fields. Simulation is useful for assessing the properties and behaviors of different models. Simulations can either be carried out directly or by means of the spectral representation of covariance functions. Spectral methods can be more efficient for large data sets and give insight into a probabilistic interpretation of the spectral representation. A variant of the spectral methods is the circulant method, which gives efficient and exact simulation for regularly spaced sites on a rectangular domain.

Gaussian random fields are the most important examples of random fields. However, with the notable exceptions of the smoothness properties in Section 2.9 and the simulation methods in Section 2.14, this chapter is mainly concerned with second moment properties and requires no assumption of Gaussianity.

2.2 Second Moment Properties

A *random field* $\{X(t) : t \in \mathbb{R}^d\}$ is a d-dimensional real-valued spatial stochastic process, $d \geq 1$. Each vector $t = (t[1], \ldots, t[d])$ is called a *site* of the process. If the random field has finite second moments, we can define the *mean function*

$$\mu(t) = E\{X(t)\}, \qquad t \in \mathbb{R}^d, \tag{2.1}$$

and the *covariance function*

$$\sigma(s,t) = \text{cov}\{X(s), X(t)\} = E\{[X(s) - \mu(s)][X(t) - \mu(t)]\}, \quad s, t \in \mathbb{R}^d, \tag{2.2}$$

which is symmetric in its arguments, $\sigma(s,t) = \sigma(t,s)$.

An important simplification occurs when the distribution of the random field is invariant under an arbitrary shift of the sites. In this case, the random field is called *strictly stationary*. Thus, the random field is strictly stationary if and only if for all $n \geq 1$ and for all choices of sites $t_j \in \mathbb{R}^d$, $1 \leq j \leq n$, and for all shift vectors $h \in \mathbb{R}^d$, the n-dimensional distribution of $\{X(t_j) : 1 \leq j \leq n\}$ is the same as that of $\{X(t_j + h) : 1 \leq j \leq n\}$. Remember our convention that the components of t_j are written using square brackets, $t_j = (t_j[1], \ldots, t_j[d])$.

If a random field is strictly stationary, the first two moments can be simplified to

$$\mu(t) = \mu, \text{ say, not depending on } t, \tag{2.3}$$

$$\sigma(t, t+h) = \sigma(h), \text{ say, not depending on } t. \tag{2.4}$$

Note that $\sigma(h)$, now a function of a single argument $h \in \mathbb{R}^d$, is an *even* function

$$\sigma(h) = \sigma(-h).$$

A less restrictive condition on a random field than *strict stationarity* is the condition of *weak* or *second-order stationarity*, in which we merely require that the first two moments be invariant under shifts, that is, (2.3) and (2.4) should hold. In this book, we use the term "stationarity" as a synonym for weak stationarity.

Stationary random fields form the fundamental models in our analysis of spatial data, and the purpose of this chapter is to explore their mathematical properties. In Chapter 3, the concept of a stationary random field is extended to define an *intrinsic* random field (that is, a random field with stationary increments) and a *generalized* random field.

For a stationary random field with a covariance function $\sigma(h)$, it is always assumed that

$$\sigma(h) \text{ is continuous for all } h \in \mathbb{R}^d, \tag{2.5}$$

and usually assumed that

$$\sigma(h) \to 0 \text{ as } |h| \to \infty, \tag{2.6}$$

where $|h| = (h[1]^2 + \cdots + h[d]^2)^{1/2}$ is the Euclidean norm of h. Thus, observations at nearby sites will be highly correlated with one another and observations at distant sites will be nearly uncorrelated.

Sometimes when estimating $\sigma(h)$ from a data set, an apparent discontinuity appears at $h = 0$. This behavior is known as a "nugget effect," (Section 1.4.1) and can be due to either measurement error or to small-scale variability in the data. Hence when fitting covariance models to data, it is often necessary to augment a continuous covariance function by adding a nugget effect (Chapter 5).

An important class of random fields is the class of *Gaussian random fields*. A random field $\{X(t)\}$ is Gaussian if for all $n \geq 1$ and all choices of sites t_j, $1 \leq j \leq n$, the distribution of $(X(t_j) : 1 \leq j \leq n)$ is multivariate normal, with mean vector $(\mu(t_j) : 1 \leq j \leq n)$ and covariance matrix $(\sigma(t_i, t_j) : 1 \leq i, j \leq n)$. The multivariate normal distribution is completely specified in terms of its first two moments. Hence, the distributional properties of a Gaussian random field are determined by its mean and covariance functions. In particular, a stationary Gaussian random field will also be strictly stationary.

If $\{X(t) : t \in D\}$ is a Gaussian process (GP) on a domain $D \subset \mathbb{R}^d$, write

$$X(\cdot) \sim GP(\mu(\cdot), \sigma(\cdot, \cdot), D) \tag{2.7}$$

for a general covariance function, and write

$$X(\cdot) \sim GP(\mu(\cdot), \sigma(\cdot), D) \tag{2.8}$$

for a stationary covariance function.

2.3 Positive Definiteness and the Spectral Representation

In this section, we examine conditions that a function $\sigma(s, t)$ must satisfy to represent the covariance function of some random field. For a stationary random field, we are able to obtain an explicit characterization.

First, we recall the definition of positive definiteness for matrices and functions.

Definition 2.3.1 An $n \times n$ symmetric matrix $B = (b_{ij})$ is called *positive semidefinite* if for all $n \times 1$ column vectors $\boldsymbol{a} = (a_j)$,

$$\boldsymbol{a}^T B \boldsymbol{a} = \sum a_i a_j b_{ij} \geq 0. \tag{2.9}$$

If the inequality is strict for all vectors $\boldsymbol{a} \neq \boldsymbol{0}$, the matrix is called *positive definite.*

Similarly, a symmetric function of two arguments $\sigma(s, t)$, $s, t \in \mathbb{R}^d$, is called *positive definite* or *positive semidefinite* if for all $n \geq 1$ and all choices of distinct sites, t_j, $1 \leq j \leq n$, an $n \times n$ matrix B with entries $b_{ij} = \sigma(t_i, t_j)$ is positive definite or positive semidefinite, respectively. For an even function of one argument $\sigma(h)$, $h \in \mathbb{R}^d$, positive definiteness and positive semidefiniteness are defined similarly with $b_{ij} = \sigma(t_i - t_j)$.

Positive definiteness is important in the study of random fields for the following reason.

Theorem 2.3.1 *If $\sigma(s, t)$ is the covariance function of a random field $\{X(t) : t \in \mathbb{R}^d\}$, then $\sigma(s, t)$ is positive semidefinite. That is, for any $n \geq 1$, any selection of sites $t_1, \ldots, t_n$ and any vector $\boldsymbol{a}$ ($n \times 1$) of coefficients, the following quadratic form is nonnegative:*

$$\sum a_i a_j \; \sigma(t_i, t_j) \geq 0. \tag{2.10}$$

Proof: The proof follows immediately from the representation

$$\sum a_i a_j \; \sigma(t_i, t_j) = \sum a_i a_j \; \mathrm{cov}\{X(t_i), X(t_j)\} = \mathrm{var}\left\{ \sum a_i X(t_i) \right\} \geq 0,$$

since any variance must be nonnegative. □

This theorem also has an important converse.

Theorem 2.3.2 *If $\sigma(s, t)$ is a symmetric positive semidefinite function, then it represents the covariance function of a random field.*

Proof: For the purposes of this proof, it is simplest to construct a *Gaussian* random field with the required covariance function and with mean 0. To carry out this construction, we call upon a theorem of Kolmogorov (see, for example, Cramér and Leadbetter, 1967, pp. 33–37), which states that to construct a stochastic process, it is sufficient to specify consistently its finite-dimensional distributions. Here "consistently" means the following. Let $\pi(t_1, \ldots, t_n)$ denote an n-dimensional probability distribution specified at sites $(t_1, \ldots, t_n)$, assumed for simplicity to have a probability density $f(x_1, \ldots, x_n)$. Let t_{n+1} be another site. Then $\pi(t_1, \ldots, t_n)$ must be the same as the marginal distribution of $\pi(t_1, \ldots, t_{n+1})$ at sites $(t_1, \ldots, t_n)$ after averaging over the values of this $(n+1)$-dimensional distribution at site t_{n+1}, or in terms of probability densities, the equation

$$f(x_1, \ldots, x_n) = \int f(x_1, \ldots, x_{n+1})\, dx_{n+1}$$

must hold.

Consider sites t_j, $1 \leq j \leq n+1$. Since $\sigma(s,t)$ is a positive semidefinite function, the matrix B with elements $b_{ij} = \sigma(t_i, t_j)$, $1 \leq i,j \leq n+1$, is positive semidefinite. Hence, there exists an $(n+1)$-dimensional multivariate normal distribution with mean 0 and covariance matrix B. Further, the marginal distribution of this $(n+1)$-dimensional distribution at sites $t_1, \ldots, t_n$ is n-dimensional normal with mean 0 and covariance matrix $\{\sigma(t_i, t_j) : 1 \leq i,j \leq n\}$ (Section A.3.4). Hence, Kolmogorov's consistency condition is satisfied and the theorem is proved. □

For technical reasons, we limit our attention in this book to covariance functions that are continuous. This assumption is made to ensure that random fields are sufficiently smooth to be useful in applications. In particular, continuity is an important assumption in the discussion of smoothness (Section 2.9), regularization (Section 2.10), and simulation (Section 2.14). Some further explanation for this assumption is given in Section 2.7.

Next, we restrict our attention to continuous covariance functions for *stationary* random fields. The following theorem gives an important characterization. For $h \in \mathbb{R}^d$, $\omega \in \mathbb{R}^d$, let $h^T\omega = h[1]\omega[1] + \cdots + h[d]\omega[d]$ denote the usual inner product.

Theorem 2.3.3 (Bochner's Theorem) *Let $\sigma(h)$, $h \in \mathbb{R}^d$, be a continuous even real-valued function. Then, $\sigma(h)$ is positive semidefinite if and only if it is the Fourier transform of a symmetric positive finite measure $F(d\omega)$ on $\mathbb{R}^d$,*

$$\sigma(h) = \int e^{i\omega^T h} F(d\omega). \tag{2.11}$$

Further, if $F(d\omega)$ has a density, $F(d\omega) = f(\omega)d\omega$, $\omega \in \mathbb{R}^d$, then $\sigma(h)$ is positive definite and $\sigma(h) \to 0$ as $|h| \to \infty$. Here $F(d\omega)$ is called the spectral measure *corresponding to $\sigma(h)$, and when it has a density, $f(\omega)$ is called the* spectral density.

Proof: Suppose $F(d\omega)$ is a nonnegative symmetric finite measure on $\mathbb{R}^d$ (so $F(d\omega) \geq 0$, $F(d\omega) = F(-d\omega)$ and $\int F(d\omega) < \infty$) and let $\sigma(h)$ be defined by (2.11).

The finiteness of $F(d\omega)$ ensures that $\sigma(h)$ is continuous, and the symmetry ensures that $\sigma(h)$ is real-valued. For any vector of coefficients $\boldsymbol{a}$, note the identity

$$\sum_{j,k=1}^{n} a_j a_k \sigma(t_j - t_k) = \int \sum_{j,k=1}^{n} a_j a_k e^{i\omega^T(t_j - t_k)} F(d\omega)$$
$$= \int |\sum_{j=1}^{n} a_j e^{i\omega^T t_j}|^2 F(d\omega), \tag{2.12}$$

which must be nonnegative if $F(d\omega)$ is nonnegative. Hence, $\sigma(h)$ must be a positive semidefinite.

Further, the principle of analytic continuation ensures that the function $\sum_{j=1}^{n} a_j e^{i\omega^T t_j}$ cannot vanish for ω in any open set in $\mathbb{R}^d$, provided the points $\{t_j\}$ are distinct and the coefficients $\{a_j\}$ are not identically 0. Hence, if $F(d\omega)$ possesses a density, the integral in (2.12) must be strictly positive. Hence, in this case, $\sigma(h)$ is a positive definite function.

The converse part of the theorem is more challenging and states that any continuous even real-valued positive semidefinite function has a representation (2.12). After standardization, this theorem becomes a statement about the class of possible characteristic functions $\sigma(h)/\sigma(0)$ for symmetric probability measures $F(d\omega)/\sigma(0)$. See, e.g., Feller (1966, p. 585). The last part of the theorem, stating that $\sigma(0) \to 0$ as $|h| \to \infty$ when the spectral measure has a density, follows from the Riemann–Lebesgue lemma; see, e.g., Feller (1966, p. 486), for the one-dimensional case. □

This theorem is important because it enables us to use our experience of characteristic functions to give examples of covariance functions; see Sections 2.5–2.7. If $\sigma(h)$ is integrable, then the spectral density $f(\omega)$ can be found by the Fourier inversion formula

$$f(\omega) = (2\pi)^{-d} \int e^{-i\omega^T h} \sigma(h) dh, \qquad \omega \in \mathbb{R}^d. \tag{2.13}$$

For some covariance models, the parameters have a simpler interpretation in the Fourier domain than in the spatial domain (especially, the autoregression lattice models in Chapter 4), thus leading to elegant spectral methods of estimation. See Section 6.5 for more details.

2.4 Isotropic Stationary Random Fields

An important class of stationary random fields possesses the property of *isotropy*. That is, the finite-dimensional distributions are invariant under rotations, so that the covariance function $\sigma(h)$ depends only on the radial component $|h|$. Thus, we can write

$$\sigma(h) = \sigma^{\#}(r), \text{ say,} \tag{2.14}$$

where $r = |h|$. We use the # notation to emphasize that σ is a function of a d-dimensional vector h, whereas $\sigma^{\#}$ depends only on the real number $r \geq 0$. We call $\sigma^{\#}(r)$ the *radial covariance function* corresponding to σ.

Let $H(d)$ denote the class of continuous radial covariance functions corresponding to d-dimensional isotropic covariance functions. The purpose of this section is to investigate how $H(d)$ varies with d. A more detailed study can be found in Kingman (1963).

Suppose $\{X(t) : t \in \mathbb{R}^d\}$ is an isotropic stationary random field on $\mathbb{R}^d$, and let Q denote a hyperplane of some dimension d', with $d' < d$. Then, $X(t)$ is still an isotropic stationary random field when restricted to Q, with the same radial covariance function. Hence, we have the following property:

$$H(d) \subset H(d') \text{ whenever } 1 \leq d' < d. \tag{2.15}$$

Thus, the class of possible radial covariance functions decreases as the dimension d increases. It is of interest to ask what functions are left in $H(d)$ as $d \to \infty$. Before answering this question, recall that a real-valued function $\varphi(v), v > 0$, is said to be *completely monotone* if it is infinitely differentiable and if its derivatives $\varphi^{(m)}(v)$ alternate in sign,

$$(-1)^m \varphi^{(m)}(v) \geq 0 \text{ for all } m \geq 0, \quad 0 < v < \infty. \tag{2.16}$$

It can be shown that a function $\varphi(v)$ is completely monotone if and only if it is the Laplace transform of a positive measure μ on $[0, \infty)$

$$\varphi(v) = \int_{[0,\infty)} e^{-v\rho} \mu(d\rho). \tag{2.17}$$

Moreover, $\varphi(0+) < \infty$ if and only if μ is finite. The existence of the integral representation (2.17) when (2.16) holds is known as Bernstein's theorem; see, e.g., Feller (1966, pp. 415–416).

The next theorem is the main result of this section.

Theorem 2.4.1 (Schoenberg, 1938) *A continuous function $\sigma^{\#}(r), r \geq 0$, lies in $H(d)$ for all $d \geq 1$ if and only if $\sigma^{\#}(r^{1/2})$ is a completely monotone function of r.*

Proof: In the course of the proof, be careful to distinguish the differential d in $d\omega$ from the dimension in $\mathbb{R}^d$. First, suppose $\sigma^{\#}(r^{1/2})$ is completely monotone; hence so is $\sigma^{\#}(\{2r\}^{1/2})$. By assumption, $\sigma^{\#}(0) < \infty$. Thus, there is a finite measure $G_0(d\rho)$ on $[0, \infty)$ with

$$\sigma^{\#}(\{2r\}^{1/2}) = \int_{[0,\infty)} e^{-\rho r} G_0(d\rho),$$

that is,

$$\sigma^{\#}(r) = \int_{[0,\infty)} e^{-r^2\rho/2} G_0(d\rho). \tag{2.18}$$

Now for any $d \geq 1$, the function $\exp(-|h|^2\rho/2)$, $h \in \mathbb{R}^d$, is the characteristic function of a d-dimensional normal distribution with mean 0 and covariance matrix ρI, where I denotes the $d \times d$ identity matrix (see, e.g., Mardia et al., 1979, p. 39). Hence, the function $\exp(-r^2\rho/2)$, $r \geq 0$, is a radial covariance function in every dimension d. Since $\sigma^{\#}(r)$ is a mixture of such functions, it too lies in $H(d)$ for all d. Thus, the first half of the theorem is proved.

Before proving the converse, we need a key result about the uniform distribution on a high-dimensional sphere. Let $\eta = (\eta[1], \ldots, \eta[d])^T$ denote a uniformly distributed random (column) vector on the unit sphere in $\mathbb{R}^d$, $d \geq 2$; thus $\eta^T\eta = 1$. Let $u \in \mathbb{R}^d$ be a fixed unit vector (so $u^Tu = 1$), and let $S = \eta^T u$ denote the projection of η onto the u-axis. Then the probability density function (pdf) of S is given by

$$n_d(s) = \begin{cases} c_d(1 - s^2)^{(d-3)/2}, & -1 < s < 1, \\ 0, & |s| \geq 1, \end{cases} \tag{2.19}$$

where $c_d = \Gamma(\frac{1}{2}d)/\{\pi^{1/2}\Gamma(\frac{1}{2}d - \frac{1}{2})\}$ and $\Gamma(\cdot)$ is the usual gamma function (Exercise 2.2). This distribution is important for our purposes, because when rescaled to $S^* = d^{1/2}S$, with pdf

$$n_d^*(s^*) = \begin{cases} d^{1/2}n_d(d^{-1/2}s^*), & -d^{1/2} < s^* < d^{1/2}, \\ 0, & |s^*| \geq d^{1/2}, \end{cases}$$

it converges to a standard normal distribution as $d \to \infty$, with pdf $n(s^*) = (2\pi)^{-1/2} \exp(-\frac{1}{2}s^{*2})$, $-\infty < s^* < \infty$. More specifically, it can be shown that as $d \to \infty$,

$$n_d^*(s^*) \to n(s^*) \text{ for each } s^* \in \mathbb{R}, \quad \text{and} \quad \sup_{s^* \in \mathbb{R}} \{n_d^*(s^*)/n(s^*)\} \to 1 \tag{2.20}$$

(see Exercise 2.3).

For the converse part of the theorem, we suppose $\sigma^{\#}(r) \in H_d$ for all d. Fix $d \geq 2$ for the moment and consider the basic spectral representation (2.11), writing $F(d\omega)$ as $F_d(d\omega)$ here for clarity. Switch to polar coordinates

$$h = ru, \quad \omega = \rho\eta, \tag{2.21}$$

where $r, \rho \geq 0$ are scalars, and u and η are unit vectors in $\mathbb{R}^d$. Since the spectral measure $F_d(d\omega)$ is isotropic, it can be expressed in the form $F_d(d\omega) = \Omega_d(d\eta)G_d(d\rho)$ where $\Omega_d(d\omega)$ denotes the uniform distribution on the unit sphere in $\mathbb{R}^d$ and $G_d(d\rho)$ is some finite measure on $\rho \geq 0$. Let $G_d^*(d\rho^*)$ denote the corresponding rescaled measure for $\rho^* = d^{-1/2}\rho$. Then (2.11) takes the form

$$\begin{aligned} \sigma^{\#}(r) = \sigma(h) &= \int e^{ih^T\omega} F_d(d\omega) \\ &= \int\int e^{ir\rho(u^T\eta)}\, \Omega_d(d\eta)\, G_d(d\rho) \end{aligned} \tag{2.22}$$

$$= \int \int e^{ir\rho s} n_d(s)\, ds\, G_d(d\rho)$$
$$= \int \int e^{ir\rho^* s^*} n_d^*(s^*)\, ds^*\, G_d^*(d\rho^*)$$
$$= A_d(r) + B_d(r), \text{say,}$$

where

$$A_d(r) = \int \int e^{ir\rho^* s^*} n(s^*)\, ds^*\, G_d^*(d\rho^*)$$
$$= \int e^{-\frac{1}{2}r^2\rho^{*2}}\, G_d^*(d\rho^*),$$

and

$$B_d(r) = \int \int e^{ir\rho^* s^*} \{n_d(s^*) - n(s^*)\}\, ds^*\, G_d^*(d\rho^*).$$

By (2.20), $n_d^*(s^*) - n(s^*)$ converges to 0 for each s^* and $|n_d^*(s^*) - n(s^*)|/n(s^*)$ is bounded uniformly over s^* as $d \to \infty$; hence by the dominated convergence theorem, $B_d(r) \to 0$ as $d \to \infty$ for each $r \geq 0$.

Note that $A_d(r)$, as a function of $\frac{1}{2}r^2$, is the Laplace transform of $G_d^*(d\rho^*)$. Since $B_d(r) \to 0$ as $d \to \infty$, $A_d(r) \to \sigma^\#(r)$. Hence, $\sigma^\#(r)$, as a function of $\frac{1}{2}r^2$, must also be the Laplace transform of a finite measure $G^*(d\rho^*)$, say, on $[0, \infty)$, with

$$\sigma^\#(r) = \int_{[0,\infty)} e^{-\frac{1}{2}r^2\rho^2}\, G^*(d\rho^*), \tag{2.23}$$

and $G_d^*(d\rho^*)$ converges weakly to $G^*(d\rho^*)$ (Feller, 1966, p. 410). The representation (2.23) completes the proof of the theorem. □

If the isotropic distribution $F_d(d\omega)$ has a density $f_d(\omega)$, then this density can depend only on the radial component $\rho = |\omega|$; that is, $f_d(\omega) = f_d^\#(\rho)$, say. After changing to polar coordinates, $G_d(d\rho)$ also has a density, given by a rescaled version of $f_d^\#(\rho)$,

$$g_d(\rho) = \pi_d \rho^{d-1} f_d^\#(\rho), \tag{2.24}$$

where

$$\pi_d = 2\pi^{d/2}/\Gamma(d/2) \tag{2.25}$$

is the surface area of the unit sphere in $\mathbb{R}^d$ (Exercise 2.1). Similarly, $G_d^*(d\rho^*)$ has a density $g_d^*(\rho^*) = d^{1/2} g_d(d^{1/2}\rho^*)$. In practice, it will usually be the case that $\sigma^\#(r) \to 0$ as $r \to \infty$ so that $G^*(d\rho^*)$ will have no mass at 0.

Next we look at explicit representations in $H(d)$. If $d \geq 2$, then from the third line of (2.22), we can write

$$\sigma^\#(r) = \int_{[0,\infty)} \left[\int_{-1}^{1} e^{ir\rho s} n_d(s) ds\right] G_d(d\rho),$$

so that

$$\sigma^\#(r) = \Gamma(d/2) \int_{[0,\infty)} (r\rho/2)^{-\lambda} J_\lambda(r\rho)\, G_d(d\rho), \tag{2.26}$$

where $\lambda = \frac{1}{2}d - 1$ (Abramowitz and Stegun, 1964, p. 360, equation 9.1.20). Equation (2.26) also holds for $d = 1$ dimension. Here

$$J_\lambda(x) = (x/2)^\lambda \sum_{k=0}^{\infty} (-x^2/4)^k / \{k!\ \Gamma(\lambda + k + 1)\}$$

is the usual Bessel function with limiting behavior

$$J_\lambda(x) \sim (x/2)^\lambda / \Gamma(\lambda + 1) \text{ as } x \to 0.$$

Note from (2.24) that when densities exist, $G_d(d\rho) = g_d(\rho)\, d\rho = \pi_d \rho^{d-1} f_d^{\#}(\rho)\, d\rho$, so that (2.26) gives an integral representation of $\sigma^{\#}(r)$ in terms of $f_d^{\#}(\rho)$. This representation is closely related to the *Hankel transform*, a version of the Fourier transform for isotropic functions. See Exercise 2.4 for more details.

For half-integer values of λ, that is, for $\lambda = \frac{1}{2}d - 1$ for odd dimensions d, $J_\lambda(x)$ can be expressed in elementary functions. In particular,

$$J_{-1/2}(x) = \sqrt{\frac{2}{\pi x}} \cos x, \quad J_{1/2}(x) = \sqrt{\frac{2}{\pi x}} \sin x$$

(Erdélyi, 1954, p. 966). The following list summarizes the representation formula (2.26) in low dimensions.

$$d = 1: \quad \sigma^{\#}(r) = \int_{[0,\infty)} \cos(r\rho)\ G_1(d\rho), \tag{2.27}$$

$$d = 2: \quad \sigma^{\#}(r) = \int_{[0,\infty)} J_0(r\rho)\ G_2(d\rho), \tag{2.28}$$

$$d = 3: \quad \sigma^{\#}(r) = \int_{[0,\infty)} (r\rho)^{-1} \sin(r\rho)\ G_3(d\rho). \tag{2.29}$$

This representation formula can be summarized as follows:

Theorem 2.4.2 *Fix $d \ge 1$. There is a one-to-one correspondence between continuous radial covariance functions $\sigma^{\#}(r)$ and finite measures $G_d(d\rho)$ on $[0,\infty)$ via the representation (2.26).*

Conversely, if $\sigma^{\#}(r)$, $r \ge 0$, is a radial covariance function in $\mathbb{R}^d$ and if $r^{d-1}\sigma^{\#}(r)$ is integrable on $(0,\infty)$, then the spectral distribution has a density that can be recovered from the inverse Fourier transform

$$f_d(\omega) = \frac{1}{(2\pi)^d} \int e^{-ih^T\omega} \sigma(h) dh$$

which reduces to

$$f_d^{\#}(\rho) = 2^{1-d} \pi^{-d/2} \int_0^\infty (r\rho/2)^{-\lambda} J_\lambda(r\rho) r^{d-1} \sigma^{\#}(r)\, dr,$$

where $\lambda = \frac{1}{2}d - 1$.

This formula reduces to the following forms in dimensions $d = 1,2,3$:

$$d = 1: \quad f_1^{\#}(\rho) = \frac{1}{\pi}\int_0^{\infty} \cos(r\rho)\, \sigma^{\#}(r)\, dr, \tag{2.30}$$

$$d = 2: \quad f_2^{\#}(\rho) = \frac{1}{2\pi}\int_0^{\infty} J_0(r\rho)\, r\, \sigma^{\#}(r)\, dr, \tag{2.31}$$

$$d = 3: \quad f_3^{\#}(\rho) = \frac{1}{2\pi^2 \rho}\int_0^{\infty} \sin(r\rho)\, r\, \sigma^{\#}(r)\, dr. \tag{2.32}$$

2.5 Construction of Stationary Covariance Functions

A valid stationary covariance function is a continuous positive semidefinite even function $\sigma(h)$, $h \in \mathbb{R}^d$, which is usually positive definite and usually satisfies $\sigma(h) \to 0$ as $|h| \to \infty$. By using standard results about positive definite functions, we can construct and manipulate such functions; see, e.g., Feller (1966, Chapter XV).

(a) *Addition, scaling, and multiplication.* If $\sigma_1(h)$ and $\sigma_2(h)$ are positive definite so are $\alpha_1\sigma_1(h) + \alpha_2\sigma_2(h)$, $\sigma_1(\beta h)$ and $\sigma_1(h)\sigma_2(h)$, where $\alpha_1 > 0,\ \alpha_2 > 0,\ \beta > 0$. In particular, multiplication of covariance functions corresponds to convolution of the underlying spectral distributions.
(b) *Limits.* If $\{\sigma_n(h)\}$ is a sequence of positive definite functions and $\sigma_n(h) \to \sigma(h)$ for each h with $\sigma(h)$ finite and continuous for all h, then $\sigma(h)$ is positive semidefinite.
(c) *Convolution.* Let the real-valued function $\varphi(t)$, $t \in \mathbb{R}^d$, be integrable ($\int |\varphi(t)|\, dt < \infty$) and square-integrable, ($\int |\varphi(t)|^2\, dt < \infty$) and set $\check{\varphi}(t) = \varphi(-t)$, and let $*$ denote convolution. Then $\sigma(h)$ defined by

$$\begin{aligned} \sigma(h) = \sigma(-h) &= (\varphi * \check{\varphi})(-h) \\ &= \int \varphi(t)\, \check{\varphi}(-h-t)\, dt \\ &= \int \varphi(t)\, \varphi(t+h)\, dt \end{aligned} \tag{2.33}$$

is positive semidefinite. (The symbol $\check{\varphi}$ is pronounced "phi-check.") To verify positive semidefiniteness, let $\tilde{\varphi}(\omega) = \int e^{it^T\omega}\varphi(t)dt$ denote the Fourier transform of φ. Since $\varphi(t)$ is square-integrable, then $\tilde{\varphi}(\omega)$ is also square-integrable. Also, the Fourier transform of $\check{\varphi}(t)$ is the complex conjugate $\overline{\tilde{\varphi}(\omega)}$. In the spectral domain, $\sigma(h)$ can be expressed as

$$\sigma(h) = \int e^{ih^T\omega} f(\omega) dw,$$

where $f(\omega) = \tilde{\varphi}(\omega)\,\bar{\tilde{\varphi}}(\omega) = |\tilde{\varphi}(\omega)|^2$ defines an integrable spectral density. Note that this representation shows that $\sigma(h)$ is continuous. This representation can also be extended to the case where $\varphi(t)$ is square-integrable but not integrable since $\int |(\tilde{\varphi}(\omega)|^2\, d\omega = (2\pi)^d \int |\varphi(t)|^2\, dt$ is still finite.

(d) *Mixtures.* As an extension of item (a) on addition and scaling, mixtures of continuous covariance functions are also continuous covariance functions. An important example is given by a scale mixture of normal characteristic functions. An integral representation is given in (2.23). By Schoenberg's Theorem 2.4.1, a scale mixture of normal characteristic functions can be characterized by the fact that $\sigma^{\#}(s^{1/2})$ is a completely monotone function of $s \geq 0$.

Using these ideas, a collection of isotropic covariance functions $\sigma(h)$ is listed in Table 2.1 in terms of the corresponding radial covariance function, $\sigma^{\#}(r)$. A parameterized family of covariance functions will be called a *scheme*, using terminology adopted from the mining literature. In practice, each covariance function will need to be augmented by a *scale* parameter $c_1 > 0$ and a *range* parameter $c_2 > 0$ to give $c_1\sigma(h/c_2)$. For simplicity, $\sigma^{\#}(r)$ in the table has been standardized so that $\sigma(0) = 1$ in each case. It is easiest to justify the positive definiteness of $\sigma^{\#}(r)$ when

Table 2.1 Some radial covariance functions.

Scheme	$\sigma^{\#}(r)$	Parameters	Dimensions
1. Constant	1		All d
2. Spherical	$\begin{cases} 1 - \frac{3}{2}r + \frac{1}{2}r^3, & 0 \leq r \leq 1 \\ 0, & r > 1 \end{cases}$	$r \leq 1$	$d \leq 3$
3. Restricted-power	$\begin{cases} (1-r)^k, & 0 \leq r \leq 1 \\ 0, & r > 1 \end{cases}$	$k \geq 2$	$d \leq 3$
4. Stable	$\exp(-r^{\alpha})$	$0 < \alpha \leq 2$	All d
5. Gaussian	$\exp(-r^2)$		All d
6. Exponential	e^{-r}		All d
7. Matérn	$\frac{2}{\Gamma(\nu)}(r/2)^{\nu}K_{\nu}(r)$	$\nu > 0$	All d
8. t-Density	$(1 + r^2)^{-\alpha}$	$\alpha > 0$	All d
9. Shell	$(r/2)^{-d/2+1}\Gamma(\frac{d}{2})J_{d/2-1}(r)$		All d
10. Ball	$(r/2)^{-d/2}\Gamma(\frac{d}{2}+1)J_{d/2}(r)$		All d
11. Annulus	$\frac{(r/2)^{-d/2}}{b^d - a^d}\Gamma(\frac{d}{2}+1) \times \{b^{d/2}J_{d/2}(br) - a^{d/2}J_{d/2}(ar)\}$	$0 \leq a < b$	All d
12. Damped sine	$\frac{1}{r}\sin r$		$d \leq 3$

Further details are given in Section 2.7.

Schoenberg's Theorem 2.4.1 is applicable; i.e. when it can be shown that $\sigma^{\#}(r^{1/2})$ is completely monotone or equivalently that $\sigma^{\#}(r^{1/2})$ is the Laplace transform of a distribution on $[0, \infty)$.

The next two sections discuss these isotropic covariance functions in more detail. The most important example, the *Matérn scheme* is covered first and then the remaining examples.

2.6 Matérn Scheme

The *Matérn scheme* (Matérn, 1960, 1986) is the single most important scheme in spatial statistics. The covariance function is given by

$$\sigma(h) = \frac{2\sigma^2}{\Gamma(\nu)}(\kappa|h|/2)^{\nu} K_{\nu}(\kappa|h|), \quad h \in \mathbb{R}^d, \tag{2.34}$$

where $K_{\nu}(\cdot)$ is a modified Bessel function of the second kind. Here, $\sigma^2 > 0$ is a scale parameter, $1/\kappa > 0$ is a range parameter and the index $\nu > 0$ governs the smoothness. As $1/\kappa$ increases, the covariance function decays more slowly in $|h|$, and as ν increases the realizations of the random field get smoother.

The covariance function is scaled so that $\lim_{|h|\to 0}\sigma(h) = \sigma^2$. In addition, $\sigma(h)$ decays monotonically to 0 as $|h| \to \infty$. Equation (2.34) defines a valid positive definite covariance function, and hence defines the covariance function for a stationary random field, in all dimensions $d \geq 1$. Table 2.1 gives (2.34) as a radial covariance function $\sigma^{\#}(r) = \sigma(|h|)$ with the range parameter $1/\kappa = 1$ and the scale parameter $\sigma^2 = 1$.

The positive definiteness of (2.34) can be confirmed in two ways. The first approach is to derive the spectral density

$$f(\omega) = \sigma^2 \frac{\Gamma(\nu + \frac{d}{2})}{\Gamma(\nu)\pi^{d/2}} \frac{\kappa^{2\nu}}{(|\kappa|^2 + |\omega|^2)^{\nu + \frac{d}{2}}}, \tag{2.35}$$

and to note that the spectral density is everywhere positive. See Exercise 2.4 for more details. When $\sigma^2 = 1$ and $\kappa = 1$, (2.35) is a multivariate t-density (e.g. Mardia et al., 1979, p. 57). Note the formula for the spectral density depends on the dimension d. The second approach is to show that when $\sigma^2 = 1$ and $\kappa = 1$, the function $\sigma^{\#}(s^{1/2})$, $s > 0$, represents the Laplace transform of a valid probability distribution on $(0, \infty)$, in this case, the reciprocal of a gamma random variable with index ν. See Exercise 2.9 for more details.

In spite of the presence of a Bessel function, the Matérn covariance function offers an attractive balance between tractability and flexibility. Here are some key properties.

(a) When ν is a half-integer value, the K Bessel functions can be written in terms of elementary functions. The most notable example is the exponential scheme when $\nu = 1/2$. Further examples, taken from Abramowitz and Stegun (1964, p. 444), are summarized in Table 2.2. A plot of Matérn covariance functions for several different indices ν, with the κ values chosen so that the covariance functions all take the value $1/2$ at $r = |h| = 1$, have been plotted in Figure 2.1. Notice that as ν increases, $\sigma^{\#}(r)$ becomes smoother at $r = 0$ and decays more quickly for large r.

(b) The Matérn scheme arises as the covariance function for the solution of a certain stochastic partial differential equation; see Chapter 4. This approach was used by Whittle (1954, 1956) to motivate the covariance function in $d = 2$ dimensions with $\nu = 1$.

(c) Several interesting limits arise by letting the parameters vary. As $\nu \to \infty$, with $\kappa = 2\sqrt{\nu}$, the Matérn scheme converges to a Gaussian scheme

$$\frac{2}{\Gamma(\nu)}(\kappa|h|/2)^{\nu} K_{\nu}(\kappa|h|) \to \exp(-|h|^2/2), \quad h \in \mathbb{R}^d. \tag{2.36}$$

See Exercise 2.9.

Table 2.2 Special cases of the Matérn covariance function in (2.34) for half-integer $\nu < \infty$ with scale parameter $\sigma^2 = 1$.

Index ν	$\sigma^{\#}(r)$
$\nu = p + 1/2$	$\exp(-r)\frac{\Gamma(p+1)}{\Gamma(2p+1)} \sum_{i=1}^{p} \frac{(p+i)!}{i!(p-1)!}(2r)^{p-i}$
$\nu = 1/2$	$\exp(-r)$
$\nu = 3/2$	$\exp(-r)\{1 + r\}$
$\nu = 5/2$	$\exp(-r)\{1 + r + \frac{1}{3}r^2\}$
$\nu \to \infty$	$\exp(-r^2)$

For $\nu < \infty$, the range parameter is fixed, $\kappa = 1$. For the limiting case $\nu \to \infty$, the range parameter $\kappa = 2\sqrt{\nu}$ depends on ν; see (2.36).

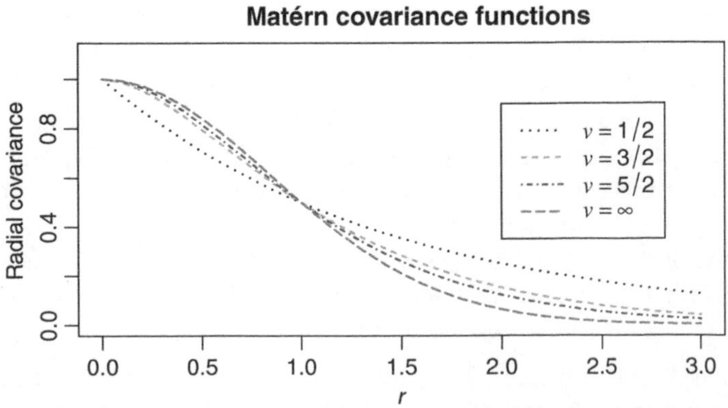

Figure 2.1 Matérn covariance functions for varying index parameters. The range and scale parameters have been chosen so that the covariance functions match at lags $r = |h| = 0$ and $r = 1$.

(d) Other limits leave the setting of stationary ordinary random fields for the *intrinsic* and/or *generalized* random fields of Chapter 3. For example, as $\nu \to 0$ the Matérn scheme converges to the covariance function of a generalized stationary random field (Section 3.8).
(e) For fixed $\nu > 0$, let $\kappa \to 0$. With appropriate scaling $\sigma^2 = 1/\kappa^{2\nu}$, the limiting process is an intrinsic and self-similar (Section 3.10.)
(f) If both $\nu \to 0$ and $\kappa \to 0$, the result is the de Wijsian scheme, which is both intrinsic and generalized.
(g) The Matérn scheme arises as the continuous limit of certain conditional autoregressive (CAR) models on the lattice. See Chapter 4.

A history of the Matérn scheme is given in Guttorp and Gneiting (2006).

2.7 Other Examples of Isotropic Stationary Covariance Functions

This section gives some further details about the radial covariance functions in Table 2.1, including spectral representations where available in concise form.

1. *Constant scheme.* This is a degenerate scheme, and arises from the spectral measure $F(d\omega)$ with mass 1 at the origin and 0 elsewhere. The spectral measure can also be written in terms of the Dirac delta (generalized) function as $F(d\omega) = \delta_0(\omega)$.
2. *Spherical scheme (Matheron, 1965, pp. 56–57).* This is obtained by convolution in $\mathbb{R}^3$ using
$$\varphi(h) = I\left[|h| < \frac{1}{2}\right],$$
where $I[\cdot]$ is the indicator function, so that $\varphi(h)$ equals 1 inside a disk of radius 1 about the origin and 0 elsewhere. This scheme is popular because of its simple algebraic form and its *finite range* ($\sigma^{\#}(r) = 0$ for $r > 1$). See Exercise 2.6.
3. *Restricted-power scheme (Mardia and Watkins,* 1989). This is similar to the spherical scheme but is a smoother function at $r = 1$ when $k > 2$. This smoothness has implications for maximum likelihood estimation; Section 5.5.3 gives some further details. The validity of the restricted-power scheme is most easily checked by showing the spectral density is positive; see Exercise 2.7.
4. *Stable scheme.* Since $\sigma^{\#}(s^{1/2}) = \exp(-s^{\alpha/2})$ is the Laplace transform of the positive stable law of index $\alpha/2$, (Feller, 1966, p. 424), for all $0 < \alpha \leq 2$, it follows that $\sigma^{\#}(r) = \exp(-r^{\alpha})$ is a valid radial covariance function in all dimensions. Further, $\sigma(h) = \exp(-|h|^{\alpha})$ is the characteristic function of the symmetric stable law on $\mathbb{R}$ of index α.
5. The *"Gaussian" or squared-exponential scheme* (which should not be confused with a Gaussian process) is also a special case ($\alpha = 2$) of the stable scheme. However, in some sense $\sigma(h)$ is "too smooth" because it is an analytic function

of h and hence is infinitely differentiable. In consequence, it turns out that with probability 1, the realizations $\{X(t)\}$ from any stationary random field following a Gaussian scheme can be taken to be infinitely differentiable analytic functions. Hence, the values of $X(t)$ are completely determined throughout $\mathbb{R}^d$ by the values on any open set. This property can lead to numerical instability in the covariance matrix for data from this scheme. See Examples 2.1 and 2.4.

6. The *exponential scheme* is a special case of the stable scheme ($\alpha = 1$). It arises as the covariance function of the stationary Ornstein–Uhlenbeck Gaussian process in $d = 1$ dimension; see, e.g., Cox and Miller (1965, p. 228). It is also a special case of the Matérn scheme.
7. The *Matérn scheme*. For more details, see Section 2.6.
8. *The t-density scheme*. The name comes from the fact that, provided $\mu = 2\alpha - d > 0$, $\sigma(h)$ is proportional to the density of an isotropic scaled t-distribution in d dimensions with μ degrees of freedom (e.g. Mardia et al., 1979, p. 57). It is easily checked that $\sigma^\#(s^{1/2})$ is completely monotone for all $\alpha > 0$, so that this scheme is, in fact, valid for all $\alpha > 0$. Its main interest lies in the tail behavior $\sigma^\#(r) \propto Cr^{-2\alpha}$ as $r \to \infty$, the so-called "power law." That is, it provides a model for long-range dependence (Section 2.13). However, this scheme suffers the same degeneracy as the Gaussian scheme since $\sigma(h)$ is an analytic function. Exercise 2.4 gives the spectral density when $\mu > 0$.
9. *Shell scheme*. This scheme arises when the radial spectral measure is proportional to a point mass at $\rho = b$; i.e. $G(d\rho) = (\pi_d b^{d-1})^{-1}\delta_b(\rho)$, so that the spectral measure $F(d\omega)$ is constant over the sphere of radius b centered at the origin in $\mathbb{R}^d$. This sphere has surface area $\pi_d b^{d-1}$ so that $F(d\omega)$ has been standardized to be a probability measure. Setting $b = 1$ yields the entry in Table 2.1. The spectral representation is given in (2.26). Note that since the J Bessel functions behave qualitatively like damped sine waves for large arguments, this scheme and the schemes in items 10–12 oscillate between positive and negative values as $r \to \infty$.
10. *Ball scheme*. This scheme arises from the spectral density that is constant on the interior of a ball of radius b about the origin in $\mathbb{R}^d$

$$f(\omega) = \{\{(\pi_d/d)b^d\}^{-1}I[a < |\omega| < b], \quad \omega \in \mathbb{R}^d. \tag{2.37}$$

The volume of the ball is $(\pi_d/d)b^d$, so that f is scaled to be a probability density. Setting $b = 1$ yields the entry in Table 2.1.

11. *Annulus scheme*. This scheme gets its name from its spectral density, which is constant on an annulus,

$$f(\omega) = \{(\pi_d/d)(b^d - a^d)\}^{-1})I[a < |\omega| < b], \quad \omega \in \mathbb{R}^d. \tag{2.38}$$

The multiplying constant is the reciprocal of the volume of the annulus.

12. *Damped sine scheme.* This scheme is a special case of the ball scheme in dimension $d = 1$, or of the shell scheme in dimension $d = 3$. See also Exercise 2.5.

One function not listed in this table is the indicator function

$$\sigma_0(h) = I[h = 0] \qquad (2.39)$$
$$= \begin{cases} 1, & h = 0 \\ 0, & h \neq 0 \end{cases}$$

since it is not continuous at $h = 0$. From the perspective of this book, (2.39) is not viewed as a valid covariance function.

The function (2.39) is intended to model a situation where observations at different sites are independent. Note that this concept is not very useful for continuously indexed spaces such as $\mathbb{R}^d$, where there are an uncountable number of sites. However, there are two modifications of (2.39) that do make sense.

(a) *Nugget effect.* Given a set of observations $\{x_i : i = 1, \ldots, n\}$ at sites $\{t_i\}_1^n$, a plot of a empirical covariance function or a semivariogram may suggest that a covariance model of the form

$$\sigma(h) = A\sigma_1(h) + B\sigma_0(h)$$

might be appropriate. Here, $\sigma_1(h)$ is a valid covariance function (and so continuous at $h = 0$), $\sigma_0(h)$ is the indicator function (2.39), and $A, B > 0$. Such a model is indicated when the sample variance of the differences between pairs of observations with a given lag h does not appear to tend to 0 as $h \to 0$. This behavior is known as a *nugget effect* (Section 1.4.1) and can be modeled using an observational error model; see Section 5.2.3. From a theoretical point of view, the observational error model is legitimate because the independent measurement errors exist only at the n sites where observations are made, not everywhere in $\mathbb{R}^d$.

(b) *White noise.* A basic building block in elementary probability theory is the concept of an infinite sequence of independent identically distributed Gaussian random variables. The most useful analogue of this concept in the random process setting is *white noise.* White noise is not an *ordinary* random process, but is an example of a *generalized* random process. Its covariance function is the generalized function $\sigma(h) = \delta_0(h)$, where $\delta_0(h)$ denotes the Dirac delta function centered at the origin. Generalized processes in general, and white noise in particular, are studied in Chapter 3.

Other choices of covariance functions and further details can be found in a variety of sources. These include Banerjee et al. (2015, Ch. 2), Diggle and Ribeiro (2007, Ch. 3), Rasmussen and Williams (2006, Ch. 4), Sherman (2011, Ch. 3), and Wendland (2005).

Example 2.1 ***Numerical stability***
The stable scheme $\sigma^{\#}(r) = \exp\{-(r/c_2)^{\alpha}\}$ suffers from a dramatic loss of numerical stability as $\alpha \to 2$. To illustrate, consider $n = 100$ equally spaced sites in $\mathbb{R}^1$ at the integers $j = 1, \ldots, n$. The numerical stability of the covariance matrix $A = (a_{jk})$, $a_{jk} = \sigma(j-k)$, can be summarized in terms of the "condition number," given by the ratio of the largest to the smallest eigenvalue of A. For $c_2 = 4$, the condition number is given by 2.1×10^3 for $\alpha = 1.9$, a value that is a bit high but does not cause numerical problems. On the other hand, the condition number rises dramatically to 1.3×10^{16} for $\alpha = 2$, which can cause severe rounding errors.

In general terms, a high condition number reflects a high level of smoothness of the underlying stochastic process and usually indicates an unrealistic model. A more detailed explanation why $\alpha = 2$ is so much worse than $\alpha = 1.9$ will be given in Example 2.4 in Section 2.11 in terms of the spectral density. □

2.8 Construction of Nonstationary Random Fields

Much of the emphasis of previous sections has been on stationary covariance functions. In particular, Section 2.5 gave several methods of construction.

This section looks at two ways to construct a nonstationary covariance function by modifying an existing covariance function. In each case, the starting point is a random field $\{X(t)\}$ (not necessarily stationary) with mean function $\mu(t)$ and covariance function $\sigma(s,t)$.

2.8.1 Random Drift

Let $f_1(t), \ldots, f_k(t)$ be a set of k linearly independent functions of the site t. Consider a new random field

$$Y(t) = X(t) + \sum_{j=1}^{k} U_j f_j(t),$$

where the coefficient vector $\boldsymbol{U} = [U_1, \ldots, U_k]^T$ is independent of $X(t)$ and has mean $\boldsymbol{v}$ and covariance matrix A, say. Then, it is straightforward to show that $Y(t)$ has mean function $\mu_Y(t) = \mu(t) + \sum v_j f_j(t)$ and covariance function

$$\sigma_Y(s,t) = \sigma(s,t) + \boldsymbol{f}(s)^T A \boldsymbol{f}(t), \tag{2.40}$$

where $\boldsymbol{f}(t) = \left[f_1(t), \ldots, f_k(t)\right]^T$. In general, the augmented covariance function $\sigma_Y(s,t)$ will not be stationary even if the original covariance matrix $\sigma(s,t)$ is stationary. If $A = I$ is the identity matrix, the augmented covariance function takes the simpler form

$$\sigma_Y(s,t) = \sigma(s,t) + \sum_{j=1}^{k} f_j(s) f_j(t). \tag{2.41}$$

2.8.2 Conditioning

Let $t_1^*, \ldots, t_n^*$ be a fixed collection of "conditioning" sites. Condition the random field $X(t)$ to take specific values $x_1^*, \ldots, x_n^*$ at these sites. Under the additional assumption that $X(t)$ is Gaussian, the conditioned random field is also Gaussian. Using standard results about the multivariate normal distribution (Section A.3.4), it is straightforward to show that the conditional mean and variance are given by

$$\mu_c(t) = \mu(t) + \boldsymbol{\sigma}^*(t)^T\{\Sigma^*\}^{-1}(\boldsymbol{x}^* - \boldsymbol{\mu}^*),$$
$$\sigma_c(s,t) = \sigma(s,t) - \boldsymbol{\sigma}^*(s)^T\{\Sigma^*\}^{-1}\boldsymbol{\sigma}^*(t), \tag{2.42}$$

where the following matrix notation is used:

$$\Sigma^* = (\sigma_{ij}^*), \quad \sigma_{ij}^* = \sigma(t_i^*, t_j^*),$$
$$\boldsymbol{\sigma}^*(t) = (\sigma_i^*(t)), \quad \sigma_i^*(t) = \sigma(t, t_i^*),$$
$$\boldsymbol{x}^* = (x_i^*), \quad \boldsymbol{\mu}^* = (\mu(t_i^*)).$$

Even if the original random field is stationary, the conditioned random field will not be stationary.

Define the "conditioned" or "residual" random field by

$$X_c(t) = X(t) - \mu(t) - \boldsymbol{\sigma}^*(t)^T\{\Sigma^*\}^{-1}(\boldsymbol{X}^* - \boldsymbol{\mu}^*), \tag{2.43}$$

where the vector $\boldsymbol{X}^* = (X(t_i^*))$ contains the values of the original random field at the sites t_i^*. Then, the residual random field is independent of $\boldsymbol{X}^*$ with constant mean $\mu = 0$ and with covariance function $\sigma_c(s,t)$ given in (2.42). This last result provides a simple way to simulate a random field conditioned to vanish at the sites $t_1^*, \ldots, t_n^*$. Just simulate an unconditioned random field, and then compute the residual in (2.43).

2.9 Smoothness

A random field $\{X(t) : t \in \mathbb{R}^d\}$ is called *continuous in probability* at t_0 if for all $\varepsilon > 0$

$$P(|X(t_0 + h) - X(t_0)| > \varepsilon) \to 0 \text{ as } h \to 0.$$

That is, sites $t_0 + h$ close to t_0 give rise to observations that are close in probability to $X(t_0)$. Similarly, the random field is called *continuous in mean-square* at t_0 if

$$E(|X(t_0 + h) - X(t_0)|^2) \to 0 \text{ as } h \to 0.$$

By Chebyshev's inequality

$$P(|X(t_0 + h) - X(t_0)| > \varepsilon) \le E(|X(t_0 + h) - X(t_0)|^2)/\varepsilon^2;$$

hence continuity in mean-square implies continuity in probability.

Let $\{X(t)\}$ have mean function $\mu(t)$ and covariance function $\sigma(s,t)$. If $\mu(t)$ and $\sigma(s,t)$ are continuous functions, then $\{X(t)\}$ must necessarily be continuous in mean-square everywhere since

$$E[X(t+h)-X(t)]^2 = [\mu(t+h)-\mu(t)]^2 + \sigma(t+h,t).$$

It should be noted that continuity in probability is a fairly weak regularity property for a random field. In particular, it does not imply that the process has continuous realizations. A simple example in $d=1$ dimension is given by the Poisson process of parameter $\lambda > 0$. This process is constructed for $t \geq 0$ by the conditions (i) $X(0)=0$, and (ii) for every $0<s<t$, $X(t)-X(s)$ has a Poisson distribution of parameter $(t-s)\lambda$, independent of $X(s)$. It is easily checked that this process has a mean function $\mu(t)=\lambda t$ and a covariance function

$$\sigma(s,t) = \lambda \min(s,t).$$

Clearly, $\mu(t)$ and $\sigma(s,t)$ are continuous, so $\{X(t)\}$ is continuous in mean-square, and hence continuous in probability. However, $\{X(t)\}$ is an integer-valued process. Its realizations are step functions, and hence cannot be continuous.

A more stringent requirement on a process than continuity in probability is that the realizations should be continuous functions with probability 1. For a Gaussian process, a simple sufficient condition for continuous realizations in a region $E \subset \mathbb{R}^d$ is that $\mu(t)$ is continuous and that

$$\frac{1}{2}\{\sigma(t+h,t+h)+\sigma(t,t)\} - \sigma(t,t+h) \leq c|h|^{\alpha} \tag{2.44}$$

for all $t, t+h \in E$, for some constant $c>0$ and for some α, $0<\alpha\leq 2$; see, e.g., Adler (1981, p. 60), who even gives a slightly weaker sufficient condition. As α increases, (2.44) gets more stringent and the realizations of the corresponding Gaussian random field become smoother. For an isotropic stationary Gaussian random field, (2.44) takes the simpler form

$$\sigma^{\#}(0) - \sigma^{\#}(r) < cr^{\alpha} \text{ as } r \to 0. \tag{2.45}$$

If (2.45) for a Gaussian random field is strengthened slightly to

$$\sigma^{\#}(0) - \sigma^{\#}(r) \sim cr^{\alpha} \text{ as } r \to 0, \tag{2.46}$$

then it is possible to make more precise statements about the realizations. In particular, the *graph* of the random field,

$$\{(t, X(t)) : t \in \mathbb{R}^d\} \subset \mathbb{R}^{d+1},$$

has *fractal* or *Hausdorff* dimension

$$D = d + 1 - \alpha/2.$$

This result follows from e.g. Adler (1981, p. 204, Th 8.4.1). Noninteger Hausdorff dimension underlies the theory of fractals developed by Mandelbrot (1982).

For non-Gaussian random fields, stronger conditions are needed to ensure continuity. In $d = 1$ dimension, a sufficient condition for continuous realizations is that (2.44) should hold for some α with $1 < \alpha \leq 2$. (For the discontinuous Poisson process, we have $\alpha = 1$.)

However, the corresponding condition in dimensions $d > 1$ is more complicated. For simplicity, we limit our attention to stationary processes with mean 0. Assume that $\sigma(h)$, now a function of a single argument $h \in \mathbb{R}^d$, is d-times differentiable, and let $p_d(h)$ denote the polynomial in h determined by the Taylor series expansion of $\sigma(h)$ up to order d. Since $\sigma(h)$ is even in h, $p_d(h)$ is an even polynomial in h of degree d if d is even, and degree $d - 1$ if d is odd. Then, a sufficient condition for continuous realizations is that for some $\alpha > d$

$$|\sigma(h) - p_d(h)| \leq c|h|^\alpha \tag{2.47}$$

as $|h| \to 0$. In other words, except for even powers of h, $\sigma(h) \to 0$ sufficiently quickly as $|h| \to 0$. See Kent (1989), where a slightly weaker sufficient condition is also given. In $d = 1$ dimension, (2.47) reduces to (2.44).

Next, we look briefly at *differentiability in probability* for a general random field $\{X(t) : t \in \mathbb{R}^d\}$ with finite second moments. Suppose $\sigma(s, t)$ is twice continuously differentiable in s and t. Then, we can define a vector field of partial derivatives $\{(Y_1(t), \ldots, Y_d(t)\}$ such that

1. For $1 \leq j \leq d$ and each $t \in \mathbb{R}^d$, $[X(t + \varepsilon e_j) - X(t)]/\varepsilon \to Y_j(t)$ in mean-square as $\varepsilon \to 0$, where e_j is a d dimensional vector with 1 in the jth place and zero elsewhere.
2. The covariance function of $\{Y_j(t)\}$ is $\partial^2\sigma(s, t)/\partial s[j]\partial t[j]$. Here, $s[j]$ and $t[j]$ are the jth components of s and t.
3. The cross-covariance function is defined by $\text{cov}\{Y_j(s), Y_k(t)\}$, $1 \leq j, k \leq d$, and is given by
$$\text{cov}\{Y_j(s), Y_k(t)\} = \partial^2\sigma(s, t)/\partial s[j]\partial t[k].$$
4. For a stationary random field, we can express $\partial^2\sigma(s, t)/\partial s[j]\partial t[k]$ as
$$-\partial^2\sigma(h)/\partial h[j]\partial h[k] \text{ with } h = t - s.$$

2.10 Regularization

In Section 2.9, we looked briefly at the differentiability of a random field. In this section, we look at the reverse operation, that of *integration* or *regularization*. Let $V \subset \mathbb{R}^d$ denote a bounded open region with volume $|V|$. A simple example of regularization is given by

$$X_V(t) = \frac{1}{|V|} \int_V X(t + s) ds, \tag{2.48}$$

so that $X_V(t)$ denotes the average value of the random field over the region $t+V = \{t+s : s \in V\}$. In particular, V might represent the interior of a rectangle or sphere. The concept of regularization has two important applications to the statistical analysis of random fields.

(a) *Data collection*. Often one cannot observe a random field at an individual site t, but can observe only the average value of the process in a small region about t. For example, in mining, the smallest practical measurement that can be made is the average mineral content over a section of a borehole. In agricultural experiments, the basic measurement is often the average yield over a plot.
Another example is given by a digital image where the observation at each discrete site or pixel $t \in \mathbb{Z}^d$ might represent the average value of a continuous random field over a square/cube of side 1, centered at t. In addition, there is often some smearing between neighboring pixels. This smearing effect is also covered by the methods of this section; see Example 2.3.

(b) *Prediction*. In mining, one wants to predict not a single value of the process at a new site, but instead the total ore content in a block of rock.

We are now ready to give a general construction of the regularization of a random field $\{X(t) : t \in \mathbb{R}^d\}$ with mean function $\mu(t)$ and covariance function $\sigma(s,t)$. Let $\varphi(t)$ be an integrable function of t

$$\int |\varphi(t)| dt < \infty. \tag{2.49}$$

Then the *regularized* random field $\{X_\varphi(t) : t \in \mathbb{R}^d\}$ is defined by

$$X_\varphi(t) = \int X(t+s)\varphi(s)ds = (X * \check{\varphi})(t), \text{ say,} \tag{2.50}$$

where $\check{\varphi}(s) = \varphi(-s)$. We also use the notation $X(\varphi) = X_\varphi(0)$ to denote the regularization of $X(t)$ with respect to φ at $t = 0$.

The mean and covariance functions of $\{X_\varphi(t)\}$ are easily seen to be

$$E\{X_\varphi(t)\} = \int \mu(t+s)\ \varphi(s)\ ds = (\mu * \check{\varphi})(t), \tag{2.51}$$

$$\begin{aligned} \operatorname{cov}\{X_\varphi(s), X_\varphi(t)\} &= \int\int \sigma(s+u, t+w)\ \varphi(u)\ \varphi(w)\ du\ dw \\ &= \sigma_\varphi(s,t), \text{ say,} \end{aligned} \tag{2.52}$$

where u and w are integrated over $\mathbb{R}^d$. For a stationary random field, (2.51)–(2.52) reduce to

$$E\{X_\varphi(t)\} = \mu \int \varphi(s)\ ds, \tag{2.53}$$

$$\begin{aligned} \operatorname{cov}\{X_\varphi(t), X_\varphi(t+h)\} &= \int\int \sigma(h+u-w)\ \varphi(u)\ \varphi(w)\ du\ dw \\ &= (\sigma * \varphi * \check{\varphi})(h) = \sigma_\varphi(h), \text{ say,} \end{aligned} \tag{2.54}$$

Note that $\{X_\varphi(t)\}$ can be regarded as a *moving average process* with weighting function φ.

Regularization of a stationary random field can also be looked at in the spectral domain. Let $\tilde{\varphi}(\omega) = \int e^{iu^T\omega}\varphi(u)\, du$ denote the Fourier transform of $\varphi(u)$. If $F(d\omega)$ is the spectral measure for $\{X(t)\}$, then

$$|\tilde{\varphi}(\omega)|^2 F(d\omega) = F_\varphi(d\omega), \text{ say}, \quad \omega \in \mathbb{R}^d, \tag{2.55}$$

represents the spectral measure of $\{X_\varphi(t)\}$ since

$$\begin{aligned}
\int e^{ih^T\omega}\, F_\varphi(d\omega) &= \int e^{ih^T\omega} |\tilde{\varphi}(\omega)|^2\, F(d\omega) \\
&= \int\int\int e^{i(h+u-w)^T\omega} \varphi(u)\varphi(w)\, F(d\omega)\, du\, dw \\
&= \int\int \sigma(h+u-w)\, \varphi(u)\, \varphi(w)\, du\, dw.
\end{aligned}$$

The integrability of $\varphi(h)$ ensures that (2.53) is well defined and finite, and that $|\tilde{\varphi}(\omega)|$ is bounded, so that $F_\varphi(d\omega)$ is a finite measure whenever $F(d\omega)$ is. Equation (2.55) is one form of the *Parseval relation*; see, e.g., Feller (1966, Chapter XIX) and Section A.5.

Example 2.2 ***Regularization with an indicator function***
For a bounded open region $V \subset \mathbb{R}^d$, let

$$\varphi(t) = |V|^{-1} I[t \in V],$$

where $I[\cdot]$ is an indicator function. We wrote $X_\varphi(t)$ as $X_V(t)$ in (2.48). □

Example 2.3 ***Regularization with a Gaussian density***
Consider a normal density $\varphi(t) = (2\pi\alpha)^{-d/2} \exp\{-|t|^2/(2\alpha)\}$, which is a bounded integrable function with infinite support. This regularization is often used to describe smearing or blurring in image analysis. The amount of blurring increases with the variance parameter α. □

2.11 Lattice Random Fields

If $\{X(t) : t \in \mathbb{Z}^d\}$ is a random field, defined now on $\mathbb{Z}^d$ rather than $\mathbb{R}^d$, then much of the above theory remains applicable, but with some simplification. For clarity, we shall often write $X(t)$ as X_t in the lattice case, to distinguish it from the continuous case. In particular, suppose $\{X_t\}$ is stationary, with covariance function $\{\sigma_h : h \in \mathbb{Z}^d\}$. Then, $\{\sigma_h\}$ has a spectral representation

$$\sigma_h = \int_{(-\pi,\pi]^d} e^{i\omega^T h} F(d\omega) \tag{2.56}$$

for some symmetric finite measure $F(d\omega)$ on $(-\pi, \pi]^d$.

Let $\{Y(t) : t \in \mathbb{R}^d\}$ be a stationary random field on $\mathbb{R}^d$ and let $X_t = Y(t),\ t \in \mathbb{Z}^d$, denote its restriction to $\mathbb{Z}^d$. If $\{Y(t)\}$ has spectral measure $G(d\omega),\ \omega \in \mathbb{R}^d$, then $\{X_t\}$ has spectral measure

$$F(d\omega) = \sum_{k \in \mathbb{Z}^d} G(2\pi k + d\omega), \quad \omega \in (-\pi, \pi]^d, \tag{2.57}$$

where frequencies separated by a lag $2\pi k, k \in \mathbb{Z}^d$, are aliased together in the construction of $F(d\omega)$. Conversely, given a stationary process on $\mathbb{Z}^d$, there are infinitely many ways to interpolate it to give a stationary random field on $\mathbb{R}^d$.

Let $\{b_j : j \in N\}$ be a set of coefficients at a finite collection of sites $D \subset \mathbb{Z}^d$. As in Eq. (2.12), the variance of this linear combination of process values can be written as

$$\text{var}\left\{\sum b_j X_j\right\} = \int_{(-\pi,\pi]^d} \left|\sum b_j \exp(i\, j^T\omega)\right|^2 F(d\omega). \tag{2.58}$$

If $F(d\omega) = f(\omega)d(\omega)$ has a density satisfying the bounds

$$0 < c/(2\pi)^d \leq f(\omega) \leq C/(2\pi)^d, \tag{2.59}$$

then it is possible to bound the above variance. In particular,

$$\begin{aligned} \text{var}\left\{\sum b_j X_j\right\} &\leq \{C/(2\pi)^d\} \int \left|\sum b_j \exp(i\, j^T\omega)\right|^2 d\omega \\ &= C \sum_{j,k \in N} b_j b_k \int \exp\{i\,(j-k)^T\omega\}\, d\omega/(2\pi)^d \\ &= C \sum b_j^2. \end{aligned} \tag{2.60}$$

The summation in the middle line simplifies because $j - k$ has integer coordinates, so that each scaled integral equals 1 or 0 for $j = k$ or $j \neq k$, respectively. Similarly, a lower bound can be obtained, $\text{var}\{\sum b_j X_j\} \geq c \sum b_j^2$.

These bounds can be used to study the *condition number* of a covariance matrix. Consider a collection of sites on a rectangular array $D = \{j : 1 \leq j[\ell] \leq n[\ell],\ \ell = 1, \ldots, d\}$. Set Σ to be the $|D| \times |D|$ covariance matrix of $\{X_j : j \in D\}$, where $|D| = \prod n[\ell]$. Define the condition number of a matrix by $\kappa(\Sigma) = \lambda_{\max}(\Sigma)/\lambda_{\min}(\Sigma)$, the ratio between the largest and smallest eigenvalues. In view of (2.60), we obtain the bound

$$\kappa(\Sigma) \leq \max\{f(\omega)\}/\min\{f(\omega)\}, \tag{2.61}$$

where the max and min are taken over $\omega \in [-\pi, \pi]^d$. The bound can be quite tight when the $n[\ell]$ are large. The condition number is useful for assessing the stability of various calculations on matrices. In particular, if the condition number is large, rounding errors can cause numerical problems.

Example 2.4 ***Numerical stability revisited***
Consider again the stable scheme of Example 2.1 as $\alpha \to 2$ with the range parameter $c_2 = 4$. For $\alpha = 2$, the spectral density for $\omega \in \mathbb{R}^d$ has the explicit form

$$f^{(2)}(\omega) = (c_2^2/4\pi)^{d/2} \exp\{-c_2^2|\omega|^2/4\}, \tag{2.62}$$

which shows squared-exponential decay as $|\omega| \to \infty$. For $\alpha < 2$, it is only possible to give an expression for the tail behavior of the spectral density

$$f^{(\alpha)}(\omega) \sim K_{d,\alpha} c_2^{-\alpha} |\omega|^{-d-\alpha} \quad \text{as } |\omega| \to \infty, \tag{2.63}$$

a power-law decay rate, where $K_{d,\alpha}$ is a constant. When restricting the process to the integer lattice with $c_2 = 4$, these expressions remain substantially valid for the corresponding spectral density on $(-\pi, \pi)^d$ since the only prominent contributions in (2.57) come from the term $k = 0$. It is this dramatic difference in the decay rates that explains the strong difference in the condition numbers reported in Example 2.1.

This example also provides an opportunity to assess the usefulness of the approximation (2.61). The exact spectral density is available only for $\alpha = 2$ for which the condition number from Example 2.1 is 1.3×10^{16}. The upper bound from (2.61) with $c_2 = 4$ is $\exp(c_2^2\pi^2/4) = \exp(4\pi^2) = 1.4 \times 10^{17}$, which has roughly the right order of magnitude. □

If φ_h is a summable function on $\mathbb{Z}^d$ (i.e. $\sum |\varphi_h| < \infty$) with the Fourier transform

$$\tilde{\varphi}(\omega) = \sum_{h \in \mathbb{Z}^d} \varphi_h e^{ih^T\omega}, \quad \omega \in (-\pi, \pi]^d,$$

then $X_{\varphi,t} = \sum X_{t+h}\, \varphi_h$, $t \in \mathbb{Z}^d$, has spectral measure given by a version of (2.55) restricted to the torus, namely,

$$F_\varphi(d\omega) = |\tilde{\varphi}(\omega)|^2 F(d\omega), \quad \omega \in (-\pi, \pi]^d. \tag{2.64}$$

This result is important in the construction of moving average and autoregression processes; see Chapter 4.

It does not make sense to talk about continuity and differentiability for lattice processes. Similarly, the concept of isotropy is not meaningful in its original form. If R is a $d \times d$ orthogonal matrix, then the vector Rh does not necessarily have integer-valued components when h does. However, we can talk of a restricted sort of isotropy called *permutation symmetry* in which $\sigma_h = \sigma_{Rh}$ for all $d \times d$ permutation matrices, that is, for all orthogonal matrices R with entries that are ± 1 or 0. For example, in $d = 2$ dimensions, a permutation-symmetric covariance function takes the same values at $(\pm h_1, \pm h_2)$, and $(\pm h_2, \pm h_1)$, a list of up to eight vector lags.

A weaker concept is that of *reflection symmetry* for which the value of covariance function σ_h depends only on the absolute values $|h[1]|, \ldots, |h[d]|$. Thus, in $d = 2$ dimensions, a reflection-symmetric covariance function takes the same values at $(\pm h_1, \pm h_2)$, a list of up to four vector lags. Of course, all lattice covariance functions always satisfy the basic symmetry property that $\sigma_h = \sigma_{-h}$.

2.12 Torus Models

So far we have considered models on $\mathbb{R}^d$ or $\mathbb{Z}^d$. However, for mathematical simplicity, especially in the Fourier domain, it is also convenient to consider models with periodic boundary conditions, i.e. on the torus. We consider separately the continuous and the lattice settings.

Models on the torus are important because of their mathematical tractability in the spectral domain. Further, although the use of periodic boundary conditions is often unrealistic in practice, torus models can still provide useful and powerful approximations.

2.12.1 Models on the Continuous Torus

Consider a rectangular region in $\mathbb{R}^d$

$$R = \{t \in \mathbb{R}^d : 0 \le t[\ell] \le 2\pi,\ \ell = 1, \ldots, d\}, \tag{2.65}$$

where we assume periodic boundary conditions; i.e. $t[\ell] = 0$ is assumed to be the same point as $t[\ell] = 2\pi$. This space is a circle in $d = 1$ dimension and a torus in higher dimensions.

Two points $s, t \in \mathbb{R}^d$ represent the same point on the torus R if their components are the same, up to multiples of 2π. Use the notation

$$s = t \text{ Mod } 2\pi \tag{2.66}$$

to mean

$$s[\ell] = t[\ell] \bmod 2\pi, \quad \ell = 1, \ldots, d, \tag{2.67}$$

where the uppercase "M" in Mod indicates the comparison of two d-vectors.

In this setting, a stationary covariance function must be periodic, i.e. $\sigma(h) = \sigma(h + 2\pi k)$ for all $k \in \mathbb{Z}^d$, where k is a multi-index, $k = (k[1], \ldots, k[d])$. That is, $\sigma(h)$ depends only on h Mod 2π. In particular, in $d = 1$ dimension, $\sigma(h)$ must satisfy $\sigma(h) = \sigma(-h) = \sigma(2\pi - h)$.

In the spectral representation (2.11), the spectral measure is concentrated on the integers and the integral reduces to a summation

$$\sigma(h) = \sum_{k \in \mathbb{Z}^d} b_k \cos(h^T k), \tag{2.68}$$

where $b_k \ge 0$, $b_k = b_{-k}$, $\sum b_k = \sigma(0) < \infty$.

Table 2.3 Some examples of stationary covariance functions $\sigma(h) = b_0 + 2\sum_{k=1}^{\infty} b_k \cos kh$ on the circle, together with the terms b_k in their Fourier series.

Name	$\sigma(h)$	b_k
Cosine	$\cos(h)$	$\frac{1}{2}I[k=1]$
Wrapped normal	$(2\pi v)^{-1/2}\sum_{j=-\infty}^{\infty}\exp\{-(h+2\pi j)^2/(2v)\}$	$\exp(-\frac{1}{2}k^2 v)/(2\pi)$
von Mises	$\{2\pi I_0(\kappa)\}^{-1}\exp(\kappa\cos h)$	$I_k(\kappa)/\{2\pi I_0(\kappa)\}$
Wrapped Cauchy	$(1-\lambda^2)/(1+\lambda^2-2\lambda\cos h)$	$\lvert\lambda\rvert^k/(2\pi)$

Here, $I_k(\kappa)$ denotes a modified Bessel function of the first kind and $I[\cdot]$ is an indicator function. The covariance functions have been scaled to integrate to 1 over the circle. The parameters are $v > 0,\ \kappa > 0$ and $0 < |\lambda| < 1$, respectively.

The circular case $d = 1$ is of particular interest, e.g. to model periodic phenomena in time. Some examples of positive definite covariance functions on the circle using popular directional densities are given in Table 2.3 (see, e.g., Mardia and Jupp, 2000). In this table, the covariance functions have been scaled to integrate to 1 over the circle. They need to be rescaled to $\sigma(h)/\sigma(0)$ to represent a random field with a marginal variance equal to 1.

2.12.2 Models on the Lattice Torus

Consider a rectangular region in $\mathbb{Z}^d$,

$$T = \{t \in \mathbb{Z}^d : 0 \le t[\ell] \le n[\ell] - 1,\ \ell = 1, \dots, d\}, \tag{2.69}$$

of size $|N| = n[1] \times \cdots \times n[d]$, where $N = (n[1], \dots, n[d])$. If for each component ℓ, we regard $t[\ell] = n[\ell] - 1$ as adjacent to $t[\ell] = 0$, then T can be regarded as a d-dimensional *lattice torus*.

As in (2.66)–(2.67), it is useful to adapt modulo notation to compare two vectors $s, t \in \mathbb{Z}^d$. Let

$$s = t \text{ Mod } N \tag{2.70}$$

be shorthand for

$$s[\ell] = t[\ell] \text{mod } n[\ell], \quad \ell = 1, \dots, d. \tag{2.71}$$

A random field $\{X_t\}$ on the lattice torus T is said to be *stationary* if $\text{cov}\{X_t, X_s\} = \sigma(s-t)$ say, depends only on $(s-t)\text{Mod } T$. A stationary random field on the lattice torus can be viewed as a special case of a lattice model on $\mathbb{Z}^d$, where the spectral measure $F(d\omega)$ in (2.56) is concentrated on a discrete set of $|T|$ frequencies $\omega_j \in \mathbb{R}^d$, indexed by $j = (j[1], \dots, j[d]) \in T$, where ω_j has components

$$\omega_j[\ell] = 2\pi j[\ell]/n[\ell], \quad j = 0, \dots, n[\ell] - 1,\ \ell = 1, \dots, d.$$

Alternatively, the process can be viewed as a discretized version of the continuous torus process in Section 2.12.1 where the Fourier series in (2.68) is limited to a finite number of nonzero coefficients.

If $\{X_t\}$ is regarded as a $|N|$-dimensional vector with $|N| \times |N|$ covariance matrix A, say, then it can be shown that the $|N|$ values of the spectral density $f(\omega_j)$ are precisely the eigenvalues of A. The matrix A is an example of a block circulant matrix. See Section A.7 for a detailed analysis of the eigenvalues and eigenvectors of such matrices.

Thus, through the lattice torus we see a link between the eigenstructure of symmetric matrix and the spectral representation (2.11).

2.13 Long-range Correlation

Let $\{X_1, X_2, \dots\}$ be independent, identically distributed random variables with mean μ and variance σ^2, and with successive sample means $\bar{X}_n = n^{-1}\sum_{i=1}^{n} X_i$. Then $\text{var}(\bar{X}_n) = \sigma^2/n$ decays at rate n^{-1} as $n \to \infty$. If instead of independence, we allow a "weak" dependence between the X_i, we typically find that $\text{var}(\bar{X}_n) \sim A/n$ as $n \to \infty$ for some $A > \sigma^2$. That is, $\text{var}(\bar{X}_n)$ still decays at rate n^{-1}, but the proportionality constant A increases, reflecting the fact that interdependence between the observations tends to increase the variance of the sample mean. If we now go further and allow a "strong" dependence between the X_i, we typically find that $\text{var}(\bar{X}_n) \sim B/n^{\beta}$ for some $B > 0$ where $0 < \beta < 1$. Thus, $\text{var}(\bar{X}_n)$ still tends to 0, but at a slower rate than before. This situation is known as long-range correlation or long-range dependence. A detailed study of the one-dimensional case is given in Beran (1994).

The purpose of this section is to explore these features in more detail in the spatial setting. Let $\{X(t) : t \in \mathbb{R}^d\}$ be a stationary random field with covariance function $\sigma(h)$. For any bounded open set $V \subset \mathbb{R}^d$ with volume $|V|$, let

$$\bar{X}(V) = |V|^{-1} \int_V X(t)\, dt \tag{2.72}$$

denote the sample mean of the random field as t varies continuously through V. Clearly,

$$\text{var}\{\bar{X}(V)\} = |V|^{-2} \int_V \int_V \sigma(s-t)\, ds\, dt. \tag{2.73}$$

Definition 2.13.1 The covariance function $\sigma(h)$ is said to exhibit *short-range correlation* if

$$\int |\sigma(h)|\, dh < \infty, \tag{2.74}$$

and to exhibit *long-range correlation* otherwise. Further, $\sigma(h)$ is said to exhibit a *power-law decay* of order β for large lags if

$$\sigma(h) \sim a|h|^{-\beta} \text{as } |h| \to \infty.$$

Suppose $\sigma(h)$ exhibits a *power-law decay* of order β. Then checking the finiteness of (2.74) shows that there is short-range correlation when $\beta > d$ and long-range correlation when $0 < \beta \le d$.

For scalar $\lambda > 0$ let $\lambda V = \{\lambda v : v \in V\}$ denote the dilation of V by a factor λ with volume $|\lambda V| = \lambda^d |V|$. The behavior of $\text{var}\{\bar{X}(\lambda V)\}$ as $\lambda \to \infty$ under short-range and long-range correlation was studied by Whittle (1956), and is summarized in the following theorem. Here the volume $|\lambda V|$ plays the role of the sample size n above. All of the covariance functions in Table 2.1 exhibit short-range correlation except for the t-density scheme $\sigma(h) \propto (1 + |h|^2)^{-\alpha}$; this scheme exhibits short-range correlation for $\alpha > d/2$ and long-range correlation for $\alpha \leq d/2$.

Theorem 2.13.1 *Let $\{X(t) : t \in \mathbb{R}^d\}$ be a stationary random field with covariance function $\sigma(h)$, and let V be a bounded open set.*

(a) If $\sigma(h)$ displays short-range correlation, then as $\lambda \to \infty$,

$$\text{var}\{\bar{X}(\lambda V)\} \sim A/\lambda^d, \quad A = |V|^{-1} \int_{\mathbb{R}^d} \sigma(h) dh, \tag{2.75}$$

provided $A > 0$.

(b) If $\sigma(h) \sim a|h|^{-\beta}$ as $|h| \to \infty$ for some $0 < \beta < d$ as $\lambda \to \infty$, then $\sigma(h)$ displays long-range correlation. In particular,

$$\text{var}\{\bar{X}(\lambda V)\} \sim B/\lambda^\beta, \quad B = a\,|V|^{-2} \int_V \int_V |s - t|^{-\beta}\, ds\, dt. \tag{2.76}$$

(c) If $\sigma(h) \sim a|h|^{-d}$ as $|h| \to \infty$, then $\sigma(h)$ also displays a limiting version of long-range correlation with $\beta = d$. In this case, as $\lambda \to \infty$,

$$\text{var}\{\bar{X}(\lambda V)\} \sim \frac{C \log \lambda}{\lambda^d}, \quad C = a\,\pi_d/|V|. \tag{2.77}$$

Proof:

(a) Let

$$\begin{aligned}
\varphi(t, \lambda) &= \lambda^d \int_V \sigma(\lambda(s - t))\, ds \\
&= \int_{\lambda(V-t)} \sigma(h)\, dh,
\end{aligned}$$

after substituting $h = \lambda(s - t)$. If t lies in the interior of V, then the set $\lambda(V - t)$ will eventually fill $\mathbb{R}^d$ as $\lambda \to \infty$, and so $\varphi(t, \lambda) \to |V|A$. Also $|\varphi(t, \lambda)| \leq \int_{\mathbb{R}^d} |\sigma(h)|\, dh < \infty$ for all t and λ. Thus, since $|\lambda V| = \lambda^d |V|$,

$$\begin{aligned}
\lambda^d \text{var}\{\bar{X}(\lambda V)\} &= \lambda^d |\lambda V|^{-2} \int_{\lambda V} \int_{\lambda V} \sigma(s - t)\, ds\, dt \\
&= \lambda^d |V|^{-2} \int_V \int_V \sigma(\lambda(s - t))\, ds\, dt \\
&= |V|^{-2} \int_V \varphi(t, \lambda)\, dt \to A
\end{aligned}$$

as $\lambda \to \infty$ by the bounded convergence theorem.

(b) First note that $\int_V \int_V |s-t|^{-\beta}\, ds\, dt < \infty$ for any $0 < \beta < d$ and any bounded open set V. Also note that we can bound $|\sigma(h)| \le K|h|^{-\beta}$ for some constant K. Hence

$$\lambda^\beta \operatorname{var}\{\bar{X}(\lambda V)\} = \lambda^\beta |V|^{-2} \int_V \int_V \sigma(\lambda(s-t))\, ds\, dt \to B$$

as $\lambda \to \infty$ by the dominated convergence theorem, since $\lambda^\beta \sigma(\lambda(s-t)) \le K|s-t|^{-\beta}$ and for $s \ne t$, $\lambda^\beta \sigma(\lambda(s-t)) \to a|s-t|^{-\beta}$.

(c) This case is a limiting form of (b) with $\beta = d$. The covariance function just barely exhibits long-range correlation and $\operatorname{var}\{\bar{X}(V)\}$ tends to 0 at a rate just less than λ^{-1}. In this case, it can be shown that if we set

$$\psi(t,\lambda) = [\log \lambda]^{-1} \int_{\lambda(V-t)} \sigma(h)\, dh,$$

then, provided t lies in the interior of V,

$$\psi(t,\lambda) \to a\, \pi_d \text{ as } \lambda \to \infty,$$

and $\psi(t,\lambda)$ is bounded above by a suitable constant for all t and $\lambda \ge 2$. Hence, it follows that

$$\begin{aligned}
\frac{\lambda^d}{\log \lambda} \operatorname{var}\{\bar{X}(\lambda V)\} &= \lambda^d [\log \lambda]^{-1} |V|^{-2} \int_V \int_V \sigma(\lambda(s-t))\, ds\, dt \\
&= |V|^{-2} \int_V \psi(t,\lambda)\, dt \\
&\to a\, \pi_d / |V| \quad \text{as } \lambda \to \infty.
\end{aligned}$$

□

Example 2.5 ***Mercer–Hall uniformity trial***

The analysis of the Mercer–Hall uniformity trial in Example 1.14 fits into the framework of this section with dimension $d = 2$. In the notation of this section, V represents a single plot, λV represents a block of plots, and $\bar{X}(\lambda V)$ represents the observed yield per unit area in a block. Since the area of λV is proportional to λ^2, the estimated slope 0.49 in that example corresponds to the power-law relationship $\operatorname{var}\{\bar{X}(\lambda V)\} \propto \lambda^{-\hat{\beta}}$ with $\hat{\beta} = 2 \times 0.49 = 0.98$. This result is compatible with a power-law decay in an isotropic covariance function $\sigma^\#(r) \propto r^{-\beta}$ for large r. Since $0.98 < d = 2$, this parameter indicates long-range correlation. □

Example 2.6 ***Long-range dependence in other uniformity trials***

Whittle (1956, 1962) summarized the results of Smith (1938) on a collection of uniformity trials; see also Whittle (1986, p. 435). It was found that the power-law behavior is widespread. The fitted values of $\hat{\beta}$ lay in the range [0.3, 1.6] with a major peak in the frequencies at $\beta = 1$ and smaller peaks at $\beta = 0.5$ and $\beta = 1.5$. All these choices for $\hat{\beta}$ are smaller than $d = 2$, again indicating long-range correlation. Whittle also proposed a spatial–temporal model to generate a power-law behavior with $\beta = 1$. □

2.14 Simulation

2.14.1 General Points

In this section, we consider the problem of simulating a Gaussian random field $\{X(t)\}$ with mean function $\mu(t)$ and covariance function $\sigma(s,t)$. Given sites $t_1, \ldots, t_n$, the objective is to simulate a realization of $\boldsymbol{X} = (X(t_1), \ldots, X(t_n))^T$.

There are three main approaches:

(a) the direct approach
(b) the spectral approach
(c) the circulant approach

The spectral and circulant approaches are limited to stationary random fields. Within the spectral approach, we describe three different methods. The spectral methods have the advantage that the values of the sites $t_1, \ldots, t_n$ do not need to be specified in advance, but have the disadvantage that only an approximation to the distribution of the desired random vector is produced. A variant of the spectral approach is the circulant approach, which does produce exact simulations. For some further details and other methods of simulation, see, e.g., Christakos (1992) and Chilés and Delfiner (2012). An extension to multivariate Gaussian random fields is treated in Emery et al. (2016).

Once a Gaussian random field has been simulated, it is straightforward to modify it to produce a random field conditioned to vanish at a specified set of sites. See Section 2.8.

2.14.2 The Direct Approach

Define an $n \times n$ matrix $\Sigma = (\sigma_{ij})$ with entries $\sigma_{ij} = \sigma(t_i, t_j)$ and a vector $\boldsymbol{\mu} = (\mu_1, \ldots, \mu_n)^T$. Then simulate from a multivariate normal distribution $N_n(\boldsymbol{\mu}, \Sigma)$ as follows. Let R be a "square root" of Σ satisfying $RR^T = \Sigma$. One choice for R is the symmetric square root of Σ. (Recall that if Σ has a spectral decomposition $\Sigma = \Gamma\Lambda\Gamma^T$, where Γ is an $n \times n$ orthogonal matrix whose columns contain the eigenvectors and $\Lambda = \text{diag}(\lambda_i)$ is a diagonal matrix whose diagonal elements are strictly positive eigenvalues, then $R = \Gamma\Lambda^{1/2}\Gamma^T$, where $\Lambda^{1/2} = \text{diag}(\lambda_i^{1/2})$.) Another choice for R is given by the Cholesky decomposition $\Sigma = RR^T$ where R is lower triangular. From a numerical point of view, the Cholesky decomposition is preferred to the singular value decomposition because it is somewhat quicker and does not involve iteration (Golub and Van Loan, 1989). The Cholesky decomposition has further speed advantages if either Σ is sparse or Σ^{-1} is sparse (Rue, 2001). The former can arise when the covariance function has a bounded range, such as the spherical scheme and the restricted-power scheme (Section 2.7).

The latter can arise for approximations to CAR and simultaneous autoregression (SAR) models on a finite grid. CAR and SAR models are discussed in Chapter 4 and sparse approximations for the inverse covariance matrix are given in Section A.11.

Once R has been calculated, simulate a vector $\boldsymbol{U} = (U_1, \ldots, U_n)^T$ of independent $N(0,1)$ random variables and set $\boldsymbol{X} = \boldsymbol{\mu} + R\boldsymbol{U}$. Then $\boldsymbol{X}$ has a multivariate normal distribution with mean $\boldsymbol{\mu}$ and covariance matrix $RR^T = \Sigma$, as required.

2.14.3 Spectral Methods

For the spectral methods that follow, suppose $\sigma(h)$ is a stationary covariance function with spectral representation $\sigma(h) = \int e^{ih^T\omega} f(\omega)\, d\omega$ where for simplicity we assume the spectral density f is continuous. For the most part, we deal with continuously indexed processes for which the lag h ranges over $\mathbb{R}^d$ and the frequency ω in the spectral integral also ranges over $\mathbb{R}^d$. Where relevant we discuss the simplifications available for lattice processes, for which h ranges over $\mathbb{Z}^d$ and the spectral integral is over $(-\pi, \pi)^d$.

Before giving the methods of simulation, we describe how the spectral representation can be used to motivate a harmonic representation of a random field in terms of a superimposition of random cosine waves at different frequencies with random phases, as follows. For simplicity, take $\mu(t) \equiv 0$.

Let $H(d\omega) = A(d\omega) + iB(d\omega)$ be a random complex Gaussian measure. Except for the symmetry property

$$H(d\omega) = \bar{H}(-d\omega), \quad \text{i.e.} \quad A(d\omega) = A(-d\omega),\ B(d\omega) = -B(-d\omega), \tag{2.78}$$

the building blocks $A(d\omega)$ and $B(d\omega)$ of $H(d\omega)$ are assumed to be independent satisfying the following properties:

(a) Independence of the real and imaginary parts.
(b) Independence at different frequencies $\omega \neq \pm\omega'$.
(c) H has no atom at the origin $\omega = 0$.
(d) $EH(d\omega) = EA(d\omega) + iEB(d\omega) = 0, \quad \text{var}(A(d\omega)) = \text{var}(B(d\omega)) = \frac{1}{2} f(\omega)d\omega$, $E(A(d\omega)B(d\omega)) = 0$.

In Property (d), the notation means that if D is a region in $\mathbb{R}^d$, with $D \cap (-D) = \emptyset$, then $\int_D A(d\omega) \sim N(0, \frac{1}{2}\int_D f(\omega)d\omega)$, with a similar interpretation for $B(d\omega)$. Define a random field by

$$X(t) = \int e^{it^T\omega} H(d\omega) = \int [\cos(t\omega)\, A(d\omega) - \sin(t\omega)\, B(d\omega)], \tag{2.79}$$

where the integral is over $\omega \in \mathbb{R}^d$. The symmetry property (2.78) guarantees that the imaginary part of the integral vanishes so that $X(t)$ is real-valued. The random field $\{X(t)\}$ has mean 0 and a covariance function

$$E\{X(t)X(t+h)\} = E\left\{\int\int \left[\cos(t^T\omega)A(d\omega) - \sin(t^T\omega)B(d\omega)\right]\right.$$
$$\left.\left[\cos((t+h)^T\omega')A(d\omega') - \sin((t+h)^T\omega')B(d\omega')\right]\right\}$$

$$= 2E \int \left\{ \left[\cos(t^T\omega)\cos((t+h)^T\omega)A^2(d\omega) \right.\right.$$
$$\left.\left. + \sin(t^T\omega)\sin((t+h)^T\omega)B^2(d\omega)\right]\right\}$$
$$= \int \left[\cos(t^T\omega)\cos((t+h)^T\omega) + \sin(t^T\omega)\sin((t+h)\right] f(\omega)\, d\omega$$
$$= \int \cos(h^T\omega) f(\omega)\, d\omega = \sigma(h),$$

as required. When simplifying the double integral, note that nonzero expectations arise only when $\omega = \omega'$ or $\omega = -\omega'$.

The random measures $A(d\omega)$ and $B(d\omega)$ are examples of random generalized random fields. Generalized random fields are discussed in more detail in Chapter 3, and the theory there can be used to give a rigorous discussion of the derivation given here.

The representation (2.79) motivates the following method of simulating a random field with the correct covariance structure. Let U, V, and W be three independent random quantities where

(a) U is a uniform random variable on $[0,2\pi)$;
(b) W is a symmetric random vector on $\mathbb{R}^d$ with density $g(\omega)$, say, satisfying $g(\omega) = g(-\omega)$, $g(\omega) > 0$ for all ω;
(c) V is a random variable with $EV^2 = 2$.

Then, define a random field by

$$X(t) = \{f(W)/g(W)\}^{1/2}\, V\cos(t^T W + U). \tag{2.80}$$

Thus, $X(t)$ is a single random cosine wave with random phase U and random frequency W, and with a random amplitude depending on the frequency. It is easy to check that $X(t)$ has mean 0 and covariance function

$$E\{X(t)\,X(t+h)\} = E\left\{V^2[f(W)/g(W)]\cos(t^T W + U)\cos((t+h)^T W + U)\right\}.$$

To evaluate this quantity, write

$$\cos((t+h)^T W + U) = \cos((t+h)^T W)\cos U - \sin((t+h)^T W)\sin U,$$

and similarly for $\cos(t^T W + U)$. Recall that $E(\cos^2 U) = E(\sin^2 U) = \frac{1}{2}$, $E(\cos U \sin U) = 0$, and average first over U to get

$$\begin{aligned} E\{X(t)X(t+h)\} &= \frac{1}{2}E\left\{V^2\left[f(W)/g(W)\right]\left[\cos(t^T W)\cos((t+h)^T W) \right.\right. \\ &\qquad \left.\left. + \sin(t^T W)\sin((t+h)^T W)\right]\right\} \\ &= \frac{1}{2}E\left\{V^2\left[f(W)/g(W)\right]\cos(h^T W)\right\} \\ &= \frac{1}{2}\int 2\left[f(\omega)/g(\omega)\right]\cos(h^T\omega)\, g(\omega)\, d\omega \\ &= \int \cos(h^T\omega) f(\omega)\, d\omega = \sigma(h), \end{aligned} \tag{2.81}$$

which is the desired covariance function. However, this random field is not Gaussian. In particular, if $t^T W = s^T W$, then $X(t) = X(s)$; that is, $X(t)$ is constant on the hyperplanes perpendicular to W. To construct a Gaussian random field, it is necessary to combine independent copies of objects such as (2.80) and to apply the central limit theorem. Three variations of this methodology are described as follows:

(a) *Random frequencies, random amplitudes.* Define $X_i(t)$, $i = 1, \ldots, n$ to be independent copies of (2.80) and set

$$X^{(n)}(t) = \sum_{i=1}^{n} X_i(t)/\sqrt{n}. \tag{2.82}$$

For each n, $X^{(n)}(t)$ has the desired covariance structure $\sigma(h)$, and as $n \to \infty$, $X^{(n)}(t)$ converges to a Gaussian process by a functional version of the central limit theorem. To apply the functional central limit theorem, we need to ensure

(i) the finite-dimensional distributions of the approximating random field converge to those of the limit, which follows by the ordinary central limit theorem in this example and

(ii) "tightness," a property which can be ensured by adding a condition such as

$$\int [1 + |\omega|^{\alpha}]\, f(\omega)\, d\omega < \infty \tag{2.83}$$

for some $\alpha > 0$.

The results of e.g. Fernique (1978) and Araujo and Giné (1980, pp. 172–173) can be used to verify tightness under (2.83) (and even under slightly weaker conditions). The relation between (2.83) and the smoothness of $\sigma(h)$ as $h \to 0$ is explored in Exercise 2.10. It should be noted that tightness is important only for random fields on $\mathbb{R}^d$; it is irrelevant for random fields on $\mathbb{Z}^d$.

In practice, all of the stationary covariance functions in common use on $\mathbb{R}^d$ satisfy (2.83). The functional version of the central limit theorem guarantees that quantities such as the maximum of the simulated process over a bounded region (which depends on the random field at an infinite number of sites) converges in distribution to the corresponding quantity for the limiting process.

(b) *Random frequencies, fixed amplitudes.* The simplest version of the general method in (a) is given by taking $g(\omega) = f(\omega)/\sigma(0)$, the spectral density scaled to be a probability density, and letting $V = 2$ be constant in (2.80).

(c) *Fixed frequencies, random amplitudes.* This method involves two levels of approximation. Pick a large number M such that most of the spectral density lies in the domain $(-M, M)^d$ (for a lattice random field, just take $M = \pi$). Next choose an integer N (with N/M also large) and approximate the domain $(-M, M)^d$ by a discrete lattice with $(2N)^d$ points of the form

$$\omega_j = \frac{M}{N}\left(j[1] + \frac{1}{2}, \ldots, j[d] + \frac{1}{2}\right), -N \le j[\ell] < N, \quad \ell = 1, \ldots, d,$$

with multi-index $j = (j[1], \ldots, j[d])$.

Let V_j be an exponential random variable with mean $\{2f(\omega_j)\}^{1/2}$ for each j, and let $\{U_j\}$ be a collection of independent uniform random variables on $[0,2\pi)$, so that $V_j \cos U_j$ and $V_j \sin U_j$ are independent $N(0, \frac{1}{2}f(\omega_j))$ random variables. Note that the V_j are not identically distributed here, but depend on the frequency through ω_j. The choice of an exponential distribution ensures that $V_j \cos U_j$ and $V_j \sin U_j$ are independent Gaussian random variables. Then set

$$X^{(M,N)}(t) = \left(\frac{M}{N}\right)^{d/2} \sum_j V_j \cos(t^T \omega_j + U_j). \tag{2.84}$$

This formula can be viewed as a discrete approximation to (2.79), and $\{X^{(M,N)}(t)\}$ is a Gaussian random field with covariance function

$$\sigma^{(M,N)}(h) = \left(\frac{M}{N}\right)^{d} \sum_j f(\omega_j) \cos(h^T \omega_j). \tag{2.85}$$

This Riemann sum converges to $\sigma(h)$ for each h, so the finite dimensional distributions of $X^{(M,N)}(t)$ have the correct limit. Further, under (2.83) the results of Araujo and Giné (1980, p. 172) can be used to ensure tightness.

(d) *Turning bands method (Matheron,* 1973; *Mantoglue and Wilson,* 1982*).* Methods (a) and (b) involve a random sample of frequencies ω. Method (c) can be viewed as a systematic sample of frequencies after approximating $\mathbb{R}^d$ by $(-M,M)^d$ and letting $g(\omega)$ correspond to a uniform distribution. The turning bands method can be viewed an intermediate approach when $d \geq 2$.

Write the frequency ω in polar coordinates, $\omega = \rho u$ where $\rho > 0$ and $u = \omega/|\omega|$ is a unit vector in $\mathbb{R}^d$. In the turning bands method, u is sampled systematically and ρ is sampled randomly. Thus, let $\{u_i : i = 1, \dots, n\}$ be a fixed collection of n unit vectors approximately equally spaced on the unit sphere in $\mathbb{R}^d$. Equal spacing is easy to guarantee in $d = 2$ dimensions (the circle) but, except for special values of n, can be done only approximately in higher dimensions.

Suppose a method is already available to simulate one-dimensional stationary Gaussian processes with a given spectral density (e.g. based on (a), (b), or (c) above). Let $X_{1,i}(\cdot)$, $i = 1, \dots, n$, be a simulated one-dimensional Gaussian process with spectral densities, $f_{1,i}(\rho) = \pi_d \rho^{d-1} f(\rho u_i)$. Note that when $f(\omega)$ is isotropic, all the one-dimensional spectral densities will be the same. Combining the one-dimensional simulations together yields the representation

$$X^{(n)}(t) = n^{-1/2} \sum_{i=1}^{n} X_{1,i}(t^T u_i), \quad t \in \mathbb{R}^d.$$

Hence, a Gaussian random field with a d-dimensional index $t \in \mathbb{R}^d$ can be approximated as a sum where each term depends on a Gaussian random field with a one-dimensional index.

2.14.4 Circulant Methods

Circulant methods are well suited to the exact simulation of lattice stationary Gaussian random fields as given in Section 2.11. Let $\{\sigma_h : h \in \mathbb{Z}^d\}$ denote the covariance function with spectral density $f(\omega)$, $\omega \in (-\pi, \pi)^d$.

The method is easiest to describe in $d = 1$ dimension. Suppose we wish to simulate the process on a domain $D = \{1,2, \ldots, n\}$ for some $n > 1$. This distribution requires the covariances σ_h, $h = 0,1, \ldots, n-1$ at lags between pairs of sites $i,j \in D$. The trick is to embed D in a larger region $E = \{1,2, \ldots, m\}$ for some m, where for simplicity of presentation we assume $m > 2n-1$ is odd. Introduce on E a circulant covariance function δ_h, $0 \leq h \leq m-1$, regarded a periodic function so that $\delta_h = \delta_{h+pm}$ for all integers p, h, say. We require $\delta_h = \sigma_h$, $0 \leq |h| \leq n-1$. The simplest way to carry out this construction is to set $\delta_h = \sigma_h$, $0 \leq |h| < m/2$. Then periodicity determines the remaining values of δ_h; in particular, $\delta_{m-h} = \delta_h$, $1 \leq h < m/2$.

Let Δ denote the $m \times m$ circulant covariance matrix with (i,j)th entry δ_{i-j}. Denote its eigenvalues by λ_k, $k = 0, \ldots, m-1$. All symmetric circulant matrices have the same eigenvectors, and a convenient orthonormal matrix is given by $G = G_m^{(\mathrm{DFT,rea})}$ in Section A.7.1. Let $F = \mathrm{diag}(f_j)$ denote the corresponding eigenvalues. Then simulate a normal random vector $\boldsymbol{Y} \sim N_m(0, F)$ with independent components, and set $\boldsymbol{X} = G\boldsymbol{Y}$. Then $\mathrm{var}(\boldsymbol{X}) = GFG^T = \Delta$, and the subvector $(x_1, \ldots, x_n)$ is normally distributed with the required covariance structure. The fast Fourier transform (FFT) can be used to carry out the simulation efficiently in only $O(m \log m)$ calculations and is most efficient if m is highly composite. A number m is highly composite if it can be written as a product of powers of a small number of prime factors; e.g. $m = 2^k$, $m = 3^k$ or $m = 2^{k_1} 3^{k_2}$.

The only snag with this method is that Δ is not guaranteed to be positive semidefinite for all m; it may have some negative eigenvalues. However, Wood and Chan (1994) have shown that under mild regularity conditions the positive definiteness of Δ can be ensured for m large enough.

A similar construction can be carried out in $d \geq 2$ dimensions, with $n = (n[1], \ldots, n[d])$ and $m = (m[1], \ldots, m[d])$ being multi-indices. Again, the presentation is simplest if we require $m[l] \geq 2n[l] - 1$ with $m[l]$ odd, $l = 1, \ldots, d$. The methodology here is a straightforward generalization of the one-dimensional case described above. We simulate independent normal random variables in the Fourier domain and then, using the FFT, transform back to the original state space. Wood and Chan (1994) describe the process in detail. They describe how to deal with the complications that can arise if one or more of the $m[l]$ is even and show that under mild regularity conditions the circulant covariance matrix will be positive definite provided all the $m[l]$ are large enough.

Exercises

2.1 Using the following hints, show that for $d \geq 1$, $\pi_d = 2\pi^{d/2}/\Gamma(d/2)$ is the surface area of the unit sphere in $\mathbb{R}^d$.

(a) ($d = 1$). Note $\pi_1 = 2$ by direct observation.

(b) ($d = 2$). Using the fact that the length of a curve $y = y(x)$ between $x = a$ and $x = b$ is given by

$$\int_a^b \{1 + (dy/dx)^2\}^{1/2} dx,$$

express the perimeter of the unit circle as

$$\pi_2 = 2\pi = 2\int_{-1}^{1} (1 - s^2)^{-1/2}\, ds.$$

(c) ($d \geq 3$). Suppose by induction that π_{d-1} takes the required form. Express π_d as the volume of a surface of rotation about the $x[d]$ axis in $\mathbb{R}^d$ to get

$$\pi_d = \int_{-1}^{1} \{\pi_{d-1}(1 - s^2)^{(d-2)/2}\}(1 - s^2)^{-1/2}\, ds.$$

Use the integral for the beta function

$$\int_0^1 v^{\alpha-1}(1 - v)^{\beta-1} dv = B(\alpha, \beta) = \Gamma(\alpha)\Gamma(\beta)/\Gamma(\alpha + \beta)$$

for $\alpha, \beta > 0$, to simplify this expression.

2.2 The purpose of this exercise is to verify the density in (2.19). Without loss of generality let $u = (1,0,\ldots,0)^T$, so $\eta^T u = \eta[1]$. Using Exercise 2.1(c), show that

$$P(\eta[1] < s) = \frac{\pi_{d-1}}{\pi_d} \int_{-1}^{s} (1 - \eta^2)^{(d-3)/2}\, d\eta.$$

Differentiate with respect to s to get the required density. See also Mardia and Jupp (2000, p. 167).

2.3 Verify the limiting behavior of Eq. (2.20). Hint: Use the elementary inequality $\log(1 - x) \leq -x$ for $0 < x < 1$, the elementary limit $(1 + x/\delta)^\delta \to e^x$ as $\delta \to \infty$, and Stirling's formula

$$\Gamma(\delta) \sim e^{-\delta}\delta^{\delta-\frac{1}{2}}(2\pi)^{\frac{1}{2}}.$$

2.4 Given a function $g(\rho)$, $\rho > 0$, the *Hankel transform* of order $\lambda > -1$ is defined by

$$\mathcal{H}_\lambda(g)(r) = \int_0^\infty (\rho r)^{1/2} J_\lambda(\rho r) g(\rho)\, d\rho$$

(Erdélyi, 1954, p. 3). In the same way that the Fourier transform links a spectral density on $\mathbb{R}^d$ to a covariance function on $\mathbb{R}^d$, we shall see that the Hankel transform links a radial spectral density on $(0, \infty)$ to a radial covariance function on $(0, \infty)$. Indeed, $\mathcal{H}_\lambda$ is directly analogous to $(2\pi)^{-d/2}\mathcal{F}$, where $\mathcal{F}$ denotes the Fourier transform in $\mathbb{R}^d$ and $\lambda = (d-2)/2$.

(a) Let $\sigma(h)$ be an isotropic covariance function in $\mathbb{R}^d$ with spectral density $f(\omega)$. Set $\sigma^\#(r) = \sigma(h)$ for $r = |h|$ and $f^\#(\rho) = f(\omega)$ for $\rho = |\omega|$. Then, $\sigma(h)$ and $f(\omega)$ are linked by the Fourier transform

$$\sigma(\omega) = \int \exp(ih^T\omega)\, f(\omega)\, d\omega = \int \cos(h^T\omega)\, f(\omega)\, d\omega.$$

Set

$$\tilde{\sigma}^\#(r) = r^{\lambda+\frac{1}{2}}\sigma^\#(r), \quad \tilde{f}^\#(\rho) = \rho^{\lambda+\frac{1}{2}}f^\#(\rho).$$

Show that

$$\tilde{\sigma}^\#(r) = (2\pi)^{d/2}\mathcal{H}_\lambda(\tilde{f}^\#)(r),$$

where the notation means that $\mathcal{H}_\lambda(\tilde{f}^\#)$ is a function evaluated at the argument r. That is, the Hankel transform takes a reweighted radial spectral density to a reweighted radial covariance function. See Eq. (2.26).

(b) Starting with the inverse Fourier transform

$$f(\omega) = (2\pi)^{-d} \int \exp(-ih^T\omega)\sigma(\omega)\, dh,$$

show that the Hankel transform is "self-reciprocal." That is, subject to suitable integrability conditions,

$$\tilde{f}^\#(\rho) = (2\pi)^{-d/2}\mathcal{H}_\lambda(\tilde{\sigma}^\#)(\rho).$$

(c) Using Erdélyi (1954, p. 24, equation (120)), show that for $\lambda > -1$, $\nu > 0$, the function

$$\varphi(\rho) = \rho^{\lambda+1/2}/(1+\rho^2)^{\nu+\lambda+1}$$

has the Hankel transform

$$\psi(r) = \mathcal{H}_\lambda(\varphi)(r) = \frac{r^{\nu+\lambda+\frac{1}{2}}K_\nu(r)}{2^{\nu+\lambda}\Gamma(\nu+\lambda+1)}.$$

Use this result and Eq. (2.26) to deduce that in all dimensions $d \geq 1$ the Matérn function

$$\sigma^\#(r) = \frac{2}{\Gamma(\nu)}(r/2)^\nu K_\nu(r)$$

is a valid radial covariance function in $\mathbb{R}^d$ with radial spectral density

$$f^\#(\rho) = \frac{\Gamma(\nu+\frac{d}{2})}{\Gamma(\nu)\pi^{d/2}}\frac{1}{(1+\rho^2)^{\nu+\frac{d}{2}}}.$$

Since $f^\#(\rho)$ is positive for all $\rho > 0$, deduce that $\sigma^\#(r)$ is positive definite.

(d) Using the fact that $K_\nu(r) > 0$ for all $\nu > 0$ and $r > 0$, deduce that the t-density scheme in Table 2.1 is positive definite when $\alpha > d/2$. (The construction can be extended to include all $\alpha > 0$.)

2.5 (Matérn, 1960, p. 16) Using Eq. (2.26) show that an isotropic radial covariance function $\sigma^{\#}(r),\ r \geq 0$, in d dimensions is bounded below by

$$\inf_{r\geq 0} \{\sigma^{\#}(r)\}/\sigma^{\#}(0) \geq \Gamma(\lambda+1)\inf_{x>0} \{(x/2)^{-\lambda} J_\lambda(x)\} = L_d, \text{ say,}$$

where $\lambda = \frac{1}{2}(d-2)$. Show that the shell scheme, for which the radial spectral measure $F^{\#}(d\rho)$ is a point mass (item 9 in Section 2.7), attains this lower bound. The first few values of this lower bound are $L_1 = -1$, $L_2 = -0.403$, $L_3 = -0.217$, $L_4 = -0.132$; and $\lim_{d\to\infty} L_d = 0$. Thus, only limited amounts of negative autocorrelation are allowed in isotropic random fields in $\mathbb{R}^d$, for $d > 1$.

2.6 For $t \in \mathbb{R}^d$ let $\varphi(t) = I[|t| < \frac{1}{2}]$ where I is an indicator function. Let $\sigma(h) = (\varphi * \check{\varphi})(h)$ as in Eq. (2.33). Show that $\sigma^{\#}(r) = \sigma(|h|)$ takes the following form in dimensions $d = 1,2,3$:

$$(d=1)\quad \sigma(r) \propto (1-r)I[r \leq 1],$$

$$(d=2)\quad \sigma(r) \propto \{1-(2/\pi)r(1-r^2)^{1/2}-(2/\pi)\sin^{-1} r\}I[r<1],$$

$$(d=3)\quad \sigma(r) \propto \left\{1-\frac{3}{2}r+\frac{1}{2}r^3\right\} I[r<1].$$

The function for $d = 3$ is the "spherical scheme" of Table 2.1. The function for $d = 2$ is the analogous construction in two dimensions and is sometimes known as the "circular scheme." The function for $d = 1$ is sometimes known as the "tent" scheme, since, when plotted as a function of h, with $r = |h|$, it has a tent-like shape between $h = -1$ and $h = 1$ with a peak at $h = 0$. Note that the tent scheme can also be viewed as a special case of the restricted-power scheme in Table 2.1 with $k = 1$, but it is a valid covariance function only in dimension $d = 1$. Chilés and Delfiner (2012, pp. 85–88) discuss higher dimensional versions.
Hint: For $d \geq 2$, $\sigma^{\#}(r)$ is given by the volume in the intersection of two spheres of radius $\frac{1}{2}$ and distance r apart. Since the volume of a $(d-1)$-dimensional sphere of radius r equals $\{\pi^{\frac{1}{2}(d-1)}/\Gamma(\frac{1}{2}d+\frac{1}{2})\}r^{d-1} = \pi_{d-1}r^{d-1}/(d-1)$ (to verify this formula, differentiate with respect to r at $r = 1$ to get the surface area π_{d-1} of the sphere), $\sigma^{\#}(r)$ can be expressed as

$$\sigma^{\#}(r) = \{2\pi_{d-1}/(d-1)\} \int_{\frac{1}{2}r}^{\frac{1}{2}} (\frac{1}{4}-u^2)^{(d-1)/2}\, du.$$

2.7 (Mardia and Watkins, 1989). Let $\sigma_k(h) = (1 - |h|)^k I[|h| \leq 1]$ denote the restricted-power scheme of Table 2.1. The purpose of this exercise is to show the positive definiteness of $\sigma_k(h)$ for integer $k \geq 2$ in $d = 3$ dimensions. This will be achieved by showing that the spectral density, $f_k(\omega)$, say, of $\sigma_k(h)$ is nonnegative for all $\omega \in \mathbb{R}^3$.

(a) Write $f_k^{\#}(\rho) = f_k(\omega)$, $\rho = |\omega|$, for the radial spectral density. Using the Fourier inversion formula, show that in $d = 3$ dimensions

$$f_k^{\#}(\rho) = \frac{1}{2\pi^2 \rho} \int_0^1 r(1-r)^k \sin(r\rho) dr.$$

Note that the integrand is positive if $0 < \rho < \pi$; hence conclude that

$$f_k^{\#}(\rho) > 0 \text{ for } 0 < \rho \leq \pi.$$

(b) Using integration by parts or otherwise, evaluate this integral for $k = 2$ and $k = 3$,

$$f_2^{\#}(\rho) = \frac{1}{\pi^2 \rho^5} \{\rho(2 + \cos\rho) - 3\sin\rho\}, \quad \rho > 0,$$

$$f_3^{\#}(\rho) = \frac{3}{\pi^2 \rho^6} \{\rho^2 - 4 + \rho\sin\rho + 4\cos\rho\}, \quad \rho > 0.$$

(c) If $\rho \geq \pi$ note that $\rho(2 + \cos\rho) - 3\sin\rho \geq \rho - 3 > 0$. Hence conclude that $f_2^{\#}(\rho) > 0$ for all $\rho > 0$.

(d) Similarly, show that $f_3^{\#}(\rho) > 0$ for all $\rho > 0$.

(e) Use the fact that positive definiteness of functions is preserved under multiplication to conclude that $\sigma_k(h)$ is positive definite for all integers $k \geq 2$ in $d = 3$ dimensions.

(f) Why does $\sigma_k(h)$ also define a positive definite function in $d = 1$ and $d = 2$ dimensions?

2.8 (Separable covariance functions; see, e.g., Martin, 1979). A covariance function $\sigma(h)$, $h \in \mathbb{R}^2$, is called *separable* if it factorizes as

$$\sigma(h) = \sigma_1(h[1])\, \sigma_2(h[2]).$$

Separable covariance functions are important because they build on our ability to model and analyze stochastic processes in one dimension.

(a) If $\sigma_1(h[1])$ and $\sigma_2(h[2])$ are positive definite functions in one dimension, show that $\sigma(h)$ is positive definite in two dimensions.

(b) If $\sigma(h)$ has spectral density $f(\omega)$ show that $\sigma(h)$ is separable if and only if $f(\omega)$ also factorizes

$$f(\omega) = f_1(\omega[1])\, f_2(\omega[2]).$$

(c) Show that $\sigma(h) = e^{-\alpha h[1]-\beta h[2]}$ defines a valid covariance function for $\alpha > 0,\ \beta > 0$. Show that $\sigma(h)$ is not isotropic.
(d) Extend the concept of separability to dimensions $d > 2$.

2.9 Start with a random variable Y following a gamma distribution with index $\nu > 0$ and scale parameter $\lambda > 0$, with density $\lambda^\nu y^{\nu-1} \exp(-\lambda y)/\Gamma(\nu),\ y > 0$. Then the reciprocal $U = 1/Y$ has density $g(u) = \lambda^\nu u^{-\nu-1} e^{-\lambda/u}/\Gamma(\nu),\ u > 0$.
(a) Using, e.g. Gradshteyn and Ryzhik (1980, p. 340, equation (9)), show that the Laplace transform of $g(u)$ is given by

$$\hat{g}(s) = \int_0^\infty g(u)e^{-su}\, du = \frac{2}{\Gamma(\nu)}(\lambda s)^{\nu/2} K_\nu(2\sqrt{\lambda s}), \quad s > 0.$$

(b) Hence, using Theorem 2.4.1 show that the function

$$\sigma(h) = c_1 \frac{2^{1-\nu}}{\Gamma(\nu)}(r/c_2)^\nu K_\nu(r/c_2), \quad r = |h|, \quad h \in \mathbb{R}^d,$$

is positive definite in all dimensions d where $c_1 > 0$ is a scale parameter and $c_2 > 0$ is a range parameter. An alternative proof of this result was given in Exercise 2.4(c).
(c) For index $\nu = 1/2$, use the identity $K_{1/2}(z) = \sqrt{\pi/(2z)}e^{-z}$ (e.g. Abramowitz and Stegun, 1964, p. 444, equation (10.2.17)) to show that the Matérn scheme reduces to the exponential scheme in Table 2.2.
(d) Using the fact that the gamma distribution has mean ν/λ and variance ν/λ^2, show that the gamma distribution converges to a point mass at $y = 1$ as $\nu \to \infty$ with $\lambda = \nu$. Hence, deduce that the Gaussian scheme appears as a limiting case of the Matérn scheme.

2.10 Let $0 < \alpha < 2$ and suppose that the spectral density satisfies the integrability condition $\int [1 + |\omega|]^\alpha f(\omega)\, d\omega < \infty$, as in (2.83). Show that $\sigma(0) - \sigma(h) = O(|h|^\alpha)$ as $h \to 0$.
Hint: Write $\sigma(0) - \sigma(h) = \int [1 - \cos(h^T\omega)]\, f(\omega)\, d\omega$ in terms of the spectral density, and use the following inequalities:
(a) $1 - \cos(x) \le \frac{1}{2}x^2$ for all $x \ge 0$, so that

$$1 - \cos(x) \le \min\left(\frac{1}{2}x^2, 2\right) \le 2 \min(x^2, 1).$$

(b) For $0 \le x \le 1, x^2 \le x^\alpha$, so that for all $x \ge 0$,

$$\min(x^2, 1) \le \min(x^\alpha, 1) \le x^\alpha.$$

There is also a converse to this result, which is harder to prove. If for some $0 < \alpha < 2$, $\sigma(0) - \sigma(h) = O(|h|^\alpha)$ as $h \to 0$, then $\int |\omega|^\beta f(\omega)\, d\omega < \infty$ for all $0 < \beta < \alpha$.

2.11 Let $\sigma(h) = \sigma^\#(r)$, $r = |h|$, be an isotropic covariance function in $\mathbb{R}^3$, with spectral density $f(\omega) = f^\#(\rho)$, $\rho = |\omega|$. In addition to the usual integrability condition $\int f(\omega)\, d\omega < \infty$, suppose also that $\int |\omega|^2 f(\omega)\, d\omega < \infty$.

(a) Show that $\sigma(h)$ is twice continuously differentiable.

(b) Show that $-\partial^2\sigma(h)/\partial h[1]^2$ is positive definite with spectral density $\omega[1]^2 f(\omega)$.

(c) Hence, show that the negative Laplacian $-\triangle\,\sigma(h) = -\sum_{j=1}^{3} \partial^2\sigma(h)/\partial h[j]^2$ is positive definite with isotropic radial spectral density $\rho^2 f^\#(\rho)$.

(d) Thus in $d = 3$ dimensions, deduce that the covariance function of the one-dimensional process needed in the turning bands algorithm in Section 2.14.3 is given by $\sigma_1(r) = -d^2\sigma^\#(r)/dr^2 - (2/r)d\sigma^\#(r)/dr$.

3

Intrinsic and Generalized Random Fields

3.1 Introduction

Chapter 2 looked at the properties of random fields $\{X(t) : t \in D\}$ as the site t ranges through a domain D; typically $D = \mathbb{R}^d$. Provided the moments are finite, there is a well-defined mean function $\mu(t) = E\{X(t)\}$ and a well-defined covariance function $\sigma(s, t) = \text{cov}\{X(s), X(t)\}$ for all $s, t \in D$. From this chapter's perspective, it is helpful to refer to these random fields as *ordinary* in contrast to the *generalized* random fields introduced later.

In addition, Chapter 2 emphasized the case of stationary random fields for which $\mu(t) = \mu$ is constant and $\sigma(s, t) = \sigma(s - t)$ depends only on the lag $s - t$. Stationarity is an important assumption to ensure the tractability of the covariance function; it simplifies the interpretation and provides a framework for the estimation of any unknown parameters.

However, the class of ordinary stationary random fields is not sufficiently general for all of our applications. It is useful to extend this class in two ways: from stationary to *intrinsic* random fields and from ordinary to *generalized* random fields.

Intrinsic random fields are studied in Sections 3.2–3.4. An intrinsic random field can also be described as a random field with *stationary increments*. For an intrinsic random field, the focus is on certain *increments* of the random field. In the simplest version (intrinsic order $k = 0$), a typical increment takes the form of a difference, e.g., $X(t + h) - X(t)$. In $d = 1$ dimension, Brownian motion is the most well-known example, which is discussed in Section 3.2. A natural tool to study intrinsic random fields of order $k = 0$ is the *semivariogram*, and its properties are studied in Section 3.3. For all orders $k \geq 0$, the class of intrinsic random fields is important because it is larger than the class of stationary random fields but retains much of the tractability.

The price to be paid for the intrinsic assumption is that it only provides a partial specification of the covariance function. The mean and variance are not specified for all linear combinations of the random field, but only for certain increments.

Spatial Analysis, First Edition. John T. Kent and Kanti V. Mardia.

Methods to complete the specification of the mean and covariance function for all linear combinations are discussed in Sections 3.2 and 3.5.

On the other hand, in a generalized random field (Section 3.6), the realizations are too rough to be ordinary functions. Instead, the realizations are generalized functions in the sense of Schwartz (Gel'fand and Shilov, 1964, 1968; Gel'fand and Vilenkin, 1964). The only way to investigate such random fields is through regularizations $X(\varphi)$ for suitably smooth test functions φ. The most widely known generalized stationary random field is Gaussian white noise, which forms the continuous analog of a sequence of independent identically distributed normal random variables.

The spectral representation for the covariance function of a stationary random field (Bochner's Theorem 2.3.3) can be extended to cover random fields that are intrinsic or generalized or both (Sections 3.7–3.8). In the spectral domain, intrinsic random fields have the property that the spectral density is too large at *low* frequencies to be integrable; conversely, generalized stationary random fields have the property that the spectral density is too large at *high* frequencies to be integrable. Regularization of the realizations of an intrinsic or generalized random field can be understood most clearly in the spectral domain and is explored in Section 3.9. The important concept of *self-similarity* for Gaussian random fields (Section 3.10) can be given a satisfactory treatment only in the setting of generalized and intrinsic random fields.

The main focus of this book is on random fields with finite second moments. Further, the emphasis in both Chapter 2 and this chapter is on just the first and second moments of the underlying random field rather than the full set of finite-dimensional distributions. Moreover, the most important application of the theory is to *Gaussian random fields*, which are fully determined by their first two moments. The assumption of Gaussianity becomes important for the specification of autoregressive models (Chapter 4) and for statistical inference (Chapters 5–7). For a more detailed investigation of the properties of Gaussian random fields, see, e.g., Adler (1981), Adler and Taylor (2007).

3.2 Intrinsic Random Fields of Order $k = 0$

This section and Section 3.3 focus on an intrinsic random field of order $k = 0$. Higher order intrinsic random fields are discussed in Section 3.4.

The motivation behind an interest in intrinsic random fields comes from the following observation. In some applications, it is found that the data appear to be modeled by a random field $\{X(t) : t \in \mathbb{R}^d\}$ for which the semivariogram

$$\gamma(h) = \frac{1}{2}\text{var}\{X(t+h) - X(t)\}$$

grows indefinitely with $|h|$. For such a process, the overall variability grows with the size of the region being examined (e.g. Example 1.12). It is useful to extend the notion of a stationary process to deal with this possibility.

Definition 3.2.1 Let $\{X(t) : t \in \mathbb{R}^d\}$ be a random field. If the first two incremental moments take the form

$$E\{X(t+h) - X(t)\} = \nu(h), \text{ say,}$$
$$\frac{1}{2}\text{var}\{X(t+h) - X(t)\} = \gamma(h), \text{ say,} \quad t, h \in \mathbb{R}^d,$$

where $\nu(h)$ and $\gamma(h)$ are continuous functions of h, not depending on t, then $\{X(t)\}$ is said to be an *intrinsic random field*. The name *random field with stationary increments* is also used. In addition, $\nu(h)$ is called the *incremental mean function* and $\gamma(h)$ is called the *semivariogram*, respectively. The function $2\gamma(h)$ is known as the *variogram*.

The property of being intrinsic depends only on the second moment properties of the increments. If, in addition, for all $n \geq 1$ and all choices of sites $t_0, \ldots, t_n \in \mathbb{R}^d$, the distribution of the n-dimensional vector of increments

$$\{X(t_1) - X(t_0), \ldots, X(t_n) - X(t_0)\}$$

depends only on the site differences $t_1 - t_0, \ldots, t_n - t_0$, then $\{X(t)\}$ is said to be *strongly intrinsic*. In particular, a Gaussian process that is intrinsic is automatically strongly intrinsic.

The two most important nonstationary intrinsic processes in $d = 1$ dimension are *Brownian motion (with drift)* and the *Poisson process*. Brownian motion is a Gaussian process $\{X(t) : t \in \mathbb{R}^1\}$ defined by the following properties:

(a) $X(0) = 0$.
(b) For $s < t$, $X(t) - X(s)$ is normally distributed with mean $(t-s)\mu$ and variance $(t-s)\sigma^2$.
(c) For $s < t \leq u < v$, $X(t) - X(s)$ is independent of $X(v) - X(u)$.

From (b) we see that $\nu(h) = \mu h$ and $\gamma(h) = \frac{1}{2}\sigma^2|h|$ for Brownian motion. Often the intrinsic mean is assumed to vanish, $\mu = 0$.

The Poisson process has a similar characterization with (b) replaced by the following condition:

(b′) For $s < t$, $X(t) - X(s)$ has a Poisson distribution with parameter $(t-s)\lambda$. Hence, $\nu(h) = \lambda h$ and $\gamma(h) = \frac{1}{2}\lambda|h|$ for the Poisson process.

Since the Poisson distribution is concentrated on the nonnegative integers, the realizations from a Poisson process will be nondecreasing step functions. By convention, the realizations are usually taken to be right-continuous, though the choice of convention makes no difference to the second moment properties.

Note that the semivariogram takes the same form for Brownian motion and the Poisson process. Also both processes are *strictly* intrinsic, and from (c), both processes have *independent increments.*

Clearly, a stationary random field on $\mathbb{R}^d$ with mean μ and covariance function $\sigma(h)$ is also intrinsic with incremental mean $\nu(h) = 0$ and with semivariogram $\gamma(h) = \sigma(0) - \sigma(h)$. Conversely, if the semivariogram of an intrinsic process satisfies $\gamma(h) \to c$, say, as $|h| \to \infty$, then $\sigma(h) = c - \gamma(h)$ defines a valid covariance function; see Exercise 3.1. In contrast, an intrinsic random field for which $\gamma(h) \to \infty$ as $|h| \to \infty$ is not stationary.

To characterize the incremental mean function $\nu(t)$ for an intrinsic random field on $\mathbb{R}^d$, we note the identity

$$\begin{aligned}\nu(t+h) &= E\{X(t+h) - X(t) + X(t) - X(0)\}\\ &= E\{X(t+h) - X(t)\} + E\{X(t) - X(0)\}\\ &= \nu(h) + \nu(t), \quad t, h \in \mathbb{R}^d.\end{aligned}$$

Since $\nu(h)$ is assumed continuous and $\nu(0) = 0$, it follows that $\nu(h)$ is a linear function of h

$$\nu(h) = \beta^T h \tag{3.1}$$

for some $\beta \in \mathbb{R}^d$. Thus, a linear drift is allowed in an intrinsic random field, in contrast to the constant mean of a stationary random field.

Next we look at further simple properties of intrinsic random fields. Start with the identity

$$\begin{aligned}2\{X(t) - X(s)\}\{X(v) - X(u)\} =& \{X(t) - X(u)\}^2 + \{X(s) - X(v)\}^2\\ &- \{X(s) - X(u)\}^2 - \{X(t) - X(v)\}^2\end{aligned} \tag{3.2}$$

for all $s, t, u, v \in \mathbb{R}^d$. After centering the increments to have mean 0 and taking expectations, we can calculate the covariance between two increments

$$\mathrm{cov}\{X(t) - X(s),\ X(v) - X(u)\} = \gamma(t-u) + \gamma(s-v) - \gamma(s-u) - \gamma(t-v). \tag{3.3}$$

The difference $X(t_1) - X(t_2)$ can be viewed as an *elementary* increment or contrast. A more general definition of an increment is as follows.

Definition 3.2.2 A linear combination $\sum_{i=1}^n a_i X(t_i)$ or, alternatively, a list of coefficients and sites $\{(a_i, t_i)\}_{i=1}^n$ is called an *increment* (of order $k = 0$) if the coefficients satisfy $\sum a_i = 0$.

From (3.3), it is straightforward to show that

$$\operatorname{var}\left\{\sum a_i X(t_i)\right\} = -\sum a_i a_j \gamma(t_i - t_j) \geq 0 \tag{3.4}$$

for all increments; see Exercise 3.2. In Section 3.4, increments of higher order $k > 0$ will be introduced.

Let $\psi(s)$ denote a continuous function of $s \in \mathbb{R}^d$, assumed for simplicity to have compact support, and suppose it is an increment function, i.e. $\int \psi(s)\, ds = 0$. Recall that a function $\varphi(t),\ t \in \mathbb{R}^d$, has compact support if there is a constant C (depending on φ) such that $\varphi(t) = 0$ for all $|t| \geq C$. A continuous version of (3.4) can be formulated for all increment functions

$$\operatorname{var}\left\{\int X(s)\psi(s)\, ds\right\} = -\int \psi(s)\psi(t)\gamma(s-t)\, ds\, dt. \tag{3.5}$$

A function $-\gamma(h)$ with the nonnegativity property (3.4) for all increments (or equivalently (3.5) for all increment functions) is said to be *conditionally positive semidefinite* of order $k = 0$; if (3.4) is strictly positive whenever the $\{a_i\}$ are not all 0, and the $\{t_i\}$) are distinct sites, then $-\gamma(h)$ is called *conditionally positive definite.* The following theorem (Matheron, 1971, pp. 56–57) demonstrates the usefulness of this concept.

Theorem 3.2.1 *A continuous even function $\gamma(h)$, $h \in \mathbb{R}^d$, with $\gamma(0) = 0$ represents the semivariogram of an intrinsic random field if and only if $-\gamma(h)$ is conditionally positive semidefinite of order $k = 0$.*

Proof: We saw above that if $\gamma(h)$ is the semivariogram of an intrinsic random field, then $-\gamma(h)$ is conditionally positive semidefinite. Conversely, suppose $-\gamma(h)$ is conditionally positive semidefinite, and let us try to construct a process for which $\gamma(h)$ is the semivariogram. Since the semivariogram only specifies the variance of the increments of a random field, we are free to specify additionally the behavior of the random field at a single given site; e.g., we can set $X(t_0) = 0$ at a site t_0, so that $\operatorname{var}\{X(t_0)\} = 0$. Set

$$\sigma(s,t) = \gamma(s - t_0) + \gamma(t - t_0) - \gamma(t - s) - \gamma(t_0 - t_0). \tag{3.6}$$

The last term is not needed since $\gamma(t_0 - t_0) = \gamma(0) = 0$, but has been included to ensure consistency with similar results for intrinsic random fields of higher order. Note that $\sigma(t_0, t_0) = 0$. It is not difficult to check that $\sigma(s,t)$ is positive semidefinite (Exercise 3.3), and hence from Theorem 2.3.2 it represents the covariance function of a (nonstationary) random field, $\{X(t)\}$, say. Further, after some simplification, it can be checked that this random field has semivariogram

$$\begin{aligned}\frac{1}{2}\text{var}\{X(t+h)-X(t)\} &= \frac{1}{2}\{\sigma(t+h,t+h)-2\sigma(t,t+h)+\sigma(t,t)\}\\ &= \gamma(h),\end{aligned}$$

as expected. □

Next we discuss the *regularization* of an intrinsic random field $\{X(t)\}$ with semivariogram $\gamma(h)$, following the construction in Section 2.10. Let $\varphi(s)$ denote a *test function*, which is bounded and has compact support. Define the regularized version of $X(t)$ by

$$X_\varphi(t) = \int X(s)\,\varphi(s-t)\,ds, \quad t \in \mathbb{R}^d. \tag{3.7}$$

Then $\{X_\varphi(t)\}$ is also an intrinsic random field whose semivariogram is easily seen to be

$$\gamma_\varphi(h) = \frac{1}{2}\text{var}\{X_\varphi(h)-X_\varphi(0)\} = \frac{1}{2}\text{var}\left\{\int X(s)\,\psi_h(s)\,ds\right\},$$

where $\psi_h(s) = \varphi(s-h) - \varphi(s)$ is an increment function. Hence, from Eq. (3.5),

$$\begin{aligned}\gamma_\varphi(h) &= -\frac{1}{2}\int\int \psi_h(s)\,\psi_h(t)\,\gamma(s-t)\,ds\,dt\\ &= \int\int \varphi(s)\,\varphi(t-h)\,\gamma(s-t)\,ds\,dt - \int\int \varphi(s)\,\varphi(t)\,\gamma(s-t)\,ds\,dt. \end{aligned}\tag{3.8}$$

Note that the second term here is constant as h varies; its presence ensures $\gamma_\varphi(0) = 0$.

Let $\{X(t)\}$ be an intrinsic random field. Since the semivariogram just specifies the variances of the increments of the process, we may replace $\{X(t)\}$ by $\{X(t) + U\}$ for any single random variable U, without affecting $\gamma(h)$. There are three common ways to think about this indeterminacy.

(a) *Registration*. The intrinsic random field $\{X(t)\}$ can be *registered* by specifying its value at a specific site t_0, e.g. $X(t_0) = x$, say. The registered random field can be written as $X(t) = x + (X(t) - X(t_0))$, the sum of a fixed value and a random increment. Registration removes the indeterminacy in the specification of the first two moments of $\{X(t)\}$. The result is a nonstationary random field with mean function $\mu(t) = x + \nu(t - t_0)$, where $\nu(\cdot)$ is given in (3.1) and with the covariance function $\sigma(s, t)$ given in (3.6).

(b) *Improper marginal distributions*. If $\gamma(h) \to c < \infty,\ 0 < c < \infty$ as $|h| \to \infty$, and $\nu(h) = 0$ in (3.1), we may regard $\{X(t)\}$ as a stationary random field. This interpretation is still available in a limiting sense if $\gamma(h) \to \infty$ as $|h| \to \infty$, by regarding $\{X(t)\}$ as an *improper* stationary random field with an improper marginal uniform distribution for $X(t) \in \mathbb{R}$ at each site $t \in \mathbb{R}^d$. The improper marginal uniform distribution can also be viewed as a limiting $N(0, \sigma^2)$ distribution as $\sigma^2 \to \infty$. If a random field model is used as a prior distribution in a Bayesian analysis, then this approach amounts to the use of an improper prior distribution. Section 7.12.3 discusses the example of Bayesian kriging.

(c) *Restriction to increments.* We may limit our interest in the intrinsic random field $\{X(t)\}$ to statements about increments of the random field, $\sum a_i X(t_i)$ where $\sum a_i = 0$. Then the underlying indeterminacy is irrelevant. This point of view is the most useful way to view intrinsic random fields and is the approach we shall generally follow. See, e.g., Sections 7.4 and 7.5 on ordinary and universal kriging (but note the adjective "ordinary" has a different meaning in kriging than when describing a random field).

For later use, we mention an important property about $\gamma(h)$. In many formulas, it is possible to replace $\gamma(h)$ by $\gamma(h) + c$, for any real constant c, without affecting the validity of the formula. Examples include Eqs. (3.4) and (3.5); see Exercise 3.4. The reason is that such formulas involve increments, and the constant c disappears after summing (or integrating) over the coefficients.

Definition 3.2.3 Given a semivariogram $\gamma(h)$, define the *intrinsic covariance function* $\sigma_I(h)$ by

$$\sigma_I(h) = -\gamma(h) + c, \tag{3.9}$$

where c is an arbitrary real constant.

Thus, $\sigma_I(h)$ is an equivalence class of functions rather than a single function. The variogram can be recovered from the intrinsic covariance function by $\gamma(h) = \sigma_I(0) - \sigma_I(h)$, and this formula does not depend on c.

One reason for introducing the intrinsic covariance function is to extend the quadratic form result in Eq. (2.10), which holds for all sites $t_1, \ldots, t_n$ and all coefficient vectors $\boldsymbol{a}$. In the intrinsic case, it is necessary to restrict the choice of coefficient vectors to increments.

Theorem 3.2.2 *If $\sigma_I(h)$ is the intrinsic covariance function of an intrinsic random field $\{X(t) : t \in \mathbb{R}^d\}$ of order 0, then $\sigma_I(h)$ is conditionally positive semidefinite of order 0. That is, for any $n \geq 1$, any selection of sites $t_1, \ldots, t_n$ and any increment vector $\boldsymbol{a}$ $(n \times 1)$ of coefficients of order 0, i.e. $\sum a_i = 0$, the following quadratic form is nonnegative:*

$$\sum_{i,j=1}^{n} a_i a_j \sigma_I(t_i - t_j) \geq 0. \tag{3.10}$$

Proof: This equation is just a restatement of Eq. (3.4), so the proof is obvious. In particular, the quadratic form does not depend on the choice of c in (3.9). □

Another reason for introducing the intrinsic covariance function is that it can be generalized naturally to higher order intrinsic random fields. In Section 3.4, the concepts of increments, intrinsic random fields, and conditional positive definiteness of order $k > 0$ are defined and investigated.

Strictly speaking, the increments of this section should be referred to as *zeroth-order* increments, i.e. linear combinations of order $k = 0$. When used without qualification, these phrases will refer to the simplest case of order $k = 0$, as we have done in this section.

3.3 Characterizations of Semivariograms

The class of conditionally positive semidefinite functions can be characterized by the following theorem (Matheron, 1975, p. 95).

Theorem 3.3.1 *Let $g(h)$ be a real-valued continuous symmetric function of $h \in \mathbb{R}^d$. Then the following conditions are equivalent.*

(a) $g(h)$ is conditionally positive semidefinite of order 0.
(b) $\exp\{\alpha\, g(h)\}$ is positive semidefinite for all $\alpha > 0$.
(c) $g(h)$ has a spectral representation

$$g(h) = g(0) + \int_{\mathbb{R}^d\setminus\{0\}} \{\cos(h^T\omega) - 1\}F(d\omega) + \frac{1}{2}h^T A h, \tag{3.11}$$

where $\{|\omega|^2/(1 + |\omega|^2)\}F(d\omega)$ is a symmetric integrable measure on $\mathbb{R}^d\setminus\{0\}$, and A is a $d \times d$ positive semidefinite matrix.

Proof: *(b)* ⇒ *(a)*. If $\exp\{\alpha\, g(h)\}$ is positive semidefinite for some $\alpha > 0$, then it is also conditionally positive semidefinite, hence so is

$$[\exp\{\alpha g(h)\} - 1]/\alpha,$$

using the definition in (3.4). As this statement remains true for all $\alpha > 0$, taking the limit as $\alpha \to 0$ shows that $g(h)$ is conditionally positive semidefinite.

(a) ⇒ *(b)*. Let $g(h)$ be conditionally positive semidefinite. We saw above (see Eq. (3.6) with $\gamma(h) = g(0) - g(h)$) that $\sigma(s, t)$ defined by

$$\sigma(s, t) = g(0) + g(s - t) - g(s) - g(t)$$

is a positive semidefinite function of two variables. Hence $1 + \alpha\, \sigma(s, t)$ is positive semidefinite for $\alpha > 0$. Schur's theorem tells us that a product of positive semidefinite (positive definite) functions is again positive semidefinite (positive definite); see Section A.3.5. Hence the product,

$$\{1 + \alpha\, \sigma(s,t)/n\}^n,$$

is positive semidefinite for any fixed $\alpha > 0$. Letting $n \to \infty$ shows that $\exp\{\alpha\, \sigma(s,t)\}$ is positive semidefinite.

Next, for any function $f(s)$ it is straightforward to show that $f(s)f(t)$ is positive semidefinite in s and t (Exercise 3.6). Letting $f(s) = \exp\{\alpha\, g(s) - \frac{1}{2}\alpha\, g(0)\}$ and multiplying $f(s)f(t)$ by $\exp\{\alpha\, \sigma(s,t)\}$ shows that $\exp\{\alpha\, g(s-t)\}$ is a positive semidefinite function in s and t; that is, $\exp\{\alpha\, g(h)\}$ is a positive semidefinite function of h.

(b) ⇔ (c). The class of continuous functions $\kappa(h)$ such that $\kappa(0) = 1$ and $\kappa(h)^\alpha$ is positive semidefinite for all $\alpha > 0$ is precisely the class of characteristic functions of *symmetric infinitely divisible probability distributions* on $\mathbb{R}^d$, and such functions are characterized by the Lévy–Khintchine integral representation for $\log \kappa(h)$ (Feller, 1966, p. 559). In terms of $g(h) - g(0) = \log \kappa(h)$, this representation is the same as (3.11). □

In practice, we are usually only interested in semivariograms $\gamma(h) = g(0) - g(h)$ for which $A = 0$ in (3.11). It can then be shown in this case that $\gamma(h)/|h|^2 \to 0$ as $h \to \infty$; see Exercise 3.7. Also in practice $F(d\omega)$ usually has a density, $F(d\omega) = f(\omega)\, d\omega$, which ensures that $\gamma(h)$ is conditionally positive definite and not just conditionally positive semidefinite. For an intrinsic process in dimensions $d = 1,2$ that is not stationary, the spectral density usually has a nonintegrable singularity $f(\omega) = O(|\omega|^{-2})$ as $\omega \to 0$. But since $1 - \cos(h^T\omega) = O((h^T\omega)^2) = O(|\omega|^2)$, the integrand in (3.11) remains bounded as $\omega \to 0$.

If $\gamma(h)$ is isotropic depending only on $r = |h|,\ h \in \mathbb{R}^d$, write $\gamma(h) = \gamma^\#(r)$ and call $\gamma^\#(r)$ a *radial semivariogram*. Just as in Section 2.4, we can ask which functions $\gamma^\#(r)$ represent radial semivariogram functions in *all* dimensions d.

Theorem 3.3.2 *A continuous function $\gamma^\#(r)$ with $\gamma^\#(0) = 0$ represents a radial semivariogram in all dimensions d if and only if $\gamma^\#(r)$ is nonnegative and the derivative $d\gamma^\#(r^{1/2})/dr$ is completely monotone in r for $r > 0$.*

Proof: From Theorem 3.2.1, $\gamma^\#(r)$ is a radial semivariogram in all dimensions if and only if $\exp\{-\alpha\, \gamma^\#(r)\}$ is a radial covariance function in all dimensions for all $\alpha > 0$, which by Theorem 2.4.1 is true if and only if $\exp\{-\alpha\, \gamma^\#(r^{1/2})\}$ is a completely monotone function of r, for all $\alpha > 0$. This last property holds if and only if $\exp\{-\gamma^{\#}(r^{1/2})\}$ is the Laplace transform of an infinitely divisible probability distribution on $[0, \infty)$. Such functions are characterized by the fact that $\gamma^{\#}(r^{1/2})$ is a nonnegative function of r with completely monotone derivative (Feller, 1966, p. 425). □

Every radial covariance function in Table 2.1 determines a radial semivariogram via $\gamma^{\#}(r) = 1 - \sigma^{\#}(r)$. Further, by checking the conditions of the above theorem, we can immediately read off two new families of radial semivariograms, each valid in all dimensions:

$$\gamma^{\#}(r) = cr^{2\alpha} \quad \text{(power scheme)}, \tag{3.12}$$

valid for $0 < \alpha \leq 1,\ c > 0$, and

$$\gamma^{\#}(r) = c\log(1 + \beta r^2) \quad \text{(shifted logarithmic scheme)}, \tag{3.13}$$

valid for $c > 0,\ \beta > 0$.

The semivariogram $\gamma^{\#}(r) = c\,r$ ($\alpha = 1/2$ in (3.12)) is called the *linear scheme*. In $d = 1$ dimension, it arises as the semivariogram of both Brownian motion and the Poisson process; see Section 3.2. In dimensions $d > 1$, it represents the semivariogram of *Lévy Brownian motion* (see, e.g., Adler, 1981, p. 244).

The semivariogram $\gamma^{\#}(r) = c\,r^2$, which arises when $\alpha = 1$ in (3.12), or when $F(d\omega) = 0$ and A is proportional to the identity matrix in (3.11), is rather degenerate. It represents the semivariogram of a process with random linear drift

$$X(t) = t^T U, \quad t \in \mathbb{R}^d, \tag{3.14}$$

where U is a single random vector in $\mathbb{R}^d$ with mean 0 and with variance $E\{UU^T\}$ proportional to the d-dimensional identity matrix.

The power scheme (3.12), $0 < \alpha < 1$, forms a subset of the self-similar random fields. See Section 3.10 for a more extended discussion.

Figure 3.1 illustrates some typical behavior in isotropic semivariograms. The power schemes $\gamma^{\#}(r) \propto r^{1/2}$ and $\gamma^{\#}(r) \propto r$ are unbounded as $r \to \infty$ and correspond to intrinsic random fields. The exponential semivariogram $\gamma^{\#}(r) = 1 - \exp(-r)$ is bounded as $r \to \infty$ and corresponds a stationary random field. Of course, a stationary random field is a special case of an intrinsic random field.

Figure 3.2 shows an example of a nugget effect with $\lim_{r\to 0}\gamma^{\#}(r) > 0$. Note there is a discontinuity in $\gamma^{\#}(r)$ at $r = 0$ since every radial semivariogram must satisfy $\gamma^{\#}(0) = 0$ by definition. The models in this chapter are all assumed to be continuous for all lags. Hence, the presence of a nugget effect is not allowed within the framework of this chapter and will not be discussed further here. However, a nugget effect can be very important in practice. In Chapter 5, the presence of a nugget effect in data is interpreted through the effects of measurement error.

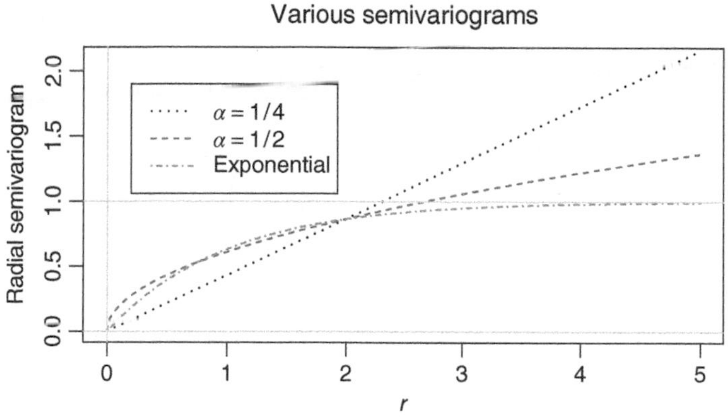

Figure 3.1 Examples of radial semivariograms: the power schemes $\gamma^{\#}(r) \propto r^{2\alpha}$ for $\alpha = 1/4, 1/2$ and the exponential scheme $\gamma^{\#}(r) = 1 - \exp(-r)$. All the semivariograms have been scaled to take the same value for $r = 2$.

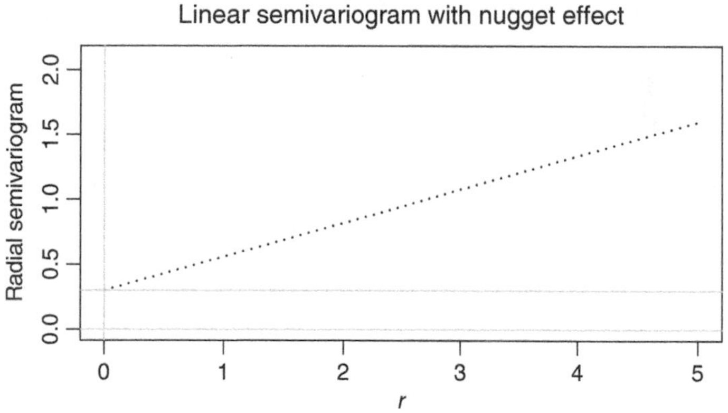

Figure 3.2 A linear semivariogram with a nugget effect: $\lim_{r \to 0} \gamma^{\#}(r) = 0.3 > 0$.

3.4 Higher Order Intrinsic Random Fields

For $k \geq 0$, let $\mathcal{H}_k$ denote the vector space of homogeneous polynomials, as a function of the site $t \in \mathbb{R}^d$. Similarly, let $\mathcal{F}_k$ denote the vector space of polynomials of degree $\leq k$, so that $\mathcal{F}_k = \mathcal{H}_0 \oplus \cdots \oplus \mathcal{H}_k$. In particular, $\mathcal{H}_0$ contains the constant function, $\mathcal{H}_1$ contains the homogeneous linear functions of t, and $\mathcal{H}_2$ contains the homogeneous quadratic functions of t.

A homogeneous polynomial of degree k in $\mathbb{R}^d$ can be written as a linear combination of monomials of the form

$$t^m = t[1]^{m[1]} \cdots t[d]^{m[d]}, \quad |m| = k, \tag{3.15}$$

where t^m is shorthand for the expansion in (3.15). Here $m = (m[1], \ldots, m[d])$ is a multi-index and $|m| = m[1] + \cdots + m[d]$. For later use, it is helpful to set $p_H(k) = \dim(\mathcal{H}_k)$ and $p_F(k) = \dim(\mathcal{F}_k)$. It can be shown that

$$p_H(k) = \dim(\mathcal{H}_k) = \binom{k+d-1}{k}, \quad p_F(k) = \dim(\mathcal{F}_k) = \binom{k+d}{k}. \tag{3.16}$$

See Exercise 3.9.

Definition 3.4.1 For any n, a collection of coefficients and sites $\{(a_i, t_i)\}_{i=1}^n$, where $n \geq 1$, is said to be a kth-*order increment* if

$$\sum_{i=1}^{n} a_i f(t_i) = 0, \quad \text{for all } f \in \mathcal{F}_k.$$

Similarly, for a random field $\{X(t)\}$, the linear combination $\sum a_i X(t_i)$ is said to be a kth-order increment of the random field.

If $\{(a_i, t_i)\}_{i=1}^n$ is a kth-order increment, then so is the shifted increment $\{(a_i, t_i + h)\}_{i=1}^n$ for any lag $h \in \mathbb{R}^d$. To see why this result is true, consider any monomial t^m with $|m| \leq k$. We need to show that the shifted increment vanishes, $\sum a_i (t_i + h)^m = 0$. To verify this claim, expand out the product and note that each term is an increment. For example, suppose $d = 2$, $k = 3$, and $\{(a_i, t_i)\}_{i=1}^n$ is a kth-order increment. Let $m = (m[1], m[2]) = (2,1)$ with $|m| = 3$. Then

$$\begin{aligned}
\sum a_i (t_i + h)^{(2,1)} &= \sum a_i (t_i[1] + h[1])^2 (t_i[2] + h[2])^1 \\
&= \sum a_i \left\{ t_i^{(2,1)} + 2h[1] t_i^{(1,1)} + h[1]^2 t_i^{(0,1)} \right. \\
&\quad \left. + h[2] t_i^{(2,0)} + 2h[1]h[2] t_i^{(1,0)} + h[1]^2 h[2] t_i^{(0,0)} \right\},
\end{aligned}$$

a sum of six increments, each of which is based on a monomial in t of degree ≤ 3. That is, each of monomials lies in $\mathcal{F}_3$ and so each of the increments vanishes.

Definition 3.4.2 A random field $\{X(t) : t \in \mathbb{R}^d\}$ is said to be an *intrinsic random field of order* k if all the kth-order increments $\sum a_i X(t_i + h)$ have a mean and variance that do not depend on h. The name *random field with stationary increments of order* k is also used.

The mean and variance of the increment can be written in the form

$$E\left\{ \sum a_i X(t_i) \right\} = \sum a_i \nu_I(t_i) \tag{3.17}$$

and

$$\operatorname{var}\left\{\sum a_i X(t_i)\right\} = \sum_{i,j=1}^{n} a_i a_j \sigma_I(t_i - t_j), \tag{3.18}$$

where $v_I(h)$ is called the *intrinsic mean function* and $\sigma_I(h)$ is called the *intrinsic covariance function*. It can be shown that the intrinsic mean function can be written as a homogeneous polynomial of degree $k + 1$

$$v_I(t) = \sum_{|m|=k+1} \beta_m t^m$$

plus an arbitrary polynomial in t of degree $\leq k$. Both the intrinsic mean function and the intrinsic variance function are best viewed an equivalence classes of functions rather than just single functions. In particular, $v_I(t)$ and $v_I(t) + t^m$, where $|m| \leq k$, define the same mean (3.17) on increments.

Similarly, $\sigma_I(h)$ is defined up to an even polynomial in h (even, without loss of generality, since covariance functions are always even functions of the lag) of degree $\leq 2k$. That is, $\sigma_I(h)$ and $\sigma_I(h) + h^m$, where h^m, where $|m|$ is even, and $|m| \leq 2k$ define the same variances on increments. To confirm that there is no contribution to the variance of an increment from the function h^m, expand out

$$(t_i - t_j)^m = (t_i[1] - t_j[1])^{m[1]} \cdots (t_i[d] - t_j[d])^{m[d]}$$

to get a representation

$$(t_i - t_j)^m = \sum_{m'} c_{m'} t_i^{m'} t_j^{m-m'} \tag{3.19}$$

for suitable constants $c_{m'}$, where the sum is over multi-indices m' such that $0 \leq m'[\ell] \leq m[\ell], \ell = 1, \ldots, d$. Hence for each m', either $|m'| \leq k$ or $|m - m'| \leq k$ (or both). Thus, if $\sigma_I(h)$ is replaced by $\sigma_I(h) + h^m$ in (3.18), the contribution of h^m to the incremental variance (3.18) becomes

$$\sum_{m'} c_{m'} \sum_{i,j} a_i a_j t_i^{m'} t_j^{m-m'} = \sum_{m'} c_{m'} \left\{\sum_i a_i t_i^{m'}\right\}\left\{\sum_j a_j t_j^{m-m'}\right\}.$$

The bounds on $|m'|$ and $|m - m'|$ imply that for each m' the sum over i and/or the sum over j must vanish. Hence, the overall contribution of h^m to the incremental variance is 0.

A function $\sigma(h)$ for which (3.18) is nonnegative for all kth order increments is said to be *conditionally positive semidefinite of order k*. This property is necessary and sufficient for $\sigma(h)$ to be a valid intrinsic covariance function. If $\sigma(h)$ is positive for nonzero kth order increments, it is said to be conditionally positive definite of order k. A spectral representation of the intrinsic covariance function is given in Section 3.8.

3.5 Registration of Higher Order Intrinsic Random Fields

Let $\nu_I(t)$ and $\sigma_I(h)$ denote an intrinsic mean function and an intrinsic variance function of order k, where $k \geq 0$ is specified. These functions specify the means and variances of the kth-order increments of an intrinsic random field $X_I(t)$, say. The subscript I indicates that these functions are equivalence classes of functions. In particular, the intrinsic mean function and the intrinsic variance function do not fully specify the first two moments of the random field, since the means and variances of nonincrements are not specified.

From this perspective, the intrinsic random field is an equivalence class of random fields, where $X_I(t)$ is only determined up to a polynomial in t of degree k. This was one of the approaches taken in Section 3.2 for $k = 0$.

Another way to think about the random field is to remove the indeterminacy by fixing its values at p sites, where $p = p_F(k)$ denotes the dimension of $\mathcal{F}_k$ in (3.16). This approach was also considered in the proof of Theorem 3.2.1 but the details are a bit more involved when $k > 0$.

Given a basis $f_1(t), \ldots, f_p(t)$ of $\mathcal{F}_k$, and given p distinct "registration" sites $t_1^*, \ldots, t_p^*$, such that the matrix

$$F = (f_{ij}), \quad f_{ij} = f_i(t_j^*)$$

is nonsingular, it is possible to define a fully specified random field $X_R(t)$ that vanishes at the registration sites and has the same incremental variances as $X_I(t)$. Then, $X_R(t)$ can be regarded as a *registered* version of the intrinsic random field.

Here are the details. First define a new basis functions $g_1(t), \ldots, g_p(t)$ by

$$\mathbf{g}(t) = \begin{bmatrix} g_1(t) \\ \vdots \\ g_p(t) \end{bmatrix} = F^{-1}\boldsymbol{f}(t) \text{ where } \boldsymbol{f}(t) = \begin{bmatrix} f_1(t) \\ \vdots \\ f_p(t) \end{bmatrix}.$$

Collect the new basis functions at the registration sites into a matrix

$$G = \begin{bmatrix} \mathbf{g}(t_1^*) & \ldots & \mathbf{g}(t_p^*) \end{bmatrix}$$

and note that $G = F^{-1}F = I$, i.e., $g_i(t_j^*) = 1$ if $i = j$ and $= 0$ otherwise.

Also define a vector function

$$\mathbf{b}(s) = (b_i(s)), \ b_i(s) = \sigma_I(s - t_i^*), \quad i = 1, \ldots, p,$$

of intrinsic covariances between a site s and the registration sites, and let

$$K = (k_{ij}), \ k_{ij} = \sigma_I(t_i^* - t_j^*), \quad i,j = 1, \ldots, p,$$

denote the intrinsic covariance matrix at the registration sites.

Define the registered process by

$$X_R(t) = X_I(t) - \sum_{i=1}^{p} g_i(t) X_I(t_i^*). \tag{3.20}$$

It can be checked that the registered random field vanishes at the registration sites, $X_R(t_j^*) = 0,\ j = 1, \ldots, p,$. Further, for each t, $X_R(t)$ is an increment of the X_I process so that it has a well-defined variance. It follows that the registered random field has a fully specified covariance function

$$\sigma_R(s,t) = \sigma_I(s,t) - \mathbf{b}(s)^T\mathbf{g}(t) - \mathbf{b}(t)^T\mathbf{g}(s) + \mathbf{g}(s)^T K\mathbf{g}(t) \tag{3.21}$$

(Exercise 3.10).

3.6 Generalized Random Fields

If $X(t)$ is an ordinary random field, $t \in \mathbb{R}^d$, then we saw in Section 2.10 how to average $X(t)$ with respect to a test function φ to construct the regularized process

$$X_\varphi(t) = \int X(s)\varphi(s-t)\,ds = \int X(s)\varphi_{(t)}(s)\,ds, \tag{3.22}$$

where $\varphi_{(t)}(s) = \varphi(s-t)$ denotes a shifted version of the test function.

It is also useful to use the notation

$$X(\varphi) = \int X(s)\varphi(s)\,ds = X_\varphi(0)$$

to give greater prominence to the test function. Then, X can be regarded either as a site-indexed random field $\{X(t) : t \in \mathbb{R}^d\}$ or as a function-indexed random field $\{X(\varphi)\}$ with respect to some suitable collection of test functions φ.

The class of function-indexed random fields is wider than the class of site-indexed random fields. Such random fields are called *generalized random fields* because their realizations are often generalized functions in the sense of Schwarz, rather than ordinary functions. For a detailed discussion of such generalized functions, see Gel'fand and Shilov (1964, 1968). The purpose of this section is to give a brief description of the properties of generalized random fields.

First, we need a suitable class of test functions. A convenient class of functions to introduce the subject is given by

$$\mathcal{K} = \{\varphi : \varphi \text{ is infinitely differentiable and has compact support}\}.$$

Less restrictive conditions on φ will be given later.

If $m = (m[1], \ldots, m[d])$ is a multi-index of nonnegative integers, denote the mth partial derivative of φ by

$$D^m\varphi(t) = \partial^{m[1]+\cdots+m[d]}\varphi(t)/\partial t[1]^{m[1]} \cdots \partial t[d]^{m[d]}.$$

Say that a sequence $\{\varphi_n(t)\}$ converges to $\varphi(t)$ in $\mathcal{K}$ as $n \to \infty$ if for each m, $D^m\varphi_n(t) \to D^m\varphi(t)$ uniformly over t.

Definition 3.6.1 A generalized function f_G is a collection of real numbers $\{f_G(\varphi) : \varphi \in \mathcal{K}\}$ with the following properties:

(a) *Linearity*. For $\alpha_1, \alpha_2 \in \mathbb{R}, \varphi_1, \varphi_2 \in \mathcal{K}$,

$$f_G(\alpha_1\varphi_1 + \alpha_2\varphi_2) = \alpha_1 f_G(\varphi_1) + \alpha_2 f_G(\varphi_2).$$

(b) *Continuity*. If the sequence $\{\varphi_n(t)\}$ converges to $\varphi(t)$ in $\mathcal{K}$, then

$$f_G(\varphi_n) \to f_G(\varphi).$$

□

Strictly speaking, we should write a generalized function in the form $\{f_G(\varphi) : \varphi \in \mathcal{K}\}$, not $\{f_G(t) : t \in \mathbb{R}^d\}$, because the values $f_G(t)$ may not even make sense. However, we shall also use the latter notation when convenient to emphasize the analogy with ordinary functions.

We shall often use the subscript "G" to distinguish generalized functions from ordinary functions. Of course, any locally integrable ordinary function can also be thought of as a generalized function, but there are many other examples as well. As an exception to this notational rule, we shall write the *Dirac delta function* centered at $h \in \mathbb{R}^d$ as $\delta_h(t)$. This is the best-known generalized function, defined through its effect on a test function by

$$\delta_h(\varphi) = \varphi(h), \tag{3.23}$$

especially for $h = 0$. Loosely speaking, $\varphi(h) = \int \delta_h(t)\, \varphi(t)\, dt$, where the delta function is visualized as a normal density centered at h with an infinitesimally small variance.

An *ordinary* random field $\{X(t) : t \in \mathbb{R}^d\}$ can be regarded as a random ordinary function. Similarly, we can define a *generalized random field* $\{X_G(t) : t \in \mathbb{R}^d\}$ to be a random generalized function. In the study of ordinary random fields, we limited our attention to random fields for which $\mu(t) = E\{X(t)\}$ and $\sigma(s,t) = \text{cov}\{X(s), X(t)\}$ are finite and continuous. Similarly, we shall limit our attention to generalized random fields for which the mean functional

$$\mu_G(\varphi) = E\{X(\varphi)\} \tag{3.24}$$

and the covariance functional

$$\sigma_G(\varphi_1, \varphi_2) = E\{[X_G(\varphi_1) - \mu_G(\varphi_1)][X_G(\varphi_2) - \mu_G(\varphi_2)]\} \tag{3.25}$$

are finite and continuous on $\mathcal{K}$.

Note that a knowledge of the variance functional $\sigma_G(\varphi, \varphi)$ for all $\varphi \in \mathcal{K}$ determines the covariance functional $\sigma_G(\varphi_1, \varphi_2)$ for all $\varphi_1, \varphi_2 \in \mathcal{K}$ from the identity

$$\sigma_G(\varphi_1 + \varphi_2, \varphi_1 + \varphi_2) = \sigma_G(\varphi_1, \varphi_1) + 2\sigma_G(\varphi_1, \varphi_2) + \sigma_G(\varphi_2, \varphi_2). \tag{3.26}$$

A generalized random field is said to be *stationary* if

$$\mu_G(\varphi) = \mu(\varphi_{(h)}),$$
$$\sigma_G(\varphi_1, \varphi_2) = \sigma_G(\varphi_{1(h)}, \varphi_{2(h)}), \quad \text{for all } h \in \mathbb{R}^d,$$

where $\varphi_{(h)}(t) = \varphi(t-h)$ denotes the φ function shifted by h. It can be shown that for a generalized stationary random field, the mean functional takes the form $\mu_G(\varphi) = \mu \int \varphi(t)\, dt$ for some scalar $\mu \in \mathbb{R}$, and that the covariance functional $\sigma_G(\varphi_1, \varphi_2)$ depends only on the convolution

$$\varphi_1 * \check{\varphi}_2(h) = \int \varphi_1(h+s)\, \varphi_2(s)\, ds, \tag{3.27}$$

where

$$\check{\varphi}(s) = \varphi(-s)$$

is the same as φ, but with the sign of the argument reversed (Gel'fand and Vilenkin, 1964, p. 167). In this case, we shall write $\sigma_G(\varphi_1, \varphi_2)$ as $\sigma_G(\varphi_1 * \check{\varphi}_2)$, a generalized function of a single argument.

A generalized random field is said to be *strictly stationary* if for every $\varphi \in \mathcal{K}$, the random variables $X(\varphi)$ and $X(\varphi_{(h)})$ have the same distribution for all $h \in \mathbb{R}^d$.

Next, we turn our attention to some properties of $\sigma_G(\varphi_1, \varphi_2)$ for a generalized, possibly nonstationary, random field.

(a) *Bilinearity and symmetry*. For all $\varphi, \varphi_1, \varphi_2 \in \mathcal{K}$ and $\alpha_1, \alpha_2 \in \mathbb{R}$,

$$\sigma_G(\alpha_1\varphi_1 + \alpha_2\varphi_2, \varphi) = \sigma_G(\varphi, \alpha_1\varphi_1 + \alpha_2\varphi_2)$$
$$= \alpha_1\sigma_G(\varphi_1, \varphi) + \alpha_2\sigma_G(\varphi_2, \varphi).$$

(b) *Positive semidefiniteness*. For all $\varphi \in \mathcal{K}$,

$$\sigma_G(\varphi, \varphi) = \text{var}\{X(\varphi)\} \geq 0.$$

(c) *Continuity*. If $\varphi_{1,n} \to \varphi_1$, and $\varphi_{2,n} \to \varphi_2$ in $\mathcal{K}$, then

$$\sigma_G(\varphi_{1,n}, \varphi_{2,n}) \to \sigma_G(\varphi_1, \varphi_2).$$

The following theorem shows that these properties characterize the class of generalized covariance functions.

Theorem 3.6.1 *Let $\sigma_G(\varphi_1, \varphi_2)$ be a symmetric bilinear continuous generalized function of $\varphi_1, \varphi_2 \in \mathcal{K}$. Then, $\sigma_G(\varphi_1, \varphi_2)$ represents the covariance functional of a generalized random field if and only if $\sigma_G(\varphi_1, \varphi_2)$ is positive semidefinite.*

The proof of this theorem requires techniques far beyond the scope of this book; see Gel'fand and Vilenkin (1964, p. 252).

In applications, two special forms of σ_G are important:

$$\sigma_G(\varphi_1, \varphi_2) = \int \varphi_1(s)\varphi_2(t) f(s,t)\, ds\, dt, \tag{3.28}$$

$$\sigma_G(\varphi_1, \varphi_2) = \int \varphi_1(s)\varphi_2(t)\delta_0(t-s)\, ds\, dt \quad \text{(white noise)}$$
$$= \int \varphi_1(t)\varphi_2(t)\, dt = (\varphi_1 * \check{\varphi}_2)(0). \tag{3.29}$$

In (3.28), $f(s,t)$ is a locally integrable function, symmetric in s and t, but often with a singularity on the diagonal, $f(s,t) \to \infty$ as $s-t \to 0$. Of course, f is required to behave so that σ_G is positive semidefinite. If $f(t, t+h) = f(h)$, say, does not depend on t, then the generalized random field is stationary, and (3.28) reduces to

$$\sigma_G(\varphi_1, \varphi_2) = \int \varphi_1(s)\varphi_2(s+h)f(h)\, ds\, dh$$
$$= \int (\varphi_1 * \check{\varphi}_2)(h)\, f(h)\, dh = \sigma_G(\varphi_1 * \check{\varphi}_2), \text{ say.}$$

Formula (3.29), known as the covariance functional of *white noise*, arises for two important strictly stationary generalized random fields: *Gaussian white noise* and the *Poisson point process*. Gaussian white noise is defined by the property that for $\varphi \in \mathcal{K}$, $X_G(\varphi)$ has a normal distribution with mean 0 and variance $\int \varphi^2(t)\, dt$. Let $\varphi(t) = I[t \in D]$ denote the indicator function for a bounded open set D of volume $|D|$. The Poisson point process is defined by the property that $X_G(\varphi)$ has a Poisson distribution with parameter $|D|$. Although $\varphi(t) = I[t \in D]$ does not lie in $\mathcal{K}$, the random functional $X_G(\varphi)$ is still well defined; see Section 3.8. Realizations from the Poisson point process can be viewed as a random collection of Dirac delta functions.

Example 3.1 A nice example of generalized random field is given by the random spectral measures $A(d\omega)$ and $B(d\omega)$ of Section 2.14. These are independent random Gaussian measures, satisfying $A(d\omega) = A(-d\omega)$ and $B(d\omega) = -B(-d\omega)$. They have mean 0 and generalized variance functions

$$\text{var}\left\{\int \chi(\omega)\, A(d\omega)\right\} = \int \{(\chi(\omega) + \chi(-\omega))/2\}^2 f(\omega)\, d\omega$$
$$\text{var}\left\{\int \gamma(\omega)\, B(d\omega)\right\} = \int \{(\gamma(\omega) - \gamma(-\omega))/2\}^2 f(\omega)\, d\omega$$

in terms of the underlying spectral density $f(\omega)$, and suitable test functions, χ and γ. If the underlying random field $X(t)$ is an ordinary stationary random field, then $f(\omega)$ is integrable, and the only requirement for this variance to be well defined is that the test functions $\chi(\omega)$ and $\gamma(\omega)$ be bounded and measurable. In particular, setting $\chi(\omega) = \cos(t^T\omega)$ and $\gamma(\omega) = \sin(t^T\omega)$ yields the spectral representation in Eq. (2.79). Further discussion of this representation for generalized and intrinsic random fields is given in Section 3.8. □

3.7 Generalized Intrinsic Random Fields of Intrinsic Order $k \geq 0$

It is possible for a random field to be both generalized and intrinsic. For integer $k \geq -1$, say that a test function $\psi(t)$ is a *kth-order increment function* if

$$\int |t|^k \psi(t)\, dt < \infty \quad \text{and} \quad \int t^m \psi(t)\, dt = 0 \text{ for all } |m| \leq k. \tag{3.30}$$

Here $m = (m[1], \ldots, m[d])$ is a multi-index with $m[\ell] \geq 0, |m| = \sum m[\ell]$, and $t^m = t[1]^{m[1]} \cdots t[d]^{m[d]}$. If $k = -1$, no constraint is imposed on ψ; if $k = 0$, there is just a single constraint $\int \psi(t)\, dt = 0$.

Let

$$\mathcal{K}_k = \{\psi \in \mathcal{K} : \ \psi \text{ is a } k\text{th-order increment function}\},$$

with $\mathcal{K}_{-1} = \mathcal{K}$. Thus, $\mathcal{K}_k$ consists of the functions in $\mathcal{K}$ that are kth-order increments.

Definition 3.7.1 A generalized random field $X_{IG}(\cdot)$ is said to be *intrinsic* of order k if the mean and covariance functionals in (3.24)–(3.25) satisfy

$$\mu_{IG}(\psi) = \mu_{IG}(\psi_{(h)}), \quad \sigma_{IG}(\psi, \psi) = \sigma_{IG}(\psi_{(h)}, \psi_{(h)}) \tag{3.31}$$

for all kth-order increment functions $\psi(t) \in \mathcal{K}_k$, where $\psi_{(h)}(t) = \psi(t - h)$.

That is, the mean and covariance functions are stationary when restricted to kth-order increment functions. By convention, we regard a generalized stationary random field as a generalized intrinsic random field of order $k = -1$. Just as for ordinary intrinsic random fields, the mean functional $\mu_{IG}(t)$ can be regarded as an equivalence class of polynomials in t of degree $k + 1$, where only the homogeneous terms of degree $k + 1$ are identifiable. Similarly, $\sigma_{IG}(\psi, \psi)$ can be regarded as a generalized function of a single functional argument $(\psi * \check{\psi})(h)$, where $\sigma_{IG}(\psi * \check{\psi})$ is determined only up to an even-order polynomial in h of degree $2k$.

3.8 Spectral Theory for Intrinsic and Generalized Processes

In Section 2.3, we showed that the covariance function of an ordinary stationary random field can be represented as the Fourier transform of a positive, symmetric, integrable measure $F(d\omega)$, $\omega \in \mathbb{R}^d$,

$$\sigma(h) = \int_{\mathbb{R}^d} e^{ih^T\omega} F(d\omega) = \int \cos(h^T\omega) F(d\omega). \tag{3.32}$$

If $\varphi \in \mathcal{K}$ has the Fourier transform

$$\tilde{\varphi}(\omega) = \int e^{ih^T\omega}\varphi(h)dh,$$

then integrating out either ω or h, respectively, in the joint integral

$$\int e^{ih^T\omega}\varphi(h)\ dh\ F(d\omega),$$

yields the Parseval formula

$$\int \sigma(h)\varphi(h)dh = \int \tilde{\varphi}(\omega)F(d\omega), \tag{3.33}$$

where both integrals are over $\mathbb{R}^d$.

If $\varphi(h)$ is replaced by $(\varphi * \check{\varphi})(h)$, with the Fourier transform $|\tilde{\varphi}(\omega)|^2$, then (3.33) becomes

$$\int \sigma(h)(\varphi * \check{\varphi})(h)dh = \int |\tilde{\varphi}(\omega)|^2 F(d\omega). \tag{3.34}$$

Further, it can be shown that the validity of (3.34) for all test functions $\varphi \in \mathcal{K}$ implies the validity of (3.32).

In the setting of generalized stationary random fields, a direct spectral representation such as (3.32) is no longer available. Instead, it is necessary to work with the Parseval relation (3.34). Further, if attention is restricted to increment functions, the spectral representation can be extended to generalized intrinsic random fields as well. The order satisfies $k \geq -1$, the case $k = -1$ being the stationary case. To state the main theorem, some notation is defined first.

Given $k \geq 0$ and a smooth function φ, it is convenient to define a vector of partial derivatives $\nabla^{\{k\}}\varphi(t)$ by

$$\nabla^{\{k\}}\varphi(t) = \{D^m\varphi(t) : |m| = k\}, \tag{3.35}$$

the set of the mixed partial derivatives of φ at t of order k, written in some specified order. This vector has length

$$p_H(k) = \binom{d+k-1}{k},$$

which is the same as the dimension of $\mathcal{H}_k$ in (3.16). See, e.g., Feller (1968, p. 38) or Exercise 3.9. In particular, $\nabla^{\{0\}}\varphi(t) = \varphi(t)$ is a scalar function, $\nabla^{\{1\}}\varphi(t)$ is the usual gradient vector of dimension d, and $\nabla^{\{2\}}\varphi(t)$ has $d(d+1)/2$ distinct components $\partial^2\varphi(t)/\partial t_i \partial t_j, 1 \leq i \leq j \leq d$.

The following result generalizes Bochner's Theorem 2.3.3 for ordinary stationary random fields in Chapter 2. In this theorem, a superscript * denotes the complex conjugate transpose of a vector.

Theorem 3.8.1 (Gel'fand and Vilenkin, 1964, p. 188) *Let $\sigma_{IG}(\varphi * \check{\varphi})$ denote the covariance functional of a generalized intrinsic random field, with order $k \geq -1$.*

Then there is a unique positive, symmetric measure $F(d\omega)$ on $\mathbb{R}^d\backslash\{0\}$ and a unique positive semidefinite array of real numbers $A = \{a(m_1, m_2) : |m_1| = |m_2| = k+1\}$ such that

$$\sigma_{IG}(\varphi * \check{\varphi}) = \int_{\mathbb{R}^d\backslash\{0\}} |\tilde{\varphi}(\omega)|^2 \, F(d\omega) + [\nabla^{\{k+1\}}\tilde{\varphi}(0)]^* A [\nabla^{\{k+1\}}\tilde{\varphi}(0)] \tag{3.36}$$

for all $\varphi \in \mathcal{K}_k$, where for some smallest integer $q \geq 0$, F satisfies the integrability condition

$$\int \frac{|\omega|^{2k+2}}{(1+|\omega|^2)^{k+q+1}} F(d\omega) < \infty. \tag{3.37}$$

Conversely, given a nonnegative symmetric measure F on $\mathbb{R}^d\backslash\{0\}$ satisfying (3.37) for some $k \geq -1$ and $q \geq 0$, and given a positive semidefinite matrix A, then (3.36) defines the covariance functional of a generalized intrinsic random field of order k and regularity order q.

Remarks Theorem 3.8.1 unifies a number of special cases. Therefore, it is helpful to look at various aspects of the representation (3.36) in more detail.

1. The integrability condition in (3.37) can be split into two parts, for small and large ω, respectively. Since $1 \leq 1 + |\omega|^2 \leq 2$ for $|\omega| \leq 1$ and $1/2 \leq |\omega|^2/(1+\omega|^2) \leq 1$ for $|\omega| \geq 1$, this condition is equivalent to

$$\int_{|\omega|\leq 1} |\omega|^{2k+2} F(d\omega) < \infty \text{ and } \int_{|\omega|\geq 1} |\omega|^{-2q} F(d\omega) < \infty. \tag{3.38}$$

   Thus if $k \geq 0$, F is allowed to have a singularity near 0, and if $q \geq 1$, F is allowed to have a singularity near ∞.
2. Next we need some background facts on Fourier transforms. Recall $\mathcal{K}$ denotes the set of infinitely differentiable functions on $\mathbb{R}^d$ of compact support. Another useful class of functions is S, the set of infinity differentiable functions on $\mathbb{R}^d$ whose derivatives of all orders are rapidly vanishing at ∞; that is, $\varphi(u) \in S$ means that

$$|\omega|^n D^m \tilde{\varphi}(\omega) \to 0 \text{ as } |\omega| \to \infty$$

   for all integers $n \geq 0$ and all orders $m = (m[1], \ldots, m[d])$ of derivatives.
   A basic result from Fourier analysis states that the Fourier transform of S is equal to S itself. In particular, if $\varphi \in \mathcal{K}_k \subset \mathcal{K} \subset S$, then $\tilde{\varphi} \in S$, so that $\tilde{\varphi}$ is rapidly decreasing at ∞. Hence, if F satisfies (3.37), or equivalently (3.38), the second part of the integral in (3.36) over $|\omega| \geq 1$ is finite, whatever the value of $q \geq 0$.
3. To ensure that the part of the integral over $|\omega| \leq 1$ converges if F satisfies (3.37), note that $\varphi \in \mathcal{K}_k$ ensures that $\tilde{\varphi}^{(m)}(0) = 0$ for $|m| \leq k$. Hence, a Taylor series expansion yields $\tilde{\varphi}^{(m)}(\omega) = O(|\omega|^k)$ as $\omega \to 0$, which ensures that the integral in (3.36) over $|\omega| \leq 1$ converges.

4. Now that the representation integral (3.36) has been established for all test functions $\varphi \in \mathcal{K}_k$, it is useful to relax the conditions on the test functions. The key requirement is that the integral (3.36) should be finite. Here are some examples that depend on the behavior of the spectral measure F.
   - If F is integrable, then it is sufficient to assume merely that φ is integrable. See Section 2.10 where this representation was used to regularize a stationary random field.
   - If $F(d\omega) = f(\omega)d\omega$ has a density and if the density is bounded $f(\omega) \leq C$, then it is sufficient to assume φ is square-integrable. Since the Fourier transform maps square-integrable functions to themselves, this condition ensures that $\int |\tilde{\varphi}(\omega)|^2 d\omega < \infty$ and hence that the integral in (3.36) is finite. The simplest example for the spectral density is white noise for which $f(\omega) = c > 0$. A simple example for a test function is an indicator function on a bounded set such as the interior of a cube or sphere.
5. If $q = 0$ in (3.37), the resulting random field is ordinary rather than generalized. Further, for $q \geq 1$, it can be shown that

$$\sigma_G(h) = (1 - \Delta)^q S(h), \tag{3.39}$$

   where $S(h)$ is an ordinary positive definite function and $\Delta = \Sigma \partial^2/\partial h[\ell]^2$ is the Laplace operator (Gel'fand and Vilenkin, 1964, p. 165).
6. The most important choice for k is $k = -1$, corresponding to a stationary random field, which is ordinary if $q = 0$ and generalized if $q \geq 0$. Then, the second part of (3.36) reduces to a simple term $a|\tilde{\varphi}(0)|^2$, which can be absorbed within the integral if desired. In particular, if $q = 0$ as well, the representation (3.36) reduces to (3.34).
7. The case $k = 0$ corresponds to an intrinsic random field (of order 0). The second part of (3.36) reduces to $[\nabla\tilde{\varphi}(0)]^* A[\nabla\tilde{\varphi}(0)]$, where ∇ is the gradient operator and A is a positive semidefinite $d \times d$ matrix. In this case, $\sigma_{IG}(\cdot)$ is determined only up to an additive constant. If $q = 0$ as well, a natural choice for this constant leads to the semivariogram of Section 3.2, and in this case a direct spectral representation was given in Theorem 3.3.1(c). In particular, the function $g(h)$ in (3.11) can be identified with $\sigma_{IG}(\cdot)$ in (3.36), which is an ordinary function in this case. Both functions are determined only up to an additive constant by the spectral measure $F(d\omega)$ and the matrix A. The function $g(\cdot)$ determines a semivariogram function by setting $\gamma(h) = g(0) - g(h)$; effectively, the arbitrary constant in $g(\cdot)$ is chosen to ensure that the semivariogram vanishes at lag $h = 0$, i.e. $\gamma(0) = 0$.
8. For $k \geq 0$, $\sigma_{IG}(t)$ is determined by (3.36) only up to an even polynomial in t of order $2k$.
9. It is interesting to contrast the roles of k and q in (3.37). As k increases, more low frequencies are allowed to enter the spectral measure. The effect of this

is to limit identifiable aspects of the random field to increment functions of order k. On the other hand, as q increases more high-frequency components are allowed, which means that more regularization is needed in order to make observations on the random field.

10. For statistical applications, we are mainly interested in special cases of (3.36). First, we shall generally take

$$A = 0. \tag{3.40}$$

To the extent that A can be identified from a single realization of a random field, it is usually more helpful to treat it as a mean ($k = -1$) or a homogeneous polynomial trend of degree $k + 1$ ($k \geq 0$) rather than as part of the covariance function.

11. When $q = 0,\ k \geq 0$, the covariance function $\sigma(t)$ can be given a direct spectral representation by slightly modifying (3.36) to

$$\sigma(h) = \int_{\mathbb{R}^d\setminus\{0\}} \left\{ \cos(h^T\omega) - \left[\sum_{j=1}^{k}(-1)^j(h^T\omega)^{2j}/(2j)!\right] I[|h^T\omega| < c] \right\} F(d\omega)$$
$$+h^{\{k+1\}T}Ah^{\{k+1\}}. \tag{3.41}$$

Here $c > 0$ and its arbitrariness reflects the fact that $\sigma(h)$ is determined only up to an even polynomial in h of degree $2k$. Also, note that the quantity in curly brackets is bounded by $|\omega|^{2k+2}$ as $\omega \to 0$ for each h, so that the integral is well defined. Finally, $h^{\{k+1\}}$ denotes the vector of the $(k+1)$th-order monomials $h[1]^{m[1]} \cdots h[d]^{m[d]}$ for $|m| = \sum m[\ell] = k + 1$, arranged in the same order as for $\nabla^{\{k+1\}}\varphi(t)$ in (3.35). When $k = 0$, there is no need to truncate the Taylor series expansion of $\cos(h^T\omega)$ and the representation (3.11) is obtained.

3.9 Regularization for Intrinsic and Generalized Processes

Let $X(t)$ denote a random field of intrinsic order $k \geq -1$ and regularity order $q \geq 0$. Let φ be a "suitable" test function and define a regularized random field $X_\varphi(t)$ as in Eq. (3.22). If $X(t)$ has spectral density $f(\omega)$, then $X_\varphi(t)$ has spectral density

$$f_\varphi(\omega) = |\tilde{\varphi}(\omega)|^2 f(\omega).$$

Regularization is even more important for intrinsic and/or generalized random fields than it is for ordinary stationary random fields because the individual values of $X(t)$ are well defined only if $k = -1$ and $q = 0$.

The character of the regularized random field depends on the values of k and q and on the behavior of φ. Let k_φ and q_φ denote the intrinsic and regularity orders for the regularized random field. Here are some examples.

(a) A minimal requirement on φ is that $\tilde{\varphi}$ be a bounded continuous function. This is guaranteed if $\varphi \in L_1$; we can also extend the notion of a test function to a signed finite measure. In this case, $k_\varphi \leq k$ and $q_\varphi \leq q$. That is, regularization does not increase the intrinsic or regularity orders.

(b) *Stationary*. Suppose φ satisfies (a), and, in addition, has vanishing moments up to order k. Then the resulting random field is stationary ($k_\varphi = -1$), but possibly still generalized.

(c) *Ordinary*. Suppose the test function φ (not a measure here) is smooth enough to ensure that $\tilde{\varphi}(\omega)$ decays to 0 quickly enough as $|\omega| \to \infty$ that $f_\varphi(\omega)$ is integrable over $|\omega| \geq 1$. In this case, the regularized random field will be ordinary ($q_\varphi = 0$), but may still be intrinsic. A simple condition on φ, which is sufficient for any value of q, is that $\varphi \in \mathcal{S}$.
However, it is also important to allow less restrictive conditions on φ. Suppose the spectral density is bounded, $f(\omega) \leq c$, for $|\omega| \geq 1$ and suppose $\varphi \in L_1 \cap L_2$. Then $\int_{\mathbb{R}^d} |\tilde{\varphi}|^2 < \infty$, and so f_φ is integrable over $|\omega| \geq 1$. Important examples of such functions φ are indicator functions on open sets with compact closure in $\mathbb{R}^d$.

(d) *Stationary and ordinary*. If φ satisfies (a), (b), and (c), then the regularized process will be both stationary and ordinary.

3.10 Self-Similarity

A fundamental property that can be possessed by a random field $\{X(t) : t \in \mathbb{R}^d\}$ is *self-similarity*. In a certain sense, the random field is invariant if the domain $\mathbb{R}^d$ is rescaled. Many of the core models in the theory of Gaussian random fields possess the property of self-similarity. However, there is a complication because a self-similar Gaussian random field cannot be both ordinary and stationary. Instead, it must be either generalized or intrinsic (or both). The purpose of this section is give a detailed description of self-similarity for Gaussian random fields.

Definition 3.10.1 Let $\{X_{IG}(t)\}$ be a generalized intrinsic random field of order $k \geq -1$. If for some real number α, the random field $\{a^{-\alpha} X_{IG}(at) : t \in \mathbb{R}^d\}$ has the same covariance structure for all $a > 0$, then $\{X_{IG}(t)\}$ is said to be a *self-similar* random field of index α.

This property says that with appropriate scaling, the covariance structure of the random field is invariant under dilations of the sites. If $\{X_{IG}(t)\}$ has covariance function $\sigma_{IG}(h)$ corresponding to a spectral density $f(\omega)$, then $\{a^{-\alpha} X_{IG}(at)\}$ has covariance function $a^{-2\alpha}\sigma_{IG}(ah)$ corresponding to a spectral density $a^{-d-2\alpha} f(\omega/a)$.

Hence, $X_{IG}(t)$ is self-similar of index α if and only if $\sigma_{IG}(h) = a^{-2\alpha}\sigma_{IG}(ah)$ or equivalently

$$f(\omega) = a^{-d-2\alpha} f(\omega/a), \quad \text{for all } a > 0. \tag{3.42}$$

It turns out that no ordinary stationary random field with finite second moments can be self-similar. Any self-similar random field must be generalized and/or have intrinsic order $k \geq 0$. In order to characterize self-similar random fields more completely, we limit our attention to *isotropic* random fields so that $\sigma_{IG}(h) = \sigma_{IG}^{\#}(r)$, say, and $f(\omega) = f^{\#}(\rho)$, say, depend only on $|h| = r$ and $|\omega| = \rho$, respectively.

Suppose that for some real α, the spectral density satisfies the isotropic version of (3.42)

$$f^{\#}(\rho) = a^{-d-2\alpha} f^{\#}(\rho/a). \tag{3.43}$$

Setting $\rho = 1$ in (3.43) yields $f^{\#}(1/a) = a^{d+2\alpha} f^{\#}(1)$; that is, after changing notation, we have the following result.

Theorem 3.10.1 *Let $\{X_{IG}(t)\}$ be an isotropic generalized intrinsic random field of order $k \geq -1$, and suppose it is self-similar of index $\alpha \in \mathbb{R}$. Then the spectral density must be proportional to*

$$f_{\alpha,d}^{\#}(\rho) = (2\pi)^{-d} \rho^{-d-2\alpha}. \tag{3.44}$$

Thus, up to a multiplicative constant, the spectral density is uniquely determined by the index of self-similarity. After integration over the unit sphere in $\mathbb{R}^d$, the differential $d\omega$ becomes $\pi_d\, \rho^{d-1} d\rho$ in polar coordinates, where $\pi_d = 2\pi^{d/2}/\Gamma(d/2)$ is the surface area of the unit sphere in $\mathbb{R}^d$ (Section 2.4). Hence (3.44) is integrable near $\rho = 0$ only if $\alpha < 0$ and integrable near $\rho = \infty$ only if $\alpha > 0$. Thus, in view of the spectral representation theorem of Section 3.8, $\{X_{IG}(t)\}$ will be an ordinary intrinsic random field of order $k = [\alpha] \geq 0$ if $\alpha > 0$ and will be a generalized stationary random field if $\alpha < 0$. If $\alpha = 0$, the random field is generalized and intrinsic of order $k = 0$.

Taking the "formal" Fourier transform (3.44) yields the following "formal" radial covariance function

$$\sigma_{\alpha}^{\#}(r) \text{ “} = \text{” } c_{\alpha,d} r^{2\alpha}, \tag{3.45}$$

where $\sigma_{IG}^{\#}(r)$ is written as $\sigma_{\alpha}^{\#}(r)$ here to emphasize the dependence on α, and

$$c_{\alpha,d} = 2^{-2\alpha-d}\pi^{-d/2}\Gamma(-\alpha)/\Gamma\left(\alpha + \frac{1}{2}d\right). \tag{3.46}$$

The adjective "formal" is used because the interpretation of (3.45) needs care. Here are the key points and special cases. The details are explored in Exercises 3.15 and 3.16. See Table 3.1 for some particular cases.

Table 3.1 Self-similar random fields with spectral density $f^{\#}(\rho) = \rho^{-d-2\alpha}$: some particular cases

Index α	Covariance functional $\sigma^{\#}_{\alpha}(r)$	Normalization constant c	Type of process and name
$0 < \alpha < 1$	$c\, r^{2\alpha}$	$c = c_{\alpha,d} < 0$	Ordinary intrinsic fractional Brownian motion
$\alpha = 0$	$c \log r$	$c = c'_{0,d}$ $= -2^{1-d}\pi^{-d/2}/\Gamma(d/2)$	Generalized intrinsic Logarithmic or de Wijsian random field
$-\frac{d}{2} < \alpha < 0$	$c\, r^{2\alpha}$	$c = c_{\alpha,d} > 0$	Generalized stationary fractional white noise
$\alpha = -\frac{d}{2}$	$c\delta^{\#}_{0}(r)$	$c = c^{*}_{0,d} = 1$	Generalized stationary white noise

For $0 \le \alpha < 1$, $\sigma^{\#}_{\alpha}(r)$ is defined only up to a constant term. The constants $c_{\alpha,d}$, $c'_{k,d}$, and $c^{*}_{j,d}$ are given in (3.46), (3.48), and (3.49), respectively.

1. If for some integer $k \ge 0$, $k \le \alpha < k+1$, then $\sigma^{\#}_{\alpha}(r)$ is intrinsic of order k and is defined only up to an even polynomial in r of degree $2k$.
2. If $\alpha = k \ge 0$ is an integer, then it is still the case that $\sigma^{\#}_{\alpha}(r)$ is intrinsic of order k and is defined only up to an even polynomial in r of degree $2k$. However, $c_{\alpha,d} = \infty$ so that the formula for $\sigma^{\#}_{\alpha}(r)$ must be replaced by the limiting value

$$\sigma^{\#}_{\alpha}(r) = \lim_{\varepsilon\to 0} 2\varepsilon c_{\alpha+\varepsilon,d}\left(\frac{r^{2k+2\varepsilon} - r^{2k}}{2\varepsilon}\right) = c'_{k,d} r^{2k} \log r, \tag{3.47}$$

   where

$$c'_{k,d} = 2^{1-2k-d}(-1)^{k-1}\pi^{-d/2}/\{\Gamma(k+d/2)k!\}. \tag{3.48}$$

3. Note the normalizing constants alternate in sign on successive intervals between the nonnegative integers. In particular, $c_{\alpha,d} < 0$ for $0 < \alpha < 1$, $c_{\alpha,d} > 0$ for $1 < \alpha < 2$, etc. Similarly, at the integers, $c'_{0,d} < 0$, $c'_{1,d} > 0$, etc.
4. For $\alpha \le 0$, $\sigma^{\#}_{\alpha}(r)$ is a generalized covariance function, and it is useful to highlight several possibilities.
   (a) If $-d/2 < \alpha < 0$, then the generalized covariance function $\sigma^{\#}_{\alpha}(r)$ can be identified with the ordinary function $c_{\alpha,d}r^{2\alpha}$, which is integrable over the unit ball in $\mathbb{R}^d$, $|h| \le 1$, where $|h| = r$.

(b) However, if $\alpha \le -d/2$, then formula (3.45) for $\sigma_\alpha^\#(r)$ is incomplete: $\sigma_\alpha^\#(r) = c_{\alpha,d} r^{2\alpha}$ for $r > 0$, but the support of the generalized covariance function also includes the origin.
(c) In particular, if $\alpha = -d/2 - j$ for some integer $j \ge 0$, then $c_{\alpha,d} = 0$ and $\sigma_\alpha^\#(r)$ reduces to the Dirac delta function and its derivatives

$$\sigma_\alpha^\#(r) = c_{j,d}^* \delta_0^{(2j)}(h), \text{ where } c_{j,d}^* = (-1)^j. \tag{3.49}$$

Here $r = |h|$, and $\delta_0^{(2j)}(h) = (\Delta)^j \delta_0(h)$, the jth power of the Laplacian operator $\Delta = \partial^2/\partial h[1]^2 + \cdots + \partial^2/\partial h[d]^2$, applied to the Dirac delta function.

5. The self-similar covariance functions for different indices are related by $-\Delta\sigma_\alpha^\#(r) = \sigma_{\alpha-1}^\#(r)$ since the spectral densities are related by $f_{\alpha,d}(\omega) = |\omega|^2 f_{\alpha+1,d}(\omega)$.

A number of random fields in this list deserve special mention. The case $\alpha = \frac{1}{2}$ is known as *Lévy Brownian motion* (Adler, 1981, p. 244), an ordinary intrinsic random field with semivariogram $\gamma^\#(r) \propto r$. This random field is best known in $d = 1$ dimension where it is just called *Brownian motion* or the *Wiener process* (Section 3.2), and it forms the foundation of much of stochastic analysis. The generalization to $0 < \alpha < 1$ is known as *fractional Brownian motion*. The semivariogram $\gamma^\#(r) \propto r^{2\alpha}, 0 < \alpha < 1$ was mentioned earlier in (3.12).

The following notation will be useful in Chapter 7. For $\alpha > 0$, let

$$\text{IRF}_d(\alpha, k) \tag{3.50}$$

denote the self-similar Gaussian random field of index $\alpha > 0$ in d dimensions with drift space given by the polynomials in t up to degree k. Since this random field is intrinsic of order $[\alpha]$, the integer part of α, it must be the case that $k \ge [\alpha]$ in order for the random field to be well defined.

The case $\alpha = 0$ is interesting because the random field is both intrinsic and generalized. It has the property that the random field looks the same at all scales of measurement. It was first proposed by de Wijs (1951) who noted the self-similarity property in a mining application. In the mining literature, this generalized intrinsic covariance function is known as the *de Wijsian scheme*.

For $\alpha < 0$, the self-similar random fields become stationary and generalized. The case $\alpha = -d/2$ corresponds to *white noise* with constant spectral density and with stationary generalized covariance function $\sigma_{-d/2}(h) = \delta_0(h)$ given by the Dirac delta function. For $-\frac{d}{2} < \alpha < 0$, the self-similar generalized random fields are known as *fractional white noise*.

Fractional white noise is interesting because it displays long-range dependence. That is, $\sigma_\alpha(h) \propto h^{2\alpha}$ tends to 0 as $|h| \to \infty$ so slowly that $\int_{|h|\ge 1} \sigma_\alpha(h)\, dh = \infty$. It represents an idealized self-similar version of the long-range dependence for ordinary stationary processes studied in Section 2.13. (For the discussion here, it is the behavior of $\sigma_\alpha(h)$ as $|h| \to \infty$, which is of interest, not the pole in the generalized covariance function as $|h| \to 0$, which makes the random field generalized rather than ordinary.) If we let $X_\alpha(t)$ be a realization from this generalized stationary random field and set

$$\overline{X}(\lambda V) = |\lambda V|^{-1} \int_{\lambda V} X_\alpha(t)\, dt,$$

where V is a bounded open set and $\lambda > 0$, then

$$\mathrm{var}\{\overline{X}(\lambda V)\} = B\lambda^{2\alpha}, \quad B = |V|^{-2} c_{d,\alpha} \int_V \int_V |s-t|^{2\alpha}\, ds\, dt, \quad -\frac{d}{2} < \alpha < 0.$$

This result is exact for fractional white noise. For a stationary random field exhibiting long-range dependence, this result holds only asymptotically as $\lambda \to \infty$; see (2.76) in Chapter 2.

3.11 Simulation

This section is an extension of the simulation methods for stationary Gaussian random fields discussed in Section 2.14.

3.11.1 General Points

Let $X(t)$ be a zero-mean random field, either ordinary or generalized and either stationary or intrinsic. A unified formula for the variance of the functional $X(\varphi) = \int X(t)\varphi(t)\, dt$ is given by Theorem 3.8.1. Assuming that $A = 0$ and that the spectral measure has a density $F(d\omega) = f(\omega)\, d\omega)$, this representation takes the form

$$\mathrm{var}\{X(\varphi)\} = \int |\tilde{\varphi}(\omega)|^2 f(\omega) d\omega. \tag{3.51}$$

This representation is valid provided the integral on the right hand side is finite, which is guaranteed under the following conditions on the test functions.

1. If $X(t)$ is ordinary, so that f is integrable as $|\omega| \to \infty$, then $\tilde{\varphi}$ just needs to be bounded. In particular, φ can be an ordinary integrable function, or even a generalized function such as a linear combination of Dirac delta functions. Thus, if

$$\varphi(t) = \sum a_j \delta_{t_j}(t) \tag{3.52}$$

   with

$$\tilde{\varphi}(\omega) = \sum a_j \exp(i t_j^T \omega), \tag{3.53}$$

   then $X(\varphi) = \sum a_i X(t_i)$ recovers linear combinations of the values of the random field.

2. On the other hand, if $X(t)$ is generalized, then $\varphi(\omega)$ is restricted to being an ordinary function.
3. If $X(t)$ is intrinsic of order k, then $\varphi(t)$ must lie in $\mathcal{K}_k$ and satisfy (3.30). When $X(t)$ is ordinary and $\varphi(t)$ takes the form of (3.52), then (3.30) takes the form $\sum a_j t_j^m = 0$ for all $m \le k$. The stationary case corresponds to $k = -1$, with no constraints on φ.
4. For example, for an ordinary intrinsic random field with $k = 0$, a common choice for φ is $\varphi(t) = \delta_{t_0} - \delta_0$, so that the representation (3.51) specifies the difference $X(t_0) - X(0)$ between the value of the random field at an arbitrary site $t = t_0$ and its value at $t = 0$.

With this unified framework, there are two natural simulation methods for a Gaussian random field that is possibly generalized and/or possibly intrinsic of intrinsic order $k \ge -1$. These methods generalize Section 2.14. Let $\psi_1, \ldots, \psi_n$ be a collection of test functions, assumed for simplicity to lie in $\mathcal{K}_k$. We wish to simulate $X(\psi_1), \ldots, X(\psi_n)$.

3.11.2 The Direct Method

The covariance matrix of these random variables can be determined from (3.51). In particular, note that $\text{cov}\{X(\psi_1), X(\psi_2)\}$ can be determined from the variance formula $\text{var}\{X(\psi_1 + \psi_2)\} = \text{var}\{X(\psi_1)\} + 2\,\text{cov}\{X(\psi_1),\ X(\psi_2)\} + \text{var}\{X(\psi_2)\}$. The simulation method in Section 2.14.2 can then be followed exactly.

3.11.3 Spectral Methods

In our unified framework, the spectral representation analogous to (2.79) takes the form

$$X(\psi) = \int \tilde{\psi}(\omega)\, H(d\omega). \tag{3.54}$$

The *random frequencies, random amplitudes* method is the only spectral method that generalizes easily from Section 2.14.3 to the present context. It takes the following form here. Let U, V, and W be three independent random quantities where

(a) U is a uniform random variable on $[0,2\pi)$.
(b) W is a symmetric random vector on $\mathbb{R}^d$ with probability density $g(\omega)$, say, satisfying $g(\omega) = g(-\omega)$, $g(\omega) > 0$ for all ω.
(c) V is a random variable with $EV^2 = 2$.

Then, define a random field by

$$X(t) = \{f(W)/g(W)\}^{1/2}\, V \cos(t^T W + U). \tag{3.55}$$

so that the functional $X(\psi)$ takes the form

$$X(\psi) = \{f(W)/g(W)\}^{1/2}\, V \int \cos(t^T W + U)\, \psi(t)\, dt. \tag{3.56}$$

Thus, $X(t)$ is a single random cosine wave with random phase U and random frequency W, and with a random amplitude depending on the frequency. It is easy to check that $X(\psi)$ has mean 0 and variance

$$\begin{aligned}
E\{X(\psi)^2\} &= E\left\{\frac{f(W)}{g(W)}V^2 \int \cos(t^T W + U)\, \cos(s^T W + U)\, \psi(t)\, \psi(s)\, ds\, dt\right\}\\
&= E\left\{\frac{f(W)}{g(W)} \int [\cos t^T W \cos s^T W + \sin t^T W \sin s^T W]\, \psi(t)\, \psi(s)\, ds\, dt\right\}\\
&= E\left\{\frac{f(W)}{g(W)}([\mathrm{Re}\, \tilde{\psi}(W)]^2 + [\mathrm{Im}\, \tilde{\psi}(W)]^2)\right\}\\
&= \int \frac{f(\omega)}{g(\omega)} |\tilde{\psi}(\omega)|^2 g(\omega)\, d\omega\\
&= \int |\tilde{\psi}(\omega)|^2 f(\omega)\, d\omega,
\end{aligned}$$

as required. The derivation is similar to Eq. (2.81). The second line is obtained from the first by writing $\cos(t^T W + U) = \cos t^T W \cos U - \sin t^T W \sin U$ and $\cos(s^T W + U) = \cos s^T W \cos U - \sin s^T W \sin U$, and then averaging over V and U. Note that $E\{\cos^2 U\} = E\{\sin^2 U\} = 1/2$, $E\{\cos U \sin U\} = 0$, and $E\{V^2\} = 2$. To simulate $X(\cdot)$ for a set of test functions $\psi_j(t)$, $j = 1, \ldots, m$, the same simulated random variables should be used.

Thus, this random field has the right covariance structure, but it is not Gaussian. In order to construct a Gaussian random field, it is necessary to add independent copies $X_i(\psi_j)$ of (3.56) together to get

$$X^{(n)}(\psi_j) = \sum_{i=1}^{n} X_i(\psi_j)/\sqrt{n}. \tag{3.57}$$

For each n, the random vector $\{X^{(n)}(\psi_j) : j = 1, \ldots, m\}$ has the desired covariance structure, and as $n \to \infty$, this random vector converges to a multivariate Gaussian distribution.

3.12 Dispersion Variance

Let $\{X(t) : t \in \mathbb{R}^d\}$ be an ordinary stationary random field with mean μ and autocovariance function $\sigma(h)$. The marginal variance $\sigma(0) = E\{[X(t) - \mu]^2\}$ measures the expected variability of $X(t)$ about the population mean μ. However, in some applications it is of more interest to look at the variability about the *sample mean* in a bounded open region $V \subset \mathbb{R}^d$,

$$\overline{X}(V) = |V|^{-1} \int_V X(t)\, dt. \tag{3.58}$$

Start with the sample variance of the random field as t varies through V

$$|V|^{-1} \int_V [X(t) - \overline{X}(V)]^2\, dt, \tag{3.59}$$

which is a continuous version of the sample variance in ordinary statistics. The expected value of (3.59) is called the *dispersion variance*

$$\sigma(0|V) = |V|^{-1} E \left\{ \int_V [X(t) - \overline{X}(V)]^2\, dt \right\}. \tag{3.60}$$

The notation $0|V$ is meant to suggest a point distributed uniformly in V. The dispersion variance is related to the marginal variance by

$$\sigma(0|V) = \sigma(0) - \operatorname{var}\{\overline{X}(V)\} \tag{3.61}$$

(Exercise 3.17). Note that $\sigma(0|V)$ is invariant under translations of V; that is, $\sigma(0|V) = \sigma(0|(V+t))$, where $V + t = \{v + t : v \in V\}$.

Next we extend the notion of dispersion variance to a pair of regions $V_1 \subset V_2$, by defining

$$\begin{aligned} \sigma(V_1|V_2) &= \sigma(0|V_2) - \sigma(0|V_1) \\ &= \operatorname{var}\{\overline{X}(V_1)\} - \operatorname{var}\{\overline{X}(V_2)\}. \end{aligned} \tag{3.62}$$

The definition of $\sigma(V_1|V_2)$ is statistically meaningful only if $\sigma(V_1|V_2) \geq 0$. The simplest setting in which to guarantee this nonnegativity is when V_2 can be partitioned into a union of copies of V_1. For example, suppose V_1 is a rectangular region and $V_2 = \lambda V_1$, where $\lambda = n$ is an integer, so that we can write $V_2 = \bigcup(t_i + V_1)$, a union of $N = n^d$ copies of V_1 for a suitable sequence of sites $t_i \in \mathbb{R}^d, i = 1, \ldots, N$. In this case, the dispersion variance can be given an explicit interpretation analogous to the usual decomposition of sums of squares in the analysis of variance. Let

$$Z = \overline{X}(V_2), \quad Y_i = \overline{X}(t_i + V_1), \quad i = 1, \ldots, N.$$

Then, it can be shown that

$$|V_2|^{-1} \int_{V_2} [X(t) - Z]^2\, dt = N^{-1} \sum_{i=1}^{N} \left\{ |V_1|^{-1} \int_{t_i+V_1} [X(t) - Y_i]^2 dt + [Y_i - Z]^2 \right\}. \tag{3.63}$$

After taking expectations,

$$\sigma(0|V_2) = \sigma(0|V_1) + N^{-1} E \sum_{i=1}^{N} [Y_i - Z]^2, \tag{3.64}$$

so that

$$\sigma(V_1|V_2) = N^{-1}E\sum_{i=1}^{N}[Y_i - Z]^2. \tag{3.65}$$

If V_1 is called a *block* and V_2 is called a *panel*, note that $\sigma(V_1|V_2)$ can be regarded as the expected *between-block, within-panel* variance.

To appreciate the importance of dispersion variance, consider a mining example. Suppose a mine contains one panel of ore that is to be extracted at the rate of one block per week over the course of one year. The mine operator is concerned that the extracted ore should be as consistent as possible over this year. The dispersion variance gives the appropriate measure of variability over this period.

Since (3.59) involves an increment of order 0 of the random field, the notion of dispersion variance can be extended to intrinsic random fields of intrinsic order 0. It can be shown (see Exercise 3.18) that the dispersion variance can be written in terms of the semivariogram as

$$\sigma(0|V) = |V|^{-2}\int_V\int_V \gamma(s-t)\, ds\, dt. \tag{3.66}$$

Finally, it should be emphasized that the nonnegativity of $\sigma(V_1|V_2)$ cannot be guaranteed for all covariance functions and all $V_1 \subset V_2$. A counterexample is given in Exercise 3.19. Thus the application of dispersion variance requires some caution. Sufficient conditions for nonnegativity include either of the following:

- V_2 is a union of copies of V_1; see (3.65) and Exercise 3.20.
- $X(t)$ forms a self-similar intrinsic random field, and $V_1 = \lambda_1 V$ and $V_2 = \lambda_2 V$ are re-scaled versions of a common underlying set; see Exercise 3.18.

Exercises

3.1 Let $\gamma(h),\ h \in \mathbb{R}^d$, be a valid semivariogram and suppose it tends to a finite limit for large lags, $\gamma(h) \to c$ as $|h| \to \infty$. Show that

$$\sigma(h) = c - \gamma(h)$$

defines a valid covariance function $\sigma(h)$ with marginal variance $\sigma(0) = c$.

Hint: Let $\{X_I(t)\}$ denote an intrinsic process with semivariogram $\gamma(h)$ and with vanishing incremental mean. Let D denote a fixed bounded open set in $\mathbb{R}^d$ (such as a disk) with volume $|D|$, and let $\lambda > 0$ be a dilation parameter. Define a (nonstationary) process $\{Y(t)\}$ by

$$Y(t) = X_I(t) - \frac{1}{|\lambda D|}\int_{\lambda D} X_I(s)ds.$$

Note that for each t, $Y(t)$ is a continuous version of an increment in Definition 3.2.2; hence the $\{Y(t)\}$ process is fully specified.
Using (3.3), the covariance function for the Y-process is given by

$$\begin{aligned}
\operatorname{cov}\{Y(t), Y(v)\} &= -\gamma(t-v) - \frac{1}{(|\lambda D|)^2}\int_{\lambda D}\int_{\lambda D}\gamma(s-u)\,du\,ds \\
&\quad + \frac{1}{(|\lambda D|)}\int_{\lambda D}\gamma(t-u)\,du + \frac{1}{(|\lambda D|)}\int_{\lambda D}\gamma(s-v)\,dv \\
&= -\gamma(t-v) - \frac{1}{(|D|)^2}\int_{D}\int_{D}\gamma(\lambda s' - \lambda u')\,du'\,ds' \\
&\quad + \frac{1}{(|D|)}\int_{D}\gamma(t-\lambda u')\,du' + \frac{1}{(|D|)}\int_{D}\gamma(s-\lambda v')\,dv' \\
&\to -\gamma(t-v) - c + c + c = c - \gamma(t-v)
\end{aligned}$$

as $\lambda \to \infty$, using the dominated convergence theorem. Note the change of variables $u = \lambda u',\ v = \lambda v'$ and bear in mind the relationship $|\lambda D| = \lambda^d |D|$.

3.2 Verify the formula

$$\operatorname{var}\left\{\sum_{i=1}^{n} a_i X_I(t_i)\right\} = -\sum_{i,j=1}^{n} a_i a_j \gamma(t_i - t_j)$$

for an intrinsic random field $\{X_I(t)\}$ of order 0 with semivariogram $\gamma(h)$, where $\boldsymbol{a}(n \times 1)$ is an increment vector of order 0, i.e. $\sum a_i = 0$.
Hint: Write $\sum a_i X_I(t_i) = \sum a_i [X_I(t_i) - X_I(t_0)]$ for any other site t_0, and using (3.3) show that

$$\begin{aligned}
\operatorname{var}\left\{\sum a_i X_I(t_i)\right\} &= \operatorname{var}\left\{\sum a_i [X_I(t_i) - X_I(t_0)]\right\} \\
&= \sum_{i,j} a_i a_j \operatorname{cov}\{[X_I(t_i) - X_I(t_0)],\ [X_I(t_j) - X_I(t_0)]\} \\
&= \sum a_i a_j \{\gamma(t_i - t_0) + \gamma(t_j - t_0) - \gamma(0) - \gamma(t_i - t_j)\} \\
&= -\sum a_i a_j \{\gamma(t_i - t_j)\}.
\end{aligned}$$

3.3 If $-\gamma(h)$ is a conditionally positive semidefinite function with $\gamma(0) = 0$, show that $\sigma(s,t) = \gamma(s) + \gamma(t) - \gamma(t-s)$ is positive semidefinite. Further show that $\sigma(s,t)$ represents the covariance function of an intrinsic random field with semivariogram $\gamma(h)$.
Hint: Given sites $t_1, \ldots, t_n$ and an $n \times 1$ coefficient vector $\boldsymbol{b}$, it is necessary to show

$$\sum_{i,j=1}^{n} b_i b_j \sigma(t_i, t_j) \geq 0.$$

Pick an additional site $t_0 = 0$ and define a new $(n+1) \times 1$ vector $\boldsymbol{b}^*$ with entries $b_0^* = -\sum_{j=1}^{n} b_j$, $b_j^* = b_j$ for $j = 1, \ldots, n$. Then $\boldsymbol{b}^*$ is an increment vector of order 0. Show that

$$\begin{aligned} 0 &\leq -\sum_{i,j=0}^{n} b_i^* b_j^* \gamma(t_i - t_j) \\ &= -b_0^{*2}\gamma(0) - 2b_0^* \sum_{j=1}^{n} b_j \gamma(t_j) - \sum_{i,j=1}^{n} b_i b_j \gamma(t_i - t_j) \\ &= 2\left(\sum_{i=1}^{n} b_i\right)\left(\sum_{j=1}^{n} b_j \gamma(t_j)\right) - \sum_{i,j=1}^{n} b_i b_j \gamma(t_i - t_j) \\ &= \sum_{i,j=1}^{n} b_i b_j \sigma(t_i, t_j), \text{ as required.} \end{aligned}$$

This argument is also valid if t_0 is one of the sites $t_1, \ldots, t_n$. For the last part, if $X(t)$ is a random field with covariance function $\sigma(s,t)$, confirm that

$$\begin{aligned} E[X(t+h) - X(t)]^2 &= \sigma(t+h, t+h) + \sigma(t,t) - 2\sigma(t+h, t) \\ &= 2\gamma(t+h) + 2\gamma(t) - 2\{\gamma(t+h) + \gamma(t) - \gamma(h)\} \\ &= 2\gamma(h), \text{ as required.} \end{aligned}$$

3.4 If a semivariogram $\gamma(h)$ is replaced by $\gamma_c(h) = \gamma(h) + c$ for any constant $c \in \mathbb{R}$, show that formulae (3.4) and (3.5) remain valid.

3.5 Let $\{X_I(t) : t \in \mathbb{R}^d\}$ be an intrinsic random field (of intrinsic order $k = 0$) with semivariogram $\gamma(h)$. Given a fixed vector $h_0 \in \mathbb{R}^d$, define a new random field by

$$Y(t) = X_I(t) - X_I(t - h_0).$$

Using (3.4) show that $\{Y(t)\}$ is stationary and that its covariance function takes the form

$$\text{cov}\{Y(t), Y(s)\} = \gamma(t - s - h_0) + \gamma(s - t - h_0) - 2\gamma(t - s).$$

3.6 If $f(s)$ is an arbitrary finite real-valued function of $s \in \mathbb{R}^d$, show that $\sigma(s,t) = f(s)f(t)$ is positive semidefinite.
Hint: For coefficients $a_1, \ldots, a_n$ and sites $t_1, \ldots, t_n$ note the identity $\sum a_i a_j f(t_i) f(t_j) = \{\sum a_i f(t_i)\}^2 \geq 0$.

3.7 If $g(h)$ satisfies the integral representation (3.11) with $A = 0$, show that $g(h)/|h|^2 \to 0$ as $|h| \to \infty$.

Hint: By assumption

$$M_{1,c} = \int_{|\omega|\le c} |\omega|^2 F(d\omega) < \infty, \quad M_{2,c} = \int_{|\omega|>c} F(d\omega) < \infty.$$

Using the elementary inequalities, $1 - \cos x \le x^2/2$, $|1 - \cos x| \le 2$, confirm that

$$\int_{|\omega|\le c} |\cos(h^T\omega) - 1| \; F(d\omega) \le |h|^2 M_{1,c},$$

$$\int_{|\omega|>c} |\cos(h^T\omega) - 1| \; F(d\omega) \le 2M_{2,c}.$$

Thus $\lim_{|h|\to\infty} |g(h)|/|h|^2 \le M_{1,c}$. But by taking c arbitrarily small, $M_{1,c}$ can be made arbitrarily small. Hence $\lim g(h)/|h|^2 = 0$.

3.8 (a) If $\sigma(s,t)$ is an ordinary continuous positive semidefinite function, show that

$$\sigma_G(\varphi,\varphi) = \int \varphi(s)\,\varphi(t)\,\sigma(s,t)\,ds\,dt$$

defines a positive semidefinite generalized bilinear functional.

(b) If $\{X(t)\}$ is an ordinary random field with covariance function $\sigma(s,t)$, show that it can also be viewed as a generalized random field $X_G(\varphi) = \int \varphi(t)\,X(t)\,dt$ with covariance functional $\sigma_G(\varphi,\varphi)$.

3.9 Show that $\mathcal{H}_k$, the space of homogeneous polynomials of degree k in $\mathbb{R}^d$, and $\mathcal{F}_k$, the space of all polynomials of degree $\le k$ in $\mathbb{R}^d$ have dimensions

$$p_H(k) = \dim(\mathcal{H}_k) = \binom{k+d-1}{k}, \quad p_F(k) = \dim(\mathcal{F}_k) = \binom{k+d}{k},$$

as stated in (3.16).

Hint: This exercise can be done using a simple counting argument (e.g. Feller, 1968, p. 38). Write down a string of $k+d-1$ characters, of which $d-1$ characters are "|" and k characters are "x." Let the number of "x" characters between successive "|" characters indicate the power of successive components of $t = (t[1], \ldots, t[d])$. For example, with $d = 4$ and $k = 5$, one such string might be $x|xx|xx|$, corresponding to the monomial $t[1]^1 t[2]^2 t[3]^2 t[4]^0$. Count the number of different arrangements of strings to get the dimension of $\mathcal{H}_k$. For $\mathcal{F}_k$, add an extra component $t[d+1] = 1$ and note that a monomial of degree k in $t[1], \ldots, t[d+1]$ corresponds to a monomial of degree $\le k$ in $t[1], \ldots, t[d]$.

3.10 The purpose of this exercise is to confirm that the registered intrinsic random field $X_R(t)$ in (3.20) has the covariance function $\sigma_R(s,t)$ in (3.21).

(a) Fix a site t. Show that the collection of coefficients $1, -g_1(t), \ldots, -g_p(t)$ at the sites $t, t_1^*, \ldots, t_p^*$ forms a kth-order increment, i.e.

$$g_i(t) - \sum_{j=1}^{p} g_j(t) g_i(t_j^*) = 0$$

for $i = 1, \ldots, p$. Hence, deduce that for each t, $X_R(t)$ is an increment of the $X_I(\cdot)$ random field.

(b) Use (3.18) to show that the variance of $X_R(t)$ can be expressed in the form (3.21) with $s = t$. Extend this result to show that the covariance between $X_R(s)$ and $X_R(t)$ can be expressed in the form (3.21) for $s \neq t$.

3.11 Let D be a bounded region in $\mathbb{R}^d$ and let $\varphi(t) = I[t \in D]$ be an indicator function on D. Show that the Fourier transform $\tilde{\varphi}(\omega)$ of φ satisfies the following:

(a) $\tilde{\varphi}(\omega)$ is bounded for all ω.

(b) $\int |\tilde{\varphi}(\omega)|^2 d\omega < \infty$.

Hint: (a) $|\tilde{\varphi}(\omega)| = |\int \exp(ih^T\omega)\varphi(h)\, dh| \le \int |\varphi(t)|\, dt \le$ volume of D.

(b) Use the identity $\int |\varphi(t)|^2\, dt = (2\pi)^{-d} \int |\tilde{\varphi}(\omega)|^2\, d\omega$ and note that $\int |\varphi(t)|^2\, dt =$ volume of D.

3.12 In one dimension, $d = 1$, consider a stationary Gaussian random field $X(t)$ with the exponential covariance function

$$\sigma(h) = \exp(-|h|).$$

Consider the indicator test function

$$\varphi(u) = I[|u| \le 1/2].$$

(a) Show that $\varphi(u)$ has the Fourier transform

$$\tilde{\varphi}(\omega) = \begin{cases} (2/\omega)\sin(\omega/2), & \omega \neq 0, \\ 1, & \omega = 0. \end{cases}$$

(b) Let $\psi(u) = \varphi * \check{\varphi}(u)$ and show that

$$\psi(u) = \begin{cases} 0, & u \le -1, \\ 1 + u, & -1 \le u \le 0, \\ 1 - u, & 0 \le u \le 1, \\ 0, & u \ge 1. \end{cases}$$

Confirm that $\psi(u)$ is a compactly supported function with a tent-like shape.

(c) Define the regularized random field $X_\varphi(t) = \int X(t+u)\varphi(u)\,du$ and the regularized covariance function

$$\sigma_\varphi(h) = \int \sigma(h+u)\psi(u)\,du. \tag{3.67}$$

By splitting the integral into three pieces, $(-1,-h)$, $(-h,0)$, $(0,1)$ if $0 < h < 1$ and two pieces $(-1,0)$, $(0,1)$ if $h > 1$, show that

$$\sigma_\varphi(h) = \begin{cases} 2 - 2|h| + e^{-1-|h|} + e^{-1+|h|} - 2e^{-|h|}, & 0 \le |h| \le 1, \\ e^{-1-|h|} + e^{-1+|h|} - 2e^{-|h|}, & |h| \ge 1. \end{cases}$$

(d) Show that $\sigma(0) - \sigma(h) \sim |h|$ and $\sigma_\varphi(0) - \sigma_\varphi(h) \sim (1-e^{-1})h^2$ as $|h| \to 0$. Hence, deduce that $\sigma_\varphi(h)$ is smoother at the origin than $\sigma(h)$.

3.13 Repeat Exercise 3.12 with the same test function $\varphi(h)$, but this time using the linear intrinsic covariance function $\sigma_I(h) = -|h|$, where $\sigma_I(h)$ is defined up to an additive constant. As before, define the regularized intrinsic covariance function $\sigma_{I,\varphi}(h)$ by (3.67).

(a) Show that

$$\sigma_{I,\varphi}(h) = \begin{cases} \frac{1}{3}|h|^3 - h^2 - \frac{1}{3}, & 0 \le |h| \le 1, \\ -|h|, & |h| \ge 1. \end{cases}$$

(b) The semivariogram for the original and regularized random fields are defined by $\gamma(h) = \sigma_I(0) - \sigma_I(h)$ and $\gamma_\varphi(h) = \sigma_{I,\varphi}(0) - \sigma_{I,\varphi}(h)$, respectively. Note they do not depend on the arbitrary additive constant in the definition of $\sigma_I(h)$. Show that $\gamma(h) \sim |h|$ and $\gamma_\varphi(h) \sim h^2$ as $|h| \to 0$. Hence, conclude that $\gamma_\varphi(h)$ is smoother at the origin than $\gamma(h)$.

3.14 Repeat Exercise 3.12 with the same test function $\varphi(h)$, but this time using the de Wijsian generalized intrinsic covariance function $\sigma_{GI}(h) = -\log h,\ h > 0$, where $\sigma_{GI}(h)$ is defined up to an additive constant. As before, define the regularized intrinsic covariance function $\sigma_{I,\varphi}(h)$ by (3.67).

(a) Show that for $h > 0$,

$$\sigma_{I,\varphi}(h) = h^2 \log h + \frac{3}{2} - \frac{1}{2}(1+h)^2 \log(1+h) - \frac{1}{2}(1-h)^2 \log|1-h|.$$

(b) Show that $\lim_{|h|\to 0} \sigma_{I,\varphi}(h) = 3/2$ is finite. Hence, deduce that $\sigma_{I,\varphi}(h)$ defines an ordinary intrinsic random field with semivariogram $\gamma_\varphi(h) = \sigma_{I,\varphi}(0) - \sigma_{I,\varphi}(h) = 3/2 - \sigma_{I,\varphi}(h)$. Show that $\gamma_\varphi(h) = h^2 \log h + O(h^2)$ as $h \to 0$. That is, $\gamma_\varphi(h)$ defines an ordinary intrinsic random field, whose semivariogram is nearly smooth enough to be twice-differentiable at the origin.

(c) Show that $\gamma_\varphi(h) = \log h + O(h^{-2})$ as $h \to \infty$. That is, the difference between $\gamma_\varphi(h)$ and $\gamma(h)$ tends to 0 as $h \to \infty$.

3.15 Consider the function

$$f_{\alpha,d}(\omega) = |\omega|^{-d-2\alpha}, \quad \omega \in \mathbb{R}^d,$$

where $\alpha \in \mathbb{R}$ is a real parameter. Then $f_{\alpha,d}$ can be viewed as the spectral density of an isotropic stochastic process on $\mathbb{R}^d$ (ordinary intrinsic if $\alpha > 0$, generalized stationary if $\alpha < 0$ and generalized intrinsic if $\alpha = 0$). In Section 3.10, it was noted that the corresponding process is self-similar for all values of α. The purpose of this exercise is to confirm the representation (3.45) for $0 < \alpha < 1$. That is, for $0 < \alpha < 1$, the radial covariance function takes the form

$$\sigma^{\#}_{\alpha,d}(r) = cr^{2\alpha}$$

up to an arbitrary additive constant, where the normalizing constant is given by $c = c_{\alpha,d}$ in Eq. (3.46).

(a) Start by considering the Matérn covariance function from Table 2.1

$$\sigma_{M,\alpha,d,a}(h) = \sigma^{\#}_{M,\alpha,d,a}(r) = \frac{(ar)^\alpha}{2^{\alpha-1}\Gamma(\alpha)} K_\alpha(ar), \quad r = |h|,$$

which defines an ordinary stationary stochastic process on $\mathbb{R}^d$ with index $\alpha > 0$ and scale parameter $a > 0$. The spectral density is

$$f_{M,\alpha,d,a}(\omega) = \frac{\Gamma\left(\frac{2\alpha+d}{2}\right) a^{2\alpha}}{\pi^{d/2}\Gamma(\alpha)} \left[a^2 + |\omega|^2\right]^{-(2\alpha+d)/2}$$

(e.g. Kotz and Nadarajah, 2004, p. 39). This spectral density can also be viewed the probability density function for a scaled version of the multivariate t-distribution with 2α degrees of freedom. In particular, since $f_{M,a,\alpha,d}$ is a probability density, its integral is 1.

Use the limiting behavior of the Bessel function $K_\alpha(z) \sim \frac{1}{2}\Gamma(\alpha)(z/2)^{-\alpha}$ as $z \to 0,\ z > 0$, to confirm that $\sigma^{\#}_{M,\alpha,d,a}(0) = 1$.

(b) To make further progress, restrict attention to the case $0 < \alpha < 1$. Then the limiting behavior of the Bessel function can be given a more refined expansion for $z > 0$,

$$K_\alpha(z) = \frac{1}{2}\left\{\Gamma(\alpha)(z/2)^{-\alpha} + \Gamma(-\alpha)(z/2)^\alpha + O(z^{2-\alpha})\right\}, \quad z \to 0.$$

[This formula can be verified from the representation of the K Bessel function in terms of the I Bessel function (Gradshteyn and Ryzhik, 1980, p. 970, eqn 8.485), together with the series expansion of the I Bessel function (Gradshteyn and Ryzhik, 1980, p. 961, eqn 8.445),

and the reflection formula for the Gamma function (Gradshteyn and Ryzhik, 1980, p. 937, eqn 8.334.3).]
Using this expansion, show that minus the semivariogram for the Matérn process takes the form

$$\int (\cos(h^T\omega) - 1) f_{M,a,\alpha,d}(\omega)\, d\omega = \sigma^{\#}_{M,\alpha,d,a}(r) - 1 = \frac{\Gamma(-\alpha)}{\Gamma(\alpha)}(ar/2)^{2\alpha} + O((ar)^{2-2\alpha}).$$

Lastly, multiply $f_{M,\alpha,d,a}$ by $\pi^{d/2}\Gamma(\alpha)a^{-2\alpha}/\Gamma(\alpha + \frac{d}{2})$ and let $a \to 0$ for fixed $r \geq 0$ to conclude that

$$\sigma^{\#}_{\alpha,d}(r) - 1 = \frac{\pi^{d/2}\Gamma(-\alpha)}{2^{2\alpha}\Gamma(\alpha + \frac{d}{2})} r^{2\alpha}.$$

Since an intrinsic covariance function (of intrinsic order $k = 0$) is defined only up to an additive constant, the required form for the intrinsic covariance function has been proved.

3.16 The purpose of this exercise is to investigate the extent to which Exercise 3.15 can be extended to other values of α.

(a) If $g(h)$ is a function of $h \in \mathbb{R}^d$, the Laplacian is defined by $\Delta g(h) = \sum_{\ell=1}^{d} \partial^2 g(h)/\partial h[\ell]^2$. If $g(h) = g^{\#}(r)$, say, with $|h| = r$, is isotropic, show that

$$\Delta g(h) = (g^{\#})''(r) + \frac{d-1}{r}(g^{\#})'(r),$$

where the dash denotes differentiation with respect to $r > 0$.

(b) Use the Laplacian formula to deduce that if $g(h) = |h|^{2\alpha}$ then $\Delta g(h) = 2\alpha(2\alpha + d - 2)r^{2\alpha-2}$, $r = |h| > 0$.

(c) If $\sigma^{\#}_{\alpha+1,d}(r) = c_{\alpha+1,d}r^{2\alpha+2}$, then show that $\Delta\sigma^{\#}_{\alpha+1,d}(r) = c_{\alpha,d}r^{2\alpha} = \sigma^{\#}_{\alpha,d}(r)$, $r > 0$. Further if $\alpha > 0$, show that this equation can be extended to $r = 0$.

(d) Hence, deduce that, at least formally, the constant c in (3.45) must take the form $c = c_{\alpha,d}$ for all noninteger values of α. The adjective "formally" means that some extra issues need to be considered for certain values of α. These issues will be explored in the following parts of the exercise.

(e) If $\alpha > 0$, then $\sigma_{\alpha,d}(h)$ is an equivalence class of functions; it equals $c_{\alpha,d}r^{2\alpha}$ plus an arbitrary even polynomial in r of degree $2[\alpha]$, where $[\alpha]$ is the integer part of α. Show that with this equivalence class interpretation,

$$\Delta\sigma_{\alpha+1,d}(h) = \sigma_{\alpha,d}(h), \ h \in \mathbb{R}^d$$

remains true.

(f) If $\alpha = k > 0$ is an integer, then in addition to the equivalence class considerations, there is the problem that $c_{\alpha,d} = \infty$. In this case, the formula for the covariance function must be modified to

$$\sigma_{k,d}(h) = c'_{k,d}|h|^{2k}\log(|h|),$$

plus an arbitrary even polynomial in h of degree $2k$, where the modified normalizing constant $c'_{k,d}$ is given in (3.48). Use the limiting argument in Section 3.10 as $\varepsilon \downarrow 0$ with $\alpha = k + \varepsilon$ to confirm that this modified form for the covariance function is valid.

(g) If $-d/2 < \alpha \le 0$, then $\sigma_{\alpha,d}(h)$ is a generalized covariance function, which can be identified with the ordinary function $\sigma_{\alpha,d}(h) = c_{\alpha,d}|h|^{2\alpha}$. In particular, show that $\sigma_{\alpha,d}(h)$ is integrable in a neighborhood of $h = 0$.

(h) However, if $\alpha \le -d/2$, the behavior of the generalized covariance function $\sigma_{\alpha,d}(h)$ is more complicated to describe. Suppose $\alpha + k$ lies in the interval $(-d/2, -d/2 + 1]$ for some integer $k \ge 1$. As before, $\sigma_{\alpha,d}(h) = c_{\alpha,d}|h|^{2\alpha}$ can be identified with an ordinary function for $h \ne 0$. However, in this case the generalized covariance function also has some support at the origin. A full interpretation of $\sigma_{\alpha,d}(h)$ is easiest to describe using of the Parseval relation. Suppose $\varphi(u) \in \mathcal{K}$ is a smooth test function with compact support, with the Fourier transform $\tilde{\varphi}(\omega)$, and write $\psi(h) = (\varphi * \check{\varphi})(h)$ with the Fourier transform $\tilde{\psi}(\omega) = |\tilde{\varphi}(\omega)|^2$. Show that the covariance functional $\sigma_{\alpha,d}$ applied to ψ takes the form

$$\begin{aligned}
\sigma_{\alpha,d}(\psi) &= \int f_{\alpha,d}(\omega)\tilde{\psi}(\omega)\, d\omega \\
&= \int \{|\omega|^{-2k} f_{\alpha,d}(\omega)\}\{|\omega|^{2k}\tilde{\psi}(\omega)\}\, d\omega \\
&= \int f_{\alpha+k,d}(\omega)\{|\omega|^{2k}\tilde{\psi}(\omega)\}\, d\omega \\
&= \int \sigma_{\alpha+k,d}(h)\Delta^k\psi(h)\, dh,
\end{aligned}$$

where Δ^k is the iterated Laplacian, and Eq. (A.37) has been used to show that the Fourier transform of $\Delta^k\psi(h)$ is given by $|\omega|^{2k}\tilde{\psi}(\omega)$. Note: If $d = 1$ and $\alpha + k \in [0,1/2)$ explain why it is necessary to confirm additionally that $\Delta^k\psi(h)$, the $(2k)$th derivative of $\psi(h)$, is an increment of order 0, i.e. $\int \Delta^k\psi(h)\, dh = 0$.

(i) White noise is an ordinary generalized process defined by a constant spectral density

$$f_{WN}(\omega) = \left(\frac{1}{2\pi}\right)^d, \quad \omega \in \mathbb{R}^d$$

with generalized covariance function given by the Dirac delta function, $\sigma_{WN}(h) = \delta_0(h)$. Up to the factor $(2\pi)^{-d}$ it is the spectral density of the self-similar process of index $\alpha = -d/2$ in (3.44). Confirm that for $\alpha = -d/2$, $c_{\alpha,d} = 0$, so that the generalized function $\sigma_{WN}(h)$ has no support outside the origin.

3.17 Let V be an open bounded region in $\mathbb{R}^d$. If $\{X(t)\}$ is a stationary random field with covariance function $\sigma(h)$, define the *(continuous) sample mean* within V by

$$\overline{X}(V) = |V|^{-1} \int_V X(t)\, dt.$$

Show that its variance is given by

$$\operatorname{var}\{\overline{X}(V)\} = |V|^{-2} \int_V \int_V \sigma(s-t)\, ds\, dt.$$

By expanding the square for the *sample continuous variance within V* in the formula for the dispersion variance

$$\sigma(0|V) = |V|^{-1} E\left\{ \int [X(t) - \overline{X}(V)]^2\, dt \right\},$$

show that

$$\sigma(0|V) = \sigma(0) - \operatorname{var}\{\overline{X}(V)\}.$$

3.18 (a) Show that the dispersion variance $\sigma(0|V)$ as defined in Eq. (3.60) continues to make sense for an intrinsic random field with semivariogram $\gamma(h)$, and is given in this case by

$$\sigma(0|V) = |V|^{-2} \int \int \gamma(s-t)\, ds\, dt.$$

To verify this formula, it is helpful to expand (3.60) as

$$\sigma(0|V) = |V|^{-3} \int_V \int_V \int_V E\{[X(t) - X(s)][X(t) - X(u)]\}\, ds\, du\, dt,$$

and to use (3.3) to simplify the result.

(b) If $\gamma(h) = |h|^{2\alpha}$ for some $0 < \alpha < 1$ and λV denotes the dilation of V by a factor $\lambda > 0$, show that

$$\sigma(0|\lambda V) = B\lambda^{2\alpha}, \quad B = |V|^{-2} \int_V \int_V |s-t|^{2\alpha}\, ds\, dt,$$

and

$$\sigma(\lambda_1 V|\lambda_2 V) = (\lambda_2^{2\alpha} - \lambda_1^{2\alpha})B > 0$$

for $0 < \lambda_1 < \lambda_2$.

Thus, $\sigma(0|\lambda V)$, the expected sample variance of the random field as the site varies continuously over the region λV increases indefinitely with λ. This behavior can be contrasted with a stationary random field where it is usually the case that $\sigma(0|\lambda V) \to \sigma(0)$ as $\lambda \to \infty$; see Chapter 2, Theorem 2.13.1.

3.19 The purpose of this exercise is to construct a counterexample showing that it is possible that $\sigma(V_1|V_2) < 0$ even when $V_1 \subset V_2$. Let $X(t) = \sqrt{2}\cos(t + \Phi)$ denote a random cosine wave in one dimension, $t \in \mathbb{R}^1$, where Φ is uniformly distributed on $[0,2\pi)$. This process is stationary with covariance function $\sigma(h) = \cos(h)$. Set $V = [0, a]$ for some $a > 0$.

(a) Show that

$$\text{var}\{\overline{X}(V)\} = 2(1 - \cos\ a)/a^2.$$

(b) In particular, note that $\overline{X}(V) = 0$ and $\text{var}\{\overline{X}(V)\} = 0$ when a is an integer multiple of 2π.

(c) Since $\text{var}\{\overline{X}(V)\}$ is not monotone decreasing in a, it is possible for the dispersion variance to be negative. For example, show that

$$\sigma([0, a) \mid [0, b)) < 0$$

for $a = 2\pi,\ b = 3\pi$.

(d) Equation (3.65) states that the dispersion variance is always nonnegative when $V_2 = [0, b]$ is a union of n disjoint copies of $V_1 = [0, a]$, i.e. $b = na,\ a > 0, n \geq 1$. Using this result, deduce the trigonometric inequalities

$$1 - \cos\ na \leq n^2(1 - \cos\ a)$$

for all real $a > 0$ and integer $n \geq 1$.

3.20 The purpose of this exercise is to show that negativity of the dispersion variance $\sigma(V_1|V_2)$ in Exercise 3.19 cannot arise under the additional assumption that V_2 can be partitioned as a union of copies of V_1. In this case, verify the continuous analysis of variance identity in (3.63). If $\{X(t)\}$ is a stationary or intrinsic random field, take expectations to get (3.64) and (3.65), and hence confirm that the dispersion variance is nonnegative, $\sigma(V_1|V_2) \geq 0$.

4

Autoregression and Related Models

4.1 Introduction

In Chapters 2 and 3, we looked at examples of possible covariance functions and semivariograms for random fields. In general, these examples are merely functions of convenient analytic form, without any probabilistic motivation. In this chapter, we look at random fields that have a natural description in terms of a probabilistic model, typically an autoregression. These models are largely restricted to random fields indexed on the integer lattice $\mathbb{Z}^d$.

The motivation for these models comes from discrete time series (e.g. Fuller, 1996) where there are three natural mechanisms for constructing a stationary Gaussian process $\{X_t : t \in \mathbb{Z}\}$, starting from a Gaussian discrete white noise process $\{\varepsilon_t : t \in \mathbb{Z}\}$, that is, a collection of independent $N(0, \sigma_\varepsilon^2)$ random variables. Here, t denotes (discrete) time.

(a) *Moving Average* of order S (MA(S)). Let

$$X_t = \sum_{s=0}^{S} b_s \varepsilon_{t-s}, \quad t \in \mathbb{Z}, \tag{4.1}$$

where $S \geq 0$ and $b_0, \ldots, b_S$ are coefficients with $b_0 > 0$. For any set of coefficients, (4.1) defines a stationary process. If we define a polynomial

$$B(z) = \sum_{s=0}^{S} b_s z^s, \quad z \in \mathbb{C}, \tag{4.2}$$

then it is usual to assume a "stability condition"

$$B(z) \neq 0 \text{ for } |z| \leq 1. \tag{4.3}$$

This stability condition ensures that the coefficients b_s are identifiable and that $\{\varepsilon_t\}$ can be interpreted as the innovation process of $\{X_t\}$

$$\varepsilon_t = X_t - E[X_t | X_s : s < t]. \tag{4.4}$$

Spatial Analysis, First Edition. John T. Kent and Kanti V. Mardia.

Without the stability condition, different sets of coefficients $\{b_s\}$ will generate the same autocovariance function for $\{X_t\}$.

The innovation Eq. (4.4) is a key result for prediction. To better understand its basic properties, let $Y_t = E[X_t|X_s : s < t]$, so that Y_t is a function of $\{X_s : s < t\}$. In particular,

$$E[Y_t|X_s : s < t] = Y_t.$$

Hence, for any particular time $r < t$, we find

$$\begin{aligned} E[X_r\varepsilon_t] &= E[E\{X_r(X_t - Y_t)|X_s : s < t\}] \\ &= E[X_r E\{X_t|X_s : s < t\} - X_r Y_t] \\ &= E[X_r Y_t - X_r Y_t] = 0. \end{aligned}$$

The first line follows by the tower law since an overall expectation is being written as the expected value of a conditional expectation (Exercise 4.1). The second line follows since X_r and Y_t are fixed given $\{X_s : s < t\}$, and hence can be factored out of the conditional expectation. Thus, ε_t is uncorrelated with X_r for all $r < t$. Intuitively, ε_t represents that part of X_t which cannot be predicted from its past values (such as X_r).

(b) *Autoregression* of order S (AR(S)). Suppose $\{X_t\}$ satisfies

$$\sum_{s=0}^{S} d_s X_{t-s} = \varepsilon_t, \tag{4.5}$$

where $S \geq 0$ and $d_0, \ldots, d_S$ are coefficients with $d_0 = 1$. If we set $D(z) = \sum d_s z^s$, then the condition

$$D(z) \neq 0, \quad \text{for } |z| \leq 1 \tag{4.6}$$

is known as an "invertibility" condition. It ensures that (4.5) defines a stationary process and that $\{\varepsilon_t\}$ can be regarded as the innovation process (4.4). In this case, Eq. (4.4) can be rephrased as

$$E[X_t|\{X_s : s < t\}] = -\sum_{s=1}^{S}(d_s/d_0)X_{t-s}, \tag{4.7}$$

so that X_t depends on its past values only through the most recent S values.

(c) *Differencing*. Let ∇ denote the first-order difference operator

$$\nabla X_t = X_t - X_{t-1}. \tag{4.8}$$

A common strategy in time-series analysis is to suppose that the first-order differences (4.8) (or possibly higher order differences) form a stationary process.

Techniques (a), (b), and (c) form the basis of the ARIMA approach to time-series modeling (AR = autoregression, MA = moving average, and I = integrated, the converse operation to differencing). Note that all three of these techniques use the fact that time is ordered; we can distinguish the "past" from the "future." However, when we move to the spatial setting, $t \in \mathbb{Z}^d$, $d \geq 2$, this distinction is no longer very meaningful. Therefore, it is useful to recast these one-dimensional models so they do not distinguish between past and future, in order to motivate similar models in higher dimensions.

For the moving average and autoregression models (4.1) and (4.5), the change is obvious; we allow the summations to include coefficients a_s and b_s for $s < 0$ as well as for $s \geq 0$. The symmetry conditions $a_s = a_{-s}$ and $b_s = b_{-s}$ are often imposed to ensure identifiability of the parameters. After some background material in Section 4.2, the moving average process is extended to $d \geq 2$ dimensions in Section 4.3. The symmetric two-sided version of the autoregression process is known as a "simultaneous autoregression" (SAR) and is extended to higher dimensions in Section 4.5. However, in the SAR model, the discrete white noise process $\{\varepsilon_t\}$ in (4.5) no longer has the interpretation of an innovation process by (4.4).

The conditioning in (4.7) on $\{X_s : s < t\}$ for the autoregression process makes explicit use of the past and does not generalize naturally to a spatial setting. We need some notation to describe a related relationship that treats the past and future symmetrically. Define a new sequence $\{c_s\}$ by $c = d * \check{d}$, where "$*$" denotes convolution and $\check{d}_h = d_{-h}$ denotes the sequence reflected about the origin; that is,

$$c_s = \sum_{h=0}^{S} d_h d_{h-s}.$$

Note that $c_s = c_{-s}$, $c_0 > 0$, and $c_s = 0$ for $s > |S|$. Then, the autoregression (4.5) can also be characterized by the property

$$\begin{aligned} E[X_t | X_{\setminus t}] &= - \sum_{\substack{s=-S \\ s \neq 0}}^{S} (c_s / c_0) X_{t-s}, \\ \text{var}[X_t | X_{\setminus t}] &= \sigma_\varepsilon^2 / c_0. \end{aligned} \tag{4.9}$$

Here, $X_{\setminus t} = \{X_s : s \in \mathbb{Z},\ s \neq t\}$ denotes the values of the process at all sites except t. See Exercise 4.4 for a proof. In other words, the conditional distribution is normal with the mean being a linear function of the S nearest neighbors and with constant conditional variance.

Thus, the autoregression model (4.5) can be given either a "unilateral" conditional representation (4.7) or a "bilateral" conditional representation (4.9). The latter representation is known as a "conditional autoregression" (CAR) and generalizes immediately to a spatial setting; see Section 4.6.

In time series, it is often convenient to transform a process by first differences to produce a stationary process. In the spatial setting, the notion of "successive sites" does not make much sense. Therefore, it is more natural to model the original process directly using an intrinsic approach as in Chapter 3. The adaptation of CAR models to include intrinsic behavior is also considered in Section 4.6.

It was mentioned earlier that the ordering of time has no completely natural analogue in higher dimensions. However, it is possible to define an ordering on $\mathbb{Z}^d$, $d \geq 2$, (indeed, many possible orderings) in a slightly artificial manner. Further, in terms of such an ordering, it is possible to create a theory of unilateral moving averages and autoregressions, which closely parallels the one-dimensional theory; see Section 4.8. Although these models are rather unnatural from a modeling point of view, they can be computationally very convenient and have some important theoretical properties. A tractable special case is given by the quadrant autoregressions (QARs) for which the dependence extends just over a quadrant; see Section 4.8.

So far, all the models considered in this chapter have been Gaussian. It is also of interest to consider alternative models, especially for integer-valued data. The general version of a CAR model is known as a Markov random field (MRF) and is considered in Section 4.9. The general version of a QAR model is known as a Markov mesh model and is considered in Section 4.10.

For the most part, we are concerned with lattice random fields $\{X_t : t \in \mathbb{Z}^d\}$, in this chapter. However, some of the constructions (such as moving averages and SARs) extend easily to continuously indexed random fields $\{X(t) : t \in \mathbb{R}^d\}$ (Sections 4.3–4.5). In addition, continuously indexed random fields can appear as limits of discrete random fields on a finely spaced lattice (Section 4.7).

4.2 Background

Here is a convenient place to summarize some basic results about Fourier transforms. See Section A.4 for more details. Start with the discrete case. That is, let $\{b_s : s \in \mathbb{Z}^d\}$ denote a set of real coefficients with the Fourier transform

$$\tilde{b}(\omega) = \sum b_s \exp(2\pi i s^T \omega), \quad \omega \in (-\pi, \pi)^d,$$

so that $\tilde{b}(\omega)$ is expressed as a Fourier series in the coefficients $\{b_s\}$. When the number of coefficients is infinite, it is necessary to impose some regularity conditions in order for the Fourier transform to be a well-defined function. Here are two standard choices.

1. If the coefficients $\{b_s\}$ are summable, $\sum |b_s| < \infty$, then $\tilde{b}(\omega)$ is a bounded continuous function of ω.

2. If the coefficients $\{b_s\}$ are square-summable, $\sum |b_s|^2 < \infty$, then $\tilde{b}(\omega)$ is square-integrable over $(-\pi, \pi)^d$ (and vice versa).

Of course, if the coefficients are summable, they are automatically square-summable.

Similar results hold in the continuous case. Let $b(s)$, $s \in \mathbb{R}^d$ be a real-valued function with the Fourier transform

$$\tilde{b}(\omega) = \int b(s) \exp(2\pi i s^T \omega)\, ds, \quad \omega \in \mathbb{R}^d.$$

Again, it is necessary to impose some regularity conditions on $b(s)$ to ensure the Fourier transform exists.

1. If the function $b(s)$ is integrable, $\int |b_s| < \infty$, then $\tilde{b}(\omega)$ is a bounded continuous function of ω.
2. If the function $\{b(s)\}$ is square-integrable, $\int |b_s|^2 < \infty$, then $\tilde{b}(\omega)$ is square-integrable over $\mathbb{R}^d$ (and vice versa).

In the continuous case, the integrability of $b(s)$ neither implies, nor is implied by, the square integrability of $b(s)$.

A key tool used to understand the stationary Gaussian models of this chapter is the Fourier transform. Start with a stationary process $\{U_t, \ t \in \mathbb{Z}^d\}$, whose covariance function $\sigma_{U;h}$ has a spectral density $f_U(\omega)$; i.e.

$$\sigma_{U;h} = \int_{(-\pi,\pi)^d} \exp\{ih^T\omega\} f_U(\omega)\, d\omega,$$

where $f_U(\omega)$ is nonnegative and integrable over $(-\pi, \pi)^d$. Next, define another stationary process $\{V_t\}$ by linearly filtering $\{U_t\}$,

$$V_t = \sum b_s U_{t-s}, \ t \in \mathbb{Z}^d, \tag{4.10}$$

where s ranges through some finite or infinite set of indices in $\mathbb{Z}^d$. The $\{V_t\}$ process can be viewed as a *filtered* version of the $\{U_t\}$ process. The covariance function of $\{V_t\}$ is given by

$$\sigma_{V;h} = \sum_{s,t} b_s b_t \sigma_{U;h+s-t}, \tag{4.11}$$

with corresponding spectral density

$$f_V(\omega) = |\tilde{b}(\omega)|^2 f_U(\omega), \quad \omega \in (-\pi, \pi)^d. \tag{4.12}$$

Here, $\tilde{b}(\omega) = \sum \{b_s \exp(is^T\omega)\}$ is the Fourier transform of the filter. This formula was used earlier Eq. (2.64).

Next consider conditions on the coefficients $\{b_s\}$ to ensure that $f_V(\omega)$ is integrable. In particular, if $f_U(\omega)$ is bounded, then a sufficient condition is that the coefficients $\{b_s\}$ are square-summable. Then $|\tilde{b}(\omega)|$ is square-integrable over $\omega \in (-\pi, \pi)^d$, i.e. $|\tilde{b}(\omega)|^2$ is integrable, and so $f_V(\omega)$ is also integrable.

A similar analysis holds in the continuous case. Let $\{U(t) : t \in \mathbb{R}^d\}$ denote a stationary (ordinary or generalized) random field, with (real or generalized) covariance function $\sigma_U(h)$, $h \in \mathbb{R}^d$, and spectral density $f_U(\omega)$, $\omega \in \mathbb{R}^d$. The covariance function $\sigma_U(h)$ satisfies

$$\sigma(h) = \int f_U(\omega)\, d\omega,$$

at least in the sense of Section 3.6. Define a new (ordinary) stationary random field $V(t)$ by

$$V(t) = \int b(s)U(t+s)\, ds$$

in terms of a filtering function $b(s)$, $s \in \mathbb{R}^d$, with spectral density

$$f_V(\omega) = |\tilde{b}(\omega)|^2 f_U(\omega).$$

In order for this construction to be well defined, it is necessary to ensure that $f_V(\omega)$ is an integrable function. Two special cases are of interest.

1. Suppose $\{U_t\}$ is an ordinary stationary random field, so that $f(\omega)$ is integrable. Then, it is sufficient to assume that $\{b(s)\}$ is summable, $\int |b(s)| < \infty$, so that $\tilde{b}(s)$ is bounded, and hence that $f_V(\omega)$ is integrable.
2. Suppose that $\{U(t)\}$ is generalized white noise, an example of a generalized random field with covariance function $\sigma_U(h) = \sigma^2\delta_0(h)$ in terms of the Dirac delta function, and with constant spectral density $f_U(\omega) = \sigma^2/(2\pi)^d$. The spectral density of V reduces to

   $$f_V(\omega) = |\tilde{b}(\omega)|^2\sigma^2/(2\pi)^d, \quad \omega \in \mathbb{R}^d.$$

   In this case, a necessary and sufficient condition to ensure the integrability of $f_V(\omega)$ is that $b(s)$ be *square-integrable*, $\int |b(s)|^2 < \infty$, which is equivalent to $\tilde{b}(\omega)$ being square-integrable.

4.3 Moving Averages

The phrase "moving average" is a way to define a stationary random field in terms of the regularization of white noise, discrete in the lattice case, and generalized in the continuous case. It is convenient to describe the lattice case and the continuously indexed cases separately.

4.3.1 Lattice Case

We first describe the lattice case. Start with a discrete white noise process, which is a collection $\{\varepsilon_t : t \in \mathbb{Z}^d\}$, say, of independent random variables with mean 0 and variance σ_ε^2. The covariance function is given by $\sigma_{\varepsilon;0} = \sigma_\varepsilon^2$, $\sigma_{\varepsilon;h} = 0$ for $h \neq 0$, with spectral density $f_\varepsilon(\omega) = \sigma_\varepsilon^2/(2\pi)^d$, $\omega \in (-\pi, \pi)^d$.

Let $\{b_s : s \in \mathbb{Z}^d\}$ be a set of known coefficients satisfying $\sum |b_s|^2 < \infty$. Define a new random field $\{Y_t\}$ by the regularization

$$\begin{aligned} Y_t &= \sum_s \varepsilon_{t+s} b_{s,} \quad t \in \mathbb{Z}^d \\ &= (\varepsilon * \check{b})_t, \end{aligned}$$

where $\check{b}_s = b_{-s}$. Note the change in sign convention from (4.1). Then, $\{Y_t\}$ is a stationary random field with mean 0 and covariance function

$$\sigma_{Y;h} = \sigma_\varepsilon^2 \sum_s b_h \, b_{h+s} = \sigma_\varepsilon^2 (b * \check{b})_h, \quad h \in \mathbb{Z}^d,$$

and with spectral density

$$f_Y(\omega) = |\tilde{b}(\omega)|^2 \sigma_\varepsilon^2/(2\pi)^d, \quad \omega \in (-\pi, \pi)^d,$$

where $\tilde{b}(\omega) = \sum \exp(ih^T\omega) b_h$ is the Fourier transform of $\{b_h\}$.

4.3.2 Continuously Indexed Case

A similar construction can be carried out for continuously indexed random fields. Let $\{\varepsilon(t) : t \in \mathbb{R}^d\}$ now denote "generalized white noise" – a generalized stationary random field with the covariance functional given in terms of the Dirac delta function, $\delta_0(h)$, by

$$\sigma_\varepsilon(h) = \sigma_\varepsilon^2 \delta_0(h), \ h \in \mathbb{R}^d,$$

and with constant spectral density

$$f_\varepsilon(\omega) = \sigma_\varepsilon^2/(2\pi)^d, \quad \omega \in \mathbb{R}^d.$$

Let $b(s)$, $s \in \mathbb{R}^d$, be a given square-integrable function and define $\{X(t) : t \in \mathbb{R}^d\}$ by

$$\begin{aligned} X(t) &= \int \varepsilon(t+s) b(s) \, ds \\ &= \varepsilon(b_t), \ t \in \mathbb{R}^d, \quad \text{say,} \end{aligned}$$

where

$$b_t(s) = b(s-t) \tag{4.13}$$

stands for the shifted version of the function $b(s)$. Then $\{X(t)\}$ is an ordinary stationary random field with covariance function

$$\sigma_X(h) = \sigma_\varepsilon^2 \int b(h+s) \, b(s) \, ds = \sigma_\varepsilon^2 (b * \check{b})(h),$$

where $\check{b}(t) = b(-t)$, and with spectral density

$$f_X(\omega) = \sigma_\varepsilon^2 |\tilde{b}(\omega)|^2/(2\pi)^d, \quad \omega \in \mathbb{R}^d, \tag{4.14}$$

where $\tilde{b}(\omega) = \int \exp(ih^T\omega)b(h)\, dh$ is the Fourier transform of $b(h)$. Note that if $b(h)$ is square-integrable in h, then $\tilde{b}(\omega)$ is square-integrable in ω, i.e. $|\tilde{b}(\omega)|^2$ is integrable. Hence $f_X(\omega)$ is integrable.

A common choice for $b(s)$ is an indicator function on a disk

$$b(s) = I[|s| \leq c].$$

In dimensions $d = 1,2,3$, this leads to the covariance functions of Exercise 2.6.

Different functions $b(s)$ satisfying $\int |b(s)|^2\, ds < \infty$ can give rise to the same spectral density $f_X(\omega)$ in (4.14). Hence, it is interesting to ask in what circumstances it is possible to determine a unique function $b(s)$ from the spectral density. For identifiability and regularity purposes in the continuous spatial setting, we restrict our attention to real-valued functions $b(h)$ satisfying

(i) $b(h) = b(-h),\ h \in \mathbb{R}^d$ (symmetry),
(ii) $\int |b(h)|e^{\kappa|h|}\, dh < \infty$ for some $\kappa > 0$, and
(iii) $\int b(h)\, dh > 0$ (positive integral).

In the lattice setting, replace $b(h),\ h \in \mathbb{R}^d$, by $b_h,\ h \in \mathbb{Z}^d$, and replace the integrals in (ii)–(iii) by sums.

Symmetry ensures that $\tilde{b}(\omega)$ is real-valued for ω real. This is very different from the one-sided condition used in time series for identifiability, where, in particular, $b(h)$ is nonzero only for $h < 0$.

The integrability assumption (ii) implies that $\tilde{b}(\omega)$ can be extended to be an analytic function of the complex vector $\omega \in \mathbb{C}^d$ for $|\mathrm{Im}(\omega)| < \kappa$, where Im stands for the imaginary part of a complex number. A positive integral (iii) means that $\tilde{b}(0) > 0$. Hence, it follows that $\tilde{b}(\omega)$ is determined from $|\tilde{b}(\omega)|^2$ in an open set about $\omega = 0$, and hence is determined for all ω, $|\mathrm{Im}(\omega)| < \kappa$, by analytic continuation. Finally, $b(h)$ can be found by the inverse Fourier transform.

4.4 Finite Symmetric Neighborhoods of the Origin in $\mathbb{Z}^d$

A finite set of sites $\mathcal{N} \subset \mathbb{Z}^d$ is called a *neighborhood of the origin* if it is symmetric, so that $s \in \mathcal{N}$ if and only if $-s \in \mathcal{N}$. By convention, the origin is not treated as an element of $\mathcal{N}$, and an augmented neighborhood can be defined by

$$\mathcal{N}_0 = \mathcal{N} \cup \{0\}.$$

Occasionally, it is helpful to take half of a neighborhood, which contains one of s or $-s$ for each $s \in \mathcal{N}$. Write a half-neighborhood as $\mathcal{N}^\dagger$ so that

$$\mathcal{N}^\dagger \cap (-\mathcal{N}^\dagger) = \emptyset, \quad \mathcal{N} = \mathcal{N}^\dagger \cup (-\mathcal{N}^\dagger). \tag{4.15}$$

Of course, the choice of half-neighborhood is not unique.

A neighborhood of the origin can be shifted to form a neighborhood of any other site t, $\mathcal{N}(t) = \mathcal{N} + t$. These neighborhoods of the origin play a fundamental role in the construction of two spatial autoregression models – SARs and CARs.

One way to think about neighbors is through a graph structure. The vertices of the graph are the sites, and two vertices are connected by an edge if they are neighbors. In practice, a neighborhood of the origin will consist of sites close to the origin.

The most important neighborhood in $\mathbb{Z}^d$ is the *first-order basic neighborhood*, for which a site t is a neighbor of the origin if all the components $t[\ell]$, $\ell = 1, \ldots, d$ of t vanish, except for one component which equals $+1$ or -1. In $d = 2$ dimensions, this neighborhood takes the form

$$\mathcal{N}^{(\text{basic},1)} = \mathcal{N}^{(1)} = \{(-1,0),\ (0,-1),\ (1,0),\ (0,1)\}. \tag{4.16}$$

The superscript "(1)" indicates that it is also possible to define higher order neighborhoods $\mathcal{N}^{(\text{basic},k)}$, $k > 1$. For example, a site t is a second-order neighbor of the origin if there exists a site u such that u is a first-order neighbor of the origin and t is a first-order neighbor of u. Figure 4.1a,c illustrates the first-order and second-order basic neighborhoods of the origin in $d = 2$ dimensions.

Another neighborhood of the origin that is sometimes of interest can be called the *first-order full neighborhood*. It includes the diagonal neighbors as well as neighbors along the coordinate axes. In $d = 2$ dimensions, it is given by

$$\mathcal{N}^{(\text{full},1)} = \mathcal{N}^{(1)} \cup \{(-1,-1),\ (-1,1),\ (1,-1),\ (1,1)\}, \tag{4.17}$$

and is plotted in Figure 4.1b.

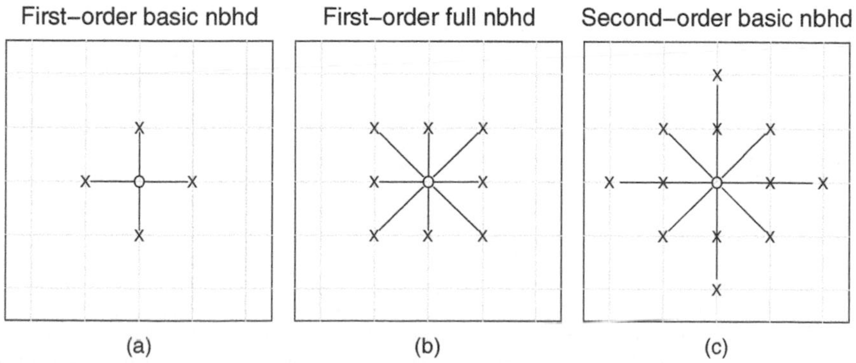

Figure 4.1 Panels (a) and (b) illustrate the first-order basic and full neighborhoods of the origin in the plane. Panel (c) illustrates the second-order basic neighborhood.

4.5 Simultaneous Autoregressions (SARs)

4.5.1 Lattice Case

Let $\{X_t : t \in \mathbb{Z}^d\}$ be a lattice stationary Gaussian random field with mean 0, and suppose there are coefficients $\{d_s : s \in \mathbb{Z}^d\}$ such that $d_0 > 0$, $\sum d_s^2 < \infty$, and

$$\sum d_s X_{t+s} = \varepsilon_t \text{ say}, \quad t \in \mathbb{Z}^d, \tag{4.18}$$

where $\{\varepsilon_t\}$ is a discrete white noise random field with variance σ_ε^2. Such a representation is called an "autoregression" for $\{X_t\}$ because X_t can be written as a linear combination of other X_s values plus an error term. By the argument in Section 4.2, the spectral density for $\{X_t\}$ must take the form

$$f_X(\omega) = \sigma_\varepsilon^2 / \{(2\pi)^d |\tilde{d}(\omega)|^2\}, \quad \omega \in (-\pi, \pi)^d, \tag{4.19}$$

where

$$\tilde{d}(\omega) = \sum d_h \exp(ih^T \omega). \tag{4.20}$$

If $\{X_t\}$ is to be a stationary random field, then (4.19) must be an integrable function. This imposes restrictions on the coefficients $\{d_s\}$. A sufficient condition for the integrability of (4.19) is that $\tilde{d}(\omega)$ is bounded away from 0, which is guaranteed if

$$d_0 > 0, \quad \sum_{s \neq 0} |d_s| < d_0. \tag{4.21}$$

If (4.21) holds, then $f_X(\omega)$ is a bounded function of $\omega \in (-\pi, \pi)^d$, and hence is integrable.

Model (4.18), as it stands, is not identifiable; that is, different sets of coefficients can give rise to the same spectral density and hence the same covariances; see Exercise 4.2. In one dimension, this indeterminacy is usually resolved by restricting attention to "unilateral autoregressions" (UARs), often with only a finite number of nonzero coefficients, for which $d_s = 0$ for $s < 0$ and for which $\{\varepsilon_t\}$ can be interpreted as the innovation process of $\{X_t\}$. This unilateral approach can be generalized to higher dimensions, though in a somewhat artificial manner; see Section 4.8.

For the rest of this section, we pursue an alternative approach based on a symmetry condition, which seems more natural in a spatial context.

Definition 4.5.1 Suppose the coefficients $\{d_s : s \in \mathbb{Z}^d\}$, with $d_0 > 0$, satisfy the symmetry condition

$$d_s = d_{-s}, \quad \text{for all } s \tag{4.22}$$

and suppose $f_X(\omega)$ in (4.19) is an integrable function over $\omega \in (-\pi, \pi)^d$. Then the stationary random field defined by (4.18) is said to be a simultaneous autoregression (SAR).

Unless otherwise stated, attention in this book will be limited to *finite* SARs, for which only a finite number of coefficients are nonzero. Let $\mathcal{N} \subset \mathbb{Z}^d$ be finite symmetric neighborhood of the origin, as in Section 4.4, with augmented neighborhood $\mathcal{N}_0 = \mathcal{N} \cup \{0\}$. Then for a finite SAR defined with respect to $\mathcal{N}$, it is assumed that $d_s = 0, s \notin \mathcal{N}_0$.

The symmetry condition implies the function

$$\tilde{d}(\omega) = \sum_{s\in\mathcal{N}_0} d_s \exp(is^T\omega) = d_0 + \sum_{s\in\mathcal{N}} d_s \cos(\omega^T s) \tag{4.23}$$

is real-valued. In the presence of (4.21), $\tilde{d}(\omega)$ is positive for all ω, and hence, can be determined from (4.19). To parameterize the scale of the random field, we can either (i) set $d_0 = 1$ and include σ_ε^2 as a parameter or (ii) set $\sigma_\varepsilon^2 = 1$ and include $d_0 > 0$ as a parameter.

The classes of models generated by the finite UARs and the finite SARs are not the same, though they do have a nonempty intersection. In general, the SAR models seem more appropriate than UARs for a spatial setting.

Example 4.1 ***(First-order basic SAR model).***
For the first-order basic neighborhood $\mathcal{N} = \mathcal{N}^{(1)}$ in Section 4.4, define coefficients d_s by

$$d_0 = 1, \quad d_s = -\beta, \quad s \in \mathcal{N}. \tag{4.24}$$

Provided $|\beta| < 1/(2d)$, $\tilde{d}(\omega) > 0$ for all ω in (4.23). The corresponding SAR model is called the *first-order basic SAR* model with spectral density

$$f_{\text{SAR-1}}(\omega) = \frac{\sigma_\varepsilon^2/(2\pi)^d}{\tilde{d}(\omega)^2} = \frac{\sigma_\varepsilon^2/(2\pi)^d}{\{1 - 2\beta\sum_{\ell=1}^d \cos h[\ell]\}^2}. \tag{4.25}$$

□

4.5.2 Continuously Indexed Random Fields

Autoregression modeling can also be extended to continuously indexed random fields, provided we take limits in a suitable way. Let $\{X(t) : t \in \mathbb{R}^d\}$ be stationary and define

$$Y(t) = \sum d_s X(t+s), \tag{4.26}$$

where s ranges over some finite collection of sites $s \in \mathbb{R}^d$. In particular, the components of s do not now need to be integer-valued. If $\{X(t)\}$ has a spectral density $f_X(\omega)$, then $\{Y(t)\}$ has spectral density

$$f_Y(\omega) = \left|\sum d_s \exp(i\omega^T s)\right|^2 f_X(\omega). \tag{4.27}$$

In particular, suppose $Y(t) = [X(t + \zeta e_1) - X(t)]/\zeta$ where $e_1 = (1,0,\ldots,0)^T$ is a unit vector along the $t[1]$–axis, and $\zeta > 0$. Letting $\zeta \to 0$ yields

$$Y(t) = \partial X(t)/\partial t[1], \quad t \in \mathbb{R}^d,$$
$$f_Y(\omega) = \omega[1]^2 f_X(\omega).$$

Of course, $\{Y(t)\}$ will exist only in a generalized function sense if $f_Y(\omega)$ is not integrable.

This relationship can be extended to general linear differential operators of the form

$$P = \sum a_m (\partial/\partial t)^m,$$

where each $m = (m[1],\ldots,m[d])$ in the (finite) sum is a multi-index and $(\partial/\partial t)^m = (\partial/\partial t[1])^{m[1]} \cdots (\partial/\partial t[d])^{m[d]}$. If $Y(t) = (PX)(t)$ then

$$f_Y(\omega) = \left|\sum (i\omega)^m a_m\right|^2 f_X(\omega).$$

In particular, if $P = \Delta = \sum \partial^2/\partial\omega[\ell]^2$ denotes the Laplacian operator, then

$$f_Y(\omega) = |\omega|^4 f_X(\omega).$$

Example 4.2 Two simple continuous SAR-type models can be constructed with the Laplacian operator. If

$$\Delta^k X(t) = \varepsilon(t),$$

for some $k \geq 1$, where $\varepsilon(t)$ is generalized white noise with constant spectral density $f_\varepsilon(\omega) = \sigma_\varepsilon^2/(2\pi)^d$, then $\{X(t)\}$ has spectral density

$$f_X(\omega) = |\omega|^{-4k}\sigma_\varepsilon^2/(2\pi)^d.$$

This spectral density is never integrable over all of $\mathbb{R}^d$, and hence never corresponds to a stationary random field. However, it is integrable as $|\omega| \to \infty$ provided $d - 1 - 4k < -1$, that is, $k > \frac{1}{4}d$. In particular, for $k = 1$, this spectral density defines an ordinary intrinsic random field in dimensions $d = 1,2,3$ of intrinsic orders 1,1,0, respectively. Links to kriging are discussed in Section 7.14.

Another random field can be defined by

$$(\kappa^2 - \Delta)^k X(t) = \varepsilon(t), \tag{4.28}$$

where $\kappa > 0$ is a scalar. In this case, the spectral density is given by

$$f_X(\omega) = (\kappa^2 + |\omega|^2)^{-2k}\sigma_\varepsilon^2/(2\pi)^d, \tag{4.29}$$

which is stationary provided $k > \frac{1}{4}d$, and corresponds to the Matérn spectral density (Tables 2.1 and 2.2). □

Discretized versions of this model (4.28) have been developed by Lindgren et al. (2011). See Section 8.8.

Example 4.3 There are also extensions to spatial-temporal models. Let $t \in \mathbb{R}^d$ denote a spatial site and let $u \in \mathbb{R}$ represent "time." A temporal-spatial model in $\mathbb{R}^{d+1}$ can be defined by

$$\partial X(t,u)/\partial u = \Delta X(t,u) - \kappa^2 X(t,u) + \varepsilon(t,u), \tag{4.30}$$

where $\Delta = \sum \partial^2/\partial t[\ell]^2$ is the Laplacian and $\varepsilon(t,u)$ is generalized white noise on $\mathbb{R}^{d+1}$. This process is called a diffusion-injection model (Whittle, 1962, 1986). The $\varepsilon(t,u)$ term represents random noise introduced into the system, the $\Delta X(t,u)$ term represents a smoothing effect through space, and the $-\kappa^2 X(t,u)$ term represents a damping effect.

From the above discussion, we see that $X(t,u)$ has spectral density

$$\begin{aligned} f(\omega,\lambda) &= \sigma_\varepsilon^2 \; |i\lambda + \kappa^2 + \sum_{\ell=1}^{d} \omega[\ell]^2 \; |^{-2}/(2\pi)^{d+1} \\ &= \sigma_\varepsilon^2 \; [(\kappa^2+\rho^2)^2 + \lambda^2]^{-1}/(2\pi)^{d+1}, \end{aligned}$$

where $\omega \in \mathbb{R}^d$, $\lambda \in \mathbb{R}$ and we set $\rho^2 = \sum \omega[\ell]^2 = |\omega|^2$. If $\{X(t,u)\}$ forms a stationary random field in space and time, then integrating $f(\omega,\lambda)$ with respect to λ gives the spectral density in $\mathbb{R}^d$ for a spatial cross section of the random field at a fixed time

$$f(\omega) = \frac{1}{2}\sigma_\varepsilon^2(\kappa^2+\rho^2)^{-1}/(2\pi)^d,$$

which is again a Matérn spectral density. □

4.6 Conditional Autoregressions (CARs)

CARs are defined only for lattice-indexed random fields. We start with the construction of stationary CARs on $\mathbb{Z}^d$ and then generalize the construction to intrinsic CARs and to CARs on the torus. Much of this material is based on a series of papers by Besag (1972, 1974, 1975, 1981), Besag and Moran (1975), Besag and Kooperberg (1995), and Besag and Mondal (2005). See also the discussion about the Hammersley–Clifford Theorem below (Theorem 4.9.3).

It is also possible to extend CARs in ways that will not be studied in detail here. For example, CARs can be constructed on irregularly spaced "sites" where the sites might represent, e.g. geographic regions in the plane; see Section 8.8 for an overview of related ideas. In addition, there is a theory of multivariate CARs (Mardia, 1988); see Section 8.5 for some other methods to model multivariate random fields.

4.6.1 Stationary CARs

Let $\mathcal{N}$ denote a finite symmetric neighborhood of the origin in $\mathbb{Z}^d$, and as in Section 4.4, write $\mathcal{N}_0 = \mathcal{N} \cup \{0\}$. In this section, we consider models for a stationary Gaussian random field $\{X_t\}$ of the form

$$E[X_t|X_{\backslash t}] = \mu + \sum_{s \in \mathcal{N}} \beta_s (X_{t+s} - \mu),$$
$$\text{var}[X_t|X_{\backslash t}] = \sigma_\eta^2, \tag{4.31}$$

where $\beta_s = -\beta_s$ and $X_{\backslash t} = \{X_s : s \neq t\}$ is shorthand for the random field at the rest of the sites other than t.

Definition 4.6.1 A stationary random field X_t, $t \in \mathbb{Z}^d$, whose conditional moments satisfy (4.31) is said to be a conditional autoregression (CAR).

Unless otherwise stated, it is always assumed that the neighborhood $\mathcal{N}$ has only a finite number of elements. The following discussion gives conditions on the coefficients $\{\beta_s\}$ to ensure that the process is well defined.

The residual process is defined by

$$\eta_t = X_t - \mu - \sum_{s \in \mathcal{N}} \beta_s (X_{t+s} - \mu). \tag{4.32}$$

Each η_t has mean 0 and variance σ_η^2. However, as shown below, the residuals are correlated. For simplicity, we set $\mu = 0$ to simplify the theoretical development.

Next we derive some equations for the covariance function $\{\sigma_h : h \in \mathbb{Z}^d\}$ for a stationary random field $\{X_t\}$ with mean 0 obeying (4.31). First, from (4.31), we can write

$$X_t = \sum_{s \in \mathcal{N}} \beta_s X_{t+s} + \eta_t, \tag{4.33}$$

where, for each t, the residual random variable η_t has mean 0 and variance σ_η^2, and η_t is independent of X_s for all $s \neq t$. Here, independence is a key property, which follows from properties of conditional expectation. In particular, multiplying (4.33) by η_t and taking the expectation yields

$$E\{X_t \eta_t\} = 0 + E\{\eta_t^2\} = \sigma_\eta^2. \tag{4.34}$$

To get some expressions for the covariance function, multiply (4.33) by X_{t+h}, $h \neq 0$, and take the expectation to give

$$\sigma_h = \sum_{s \in \mathcal{N}} \beta_s \sigma_{h-s}, \quad h \neq 0, \tag{4.35}$$

since $E\{X_{t+h}\eta_t\} = 0$ for $h \neq 0$. Similarly from (4.33) we find

$$\begin{aligned}\sigma_0 &= EX_t^2 \\ &= E\left[\sum_{s\in\mathcal{N}} \beta_s X_{t+s} + \eta_t\right]\left[\sum_{h\in\mathcal{N}} \beta_h X_{t+h} + \eta_t\right] \\ &= \sum_{s,h\in\mathcal{N}} \beta_s\beta_h\sigma_{h-s} + \sigma_\eta^2.\end{aligned}$$

Since $h \neq 0$ for $h \in \mathcal{N}$, we can simplify the sum over $s \in \mathcal{N}$ using (4.35) to give

$$\sigma_0 = \sigma_\eta^2 + \sum_{h\in\mathcal{N}} \beta_h\sigma_h = \sigma_\eta^2 + \sum_{h\in\mathcal{N}} \beta_h\sigma_{-h}. \tag{4.36}$$

Symbolically (4.35) and (4.36) can be combined as

$$\sigma_h = (\beta * \sigma)_h + \sigma_\eta^2\delta_h \tag{4.37}$$

in terms of the Kronecker delta, $\delta_0 = 1$ and $\delta_h = 0$ for $h \neq 0$. Taking inverse Fourier transforms in (4.37) yields a spectral equation

$$f_X(\omega) = \tilde{\beta}(\omega)f_X(\omega) + \sigma_\eta^2/(2\pi)^d,$$

which we can solve to give

$$f_X(\omega) = \frac{\sigma_\eta^2/(2\pi)^d}{1-\tilde{\beta}(\omega)}. \tag{4.38}$$

Here,

$$\tilde{\beta}(\omega) = \sum_{h\in\mathcal{N}} \beta_h \exp(ih^T\omega) = \sum_{h\in\mathcal{N}} \beta_h \cos(h^T\omega) \tag{4.39}$$

is real-valued since $\beta_h = \beta_{-h}$.

Further, the residual random field $\{\eta_t\}$ has spectral density

$$\begin{aligned}f_\eta(\omega) &= [1-\tilde{\beta}(\omega)]^2 f_X(\omega) \\ &= \sigma_\eta^2[1-\tilde{\beta}(\omega)]/(2\pi)^d.\end{aligned}$$

Thus, the residuals are correlated in a CAR model, in contrast to a SAR model, where the residuals are uncorrelated.

A sufficient condition for the model (4.31) to specify a valid stationary random field is that the spectral density (4.38) be bounded, which from (4.39) is true if

$$\tilde{\beta}(\omega) = \sum_{h\in\mathcal{N}} \beta_h \cos(\omega^T h) < 1, \quad \text{for all } \omega \in [-\pi,\pi)^d, \tag{4.40}$$

and which, in turn, is implied by the simpler condition

$$\sum_{h\in\mathcal{N}} |\beta_h| < 1. \tag{4.41}$$

It is interesting to note that every stationary SAR model can be given a CAR representation (Exercise 4.4). However, the converse is not true (Exercise 4.2).

Example 4.4 ***(First-order basic CAR model).***
For the first-order basic neighborhood $\mathcal{N} = \mathcal{N}^{(1)}$ in Section 4.4, define coefficients β_s by

$$\beta_s = \beta, \quad s \in \mathcal{N}. \tag{4.42}$$

Provided $|\beta| < 1/(2d)$, $\tilde{\beta}(\omega) > 0$ for all ω in (4.23). The corresponding CAR model is called the *first-order basic CAR* model with spectral density

$$f_{\text{CAR-1}}(\omega) = \frac{\sigma_\eta^2/(2\pi)^d}{1 - \tilde{\beta}(\omega)} = \frac{\sigma_\eta^2/(2\pi)^d}{1 - 2\beta \sum_{\ell=1}^{d} \cos h[\ell]}. \tag{4.43}$$

Note the difference with the corresponding SAR model. Both involve the same function in the denominator, but in (4.25) the denominator is squared. That is,

$$f_{\text{SAR-1}}(\omega) \propto \{f_{\text{CAR-1}}(\omega)\}^2. \tag{4.44}$$

This result will prove useful in the analysis of iterated models (Section 4.6.2) and in discrete approximations to continuum models (Section 8.8). □

4.6.2 Iterated SARs and CARs

Start with coefficients $\{d_s,\ s \in \mathcal{N}_0\}$ as in (4.22) and Fourier transform $\tilde{d}(\omega)$ as in (4.23). The convolution of the coefficients

$$d_h^{\otimes 2} = \sum_{s \in \mathcal{N}} d_{h-s} d_s, \quad h \in \mathcal{N}^{(2)}$$

has nonzero entries for h in the second-order neighborhood $\mathcal{N}^{(2)}$ defined in Section 4.4, and has the Fourier transform $\{\tilde{d}(\omega)\}^2$.

Extending this idea to a k-fold convolution, $k > 1$, it is possible to define k-fold SAR and CAR models with spectral densities

$$f_{\text{SAR-}k}(\omega) \propto \{\tilde{d}(\omega)\}^{-2k}, \quad f_{\text{CAR-}k}(\omega) \propto \{\tilde{d}(\omega)\}^{-k}. \tag{4.45}$$

In passing note that a SAR model of order k can be viewed as an example of a CAR model of order $2k$.

Example 4.5 Let $\mathcal{N}$ denote the first-order basic neighborhood in Section 4.4 and consider the coefficients $\{d_h\}$ defined in (4.24). In $d = 2$ dimensions, the coefficients $\{d_h\}$ can be represented graphically by

$$\begin{array}{ccc} & -\beta & \\ -\beta & 1 & -\beta \\ & -\beta & \end{array}.$$

Then, $d^{\otimes 2}$ can be represented graphically by

$$\begin{array}{ccccc} & & \beta^2 & & \\ & 2\beta^2 & -2\beta & 2\beta^2 & \\ \beta^2 & -2\beta & 1+4\beta^2 & -2\beta & \beta^2. \\ & 2\beta^2 & -2\beta & 2\beta^2 & \\ & & \beta^2 & & \end{array}$$

More generally, the coefficients $d^{\otimes k}$ can be used to define a kth-order basic CAR model with spectral density $f_{\text{CAR-}k}(\omega) \propto \{\tilde{d}(\omega)\}^{-k}$. □

4.6.3 Intrinsic CARs

The CAR model (4.31) can also be used to construct intrinsic random fields (Künsch, 1987). In the simplest case, suppose that the coefficients β_s satisfy

$$\beta_s > 0, \quad \sum_{s\in\mathcal{N}} \beta_s = 1. \tag{4.46}$$

Then

$$\begin{aligned} 1 - \tilde{\beta}(\omega) &= 1 - \sum_{s\in\mathcal{N}} \beta_s \cos(h^T\omega) \\ &= \frac{1}{2}\sum \beta_h (h^T\omega)^2 + O(|\omega|^4) \text{ as } \omega \to 0 \\ &= \omega^T A\omega + O(|\omega|^4), \text{ say,} \end{aligned}$$

where $A = \frac{1}{2}\sum_{h\in\mathcal{N}} \beta_h hh^T$. Provided the $d \times d$ matrix A is positive definite, with smallest eigenvalue λ_0 and largest eigenvalue λ_1, say, with $0 < \lambda_0 \le \lambda_1$, the spectral density can be bounded by

$$c_0|\omega|^{-2} \le f(\omega) \le c_1|\omega|^{-2} \qquad \text{as } \omega \to 0,$$

where $c_0 = \sigma_\eta^2/[(2\pi)^d\lambda_1]$ and $c_1 = \sigma_\eta^2/[(2\pi)^d\lambda_0]$. Further, the singularity at $\omega = 0$ is the only singularity of $f(\omega)$ for $\omega \in (-\pi,\pi)^d$. The integral of $f(\omega)$ about $\omega = 0$ behaves like $\int_0^1 \rho^{-2}\rho^{d-1}d\rho$, which is infinite in dimensions $d = 1,2$ and finite for $d \ge 3$. Hence, $f(\omega)$ defines an intrinsic random field of order 0 in dimensions $d = 1,2$ and defines a stationary random field in dimensions $d \ge 3$.

In particular, in dimensions $d \ge 3$, the singularity at $\omega = 0$ is integrable so that the resulting random field is stationary, but with long-range correlation. Further, in view of the identity

$$X_t - \mu = \sum_{s\in\mathcal{N}} \beta_s(X_{t+s} - \mu) + \eta_t \tag{4.47}$$

for all values of $\mu \in \mathbb{R}$ when (4.46) holds, we see that for $d \ge 3$, the CAR representation is compatible with any mean value of the random field. Therefore, the mean μ of the random field needs to be specified separately in this case. Thus, for $d \geq 3$, we shall describe this process as a "quasi-intrinsic CAR" or a QICAR (Kent and Mardia, 1988).

The theoretical details behind CAR models are explored in detail in Georgii (1988, Chapter 13). In particular, he notes that even in the simplest case with $1 - \tilde{\beta}(\omega)$ bounded away from 0, the conditional normal distributions given by (4.31) do not completely specify the means in the joint distribution; it is necessary to impose a stationarity condition as well. See Exercise 4.6.

4.6.4 CARs on a Lattice Torus

CAR models can be developed quite easily on the lattice torus

$$T = \{t \in \mathbb{Z}^d : 0 \leq t[\ell] \leq m[\ell] - 1, \ \ell = 1, \ldots, d\} \tag{4.48}$$

of size $|M| = m[1] \times \cdots \times m[d]$, where $M = (m[1], \ldots, m[d])$. Thus, T is a rectangular region with opposite sides treated as adjacent. The construction is most straightforward when the neighborhood $\mathcal{N}$ is small enough that for all $s, t \in \mathcal{N}$,

$$|s[\ell] - t[\ell]| < m[\ell]/2, \quad \ell = 1, \ldots, d.$$

A CAR model can be constructed using (4.31) provided $t + s$ is interpreted Mod M (see Section 2.12.2). From (4.38), the spectral "density" reduces to

$$f_X(\omega_j) = \frac{\sigma_\eta^2/(2\pi)^d}{1 - \tilde{\beta}(\omega_j)}, \tag{4.49}$$

where $j = (j[1], \ldots, j[d])$ and the vector ω_j has components

$$\omega_j[\ell] = 2\pi j[\ell]/m[\ell], \quad j[\ell] = 0, \ldots, m[\ell] - 1, \quad \ell = 1, \ldots, d.$$

Clearly, the values of (4.49) will be finite for all ω_j if the spectral density (4.38) for the corresponding random field on $\mathbb{Z}^d$ is bounded. The corresponding random field is a circulant version of a stationary random field for which the spectral "density" is a finite set of numbers equal to $(2\pi)^{-d}$ times the eigenvalues of the covariance matrix.

The above construction also makes sense if $\tilde{\beta}(0) = 1$ and $\tilde{\beta}(\omega_j) < 1$ for all other frequencies. In this case, (4.49) defines a first-order intrinsic CAR; the spectral mass at frequency 0 is left undefined. However, it does not make sense to define higher order intrinsic CARs on the lattice torus.

4.6.5 Finite Regions

Let $\{X(t) : t \in \mathbb{R}^d\}$ be a stationary CAR model (4.31) with mean μ. Given a rectangular finite region $D \subset \mathbb{Z}^d$ of size $|M| = m[1] \times \cdots \times m[d]$ (the same as T in (4.48) but without the periodic boundary condition), there are several ways to construct a version of the CAR model on D. Except for the marginalization approach, these methods specify a specific form for the precision matrix Ψ and then define the covariance matrix by inversion, $\Sigma = \Psi^{-1}$.

(i) *Marginalization.* Take the marginal distribution of $\{X_t : t \in D\}$. This approach specifies the covariance matrix Σ as a Toeplitz matrix based on the covariance function σ_h (see Section A.10). However, in practice this approach is computationally cumbersome because Σ will not be sparse and, in general, the covariance function can only be calculated numerically using the Fourier inversion formula.

(ii) *Conditioning on values outside D.* Starting with the stationary random field $\{X_t : t \in \mathbb{R}^d\}$, condition on $X_t = \mu,\ t \notin D$, to give a (nonstationary) random field in R. The inverse covariance matrix Ψ is Toeplitz and for large M it is sparse (see Section A.11). This approach is statistically tractable, but is somewhat unrealistic because the random field is artificially constrained to be close to μ near the boundary.

(iii) *Conditioning on values on the boundary of D.* Split D into two disjoint parts, the "interior" int(D) and the "boundary" $\partial(D)$, where $\partial(D)$ consists of a band of values near the edge of D large enough such that

$$\bigcup_{t \in \text{int}(D)} (t + \mathcal{N}) \subset D.$$

In this case, the conditional distribution of $\{X_t : t \in \text{int}(D)\}$ given $\{X_t : t \notin \text{int}(D)\}$ depends just on the observed values in $\partial(D)$. Hence, there is no need to condition on unobserved values of the process.

(iv) *Periodic boundary conditions.* Regard $D = \mathbb{Z}_M^d$ as the lattice torus (Eq. (A.4)) and use the construction in Section 4.6.4 to produce a stationary random field on the torus. The inverse covariance matrix Ψ is block circulant (Section A.7.3). The block circulant property means that this approach is computationally tractable, but the imposition of periodic boundary conditions is usually unrealistic in practice.

(v) *Reflection boundary conditions.* Reflect the data about each of the edges to get a region $D^{(2)}$ of size $|2M| = 2^d|M| = 2m[1] \times \cdots \times 2m[d]$, and impose a CAR model with circulant boundary conditions on the enlarged data set. The precision matrix $\Psi^{(2)}$ is block circulant and so can be easily inverted to give a block circulant covariance matrix $\Sigma^{(2)}$. Restricting $\Sigma^{(2)}$ to the original sites in D is called the folded circulant approximation. Further details are given in Section A.11.4.

(vi) *Extensions.* If the spectral density blows up at the origin, $f(\omega) \sim c|\omega|^{-2}$ as $\omega \to 0$, then the random field $\{X(t) : t \in \mathbb{R}^d\}$ will be either intrinsic of order 0 (in dimensions $d = 1,2$) or quasi-intrinsic (dimensions $d \geq 3$). The above constructions can be extended in a natural way to this setting. For example,

in the marginalization approach (i), if the original random field is intrinsic of order $k \geq 0$, then the marginal random field on D is also intrinsic; hence it is defined on D only up to an arbitrary polynomial in t of degree k.
For the conditioning approaches (ii) and (iii), the singularity at the origin does not matter. These approaches give a nonsingular conditional distribution for $t \in D$ or $t \in \text{int}(D)$, respectively. Note that for approach (ii), this conditional distribution depends on μ, even though in the intrinsic case the choice of μ is arbitrary. Hence, this approach is not very satisfactory in the intrinsic case.
For approaches (iv) and (v) based on block circulant matrices, the resulting random fields can be viewed as determined up to a constant term.

Rue and Held (2005) discuss the numerical issues related to the inversion of Ψ. In general, the inversion of an $|M| \times |M|$ matrix involves $O(|M|^3)$ numerical calculations. However, if Ψ is sparse, then the time can be dramatically reduced. Further, methods (iv) and (v) involve the inversion of a circulant matrix so that the fast Fourier transform can be used to reduce the time to $O(|M| \log |M|)$.

4.7 Limits of CAR Models Under Fine Lattice Spacing

If the lattice spacing gets small in a CAR model, and the parameters are varied appropriately, then the Matérn process arises in the limit. This section provides details of the calculations.

First note what happens to a general stationary lattice random field X_t, $t \in \mathbb{Z}^d$, with spectral density $f(\omega)$, $\omega \in (-\pi, \pi)^d$, if the lattice is rescaled by a factor δ. Define $X^{(\delta)}(\cdot)$ on the rescaled lattice $\delta\mathbb{Z}^d$ by

$$X^{(\delta)}(\delta t) = X_t, \quad t \in \mathbb{Z}^d.$$

Then $X^{(\delta)}(\cdot)$ has spectral density

$$f_\delta(\omega) = \delta^d f(\delta\omega), \quad \omega \in (-\pi/\delta, \pi/\delta)^d.$$

As δ gets smaller, the random field is located on a more finely spaced lattice and the domain of the spectral density gets larger.

To illustrate the limiting procedure, fix an integer $k > d/2$ and consider the spectral density of the k-fold iterated CAR model

$$f(\omega) = \frac{\tau^2/(2\pi)^d}{\{1 - \tilde{\beta}(\omega)\}^k}, \quad \omega \in (-\pi, \pi)^d,$$

where

$$1 - \tilde{\beta}(\omega) = 1 - 2\beta \sum_{\ell=1}^{d} \cos \omega[\ell], \quad 0 < \beta < 1/(2d) \tag{4.50}$$

is the spectral density for the basic first-order CAR model (Example 4.4).

In order to get a nondegenerate limit, let β and τ^2 vary with the lattice spacing δ. In particular, for fixed values of $\kappa^2 > 0$ and $\tau_0^2 > 0$, let

$$\beta = \beta(\delta) = \frac{1}{2d} - \frac{\delta^2\kappa^2}{(2d)^2}, \quad \tau^2 = \tau^2(\delta) = \delta^{2k-d}(2\pi)^d\tau_0^2/(2d)^k. \tag{4.51}$$

Then $0 < \beta < 1/(2d)$ for small enough δ, and $\beta \to 1/(2d)$ as $\delta \to 0$, where $\beta = 1/(2d)$ is the value of β for the limiting Intrinsic CAR or QICAR model.

Since $\cos\theta = 1 - \frac{1}{2}\theta^2 + O(\theta^4)$ as $\theta \to 0$, it follows that for fixed $\omega \in \mathbb{R}^d$ as $\delta \to 0$,

$$1 - \tilde{\beta}(\delta\omega) = 1 - 2d\beta + \frac{1}{2}2\beta\sum(\delta\omega[\ell]^2) + O(\delta^4) = \frac{\delta^2\kappa^2}{2d} + \frac{\delta^2|\omega|^2}{2d} + O(\delta^4)$$

and so the rescaled spectral density becomes

$$\begin{aligned} f_\delta(\omega) &= \frac{\delta^d\tau^2/(2\pi)^d}{\{1-\tilde{\beta}(\delta\omega)\}^k} I[\omega \in (-\pi/\delta, \pi/\delta)^d] \\ &= \frac{\delta^{2k}\tau_0^2/(2d)^k}{\{\delta^2\kappa^2 + \delta^2|\omega|^2 + O(\delta^4)\}^k/(2d)^k} I[\omega \in (-\pi/\delta, \pi/\delta)^d] \\ &\to \frac{\tau_0^2}{\left\{\kappa^2 + |\omega|^2\right\}^k}, \quad \omega \in \mathbb{R}^d, \end{aligned}$$

the spectral density of the Matérn model of index $\eta = k - d/2$; see Section 2.6.

This argument remains valid if the integer power $k > d/2$ is allowed to be a real power $\gamma > d/2$ in the iterated CAR model. Further, a similar result holds if $\beta = 1/(2d)$ so that $\kappa^2 = 0$. In this case, the CAR model is intrinsic (or quasi-intrinsic if $d \geq 3$) for all δ and the limiting process is a self-similar process of index η; see Section 3.10.

4.8 Unilateral Autoregressions for Lattice Random Fields

4.8.1 Half-spaces in $\mathbb{Z}^d$

In time series, an important representation of a stationary process is given by a UAR, in which the present value of the process is decomposed as a linear combination of "past" values of the process, plus an "innovation." In higher dimensions, there is no completely natural definition of "past," but using a somewhat artificial construction, a complete generalization of the one-dimensional theory can be obtained.

Definition 4.8.1 A subset $\mathcal{H} \subset \mathbb{Z}^d$ is called a "half-space" if the following conditions hold:

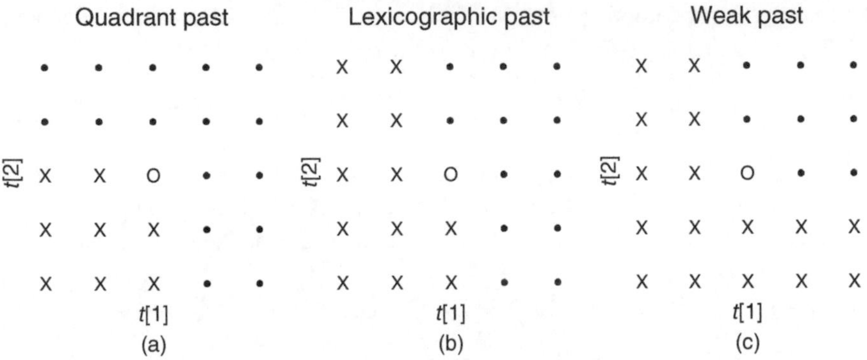

Figure 4.2 Three notions of "past" of the origin in $\mathbb{Z}^2$: (a) quadrant past ($-\mathcal{Q}$), (b) lexicographic past ($-\mathcal{L}$), and (c) weak past ($-\mathcal{B}$). In each plot, ○ denotes the origin, × denotes a site in the past, and • denotes a site in the future.

(a) for $s \neq 0$, $s \in \mathcal{H}$ if and only if $-s \notin \mathcal{H}$ (anti-symmetry)
(b) $0 \notin \mathcal{H}$
(c) if $s_1, s_2 \in \mathcal{H}$, then $s_1 + s_2 \in \mathcal{H}$ (closure under addition).

Thus, $\mathcal{H} \cup (-\mathcal{H}) \cup \{0\} = \mathbb{Z}^d$ and $\mathcal{H} \cap (-\mathcal{H}) = \emptyset$. Write $\mathcal{H}_0 = \mathcal{H} \cup \{0\}$. The "past" of a site t can be defined as the collection of sites $t - \mathcal{H} = \{t - s \in \mathbb{Z}^d : s \in \mathcal{H}\}$. Of course, the notion of "past" here depends on the choice of $\mathcal{H}$.

An important choice of $\mathcal{H}$ is the *lexicographic half-space*

$$\mathcal{L} = \{s \in \mathbb{Z}^d : s \neq 0 \text{ and the first nonzero component } s[\ell] \text{ of } s \text{ is positive}\}. \tag{4.52}$$

In $d = 1$ dimension, $\mathcal{L}$ reduces to positive integers. In $d = 2$ dimensions,

$$\mathcal{L} = \{s \in \mathbb{Z}^2 : s[1] > 0, \text{or } s[1] = 0 \text{ and } s[2] > 0\}. \tag{4.53}$$

See Figure 4.2(b) for an illustration of $-\mathcal{L}$. The half-space $\mathcal{L}$ gives rise to *lexicographic order* on $\mathbb{Z}^d$, which is the same as the order used for words in a dictionary.

4.8.2 Unilateral Models

Next, let $\{X_t\}$ be a zero-mean Gaussian stationary random field on $\mathbb{Z}^d$. Given a choice of half-space $\mathcal{H}$, define the "innovation random field" $\{\varepsilon_t\}$ by

$$\varepsilon_t = X_t - E\{X_t | X_{t-s} : s \in \mathcal{H}\}.$$

It is easy to check that $\{\varepsilon_t\}$ is a zero-mean process, $E(\varepsilon_t) = 0$, and has the following properties:

(a) ε_t depends on $\{X_{t-s} : s \in \mathcal{H}_0\}$.

(b) $E\{\varepsilon_s \varepsilon_t\} = 0$ for $s \neq t$, so that $\{\varepsilon_t\}$ is a white noise stationary random field.
(c) $E\{\varepsilon_t X_u\} = 0$ for $t - u \in \mathcal{H}$, so that the innovation ε_t at site t is independent of the past of the random field $\{X_t\}$.
(d) $E\{\varepsilon_t X_t\} = \text{var}\{\varepsilon_t\}$.

Indeed, properties (a)–(d) characterize the innovation random field. Intuitively, ε_t represents the "new" information in X_t not contained in the "past" $\{X_{t-s} : s \in \mathcal{H}\}$.

Theorem 4.8.1 *Let $\{X_t\}$ be a Gaussian stationary random field on $\mathbb{Z}^d$ with spectral density $f(\omega)$, $\omega \in (-\pi, \pi)^d$. Given a half-space $\mathcal{H}$, set $\mathcal{H}_0 = \mathcal{H} \cup \{0\}$, and let $\{\varepsilon_t\}$ denote the innovation random field. Then, the following properties hold.*

(a) The variance of ε_t, σ_ε^2, say, is given by

$$\sigma_\varepsilon^2 = (2\pi)^d \exp\left\{ (2\pi)^{-d} \int_{(-\pi,\pi)^d} \log f(\omega)\, d\omega \right\}. \tag{4.54}$$

(b) If $\sigma_\varepsilon^2 > 0$, then $\{X_t\}$ has a unilateral infinite moving average representation

$$X_t = \sum_{s \in \mathcal{H}_0} b_s\, \varepsilon_{t-s}, \quad t \in \mathbb{Z}^d, \tag{4.55}$$

where $b_0 = 1$, $\sum b_s^2 < \infty$.

(c) If $\int f^{-1}(\omega)\, d\omega < \infty$, then $\sigma_\varepsilon^2 > 0$ and there is also a unilateral autoregression representation

$$\varepsilon_t = \sum_{s \in \mathcal{H}_0} a_s\, X_{t-s}, \quad t \in \mathbb{Z}^d, \tag{4.56}$$

where $a_0 = 1$, $\sum a_s^2 < \infty$.

Proof: A complete proof is beyond the scope of this book; see Helson and Lowdenslager (1958) for more details. For our purposes, we make the simplifying assumption that $\log f(\omega)$ has an absolutely convergent Fourier expansion

$$\log f(\omega) = \sum_{s \in \mathbb{Z}^d} d_s \exp\{i s^T \omega\} \tag{4.57}$$

with $\sum |d_s| < \infty$. Note that $d_s = d_{-s}$ since $f(\omega) = f(-\omega)$. This simplifying assumption implies $\log f(\omega)$ is a bounded continuous function; hence so are $f(\omega)$ and $1/f(\omega)$. In particular, $\int f^{-1} < \infty$ and $\int \log f > -\infty$. Also note that by the Fourier inversion formula,

$$d_0 = (2\pi)^{-d} \int_{(-\pi,\pi)^d} \log f(\omega)\, d\omega. \tag{4.58}$$

Although it is often straightforward in principle to establish the existence of an expansion (4.57), it may be very difficult in dimensions $d \geq 2$ to explicitly calculate the coefficients.

Define functions with one-sided Fourier expansions by

$$g_1(\omega) = \sum_{s\in\mathcal{H}} d_s \exp\{is^T\omega\}, \quad g_2(\omega) = \sum_{s\in-\mathcal{H}} d_s \exp\{is^T\omega\}, \tag{4.59}$$

so that $\log f(\omega) = d_0 + g_1(\omega) + g_2(\omega)$ and $g_2(\omega) = \overline{g_1(\omega)}$ is the complex conjugate of $g_1(\omega)$. Setting $f_+(\omega) = \exp\{g_1(\omega)\}$, $f_-(\omega) = \exp\{-g_1(\omega)\} = 1/f_+(\omega)$, we see that

$$f(\omega) = e^{d_0}|f_+(\omega)|^2 = e^{d_0}|f_-(\omega)|^{-2}. \tag{4.60}$$

Because $\sum |d_s| < \infty$ and $0 \notin \mathcal{H}$, we can evaluate the power series expansion for $\exp\{\pm g_1(\omega)\}$ to give one-sided Fourier expansions for $f_\pm(\omega)$

$$f_+(\omega) = \sum_{s\in\mathcal{H}_0} b_s \exp\{is^T\omega\}, \quad f_-(\omega) = \sum_{s\in\mathcal{H}_0} a_s \exp\{is^T\omega\}, \text{say}, \tag{4.61}$$

where the coefficients satisfy $a_0 = b_0 = 1$, $\sum |a_s| < \infty$, $\sum |b_s| < \infty$.

Define a random field $\{Y_t\}$ by

$$Y_t = \sum_{s\in\mathcal{H}_0} a_s X_{t-s},$$

with spectral density $f_Y(\omega) = f(\omega)|f_-(\omega)|^2 = e^{d_0}$, using (4.60). Hence, $\{Y_t\}$ is a white noise process with variance $(2\pi)^d e^{d_0}$. Since $f_+(\omega)f_-(\omega) \equiv 1$, it follows that

$$X_t = \sum_{s\in\mathcal{H}_0} b_s Y_{t-s}.$$

Hence, the conclusion of the theorem follows for the white noise process $\{Y_t\}$.

Our last task is to identify $\{Y_t\}$ with the innovation process $\{\varepsilon_t\}$. This is straightforward. Due to the invertibility of the relationship between $\{Y_t\}$and $\{X_t\}$, we have

$$\begin{aligned}
\varepsilon_t &= X_t - E[X_t | X_{t-s} : s \in \mathcal{H}] \\
&= X_t - E[X_t | Y_s : s \in \mathcal{H}] \\
&= X_t - E\left[\sum_{s\in\mathcal{H}_0} b_s Y_{t-s} | Y_{s'} : s' \in \mathcal{H}\right] \\
&= X_t - \sum_{s\in\mathcal{H}} b_s Y_{t-s} = Y_t,
\end{aligned}$$

since for the white noise $\{Y_t\}$, $E[Y_t | Y_{t-s} : s \in \mathcal{H}] = 0$. Thus, the theorem is proved. □

From the UAR representation (4.56), we obtain

$$E[X_t | X_{t-s} : s \in \mathcal{H}] = -\sum_{s\in\mathcal{H}} a_s X_{t-s}. \tag{4.62}$$

Indeed, (4.62) is often used as a starting point to define a stationary random field. However, a specified set of square-integrable coefficients $\{a_s,\ s \in \mathcal{H}_0\}$ will define a valid UAR representation, provided

$$\left[\sum_{s\in\mathcal{H}_0} a_s \exp\{i\omega^T s\}\right]^{-1} = [f_-(\omega)]^{-1}, \tag{4.63}$$

is square-integrable; see condition (c) in Theorem 4.8.1. A simple sufficient condition to ensure (4.63) is

$$\sum_{s\in\mathcal{H}} |a_s| < a_0 = 1. \tag{4.64}$$

Of course, it should be remembered that the values of the coefficients $\{a_s\}$ depend on the particular half-space used.

4.8.3 Quadrant Autoregressions

The *upper quadrant* in $\mathbb{Z}^d$, defined by

$$\mathcal{Q} = \{t \in \mathbb{Z}^d : t \neq 0 \text{ and } t[\ell] \geq 0, \quad \ell = 1, \dots, d\}, \tag{4.65}$$

is a subset of $\mathcal{L}$; see Figure 4.2c. Set $\mathcal{Q}_0 = \mathcal{Q} \cup \{0\}$. A UAR with respect to $\mathcal{L}$ is called *quadrant autoregression (QAR)* when the coefficients a_s in (4.62) vanish for $s \notin \mathcal{Q}$; that is

$$E[X_t | X_{t-s},\ s \in \mathcal{L}] = -\sum_{s\in\mathcal{Q}} a_s X_{t-s}. \tag{4.66}$$

In this case, a less restrictive sufficient condition than (4.64) can be given on the coefficients $\{a_s\}$ to ensure a valid UAR representation. This condition generalizes the stability condition usually used in time series.

Define a function of the complex variables $z = (z[\ell],\ \ell = 1, \dots, d)$ by

$$A(z) = \sum_{s\in\mathcal{Q}_0} a_s z^s,\ a_0 = 1. \tag{4.67}$$

If the power series for $A(z)$ converges for $\{z : |z[\ell]| < 1 + \kappa,\ \ell = 1, \dots, d\}$, for some $\kappa > 0$ and if $A(z) \neq 0$ in this region, then the coefficients $\{a_s,\ s \in \mathcal{Q}_0\}$ define a valid UAR representation. Here z^s is shorthand for $z[1]^{s[1]} \cdots z[d]^{s[d]}$. The conditioning property (4.66) can be strengthened to

$$E[X_t | X_{t-s} : s \in \mathcal{B}] = -\sum_{s\in\mathcal{Q}} a_s X_{t-s}, \tag{4.68}$$

where

$$\mathcal{B} = \{s \in \mathbb{Z}^d : s[\ell] \geq 0 \text{ for some } \ell = 1, \dots, d, \text{and } s \neq 0\}$$

(see Exercise 4.8). Here $(-\mathcal{B})$ can be viewed as the "big past" of the origin, in contrast to the quadrant or "little" past $(-\mathcal{Q})$ and the lexicographic past $(-\mathcal{L})$; see

Figure 4.2 for an illustration in $d = 2$ dimensions. Thus, for a QAR, the regression of X_t on either the "big" past or the "lexicographic" past yields terms only in the "quadrant" past.

The class of QAR models is attractive from a computational point of view, but feels unnatural in practice because of the arbitrary notion of "past." Further, this class of models lacks symmetry; replacing $t[\ell]$ by $-t[\ell]$ for a proper subset of the components $\ell = 1, \ldots, d$, changes the class of QAR models, even when an infinite number of coefficients are allowed (Exercise 4.9).

Example 4.6 In practice, QAR models are most useful when the number of terms in (4.68) is finite. Let $\mathcal{M} \subset \mathcal{Q}$ denote a finite one-sided neighborhood of the origin. Perhaps the simplest example in Z^2 is the 3-site neighborhood

$$\mathcal{M} = \{(1,0),\ (0,1),\ (1,1)\},$$

yielding the model

$$E[X_t | X_{t-s},\ s \in \mathcal{B}] = a\, X_{t-(1,0)} + b\, X_{t-(0,1)} + c\, X_{t-(1,1)}.$$

If $c = -ab$, then the model reduces to a separable AR(1) × AR(1) model (Martin, 1979). See Exercise 4.10. □

4.9 Markov Random Fields (MRFs)

4.9.1 The Spatial Markov Property

The models of Sections 4.6–4.8 involve the conditional expectation of the Gaussian stationary random field $\{X_t\}$ at one site given its values at certain other sites. In this section, we extend the principles behind the CAR models of Section 4.6 to non-Gaussian distributions and to allow other discrete index spaces. Indeed, the CAR model in Section 4.6.1 is an example of a Gaussian Markov random field (GMRF).

In Section 4.6, it was simplest to work with random fields on the infinite lattice $\mathbb{Z}^d$ in order to use spectral representations. In contrast, it is simplest here to replace $\mathbb{Z}^d$ by a finite domain T. Recall that a random field $\{X_t : t \in T\}$ is just a collection of random variables defined on T, where the elements of T are called *sites*. In many examples, T will be a subset of $\mathbb{Z}^d$, but this is not necessary.

Let $\mathcal{X}$ denote the state space of each random variable X_t. In previous sections, we used $\mathcal{X} = \mathbb{R}$, the real line. Here, $\mathcal{X}$ will often be discrete. We shall use lowercase letters, e.g. x_t, to denote possible values of X_t. For discrete $\mathcal{X}$, a formula such as $p_t(x_t)$ denotes the probability $P(X_t = x_t)$, whereas for continuous $\mathcal{X}$, $p_t(x_t)$ stands for probability density. Also for convenience we shall suppose that $\mathcal{X}$ contains a state labeled "0". This condition can always be achieved by relabeling the states if necessary.

For each $t \in T$ let $\mathcal{N}(t) \subset T$ denote a "neighborhood" of t (by convention $t \notin \mathcal{N}(t)$). The neighborhood structure is assumed to be *balanced*; that is, s is a neighbor of t if and only if t is a neighbor of s ($s \in \mathcal{N}(t)$ if and only if $t \in \mathcal{N}(s)$ for all $s, t \in T$).

Example 4.7 Let T denote a finite rectangular region in $\mathbb{Z}^d$. Let $\mathcal{N} \subset \mathbb{Z}^d$ be any finite symmetric neighborhood of the origin; that is, (i) $0 \notin \mathcal{N}$ and (ii) $t \in \mathcal{N}$ if and only if $-t \in \mathcal{N}$. Then set

$$\mathcal{N}(t) = (t + \mathcal{N}) \cap T = \{t + s : s \in \mathcal{N} \text{and } t + s \in T\}$$

to be the translate of $\mathcal{N}$ with obvious adjustments near the boundary of T. Common choices for $\mathcal{N}$ were given in Section 4.4. □

Some concise notation will be very useful in this section. Let X_T denote the full vector of values $(X_t,\ t \in T)$. Recall T is assumed finite. More generally, for any nonempty subset $\Lambda \subset T$, let X_Λ denote the subvector of values $(X_t, t \in \Lambda)$, with $|\Lambda|$ equal to the number of sites in Λ. Let $p_\Lambda(x_\Lambda)$ denote the joint probability of X_Λ. Also, define x_T^Λ by

$$x_t^\Lambda = x_t,\ t \in \Lambda; \quad x_t^\Lambda = 0,\ t \notin \Lambda \tag{4.69}$$

for $\Lambda \subset T$. Let 0_T denote the vector, all of whose components are 0.

Throughout this section, we limit our attention to joint probability distributions satisfying the "positivity" condition

$$p_T(x_T) > 0 \qquad \text{for all } x_T \in \mathcal{X}^T. \tag{4.70}$$

Definition 4.9.1 A random field $\{X_t : t \in T\}$ is said to be a *MRF* (with respect to the neighborhood structure $\{\mathcal{N}(t)\}$) if the full conditional probabilities satisfy *spatial Markov property*

$$p_t(x_t | x_{\setminus t}) = p_t(x_t | x_s : s \in \mathcal{N}(t)), \tag{4.71}$$

where $x_{\setminus t} = \{x_s : s \neq t\}$. That is the conditional probability at site t, given the values of the random field at all other sites, depends only on the neighboring sites in $\mathcal{N}(t)$.

In the Gaussian setting, the spatial Markov property was used to define the CAR models in Section 4.6.

The full conditional probabilities play a fundamental role here. The following background result shows that the full conditional probabilities determine the joint probabilities, whether or not the Markov assumption holds.

Theorem 4.9.1 Brook Expansion (Brook, 1964). *Let T be a finite index set of sites, with size $|T| = n,\ n \geq 2$, and let $p_T(x_T)$ be a joint probability function (discrete case) or probability density (continuous case) on a state space χ^T. Assume the positivity condition (4.70). Then the full conditional probabilities*

$$p_t(x_t|x_{\backslash t}), \quad t \in T, \tag{4.72}$$

determine the joint probabilities $p_T(x_T)$.

Proof: Label the sites $t = 1, \ldots, n$ say, in some arbitrary order. Given two vectors x_T and y_T, let

$$A_t = \frac{p_t(x_t|x_1, \ldots, x_{t-1}, y_{t+1}, \ldots, y_n)}{p_t(y_t|x_1, \ldots, x_{t-1}, y_{t+1}, \ldots, y_n)}, \qquad t = 1, \ldots, n.$$

Thus, A_t is a ratio of two full conditional probabilities at site t given the rest of the sites. These conditional probabilities are evaluated at x_t and y_t, respectively, and the common conditioning values are based on the components of x_T for sites with index less than t and the components of y_T for sites with index greater than t. The positivity condition (4.70) is needed to ensure that the conditional probabilities are nonzero.

For example, consider $A_n = p_n(x_n|x_1, \ldots, x_{n-1})/p_n(y_n|x_1, \ldots, x_{n-1})$. Write $p_T(x_T) = p(x_T)$. The definition of conditional probability implies

$$p(x_T) = p_n(x_n|x_1, \ldots, x_{n-1}) p_{T\backslash\{n\}}(x_1, \ldots, x_{n-1})$$
$$p(x_1, \ldots, x_{n-1}, y_n) = p_n(y_n|x_1, \ldots, x_{n-1}) p_{T\backslash\{n\}}(x_1, \ldots, x_{n-1})$$

and combining these equations yields

$$p_T(x_T) = A_n\, p_T(x_1, \ldots, x_{n-1}, y_n).$$

Proceeding by induction it follows that

$$\begin{aligned} p_T(x_T) &= A_n\, p_T(x_1, \ldots, x_{n-1}, y_n) \\ &= A_n A_{n-1}\, p_T(x_1, \ldots, x_{n-2}, y_{n-1}, y_n) \\ &= \cdots = \left(\prod_{t=1}^{n} A_t\right) p_T(y_T). \end{aligned} \tag{4.73}$$

Hence, the ratios of joint probabilities $p_T(x_T)/p_T(y_T)$ depend only on the full conditional probabilities. Thus, the joint probabilities are determined up to a multiplicative constant. Further, this constant is determined since the probabilities must add or integrate to 1. □

In many applications, the probability function $p(x_T)$ is not specified directly. Instead, the starting point is a function $Q(x_T)$, which determines $p(x_T)$ through the equation

$$p_T(x_T) = C^{-1} \exp\{Q(x_T)\}, \tag{4.74}$$

where the normalization constant is given by

$$C = \sum \exp\{Q(x_T)\} \tag{4.75}$$

(or a multivariate integral if $\mathcal{X}$ is a continuous space), and the sum is over all possible values of x_T. In this approach, $Q(\cdot)$ is allowed to be an arbitrary real-valued function such that (4.75) is finite.

For a given probability function $p(x_T)$, the function $Q(\cdot)$ is determined up to an additive constant. In statistical mechanics, $-Q(\cdot)$ is known as the "potential function" (as a function of x_T), so $Q(\cdot)$ can be called the "negative potential function." Further, C is known as the "partition function" (as a function of any parameters present in the distribution of x_T).

As it stands, there is too much flexibility in the choice of the function $Q(x_T)$ to produce tractable statistical models. The specification of useful choices for $Q(x_T)$ will occupy the rest of this section. The first step is to give a general "subset expansion" for $Q(x_T)$. The second step is to introduce the concept of a neighborhood to remove most of the terms in this expansion.

4.9.2 The Subset Expansion of the Negative Potential Function

In order to introduce constraints on the function $Q(x_T)$, the following "subset expansion" is useful. Pick out an arbitrary element of the state space χ and label it by the value 0.

Theorem 4.9.2 (Subset expansion). *Any real-valued finite function $Q(x_T)$ can be written uniquely in a subset expansion of the form*

$$Q(x_T) = \text{const} + \sum_{\Lambda \subset T} G_\Lambda(x_\Lambda), \tag{4.76}$$

where the summation is over all nonempty subsets Λ of T. Each function $G_\Lambda(\cdot)$ is allowed to depend only on x_Λ and is required to satisfy

$$G_\Lambda(x_\Lambda) = 0 \text{ if } x_t = 0, \quad \text{for at least one } t \in \Lambda. \tag{4.77}$$

Proof: Define $G_\Lambda(\cdot)$ by

$$G_\Lambda(x_\Lambda) = \sum_{\Gamma \subset \Lambda} (-1)^{|\Lambda|-|\Gamma|} \, [Q(x_T^\Gamma) - Q(0_T)], \tag{4.78}$$

where the sum is over all nonempty subsets Γ of Λ, and the notation x_T^Γ is defined in (4.69). Clearly, the right-hand side of (4.78) depends just on x_Λ. All that remains is to verify (4.77), (4.76) and uniqueness, which we do in that order.

To verify (4.77), let Λ be a subset of T, let $t \in \Lambda$ be a site in Λ, and suppose $x_t = 0$. If we split Λ into those subsets containing or not containing t, we get

$$G_\Lambda(x_\Lambda) = \sum_{\Delta \subset \Lambda \backslash t} \{(-1)^{|\Lambda|-|\Delta|} [Q(x_T^\Delta) - Q(0_T)] + (-1)^{|\Lambda|-|\Delta \cup \{t\}|} [Q(x_T^{\Delta \cup \{t\}}) - Q(0_T)]\}. \tag{4.79}$$

Since $x_T^\Delta = x_T^{\Delta \cup \{t\}}$ if $x_t = 0$, all the terms in this expression vanish, so $G_\Lambda(x_\Lambda) = 0$.

To verify (4.76) note that

$$\begin{aligned}\sum_{\Lambda \subset T} G_\Lambda(x_\Lambda) &= \sum_{\Lambda \subset T} \sum_{\Gamma \subset \Lambda} (-1)^{|\Lambda|-|\Gamma|} \left[Q(x_T^\Gamma) - Q(0_T)\right] \\ &= \sum_{\Gamma \subset T} \left[Q(x_T^\Gamma) - Q(0_T)\right] \sum_{\Lambda \supset \Gamma} (-1)^{|\Lambda|-|\Gamma|}.\end{aligned} \tag{4.80}$$

For fixed Γ, let $m = |T| - |\Gamma| \geq 0$. Then given a value of k, $0 \leq k \leq m$, there are $\binom{m}{k}$ possible subsets Λ in this summation with $k = |\Lambda| - |\Gamma|$. Thus

$$\begin{aligned}\sum_{\Lambda \supset \Gamma} (-1)^{|\Lambda - \Gamma|} = \sum_{k=0}^{m} (-1)^k \binom{m}{k} &= 0 \text{ if } m > 0 \\ &= 1 \text{ if } m = 0,\end{aligned}$$

since this is the binomial expansion of $(1-1)^m$. Hence, all the terms in (4.80) vanish except for $\Gamma = T$. Thus,

$$\sum_{\Lambda \subset T} G_\Lambda(x_\Lambda) = Q(x_T) - Q(0_T)$$

as required. This argument is a variant of the familiar inclusion–exclusion principle.

Finally, to prove uniqueness we use induction. If $x_T = x_T^{\{t\}}$ is nonzero for at most one site t, then the right-hand side of (4.76) reduces to a constant plus $G_{\{t\}}(x_t)$, a term based on a singleton subset, so that

$$G_{\{t\}}(x_t) = G_{\{t\}}(x_t) - G_{\{t\}}(0) = Q(x_T) - Q(0_T),$$

a special case of (4.78). Similarly, for a doubleton subset $\{s, t\}$, if $x_T = x_T^{\{s,t\}}$, the right-hand side of (4.76) reduces to terms involving the singleton subsets $\{s\}$, $\{t\}$ and the doubleton subset $\{s, t\}$. Hence, we can define $G_{\{s,t\}}(x_{\{s,t\}})$ in terms of known quantities, which again reduces to a special case of (4.78). By successively increasing the size of the subsets by one at each stage, we can eventually define $G_\Lambda(x_\Lambda)$ for all subsets Λ. □

If the state space $\mathcal{X}$ is finite, then a valid joint probability distribution $p_T(x_T)$ is specified whenever the $G_\Lambda(\cdot)$ are arbitrary functions satisfying (4.77). However, if $\mathcal{X}$ is infinite (or is a continuous space), then the G_Λ must be restricted so that the normalizing constant is finite, $C < \infty$ in (4.75).

4.9.3 Characterization of Markov Random Fields in Terms of Cliques

Given a neighborhood structure $\{\mathcal{N}(t)\}$ on T, a nonempty subset $\Lambda \subset T$ is known as a "clique" if all the sites in Λ are neighbors of one another. By convention singleton sets, e.g. $\{t\}$, are also regarded as cliques. Cliques for the first-order basic neighborhood and the first-order full neighborhood in $d = 2$ dimensions are illustrated in Figures 4.3 and 4.4. See Exercise 4.13 for verification.

Equations (4.74)–(4.76) can be rewritten in the form

$$\log\{p_t(x_t|x_{\backslash t})/p_t(0|x_{\backslash t})\} = \sum_{\Lambda\subset T:t\in\Lambda} G_\Lambda(x_\Lambda). \tag{4.81}$$

To verify (4.81), note that the ratio of conditional probabilities is the same as the ratio of joint probabilities since the conditioning sets are the same. After expanding the log ratio of joint probability functions using (4.76), many of the terms either vanish or cancel to get (4.81). For more details, see Exercise 4.14.

Next, we show that only the subsets corresponding to cliques need to be considered for an MRF.

Theorem 4.9.3 (Hammersley–Clifford Theorem; see Besag (1974) and Clifford (1990)). *Consider a random field on a state space χ^T satisfying the positivity constraint (4.70). Let $Q(x_T)$ denote the negative potential function in (4.74) with subset expansion (4.76). Let $\{\mathcal{N}(t)\}$ be a neighborhood structure on T. Then, the random field is a MRF with respect to this neighborhood structure if and only if*

$$G_\Lambda(\cdot) = 0 \text{ whenever } \Lambda \text{ is not a clique.} \tag{4.82}$$

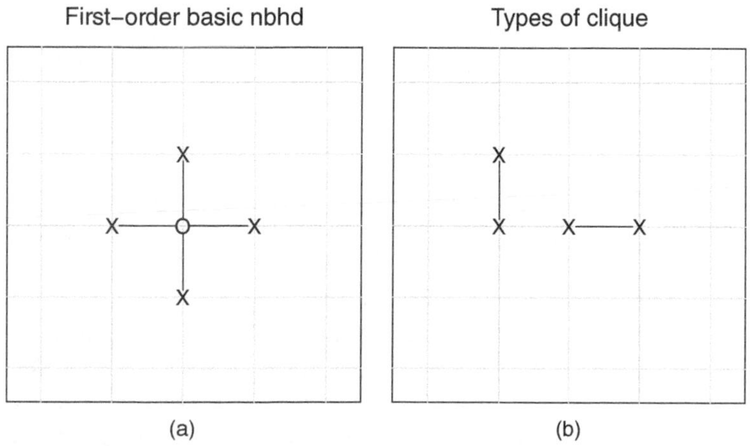

Figure 4.3 (a) First-order basic neighborhood (nbhd) of the origin ○ in $d = 2$ dimensions. Neighbors of the origin are indicated by ×. (b) Two types of clique in addition to singleton cliques: horizontal and vertical edges.

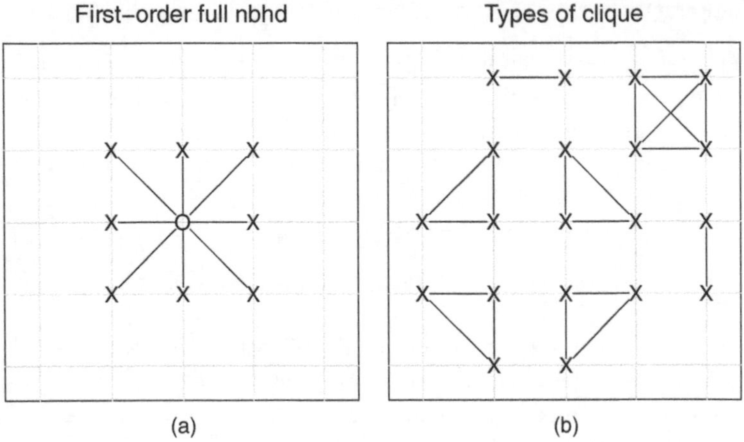

Figure 4.4 (a) First-order full neighborhood (nbhd) of the origin ○ in $d = 2$ dimensions. Neighbors of the origin are indicated by ×. (b) Seven types of clique in addition to singleton cliques: horizontal and vertical edges, four shapes of triangle and a square.

Proof: To verify this theorem, suppose first that (4.82) holds. Let u denote a site that is not a neighbor of t. Then, $u \notin \Lambda$ for any subset Λ appearing on the right-hand side of (4.81) for which $G_\Lambda(x_\Lambda) \neq 0$. Therefore, (4.81), and hence the conditional probability of x_t given $x_{\setminus t}$, does not depend on x_u. Hence, the random field is a MRF.

Conversely, suppose that the distribution of x_T forms a MRF, and consider subsets Λ of T that are not cliques. We shall carry out induction on the size of Λ.

The smallest possible size is $|\Lambda| = 2$. Thus, Λ contains two sites t and u, say, which are not neighbors. By the MRF property, the left-hand side of (4.81) cannot depend on x_u.

Let x_T be a vector for which $x_T = x_T^{\{t,u\}}$, so that all but two of its elements are 0. The right-hand side of (4.81) reduces to two possibly nonzero terms, $G_{\{t\}}(x_t) + G_{\{t,u\}}(x_{\{t,u\}})$. However, since this expression cannot depend on x_u, it follows after setting $x_u = 0$ that $G_{\{t,u\}}(\cdot) \equiv 0$. Thus, the only nonzero term involves the singleton clique $\{t\}$.

Next, consider a nonclique set Λ of size $|\Lambda| = m$, $m \geq 3$. Let t and u be two sites in Λ that are not neighbors, and consider a vector x_T for which $x_T = x_T^\Lambda$. The right-hand side of (4.81) reduces to a sum of terms involving sets Γ for which $\{t\} \subset \Gamma \subset \Lambda$. If $|\Gamma| < m$ and Γ is not a clique, then $G_\Gamma(\cdot) \equiv 0$ by induction. Thus, $G_\Lambda(\cdot)$ is the only remaining term on the right-hand side of (4.81) depending on x_u. However, since the left-hand side of (4.81) does not depend on x_u, neither does the right-hand side, and so, after setting $x_u = 0$, $G_\Lambda(\cdot) \equiv 0$. Thus, the proof is completed. □

Useful subclasses of MRFs can be obtained by limiting the level of interaction in (4.76) under the restrictions (4.82). The most extreme restriction prohibits any interaction; that is,

$$G_\Lambda(\cdot) \equiv 0, \quad \text{for all } |\Lambda| \geq 2.$$

In this case, the values of X_t at different sites are independent.

A more interesting class of models allows pairwise interaction and is studied in Section 4.9.4.

4.9.4 Auto-models

For statistical purposes, Besag (1974) proposed a parametric family of MRF models called "auto-models." In time series, an "autoregression" involves the regression of a random variable at time t on its past values; in the spatial setting, an "auto-model" involves the regression of a random variable at site t on its values at all other sites. In the Gaussian case, auto-models are the same as CAR models, but restricted to a finite set of sites T.

Auto-models are characterized by the properties that (i) the conditional distribution of $X_t|X_{\setminus t}$ forms an exponential family, and (ii) only pairwise interactions are involved, so that

$$G_\Lambda(\cdot) \equiv 0, \quad \text{for all } |\Lambda| \geq 3.$$

Here are some examples.

1. *Auto-logistic model.* This is a model for binary data, $\mathcal{X} = \{0,1\}$. Let

$$Q(x_T) = \sum_{t\in T} \alpha_t\, x_t + \sum_{t\in T}\sum_{s\in\mathcal{N}(t)} \beta_{st}\, x_s\, x_t, \tag{4.83}$$

where $\beta_{st} = \beta_{ts}$ for all $s, t \in T$. Thus, the conditional probability $p_t = P(X_t = 1|X_{\setminus t} = x_{\setminus t})$ is modeled by

$$\text{logit}\, p_t = \log\{p_t/(1-p_t)\} = \alpha_t + \sum_{s\in\mathcal{N}(t)} \beta_{st}\, x_s \tag{4.84}$$

(Exercise 4.15). In practice, we will often assume α_t does not depend on t, and, when $T \subset \mathbb{Z}^d$, that β_{st} depends only on $t - s$.

2. *Auto-binomial model.* Let $\mathcal{X} = \{0, \dots, m\}$ and set

$$Q(x_T) = \sum_t \left[\alpha_t x_t + \log\binom{m}{x_t}\right] + \sum_t \sum_{s\in\mathcal{N}(t)} \beta_{st}\, x_s\, x_t.$$

Thus, X_t given $X_{\setminus t} = x_{\setminus t}$ follows a binomial distribution, $B(m, p_t)$, where p_t is again given by (4.84).

3. *Auto-normal model.* Let $\mathcal{X} = \mathbb{R}$ and set

$$Q(x_T) = -\frac{1}{2\sigma_\eta^2}\left\{\sum_t x_t^2 - 2\sum_t \sum_{s\in\mathcal{N}(t)} \beta_{st}x_s x_t\right\}. \tag{4.85}$$

Use boldface to denote the random vector $\boldsymbol{X}_T$ with a possible realization $\boldsymbol{x}_T$. Then, $\boldsymbol{X}_T$ follows a multivariate normal distribution. If $T \subset \mathbb{Z}^d$ is a rectangular region and β_{st} depends only on $s - t$, say, does not depend on t, then (4.85) is an example of the CAR model on $\mathbb{Z}^d$ of Section 4.6, with x_t conditioned to equal 0 for sites t outside T (and identifying β_{st} here with β_{s-t} in (4.31)). Equation (4.85) can be written in matrix form as

$$Q(x_T) = -\frac{1}{2\sigma_\eta^2}\left\{\boldsymbol{x}_T^T B \boldsymbol{x}_T\right\},$$

where B is a $|T| \times |T|$ matrix of with elements

$$b_{tt} = 1, \quad b_{st} = b_{ts} = -\beta_{st}, \quad s \in \mathcal{N}(t).$$

Note that $\sigma_\eta^2 B^{-1}$ is the covariance matrix of $\boldsymbol{X}_T$ and that the parameters β_{st} must be chosen so that B is strictly positive definite. Sufficient conditions for strict positive definiteness were discussed in Section 4.6. Mardia (1988) has given an extension to the multivariate case. An auto-normal model is also called a GMRF and the multivariate version is termed the multivariate Gaussian Markov random field (MGMRF); for a broader review, we refer to Rue and Held (2010).

Also, note that $\sigma_\eta^{-2}B$, the inverse covariance matrix of $\boldsymbol{X}_T$, also known as the precision matrix, is sparse: $b_{st} = 0$ unless $s = t$ or s and t are neighbors. This property can be contrasted with a finite range covariance scheme for which the covariance matrix itself will be sparse.

Lastly, we comment briefly on the construction of MRFs on an infinite domain, e.g. $T = \mathbb{Z}^d$. In this case, when the interaction parameters are sufficiently large it sometimes happens that the conditional probabilities (4.71) do not provide enough information to completely specify the joint distribution of the random field. A famous example is the Ising model. This is an auto-logistic model on $\mathbb{Z}^d$ based on a first-order neighborhood. In this case, for $d \geq 2$, there is more than joint distribution for the random field on $\mathbb{Z}^d$, which is compatible with the conditional distributions. See, e.g., Georgii (1988, Section 6.2) for more details.

This situation is somewhat analogous to the attempt to construct an intrinsic CAR or QICAR on all of $\mathbb{Z}^d$ as in Section 4.6 with coefficients $a_s > 0$ adding to 1; see (4.47). When $d = 1$ or 2, there is a unique intrinsic Gaussian random field with the required conditional probabilities. However, when $d \geq 3$, there exist many Gaussian stationary random fields with the required conditional distributions (and with long-range dependence). The mean of the random field is not determined by the conditional probabilities in this case and needs to be specified separately.

4.10 Markov Mesh Models

The QAR Gaussian models of Section 4.8 allow autoregressive dependence on the "quadrant" past $t - \mathcal{Q}$ of a site t. In this section, we extend these models to a non-Gaussian setting, but limit our attention to a finite domain $T = \{t \in \mathbb{Z}^d : 1 \le t[\ell] \le m[\ell],\ \ell = 1, \dots, d\}$ of size $|T| = m[1] \cdots m[d]$. The resulting models are known as *Markov mesh* models.

Given a site $t \in T$, let $\mathcal{Q}_t = (t - \mathcal{Q}) \cap T$ denote the portion of the quadrant past lying in T. Markov mesh models are built from a set of (discrete or continuous) conditional probabilities densities

$$\{p_t(x_t | x_{\mathcal{Q}_t}),\ t \in T\}.$$

Note the notation differs from Section 4.9. Here, $p_t(\cdot|\cdot)$ denotes the conditional density at site t given the quadrant past; there $p_t(\cdot|\cdot)$ denoted the conditional density at site t given the values of the random field at all other sites.

Example 4.8 Let $d = 2$, let $\mathcal{X} = \{0,1\}$, and set

$$p_t(x_t = 1 | x_{\mathcal{Q}_t}) = e^a/(e^a + 1),$$

where a depends on the three values of the process to the left of t, below t, and diagonally to the lower left of t as follows:

$$\begin{aligned}
&a = \alpha(x_{t-(1,0)} + x_{t-(0,1)} + x_{t-(1,1)}), \quad \text{if } t[1] > 1,\ t[2] > 1,\\
&a = \beta x_{t-(1,0)}, \quad \text{if } t[1] > 1,\ t[2] = 1,\\
&a = \beta x_{t-(0,1)}, \quad \text{if } t[1] = 1,\ t[2] > 1,\\
&a = \gamma, \quad \text{if } t[1] = 1,\ t[2] = 1.
\end{aligned}$$

The last three lines give a modified definition of a when t lies on the left or lower boundary of T. Thus, the model contains three parameters that may be taken as equal or different. □

Once these conditional probabilities have been specified, they can be combined together to define a joint probability on T by

$$p(x_T) = \prod_{t \in T} p_t(x_t | x_{\mathcal{Q}_t}) \tag{4.86}$$

4.10.1 Validity

To confirm that (4.86) is a valid density, it is necessary to check that it sums to 1. List the sites t in lexicographic order using the half-space $\mathcal{H}$ in (4.52), yielding a list $\{t_1, t_2, \ldots, t_n\}$, say, where $n = |T|$. Note that $\mathcal{Q} \subset \mathcal{H}$, so that we can define a conditional probability with a larger conditioning set by setting

$$p_t(x_t | x_{\mathcal{H}_t}) = p_t(x_t | x_{\mathcal{Q}_t}),$$

where $\mathcal{H}_t = (t - \mathcal{H}) \cap T$. Then, (4.86) represents the usual decomposition of a joint probability of a vector $x = (x_1, \ldots, x_n)$ in terms of a product of conditional probabilities,

$$p(x) = \prod_{j=1}^{n} p_1(x_1) p_2(x_2|x_1) p_3(x_3|x_1, x_2) \cdots p_n(x_n|x_1, \ldots, x_{n-1}).$$

If each of the conditional probabilities sums to 1, so does the joint probability; just sum sequentially over the possible values of $x_n, \ldots, x_1$.

4.10.2 Marginalization

Markov mesh models have a convenient marginalization property. Let

$$T' = \{t \in \mathbb{Z}^d : 1 \le t[\ell] \le m[\ell]', \ \ell = 1, \ldots, d\}$$

with $m[\ell]' \le m[\ell], \ell = 1, \ldots, d$ be a subregion of T with the same lower corner as T. Then the marginal joint probability density of $x_{T'}$ is given by (4.86) but with the product taken over $t \in T'$ rather than $t \in T$. The proof is similar to that given in the above paragraph. Just sum over the possible values of x_t sequentially for the sites $t \in T \backslash T'$ listed in reverse lexicographic order.

Note that the marginal density over T' does not depend on the size of the original domain T provided $T' \subset T$. Thus, the original domain can be extended to the infinite quadrant $\{t \in \mathbb{Z}^d : t[\ell] \ge 1, \ \ell = 1, \ldots, d\}$.

4.10.3 Markov Random Fields

A Markov mesh model is a special case of a MRF on a finite region. Recall that $p(x_t|x_{\backslash t})$ can be found by looking at the joint density (4.86), treating factors that do not depend on x_t as constant, and rescaling what remains to sum to 1 over x_t. Thus,

$$p(x_t|x_{\backslash t}) \propto p_t(x_t|x_{\mathcal{Q}_t}) \prod_{s\,:\,t \in \mathcal{Q}_s} p_s(x_s|x_{\mathcal{Q}_s}). \tag{4.87}$$

In other words, the distribution of x_t given $x_{\backslash t}$ depends on the sites $\{s : s \in \mathcal{Q}_t \text{ or } t \in \mathcal{Q}_s\}$.

Example 4.9 Consider the Gaussian Markov mesh model with

$$x_t | x_{\mathcal{Q}_t} \sim N(ax_{t-(1,0)} + bx_{t-(0,1)} + cx_{t-(1,1)}, \sigma^2)$$

for interior sites t. Then after combining the linear and quadratic terms involving x_t in the exponents of the four factors in (4.87), we find

$$\begin{aligned} x_t|x_{\setminus t} \sim N\big(\{(a-bc)(x_{t+(1,0)}+x_{t-(1,0)})+(b-ac)(x_{t+(0,1)}+x_{t-(0,1)}) \\ +c(x_{t+(1,1)}+x_{t-(1,1)})-ab(x_{t+(1,-1)}+x_{t+(-1,1)})\}/\alpha, \sigma^2/\alpha^2\big), \end{aligned} \tag{4.88}$$

where $\alpha^2 = 1 + a^2 + b^2 + c^2$, so that this random field is a special case of a CAR model with neighborhood $\mathcal{N}^{(\max,1)}$ (See Exercise 4.16.). Another example based on logistic regression is given in Exercise 4.17. □

4.10.4 Usefulness

Markov mesh models are particularly convenient because it is straightforward to write down the joint density. They were originally developed by Abend et al. (1965), as a tractable generalization of a one-dimensional Markov chain. However, in the Gaussian case, these models have been, to some extent, superseded by CAR models and the approximate methods to deal with them on finite regions in Section 4.6.5. In general, in dimensions $d > 1$, the set of Markov mesh models is strictly smaller than the set of MRFs (Besag, 1972).

Exercises

4.1 (Tower rule for conditional expectations). Let $(U, V_1, V_2, \ldots, V_n)$ denote a collection of jointly distributed random variables. The notation $E[U|v_1, \ldots, v_n]$ denote the expected value of U conditional on $(V_1, \ldots, V_n)$ taking the values $(v_1, \ldots, v_n)$. The notation $E[U|V_1, \ldots, V_n] = W$, say, denotes this conditional expectation, treated as a random variable depending on $(V_1, \ldots, V_n)$. Prove the following two important properties of W.

(a) $E[W|V_1, \ldots, V_n] = W$, since W is constant given $(V_1, \ldots, V_n)$.

(b) $E\{E[W|V_1, \ldots, V_n]\} = E[W] = E[U]$, that is, the expectation of a conditional expectation is the same as the original expectation. This result for expectation is sometimes known as the *Tower law*.

4.2 Consider the following two autoregressions in one dimension:

(i) $6X_t - 5X_{t-1} + X_{t-2} = \varepsilon_t$,

(ii) $2X_t - 7X_{t-1} + 3X_{t-2} = \varepsilon_t$,

where in each case $\{\varepsilon_t\}$ is a white noise process with mean 0, variance 1.

(a) Show that the spectral density for the two autoregressions is given by the same function

$$f(\omega) = (2\pi)^{-1}|(2-e^{i\omega})(3-e^{i\omega})|^{-2}, \quad \omega \in [-\pi, \pi).$$

(b) Show that one of these choices satisfies the stability condition (4.3) and the other does not.

(c) Show that it is not possible to find a first-order SAR representation

$$aX_t - b\left[X_{t+1} + X_{t-1}\right] = \delta_t$$

or a higher order finite SAR representation with the same spectral density, where $\{\delta_t\}$ is another white noise process.

4.3 In $d = 1$ dimension, consider the SAR model

$$\sum_{s=-S}^{S} d_s X_{t-s} = \varepsilon_t,$$

where $d_0 > 0$, $d_s = d_{-s}$ and $\{\varepsilon_t\}$ is a white noise process. Assume that

$$\sum d_s e^{i\omega s} \neq 0 \text{ for all } \omega \in [-\pi, \pi).$$

Show that it is possible to find a unilateral representation

$$\sum_{s=0}^{2S} a_s X_{t-s} = \varepsilon'_t,$$

in terms of another white noise process $\{\varepsilon'_t\}$ for which the coefficients $\{a_s\}$ satisfy the stability conditions

$$\sum_{s=0}^{2S} a_s z^s \neq 0, \quad \text{for all } |z| \leq 1,$$

where z is a complex number.

4.4 Consider a d-dimensional AR model for a stationary Gaussian random field $\{X_t\}$ given by

$$\sum_{s \in K} d_s X_{t-s} = \varepsilon_t, \quad t \in \mathbb{Z}^d,$$

where $\{\varepsilon_t\}$ is white noise, and $K \subset \mathbb{Z}^d$ is a finite set. This framework includes both SARs (Section 4.5) and UARs (Section 4.8). Define a new sequence $\{a_s\}$ with terms

$$a_s = \sum_{h \in D} d_h\, d_{h-s}, \quad s \in D^{(2)},$$

where $D^{(2)} = \{h - s : \; h, s \in D\}$. Show that this random field can be rewritten as a CAR model (4.31) with coefficients

$$\beta_s = -a_s/a_0, \quad s \in D^{(2)}.$$

Hence, conclude that every AR model can be written as a CAR model. Hint: Work with the spectral density in both cases.

4.5 In one dimension, consider a stationary CAR model (4.31),

$$E[X_t|X_{\backslash t}] = \sum_{\substack{s=-S \\ s\neq 0}}^{S} \beta_s X_{t-s}, \quad \text{var}[X_t|X_{\backslash t}] = \sigma_\eta^2,$$

where $1 - \tilde{\beta}(\omega) = 1 - 2\sum_{s=1}^{S} \beta_s \cos s\omega \neq 0$ for all $\omega \in [-\pi, \pi]$. Show that this process can be given a unilateral representation

$$\sum_{s=0}^{S} d_s X_{t-s} = \varepsilon_t,$$

where ε_t is a white noise process and the roots of $\sum_{s=0}^{d} d_s z^s$, $z \in \mathbb{C}$, lie outside the unit disk.
Hint: Factorize

$$P(z) = P(z^{-1}) = 1 - \sum_{\substack{s=-S \\ s\neq 0}}^{S} \beta_s z^s$$

$$= \prod_{i=1}^{S} (1 - \alpha_i z)(1 - \alpha_i z^{-1}),$$

where the roots satisfy $0 \leq |\alpha_i| < 1$, $i = 1, \ldots, S$. Then define $\{d_s\}$ by $\sum d_s z^s = \prod(1 - \alpha_i z)$.

4.6 Nonuniqueness of the mean for a CAR model. Let $\{X_t\}$ be a stationary AR(1) process in one dimension, with mean μ, autoregression parameter $0 < \lambda < 1$, and residual variance σ_ε^2. Then $\{X_t\}$ is also a CAR model satisfying the equations

$$E[(X_t - \mu)|X_{\backslash t}] = \frac{\lambda}{1+\lambda^2}\{(X_{t-1} - \mu) + (X_{t+1} - \mu)\},$$
$$\text{var}(X_t|X_{\backslash t}\} = \sigma_\varepsilon^2/(1 + \lambda^2)$$

(see Exercise 4.4). Define a new process $Y_t = X_t + c\lambda^t$, with mean $E(Y_t) = \mu + c\lambda^t = \nu_t$, say, where c is a scalar constant.
Show that $\{Y_t\}$ satisfies the same CAR equations

$$E[(Y_t - \mu)|Y_{\backslash t}] = \frac{\lambda}{1+\lambda^2}\{(Y_{t-1} - \mu) + (Y_{t+1} - \mu)\},$$
$$\text{var}(Y_t|Y_{\backslash t}\} = \sigma_\varepsilon^2/(1 + \lambda^2).$$

Hence, deduce that if $\{Y_t\}$ is not assumed to be stationary, then the CAR equations do not determine the mean of $\{Y_t\}$.

4.7 Regarding $\mathbb{Z}^2 \subset \mathbb{R}^2$ as part of a complex plane $\mathbb{C}$, fix an angle α and set

$$\mathcal{H} = \{t \in Z^2 \, : \, t \neq 0, \ \alpha \leq \arg t < \alpha + \pi\},$$

where $\arg t$ is shorthand for $\arg(t[1] + it[2])$. Note that $\arg t$ and α are angles; if they are treated as numbers in $[0,2\pi)$, then the above angular inequality is equivalent to the numerical conditions $\alpha \leq \arg t < \alpha + \pi$ or $\alpha \leq 2\pi + \arg t < \alpha + \pi$. Show that $\mathcal{H}$ is a half-space.

4.8 Verify Eq. (4.68) to show that the three types of conditioning on the past are equivalent for a Gaussian QAR model,

$$E[X_t | (X_{t-s}, \ s \in \mathcal{B})] = \sum_{s \in \mathcal{Q}} a_s X_{t-s},$$

$$E[X_t | (X_{t-s}, \ s \in \mathcal{H})] = \sum_{s \in \mathcal{Q}} a_s X_{t-s},$$

$$E[X_t | (X_{t-s}, \ s \in \mathcal{Q})] = \sum_{s \in \mathcal{Q}} a_s X_{t-s}.$$

Hint: For simplicity work in $d = 2$ dimensions. The Gaussian zero-mean QAR model (4.66) is a special case of a UAR with respect to the lexicographic half-space $\mathcal{L}$ in (4.52). The spectral density is given by

$$f(\omega) \frac{\sigma^2}{(2\pi)^d} \left| 1 - \sum_{s \in \mathcal{Q}} a_s e^{is^T \omega} \right|^{-2},$$

and the innovation term, $\varepsilon_t = X_t - \sum_{s \in \mathcal{Q}} a_s X_{t-s}$, is uncorrelated with X_{t-s} for all $s \in \mathcal{L}$.

Set

$$\mathcal{L}' = \{s \in Z^d : s \neq 0 \text{ and the last nonzero component } s[\ell] \text{ of } s \text{ is positive}\}.$$

to be the lexicographic half-space of $\mathbb{R}^2$ obtained by interchanging the roles of $t[1]$ and $t[2]$ in the definition of $\mathcal{H}$ in (4.52). Then $f(\omega)$ also defines a UAR with respect to $\mathcal{L}'$.

The innovation term is still ε_t, as defined above, and is thus also uncorrelated with X_{t-s} for all $s \in \mathcal{L}'$. After writing $X_t = \varepsilon_t + \sum_{s \in \mathcal{Q}} a_s X_{t-s}$, the desired result follows from the following properties: (i) $\mathcal{Q} \subset \mathcal{L} \subset \mathcal{B} = \mathcal{L} \cup \mathcal{L}'$, (ii) ε_t is uncorrelated (and hence independent under the Gaussian assumption) with X_{t-s}, $s \in \mathcal{B}$, and (iii) $\sum_{s \in \mathcal{Q}} a_s X_{t-s}$ is constant given $(X_{t-s}, \ s \in \mathcal{B})$.

4.9 Consider random variables X_{ij}, $i,j \in \mathbb{Z}$, such that

(a) For each i, the random variables $\{X_{ij}, \ j \in \mathbb{Z}\}$ follow an AR(1) process with parameter λ, so that

$$E\{X_{ij} | X_{ij'} \, : \, j' < j\} = \lambda X_{i,j-1}.$$

(b) For different i the random variables are independent.

Arrange these processes on diagonal lines $t[2] = t[1] + i$ for $t \in \mathbb{Z}^2$ to give a random field

$$Y_t = X_{(t[1]-t[2],t[2])}.$$

Show that

$$E\{Y_t | Y_{t-s},\ s \in \mathcal{L}\} = E\{Y_t | Y_{t-s},\ s \in \mathcal{Q}\} = \lambda Y_{(t[1]-1,t[2]-1)}.$$

Hence, conclude that $\{Y_t\}$ is a quadrant regression.
Define a new random field $\{Y_t^*\}$ by

$$Y^*_{(t[1],\ t[2])} = Y_{(t[1],-t[2])}.$$

Show that

$$E\{Y_t^* | Y_{t-s}^*,\ s \in \mathcal{L}\} = \lambda Y^*_{(t[1]-1,\ t[2]+1)}$$

so that $\{Y_t^*\}$ does not have a quadrant autoregression representation.

4.10 Consider the QAR model for a stationary Gaussian random field in Example 4.6. If $c = -ab$, show that the spectral density takes the form

$$\begin{aligned} f(\omega) &= \sigma^2 |1 - a e^{i\omega[1]} - b e^{i\omega[2]} + ab e^{i(\omega[1]+\omega[2])}|^{-2} \\ &= \sigma^2 |1 - a e^{i\omega[1]}|^{-2} |1 - b e^{i\omega[2]}|^{-2}. \end{aligned}$$

Hence, deduce that this model is separable in $\omega[1]$ and $\omega[2]$ (see Exercise 2.8).

4.11 (Brook expansion (Brook, 1964)). Verify the expansion (4.73) and hence confirm that the full conditional probability functions in (4.72) determine the joint probability function $p_T(x_T)$.

4.12 Verify the proof of Theorem 4.9.2 for the subset expansion of a negative potential function.

4.13 The purpose of this exercise is to give some examples of cliques based on Figures 4.3 and 4.4. Let $T = \mathbb{Z}^2$ and consider the neighborhood structures generated by translates $\{\mathcal{N}_t^{(\text{basic},1)} = t + \mathcal{N}^{(\text{basic},1)},\ t \in \mathbb{Z}^2\}$ of the first-order basic neighborhood $\mathcal{N}^{(\text{basic},1)}$ and by translates of the first-order full neighborhood $\mathcal{N}^{(\text{full},2)}$ defined in Section 4.4.

(a) For the first-order basic neighborhood structure, show that the cliques are given by

(i) Singletons $\{t\},\ t \in \mathbb{Z}^2$.

(ii) Horizontal and vertical two-point segments $\{t, t + (1,0)\}$ and $\{t, t + (0,1)\}$.

(b) For the first-order full neighborhood structure, show that there are also additional cliques given by
  (i) Triangles of the form $\{t, t + (1,0), t + (0,1)\}$, $\{t, t + (1,0), t + (1,1)\}$, $\{t, t + (0,1), t + (1,1)\}$, and $\{t + (1,0), t + (0,1), t + (1,1)\}$.
  (ii) Squares $\{t, t + (1,0), t + (0,1), t + (1,1)\}$.

4.14 The purpose of this exercise is to confirm that Eqs. (4.74)–(4.76) imply (4.81). Let x_T be a possible value of the random field and define y_T by $y_t = 0$ and $y_{\backslash t} = x_{\backslash t}$. Writing $p_t(x_t|x_{\backslash t}) = p(x_t|x_{\backslash t})$, show that

$$\frac{p(x_t|x_{\backslash t})}{p(y_t|y_{\backslash t})} = \frac{p(x_t|x_{\backslash t})p_{T\backslash\{t\}}(x_{\backslash t})}{p(y_t|y_{\backslash t})p_{T\backslash\{t\}}(y_{\backslash t})}$$
$$= \frac{p(x_T)}{p(y_T)}$$
$$= \exp\left[\sum_{\Lambda\subset T}\{G_\Lambda(x_\Lambda) - G_\Lambda(y_\Lambda)\}\right].$$

If $t \notin \Lambda$, note that $x_\Lambda = y_\Lambda$, so the corresponding term disappears from the sum. If $t \in \Lambda$, show that $G_\Lambda(y_\Lambda) = 0$. Hence, confirm that the formula for $p(x_t|x_{\backslash t})/p(y_t|y_{\backslash t})$ reduces to the right-hand side of (4.81).

4.15 For the auto-logistic model (4.83), show that

$$\log\frac{p_t(1|x_{\backslash t})}{p_t(0|x_{\backslash t})} = \alpha_t + \sum_{s\in\mathcal{N}(t)} \beta_{st}\, x_t,$$

and hence deduce (4.84).

4.16 Verify the conditional distribution for $X_t|X_{\backslash t}$ in Example 4.9.

4.17 The purpose of this exercise is to show how a $\{0,1\}$-valued Markov mesh model can be recast as a Markov random field. For simplicity, restrict attention to the one-dimensional case and suppose that the joint distribution is built from the one-sided conditional distributions for $t > 0$,

$$p(x_t|x_{t-1}) = \exp(\alpha x_t + \beta x_t x_{t-1})/\{1 + \exp(\alpha + \beta x_{t-1})\}.$$

Note that $p(0|x_{t-1}) + p(1|x_{t-1}) = 1$ whether x_{t-1} equals 0 or 1. Show that the two-sided conditional probabilities are given by

$$\begin{aligned} p(x_t|x_{\setminus t}) &= p(x_t|\{x_{t-1}, x_{t+1}\}) \\ &\propto \exp\{\alpha x_t + \beta x_t(x_{t-1} + x_{t+1})\}/\{1 + \exp(\alpha + \beta x_t)\} \\ &\propto \exp\{\gamma x_t + \beta x_t(x_{t-1} + x_{t+1})\}, \end{aligned}$$

where $\gamma = \alpha + \log\{(1 + e^{\alpha})/(1 + e^{\alpha+\beta})\}$. To verify the last line, note that x_t can only take the values 0 and 1, so that the conditional distribution is determined by the odds ratio $p(1|x_{\setminus t})/p(0|x_{\setminus t})$.

5

Estimation of Spatial Structure

5.1 Introduction

In this chapter, we look at the problem of fitting a stationary or intrinsic Gaussian model to a set of spatial data. In principle, given a model, it is straightforward to write down the likelihood in terms of the underlying covariance function or semivariogram and to optimize it with respect to any unknown parameters. However, in practice, there are several complications.

(a) Lack of intuition. Black box fitting.
(b) Computational issues for large data sets.
(c) For some models, especially autoregressions, the covariance function is not known in an explicit form.

Hence, we also look at various alternatives and approximations to maximum likelihood. The available estimation procedures depend on the nature of the spatial model and the nature of the data.

Here are the main distinctions between the different types of processes and the different types of data. More details are given in Section 5.3.

(a) *Spacing of the process*. Continuously indexed on $\mathbb{R}^d$ vs. lattice-indexed on $\mathbb{Z}^d$
(b) *Specification of dependence*. Directly specified covariance function (on $\mathbb{R}^d$ or $\mathbb{Z}^d$) vs. directly specified spectral density (on $\mathbb{R}^d$ or $(-\pi, \pi)^d$) vs. autoregressive models (on $\mathbb{Z}^d$)
(c) *Regularity of the process*. Stationary (described by a covariance function) vs. intrinsic (described by a semivariogram)
(d) *Spacing of the data*. Irregularly spaced (arbitrary sites in $\mathbb{R}^d$) vs. regularly spaced (on a rectangle in $\mathbb{Z}^d$)

Here is an overview of the key topics covered in the chapter.

(a) Exploratory methods, including graphical analysis, for both regularly spaced and irregularly spaced data (Section 5.4). Such methods help to develop our

Spatial Analysis, First Edition. John T. Kent and Kanti V. Mardia.

intuition about the behavior of different models and provide informal assessments of goodness of fit.

(b) Maximum likelihood methods and approximations (Sections 5.5–5.11). Provided that the covariance function or semivariogram has an explicit form, there is an explicit formula for the likelihood function. However, in general, optimization must be done numerically and it becomes computationally infeasible for large data sets. Hence, various approximations have been explored.

(c) *Asymptotics*. An important issue for estimators is their asymptotic behavior with the increasing sample size n. It is helpful to distinguish two cases, "outfill" asymptotics (Section 5.11.2), also known as "increasing domain" asymptotics, and "infill" asymptotics (Section 5.14), also known as "fixed-domain" asymptotics. For outfill asymptotics, the spacing between the data sites remains fixed, and as the sample size increases, the data lie in a larger region. In contrast, for infill asymptotics, the data lie in a fixed region, and as the sample size increases, the data sites fill out this region more densely. Outfill asymptotics are the most relevant for practical data analysis.

Much of the material in this chapter, including the spatial linear model, is based on a series of papers by Mardia (1980, 1990), Mardia and Gill (1982), Mardia and Marshall (1984), and Mardia and Watkins (1989). In addition, there are estimation methods that are only applicable to regularly spaced or lattice data such as moment methods for unilateral autoregressions (UARs), moment methods for conditional autoregressions (CARs), and spectral methods, which are covered in Chapter 6.

5.2 Patterns of Behavior

Before fitting stationary and intrinsic Gaussian models of spatial variability to data, it is worth looking at the broad sorts of behavior that can occur. First, we examine the types of behavior for a one-dimensional semivariogram and then look at how one-dimensional semivariograms can fit together in higher dimensional cases.

5.2.1 One-dimensional Case

Let $\gamma(h)$ be a semivariogram in the one-dimensional case, $h \in \mathbb{R}^1$. To begin with, suppose that $\gamma(h)$ is nondecreasing in $|h| \geq 0$ and set $b = \lim \gamma(h)$ as $|h| \to \infty$, where $0 < b \leq \infty$. This limit b is called the *sill* of the semivariogram; see Section 1.4.1.

(a) *Finite range*. If $b < \infty$ and b is attained for a finite value $|h| = c$, say, (so that $\gamma(h) < b$ for $|h| < c$ and $\gamma(h) = b$ for $|h| \geq c$), then the semivariogram is said to have a "finite range" c. Under this condition, sites more than c units apart are uncorrelated. Further, $b - \gamma(h) = \sigma(h)$, say, defines a stationary covariance function with marginal variance $\sigma(0) = b$. Examples include the spherical scheme and the restricted power scheme (see Table 2.1).

(b) *Finite sill with infinite range.* If $b < \infty$ but $\gamma(h) < b$ for all $h \geq 0$, then the range of the semivariogram is infinite. Such a semivariogram is still a model for a stationary random field, but values of the random field at distant sites now have a low rather than zero correlation. For simplicity, suppose that $\gamma(h)$ is monotonically increasing in h. An important example is the Matérn covariance function. In this case, it is sometimes convenient to use a modified concept of range. For example, a "90% – range," $c_{0.9}$, say, is defined so that $\gamma(h) < 0.9b$ for a $|h| < c_{0.9}$ and $\gamma(h) \geq 0.9b$ for $|h| \geq c_{0.9}$. The exponential covariance function, $\gamma(h) = 1 - \exp(-|h|/\varphi)$, is a special case of the Matérn covariance function with index $\nu = 0.5$. In this case, $c_{0.9} = 2.3\varphi$ since $2.3 = \log(10) = -\log(0.1)$.

(c) *Intrinsic behavior.* If $b = \infty$, then the semivariogram arises from an intrinsic rather than a stationary random field. Typical behaviors of $\gamma(h)$ for large $|h|$ are $\gamma(h) \propto \log(h)$, or $\gamma(h) \propto |h|^\beta, 0 < \beta < 2$.

(d) *Oscillatory behavior.* In most applications, a monotone semivariogram $\gamma(h)$ will be appropriate. However, occasionally, it is important to allow an oscillatory semivariogram, e.g. $\gamma(h) = 1 - (\sin h)/h$ in dimension $d = 1$. Such examples are rarer in spatial analysis than in time series. One notable example where periodicity is important is the agricultural fertility data of Mercer and Hall (1911) of Example 1.14 in which ridges and furrows appeared in the pattern of fertility due to regularity in the plowing of the field. An analysis of this data set is given in Examples 6.1 and 6.2. Another example is the fingerprint data of Example 1.8 and Exercise 1.5.

5.2.2 Two-dimensional Case

If $\gamma(h), h \in \mathbb{R}^2$, is a semivariogram, write $\gamma(h) = \gamma(r; \theta)$ in polar coordinates where $h = (h[1], h[2]) = (r\cos\theta, r\sin\theta)$. The semivariogram is most easily studied by plotting $\gamma(r; \theta)$ vs. r for several values of θ (the four directions $\theta = 0°, 45°, 90°, 135°$ measured counterclockwise from the $h[1]$-axis, as shown in Figure 1.8, usually suffice). For fixed θ, the behavior of $\gamma(r; 0)$ can be described as in the one-dimensional case. When combining semivariograms from separate directions, several possible patterns can be noted.

(a) *Full symmetry.* A semivariogram automatically satisfies the "antipodal" symmetry property $\gamma(h[1], h[2]) = \gamma(-h[1], -h[2])$ in which both axes are reflected. If, in addition, the semivariogram satisfies $\gamma(h[1], h[2]) = \gamma(h[1], -h[2])$ in which just one axis is reflected, then the semivariogram is said to be "fully symmetric" (Cressie, 1993), or sometimes "reflection symmetric" (Martin, 1979).

(b) *Isotropy*. If $\gamma(r;\theta)$ does not depend on θ, a semivariogram is said to be "isotropic." This is the simplest type of behavior to study. An isotropic semivariogram is always fully symmetric.

(c) *Geometric anisotropy*. The simplest form of nonisotropic behavior occurs when a semivariogram can be written in the form

$$\gamma(r;\theta) = \gamma_0(c_\theta r) = \gamma_0\left(\sqrt{h^T A h}\right)$$

for some function $\gamma_0(r)$, where

$$c_\theta = \begin{bmatrix}\cos\theta & \sin\theta\end{bmatrix} A \begin{bmatrix}\cos\theta \\ \sin\theta\end{bmatrix}$$

and A is a positive definite 2×2 matrix. Then, c_θ lies between the two eigenvalues of A and attains these extreme values when θ points in the direction of the corresponding eigenvectors. Note that the semivariograms, as a function of r for different fixed values of θ, can be made to coincide by suitably dilating the r-axis by a certain amount depending on θ.

(d) *Lattice isotropy*. For semivariograms on the integer lattice, it is useful to consider a restricted form of isotropy, called "lattice isotropy," for which $\gamma(h[1], h[2]) = \gamma(h[2], h[1]), h \in \mathbb{Z}^2$, holds together with full symmetry; that is, a semivariogram is invariant under 90° rotations. As a simple example, given most easily in terms of a covariance function rather than a semivariogram, consider the geometric scheme $\sigma(h) = \exp\{-(|h[1]| + |h[2]|)/\varphi\}$, which possesses lattice symmetry (Martin, 1979). However, note that

$$\sigma(r;0^\circ) = \sigma(r;90^\circ) = e^{-r/\varphi},$$

whereas

$$\sigma(r;45^\circ) = \sigma(r;135^\circ) = e^{-r\sqrt{2}/\varphi},$$

so that $\sigma(h)$ is not isotropic. It decays away more quickly along the diagonals than along the main axes. Further, the behavior in this example cannot be captured by any semivariogram with geometric anisotropy.

(e) *Linear drift*. Let $Y(t) = X(t) + b^T t,\ t \in \mathbb{R}^2$, where $\{X(t)\}$ is stationary and $b^T = |b|(\cos\theta_0, \sin\theta_0)$, represents a linear drift in the direction θ_0. Then, $\{Y(t)\}$ has a well-defined semivariogram

$$\gamma_y(h) = \gamma_x(h) + \frac{1}{2}(b^T h)^2.$$

Hence, $\gamma_y(r;\theta_0 + 90^\circ) = \gamma_x(h)$ is bounded as $|h| \to \infty$ whereas $\gamma_y(r;\theta_0) = \gamma_x(h) + \frac{1}{2}|b|^2 r^2$ grows quadratically.

5.2.3 Nugget Effect

From this book's perspective (see Section 2.2), a population covariance function $\sigma(h)$ defined for $h \in \mathbb{R}^d$ is required to be a continuous function for all h.

In particular, $\sigma(h) \to \sigma(0)$ as $h \to 0$. Similarly, for intrinsic processes, a population semivariogram $\gamma(h)$ defined for $h \in \mathbb{R}^d$ is required to be a continuous function for all h. In particular, $\gamma(h) \to \gamma(0) = 0$ as $h \to 0$.

However, when fitting covariance functions visually to data, it can appear that $\sigma(h)$ should have a discontinuity with $0 < \lim \sigma(h) < \sigma(0)$ as $h \to 0$. Similarly, when fitting a semivariogram, it can appear that $\gamma(h)$ should have a discontinuity with $\lim \gamma(h) > 0$. This behavior was described in Section 1.4.1 as a "nugget effect."

A nugget effect is most easily described mathematically using a latent process, also known as an "errors-in-variables" model. Let $\{X(t) : t \in \mathbb{R}^d\}$ be an unobserved stationary random field with a (continuous) covariance function $\sigma_x(h)$. Suppose we observe random variables

$$Y_i = X(t_i) + \varepsilon_i, \quad i = 1, \ldots, n, \tag{5.1}$$

at a collection of sites $t_1, \ldots, t_n$, where $E(\varepsilon_i) = 0$, $\text{var}(\varepsilon_i) = \tau^2$, and the ε_i are independent of each other and of the random field $\{X(t)\}$. The ε_i may represent, for example, "measurement errors" or an influence similar to the nuggets of the mineral mentioned earlier. Then, the $n \times n$ covariance matrix $A = (a_{ij})$ of $[Y_1, \ldots, Y_n]^T$ is given by

$$a_{ij} = \sigma_x(t_i - t_j) + \tau^2 \delta_{ij}, \quad 1 \le i, j \le n,$$

where δ_{ij} is the Kronecker delta.

The nugget effect concept is useful for lattice models even though, strictly speaking, it does not make sense to talk about the continuity of σ_h, $h \in \mathbb{Z}^d$, at $h = 0$. For example, suppose $\{X_t\}$ is a stationary Gaussian process on $\mathbb{Z}^d$ with spectral density $f_x(\omega)$, $\omega \in (-\pi, \pi)^d$, and consider an observed random field

$$Y_t = X_t + \varepsilon_t,$$

where ε_t is a white noise random field (i.e. a set of i.i.d random variables) with mean 0 and with $\text{var}(\varepsilon_t) = \tau^2$, independent of $\{X_t\}$. Then, $\{Y_t\}$ has spectral density

$$f_y(\omega) = f_x(\omega) + \tau^2/(2\pi)^d, \quad \omega \in (-\pi, \pi)^d.$$

When modeling a nugget effect at sites $t_1, \ldots, t_n$, there is an important conceptual difference between $\mathbb{Z}^d$ and $\mathbb{R}^d$. In the lattice case $\mathbb{Z}^d$, measurement errors can be modeled as a sample from a white noise stationary random field $\{\varepsilon_t : t \in \mathbb{Z}^d\}$; that is, there is an error term ε_t for each site in the integer lattice, where the $\{\varepsilon_t\}$ have 0 mean, constant variance and are independent. However, in the continuous case, the measurement errors $\{\varepsilon_i, i = 1, \ldots, n\}$ are well defined only at the sites t_i. The $\{\varepsilon_i\}$ cannot be regarded as a sample from a white noise random field on all of $\mathbb{R}^d$; indeed, in the continuous case, a white noise random field makes sense only in a "generalized" sense; see Section 3.6.

In the continuous case, it is possible to regard the measurement errors $\{\varepsilon_i\}$ as a sample from an ordinary stationary random field with the covariance function

$\sigma_\varepsilon(h)$ where $\sigma_\varepsilon(0) = \tau^2$ and the range φ_ε, say, of $\sigma_\varepsilon(h)$ is smaller than any of the intersite distances; indeed, this perspective is the origin of the term "nugget effect." However, this approach is not very fruitful for modeling purposes since in this case the $\{\varepsilon_i\}$ are still independent and the data sites are too far apart to estimate anything about $\sigma_\varepsilon(h)$, other than τ^2, from the data.

5.3 Preliminaries

Before we can discuss estimation methods in detail, first it is necessary to describe the different types and layout of the data and the different formulations of the models that will be considered.

5.3.1 Domain of the Spatial Process

Recall that spatial processes in d dimensions are typically defined on either the integer lattice $\mathbb{Z}^d$ or continuous space $\mathbb{R}^d$. Some of the estimation procedures can be applied in either setting, and we write a spatial process as $X(t), t \in \mathbb{R}^d$, when we are discussing methods that can be applied to a continuously indexed case. Other methods are restricted to models on the integer lattice with regularly spaced data, and we use subscript notation $X_t, t \in \mathbb{Z}^d$, to describe a spatial process in this setting.

5.3.2 Model Specification

Models in spatial analysis can be specified in two common ways.

(a) *Direct specification.* The most direct way to specify a model is through an explicit formula for the covariance function $\sigma(h)$ for a stationary process, or its semivariogram $\gamma(h)$ for an intrinsic process.
(b) *Spectral specification.* On the other hand, other models are more simply specified by an explicit representation for the spectral density $f(\omega)$. The spectral specification is particularly useful for structural models on $\mathbb{Z}^d$. The main examples include unilateral autoregressions (UARs), simultaneous autoregressions (SARs), and conditional autoregressions (CARs) of Chapter 4.

To use a model in practice, at least one of these two formulations must be available in an explicit form.

Example 5.1 ***Exponential covariance function***
The simplest example of a stationary covariance function for a continuously indexed process is the ($d = 1$)-dimensional exponential covariance function, with direct representation

$$\sigma(h) = \sigma^2 \exp(-|h|/\varphi), \quad h \in \mathbb{R}^1.$$

in terms of the scale parameter $\sigma^2 > 0$ and the range parameter $\varphi > 0$. It is a special case of the Matérn covariance function with index $\nu = 1/2$ (Table 2.2). The spectral density also has an explicit form (with suffix "cont" for continuous)

$$f_{\text{cont}}(\omega) = \frac{1}{\pi} \frac{\sigma^2 \varphi^2}{1 + \varphi^2 \omega^2}, \qquad \omega \in \mathbb{R}^d.$$

If a continuously indexed stochastic process is restricted to integer indices, then the resulting process is known as a lattice process, and the covariance function can be written as

$$\sigma_h = \sigma^2 \lambda^{|h|}, \quad h \in \mathbb{Z},$$

where $\varphi = -1/\log \lambda$, i.e. $\lambda = \exp(-1/\varphi)$, with $0 < \lambda < 1$. The spectral density for the lattice process takes the explicit form (with suffix "disc" for discrete)

$$f_{\text{disc}}(\omega) = \sum_{k=-\infty}^{\infty} f_{\text{cont}}(\omega + 2\pi k) \tag{5.2}$$

$$= \frac{1}{2\pi} \frac{\sigma^2(1 - \lambda^2)}{1 + \lambda^2 - 2\lambda \cos \omega}, \qquad \omega \in (-\pi, \pi). \tag{5.3}$$

Note that when attention is restricted to the lattice process, values of the continuous spectral density at frequencies separated by multiples of 2π cannot be distinguished (the aliasing property) and are combined together to form the discrete spectral density in (5.2). However, the fact that the sum has a simple closed form in (5.3) is unexpected.

The lattice process can also be given two autoregression representations (Sections 4.8 and 4.6) as a UAR and as a CAR. These representations are studied in more detail in Chapter 6. □

5.3.3 Spacing of Data

The data consist of a collection of sites D in $\mathbb{R}^d$ together with real-valued observations at these sites. As discussed in Section 1.2, it is useful to distinguish two different settings:

(a) *Regular data*. In this case, $D \subset \mathbb{Z}^d$ represents a rectangular region of size $n = n_1 \times \cdots \times n_d$

$$D = \{t \in \mathbb{Z}^d : 1 \leq t[\ell] \leq n[\ell], \ell = 1, \ldots, d\}. \tag{5.4}$$

The values of the process are indicated using subscripts $\{X_t, t \in D\}$ and can be laid out in a d-way array.

(b) *Irregular data*. In this case, $D \subset \mathbb{R}^d$ is a general collection of sites

$$D = \{t_i, i = 1, \ldots, n\}. \tag{5.5}$$

The values of the process are indicated using parentheses $\{X(t_i), i = 1, \ldots, n\}$. The data can be laid out as an $n \times (d + 1)$ matrix, with one column representing the values of the process and the remaining d columns giving the coordinates of the data sites.

There are several reasons for the dual notation. First, the notation helps to remind the reader which setting is being used. Also, in the irregular case, the bracket notation avoids the need for double subscripts. In the regular case, the index notation facilitates the consideration of specific lags.

5.4 Exploratory and Graphical Methods

Given a set of spatial data, a useful first step is to plot the sample covariance function or the sample semivariogram. The semivariogram is more useful than the sample covariance function because it can pick out intrinsic as well as stationary behavior in the data. For a stationary random field, the semivariogram levels off for large lags. However, for an intrinsic random field of order 0 (IRF-0), the semivariogram can increase indefinitely for large lags. Mathematically, these two possibilities are very distinct. However, in practice, it can be hard to tell them apart on the basis of limited data. An analogous question in time-series analysis involves distinguishing between a stationary autogression moving average (ARMA) model and an autogression integrated moving average (ARIMA) model. In time series, one way to tackle the problem is to take successive differences of time series until stationarity is achieved. However, in dimensions $d \geq 2$, there is no simple analogue of "differencing."

Throughout this chapter, several data sets from Section 1.2 have been used to illustrate various estimation methods. In this section, we recall the preliminary graphical analysis of the bauxite data and the elevation data.

Example 5.2 ***Bauxite data***
The bauxite data set gives the bauxite ore grade in percentages at $n = 33$ irregularly spaced sites in the plane (constructed from Marechal and Serra, 1970). See Example 1.4 for more details. Figure 5.1 shows the bubble plot and the semivariograms; the same plots are given in Figures 1.4 and 1.11. As noted there, the bubble plot shows an approximate basin shape with the smallest value in the middle of the plot and larger values near the edges.

The semivariograms in the four principal directions do not suggest any anisotropy. A possible jump in the overall semivariogram at $h = 0$ suggests the need for a nugget term. At larger lags, the semivariograms are so noisy that it is hard to pick out any pattern. Later in the chapter (Examples 5.5 and 5.8), several models will be fitted to the data, including isotropic stationary models and models with quadratic drift. □

Example 5.3 ***Elevation data***
The topographic elevation data set of Davis (1973) gives the heights in a surveying problem at $n = 52$ irregularly spaced sites in the plane. See Example 1.3 for more details. Figure 5.2 shows the bubble plot and the semivariograms from Figures 1.2

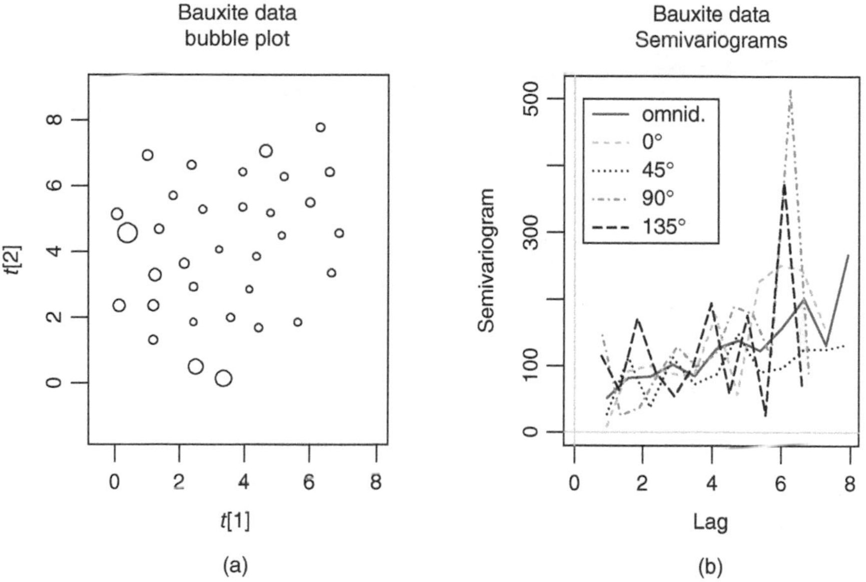

Figure 5.1 Bauxite data: Bubble plot and directional semivariograms.

and 1.10. As noted there, the bubble plot shows a valley in the top middle of the plot and a peak in the bottom middle. The semivariograms in the four principal directions indicate some possible mild anisotropy; the semivariogram in the 90° direction increases slowly, whereas the semivariogram in the 0° direction increases fast. However, an assumption of isotropy does not seem unreasonable. There is no suggestion of a nugget effect. Later in the chapter (Example 5.6), several isotropic stationary models will be fitted to the data. □

Graphical methods can also be used to estimate the parameters of a semivariogram model. One method involves matching the sample semivariogram to a population semivariogram. One of the simplest matching methods is weighted least squares (WLS). That is, choose the population parameters θ to minimize

$$f(\theta) = \sum_h w_h (g(h) - \gamma(h;\theta))^2,$$

where h ranges through the set of lags at which the semivariogram is calculated, and the w_h are preassigned weights designed to give more weight to lags for which $g(h)$ is more accurate.

This method of estimation is reviewed in Cressie (1993). As noted by Zimmerman and Stein (2010), although the WLS procedure is very popular among practitioners due to its relative simplicity, the fact that it is not based on an underlying probabilistic model for the underlying spatial process means that it is suboptimal and does not rest on as firm a theoretical footing as likelihood-based procedures.

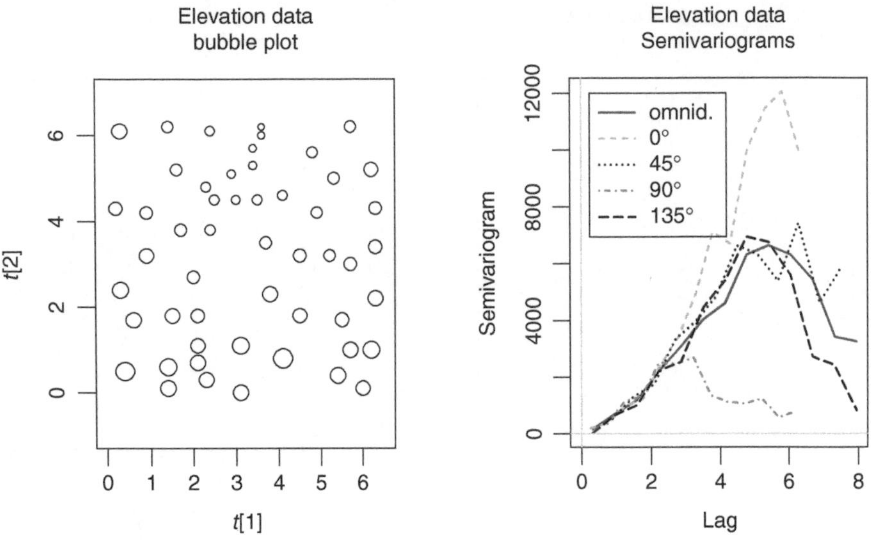

(a) (b)

Figure 5.2 Elevation data: Bubble plot and directional semivariograms.

Further, as noted by Stein (1999, Section 6.2), the act of binning the data to produce an empirical semivariogram involves information loss, especially for differentiable processes.

Moreover, when there is a drift in the data as well as spatial dependence, it can be problematic to use semivariogram-based procedures for estimating both features simultaneously. Indeed, to properly estimate the drift, it is necessary to know the spatial dependence of the residuals. And conversely, to accurately estimate the spatial dependence of the residuals, drift parameters should be known. Hence, there is a circular argument in the estimation procedure. Maximum likelihood estimates both features simultaneously, and hence avoids the circularity problem. In summary, although WLS is a method with some intuitive appeal, we will not investigate such methods in this book. This section has given a very brief introduction to graphical methods. For a more thorough description of their use for exploratory analysis and diagnostics, see, e.g., Cressie (1993), Cressie and Burden (2015), and Chilés and Delfiner (2012).

5.5 Maximum Likelihood for Stationary Models

Recall that the maximum likelihood estimator (MLE) is defined to be the parameter value which maximizes the likelihood. In this section, we consider data from a Gaussian process whose covariance function has an explicit representation and we derive the equations for the MLE of the parameter vector. Let $\{X(t) : t \in \mathbb{R}^d\}$ be a stationary Gaussian process with a constant mean μ and with a continuous covariance function $\sigma^*(h)$, where $\sigma^*(h) = \sigma^{*2}\rho^*(h)$ and $\rho^*(0) = 1$. Suppose that noisy measurements of the process are observed at a set of n sites $D = \{t_i, i = 1, \ldots, n\}$, not necessarily on a lattice,

$$Y_i = X(t_i) + \varepsilon_i, \quad i = 1, \ldots, n,$$

where the $\{\varepsilon_i\}$ are $N(0, \tau^2)$ random variables, independent of one another and of the x-process, representing a nugget effect. If $\tau^2 = 0$, the nugget effect is not present and $Y_i = X(t_i)$.

Let $\mathbf{y} = \{y_i, i = 1, \ldots, n\}$ denote the vector of observed data, with the sites listed in lexicographic order, say. Then, $\mathbf{y}$ follows a multivariate normal distribution $\mathbf{y} \sim N_n(\mu\mathbf{1}, \Sigma)$, where the covariance matrix has several equivalent representations:

$$\begin{aligned}\Sigma &= \Sigma^* + \tau^2 I = \sigma^{*2}P^* + \tau^2 I \\ &= \sigma^2((1-\psi)P^* + \psi I) = \sigma^2 P.\end{aligned} \tag{5.6}$$

In the first line $\Sigma^* = (\sigma^*_{ij}), \sigma^*_{ij} = \sigma^*(t_i - t_j)$ and $P^* = (\rho^*_{ij}), \rho^*_{ij} = \rho^*(t_i - t_j)$ are the covariance matrix and correlation matrix for the latent x-process at the data sites; σ^{*2} and τ^2 represent the variance of the x-process and the nugget variance, respectively. The second line uses the sill or overall variance, $\sigma^2 = \sigma^{*2} + \tau^2$, and the *relative nugget effect* $\psi = \tau^2/\sigma^2$; the overall correlation matrix is $P = (1-\psi)P^* + \psi I$. Note that quantities with an asterisk, such as Σ^*, P^*, and σ^{*2} refer to the latent x-process, whereas quantities without an asterisk refer to the observation vector $\mathbf{y}$. In most practical applications, there are several parameters to estimate: mean μ, variances σ^{*2}, and τ^2 (or equivalently the overall variance σ^2 and the relative nugget ψ), and any correlation parameter in the correlation function $\rho^*(h)$. For example, the exponential correlation function $\rho^*(h) = \exp(-|h|/\varphi)$ depends on the range parameter $\varphi > 0$. Let $\boldsymbol{\theta}$ be a p-dimensional parameter vector containing the covariance parameters. It is convenient to partition $\boldsymbol{\theta} = (\boldsymbol{\theta}_c^T, \sigma^2)^T$, where $\boldsymbol{\theta}_c$ contains the $p-1$ correlation parameters and the final element σ^2 contains the overall variance. For the exponential correlation function, $p = 3$ and $\boldsymbol{\theta}_c = (\varphi, \psi)^T$ when a nugget is present; $p = 2$ and $\boldsymbol{\theta}_c = \varphi$ when there is no nugget term.

5.5.1 Maximum Likelihood Estimates – Known Mean

To begin with, suppose that the mean is known and equal to 0, i.e. $\mu = 0$. Then, the log-likelihood takes the form

$$\log L(\boldsymbol{\theta}; \mathbf{y}) = -\frac{1}{2}\{\mathbf{y}^T\Sigma^{-1}\mathbf{y} + \log|\Sigma| + n\log(2\pi)\}. \tag{5.7}$$

The maximum likelihood estimate of $\boldsymbol{\theta}$ is obtained by differentiating (5.7) with respect to the elements of $\boldsymbol{\theta}$ and setting the derivatives to 0. Recall the matrix derivative formulas

$$\frac{\partial \log|\Sigma|}{\partial\theta_k} = \mathrm{tr}(\Sigma^{-1}\Sigma_k), \qquad \frac{\partial\Sigma^{-1}}{\partial\theta_k} = -\Sigma^{-1}\Sigma_k\Sigma^{-1},$$

where $\Sigma_k = \partial\Sigma/\partial\theta_k$ (e.g. Mardia et al., 1979, pp. 478–479). Then

$$\partial \log L/\partial\sigma^2 = -\frac{1}{2}\{-\mathbf{y}^T P^{-1}\mathbf{y}/\sigma^4 + n/\sigma^2\}, \tag{5.8}$$

$$\partial \log L/\partial\theta_k = -\frac{1}{2}\{-\mathbf{y}^T\Sigma^{-1}\Sigma_k\Sigma^{-1}\mathbf{y} + \mathrm{tr}(\Sigma^{-1}\Sigma_k)\}, \quad k = 1, \ldots, p-1, \tag{5.9}$$

where k indexes the correlation parameters in $\boldsymbol{\theta}_c$. Setting the derivatives to 0 and rearranging these equations yields the likelihood equations

$$\hat{\sigma}^2 = \mathbf{y}^T\hat{P}^{-1}\mathbf{y}/n, \tag{5.10}$$

$$\mathbf{y}^T\hat{\Sigma}^{-1}\hat{\Sigma}_k\hat{\Sigma}^{-1}\mathbf{y} = \mathrm{tr}(\hat{\Sigma}^{-1}\hat{\Sigma}_k), \quad k = 1, \ldots, p-1, \tag{5.11}$$

where $\Sigma^2 = \sigma^2 P$ and where a hat "^" is used to denote the MLEs and other quantities evaluated at the MLEs. These results can be used to simplify the log-likelihood.

(a) For a given value of $\boldsymbol{\theta}_c$, the maximizing value of σ^2 is given by (5.10).
(b) Substituting this value for σ^2 into the log-likelihood yields the *profile log-likelihood* for the correlation parameters $\boldsymbol{\theta}_c$

$$\log L_{\mathrm{profile}}(\boldsymbol{\theta}_c) = -\frac{1}{2}\{n + n\log(\mathbf{y}^T P^{-1}\mathbf{y}/n) + \log|P| + n\log(2\pi)\}, \tag{5.12}$$

where the correlation matrix P depends on $\boldsymbol{\theta}_c$. In general, the profile log-likelihood must be maximized numerically over $\boldsymbol{\theta}_c$. The derivative of the profile log-likelihood with respect to the elements of $\boldsymbol{\theta}_c$ can be written as

$$\partial \log L_{\mathrm{profile}}/\partial\theta_k = -\frac{1}{2}\mathrm{tr}\left\{P^{-1}P_k\left(I_n - \frac{n}{\mathbf{y}^T P^{-1}\mathbf{y}}P^{-1}\mathbf{y}\mathbf{y}^T\right)\right\}. \tag{5.13}$$

Here $P_k = \partial P/\partial\theta_k$ and $k = 1, \ldots, p-1$ ranges through the elements of $\boldsymbol{\theta}_c$.

Example 5.4 ***Exponential covariance function with nugget***
Following on from Example 5.1, recall that the overall correlation matrix can be written as

$$P = (1-\psi)P^* + \psi I,$$

where $P^* = (\rho^*_{ij})$, $\rho^*_{ij} = \exp(-|t_i - t_j|/\varphi)$, $i, j = 1, \ldots, n$. The correlation parameter vector $\boldsymbol{\theta}_c = (\varphi, \psi)$ has length $p - 1 = 2$ and contains the range parameter φ and the relative nugget effect ψ. The partial derivatives P_k, $k = 1, 2$ in this case correspond to partial derivatives with respect to φ and θ

$$\partial P/\partial\varphi = (1-\psi)P^{*\prime}, \quad \partial P/\partial\psi = I - P^*, \tag{5.14}$$

where $P^{*\prime} = (\rho^{*\prime}_{ij})$, $\rho^{*\prime}_{ij} = (|t_i - t_j|/\varphi^2)\exp(-|t_i - t_j|/\varphi)$; the dash denotes differentiation with respect to φ. In terms of the covariance matrix $\Sigma = \sigma^2 P$, the derivatives become

$$\partial\Sigma/\partial\sigma^2 = P, \quad \partial\Sigma/\partial\varphi = \sigma^2(1-\psi)P^{*\prime}, \quad \partial\Sigma/\partial\psi = \sigma^2(I - P^*). \tag{5.15}$$

□

5.5.2 Maximum Likelihood Estimates – Unknown Mean

Suppose the mean μ is included as an unknown parameter. The log-likelihood takes the form

$$\log L(\mu, \boldsymbol{\theta}; \boldsymbol{y}) = -\frac{1}{2}\{(\boldsymbol{y} - \mu\mathbf{1}_n)^T\Sigma^{-1}(\boldsymbol{y} - \mu\mathbf{1}_n) + \log|\Sigma| + n\log(2\pi)\}. \tag{5.16}$$

Recall the matrix derivative formula

$$d(\boldsymbol{u}^TA\boldsymbol{u})/d\boldsymbol{u} = 2A\boldsymbol{u},$$

where A is a symmetric matrix and $\boldsymbol{u}$ is a vector; see Section A.3.11. The derivative of the log-likelihood with respect to μ is

$$\partial \log L/\partial\mu = \mathbf{1}_n^T\Sigma^{-1}(\boldsymbol{y} - \mu\mathbf{1}_n). \tag{5.17}$$

Setting the derivative to 0 yields the *generalized least squares (GLS)* estimator

$$\hat{\mu} = \hat{\mu}(\boldsymbol{\theta}_c) = \boldsymbol{y}^T\Sigma^{-1}\mathbf{1}_n/\mathbf{1}_n^T\Sigma^{-1}\mathbf{1}_n = \boldsymbol{y}^TP^{-1}\mathbf{1}_n/\mathbf{1}_n^TP^{-1}\mathbf{1}_n \tag{5.18}$$

for a given value of $\boldsymbol{\theta}_c$.

This result can be used to simplify the log-likelihood.

(a) For a given value of $\boldsymbol{\theta}_c$, the maximizing value of μ is given by (5.18), and it does not depend on σ^2.
(b) For a given value of $\boldsymbol{\theta}_c$ and for $\mu = \hat{\mu}(\boldsymbol{\theta}_c)$, the maximizing value of σ^2 is given by (5.10), with $\boldsymbol{y}$ replaced the residual

$$\hat{\boldsymbol{w}} = \hat{\boldsymbol{w}}(\boldsymbol{\theta}_c) = \boldsymbol{y} - \hat{\mu}\mathbf{1}_n \tag{5.19}$$

and $\hat{\mu}$ given by (5.18).
(c) Substituting these values for μ and σ^2 into the log-likelihood yields the *profile log-likelihood* for the correlation parameters $\boldsymbol{\theta}_c$

$$\log L_{\text{profile}}(\boldsymbol{\theta}_c) = -\frac{1}{2}\{n + n\log(\hat{\boldsymbol{w}}^TP^{-1}\hat{\boldsymbol{w}}/n) + \log|P| + n\log(2\pi)\}. \tag{5.20}$$

Note both $\hat{\boldsymbol{w}}$ from (5.19) and P depend on $\boldsymbol{\theta}_c$.

In general, the profile log-likelihood must be maximized numerically over $\boldsymbol{\theta}_c$. Equation (5.20) and its derivatives can be simplified by using a suitable linear transformation. Let G be an $n \times (n-1)$ column orthonormal matrix, not depending on the data, such that each column is orthogonal to the constant vector of ones, so that $G^T\mathbf{1}_n = \mathbf{0}_{n-1}$ and $G^TG = I_{n-1}$. A particular choice based on the Helmert matrix is examined in Exercise 5.4. Set

$$\boldsymbol{z} = G^T\boldsymbol{y} \tag{5.21}$$

to be an $(n-1)$-vector of contrasts between the elements of $\boldsymbol{y}$. Then $\boldsymbol{z}$ has a mean $\mu G^T\mathbf{1} = \mathbf{0}$ and the covariance matrix

$$\Sigma_{GG} = G^T\Sigma G. \tag{5.22}$$

The quadratic form in (5.20) can be simplified to

$$\hat{\boldsymbol{w}}^T P^{-1} \hat{\boldsymbol{w}} = \boldsymbol{z}^T P_{GG}^{-1} \boldsymbol{z} \tag{5.23}$$

(see Section A.3.7), so that the profile log-likelihood becomes

$$\log L_{\text{profile}}(\boldsymbol{\theta}_c) = -\frac{1}{2}\left\{n + n\log(\boldsymbol{z}^T P_{GG}^{-1}\boldsymbol{z}/n) + \log|P| + n\log(2\pi)\right\}, \tag{5.24}$$

where $P_{GG} = G^T P G$. The derivatives of the profile log-likelihood with respect to the correlation parameters in $\boldsymbol{\theta}_c$ are given by

$$\frac{\partial \log L_{\text{profile,ML}}}{\partial \theta_k} = -\frac{1}{2}\left\{-n\frac{\boldsymbol{z}^T P_{GG}^{-1} P_{GG;k} P_{GG}^{-1}\boldsymbol{z}}{\boldsymbol{z}^T P_{GG}^{-1}\boldsymbol{z}} + \text{tr}(P^{-1}P_k)\right\} \tag{5.25}$$

for $k = 1, \dots, p-1$, where $p-1$ is the number of elements in $\boldsymbol{\theta}_c$. Here $P_k = \partial P/\partial \theta_k$ and $P_{GG;k} = G^T P_k G$.

5.5.3 Fisher Information Matrix and Outfill Asymptotics

The Fisher information matrix is obtained from the matrix of expected values of the second derivatives of the log-likelihood with respect to the parameters, after changing the sign. For a stationary Gaussian model $N_n(\mu\mathbf{1}_n, \Sigma)$, it takes the form

$$\mathcal{I} = \begin{bmatrix} \mathcal{I}^{(\mu)} & 0 \\ 0 & \mathcal{I}^{(\theta)} \end{bmatrix} = \begin{bmatrix} \mathbf{1}_n^T \Sigma^{-1} \mathbf{1}_n & 0 \\ 0 & A \end{bmatrix}, \tag{5.26}$$

where $A = \mathcal{I}^{(\theta)}$ is a $p \times p$ matrix with entries

$$a_{jk} = \frac{1}{2}\text{tr}(\Sigma^{-1}\Sigma_j\Sigma^{-1}\Sigma_k). \tag{5.27}$$

See Section A.12 for more details about (5.26)–(5.27).

Under suitable regularity conditions, maximum likelihood estimators are approximately normally distributed about the true parameters

$$\begin{bmatrix} \hat{\mu} \\ \hat{\theta} \end{bmatrix} - \begin{bmatrix} \mu \\ \theta \end{bmatrix} \sim N_{p+1}(\mathbf{0}, \mathcal{I}^{-1}).$$

In particular, $\hat{\mu}$ and $\hat{\boldsymbol{\theta}}$ are asymptotically independent. The regularity conditions can be described as follows:

(a) $\sigma(h; \boldsymbol{\theta})$ is twice continuously differentiable with respect to $\boldsymbol{\theta}$, and the true value of $\boldsymbol{\theta}$ lies in the interior of the parameter space.
(b) The smallest eigenvalue of the Fisher information matrix $\mathcal{I}^{(\theta)}$ in (5.26) is bounded away from 0 as $n \to \infty$.

The most important example of a covariance model that does not satisfy these conditions is a spherical scheme (see Exercise 5.1). Numerical problems with maximization of the likelihood for a spherical scheme are extensively discussed in the literature; see Warnes and Ripley (1987) and Mardia and Watkins (1989), and, for a more recent discussion, Zimmerman (2010). The failure to satisfy the regularity conditions may be a contributory factor. See also Section B.3.2.

One important situation is not entirely covered by the assumptions here. Suppose that the model includes a nugget variance $\tau^2 \geq 0$, but that the true value of τ^2 lies on the boundary $\tau^2 = 0$. In this case, the asymptotic distribution of $\hat{\tau}^2$ follows a half-normal distribution (Mardia, 1990).

The second condition can be ensured by supposing that the covariance function exhibits short-range correlation (see Eq. (2.74)) and that the data lie in an increasing domain D as the sample size n increases ("outfill" asymptotics). Short-range correlation is a property of most of the covariance functions used in practice, such as the spherical scheme and all the Matérn covariance functions (including the exponential scheme and the limiting Gaussian scheme). Further, including a nugget effect preserves the short-range correlation property. As far as the data sites are concerned, the simplest specification in the outfill setting is to suppose that they lie on an integer lattice D of size $n = |D| = n_1 \times \cdots \times n_d$, where all the $n[\ell]$ are increasing at the same rate, $\ell = 1, \ldots, d$

$$n[\ell] = c_\ell m, \text{say}, \tag{5.28}$$

with $m \to \infty$. Call this setting "regular outfill asymptotics." The assumption of an integer lattice facilitates some of the calculations and limiting results. Many of the asymptotic results also hold for irregularly spaced data sites (Mardia and Marshall, 1984). Formally, we consider an infinite sequence of sites $t_1, t_2, \ldots$ that fill an expanding region as n increases. In particular, the spacing between the data sites is not allowed to get too fine as n increases. That is, we suppose there is a constant $c > 0$ such that

$$|t_i - t_j| \geq c, \quad \text{for all } i, j \geq 1. \tag{5.29}$$

Call this setting "irregular outfill asymptotics."

For lattice data on a rectangular integer domain, it can be more straightforward to verify sufficient regularity conditions on covariance functions by moving to the spectral domain. Let $f_T(\omega; \boldsymbol{\theta})$ denote the spectral density, defined on the torus $\omega \in (-\pi, \pi)^d$.

(a) $f_T(\omega; \boldsymbol{\theta})$ is twice continuously differentiable with respect to θ.
(b) $f_T(\omega; \boldsymbol{\theta})$ is bounded away from 0 and ∞ and the first two derivatives of $f_T(\omega; \boldsymbol{\theta})$ with respect to the elements of $\boldsymbol{\theta}$ are bounded in absolute value.

The asymptotic bias of maximum likelihood estimators is discussed in Section 5.11 for the more general setting of the spatial linear model.

5.6 Parameterization Issues for the Matérn Scheme

A popular model for stationary data is the Matérn covariance function, in (2.34), which can be written in the form

$$\sigma(h) = \frac{2\sigma^2}{\Gamma(\nu)} \left(\frac{1}{2}\,|h|/\phi\right)^\nu K_\nu(|h|/\phi),$$

with three parameters: the marginal variance $\sigma^2 = \sigma(0)$, the range parameter $\varphi > 0$, and the index parameter $\nu > 0$. Except in very densely spaced data sets, it is difficult to estimate the index parameter. Hence, ν is usually treated as a known parameter. Typical values are the half-integers $\nu = 0.5, 1.5, 2.5$ because in these cases the Bessel function can be expressed in terms of elementary functions such as exponentials and powers. In particular, the choice $\nu = 0.5$ corresponds to the exponential scheme. In practice, it turns out the MLEs of σ^2 and φ can be very highly correlated when φ is large. Hence, it is useful to reparameterize σ^2, as in Section 5.14 on infill asymptotics, to

$$\sigma_e^2 = \sigma^2/\varphi^{2\nu}. \tag{5.30}$$

Then, it can be shown that the MLEs of σ_e^2 and φ are much less correlated than σ^2 and φ. This reparameterization (5.30) was suggested by Zhang (2004), motivated by arguments based on infill asymptotics; see Section 5.14.

5.7 Maximum Likelihood Examples for Stationary Models

In this section, we illustrate maximum likelihood (ML) estimation using two examples, the bauxite data and the elevation data. We focus on the Matérn scheme and focus on to several issues.

(a) Whether a nugget effect is present
(b) The effect of varying the Matérn index parameter ν
(c) Numerical issues when the range parameter is large

As discussed in Section A.12.7, if a statistical model has an extra parameter added to it, the more complicated model is preferred only if the log-likelihood increases sufficiently. If a hypothesis test is used, the maximized likelihood should be increased by a value of at least $3.84/2 = 1.92$ to be judged significant (3.84 is the upper 5% critical value of the χ_1^2 distribution). If the Akaike information criterion (AIC) criterion is used, the log-likelihood should be increased by at least 1 to justify the "expense" of the more complicated model. These rules give rough guidelines about whether a nugget effect is important.

Example 5.5 ***ML estimation for the bauxite data, stationary Matérn models***
Consider the bauxite data given earlier. In many ways, the simplest model is an exponential scheme, i.e. Matérn with index $\nu = 0.5$, possibly with a nugget effect. The first two rows of Table 5.1 give parameter estimates for this model both without and with the nugget term. For some parameters, we work with log parameters in an attempt to improve the symmetry of the confidence intervals. Standard errors are produced using the expected Fisher information matrix.

Table 5.1 Parameter estimates (and standard errors) for the bauxite data using the Matérn model with a constant mean and with various choices for the index ν.

Index ν	Nugget	$\hat{\mu}$	$\log\hat{\sigma}^2$	$\log\hat{\sigma}_e^2$	$\log\hat{\varphi}$	$\hat{\tau}^2$	$\log L$
0.5	No	13.7 (3.7)	4.73 (0.33)	4.49 (0.35)	0.24 (0.47)	0 (0)	−120.5
0.5	Yes	14.6 (4.5)	4.55 (0.63)	3.88 (1.26)	0.67 (0.96)	23 (44)	−120.2
1.5	Yes	14.8 (4.5)	4.44 (0.58)	4.07 (1.79)	0.12 (0.59)	39 (24)	−119.8
2.5	Yes	14.8 (4.5)	4.44 (0.57)	5.37 (2.38)	−0.19 (0.48)	41 (20)	−119.7

For $\nu = 0.5$, the likelihood increases by only $-120.2 + 120.5 = 0.3$, when the nugget term is added, which is not significant. Hence, we choose the model without the nugget effect. Note that the inclusion of the nugget term is accompanied by an increase in the estimate of φ ($\log\varphi$ increases from 0.24 to 0.67). The reason is that the nugget term implies greater independence between the observations. This effect is counterbalanced by increasing the range parameter, which implies increasing dependence between the observations. Whether a solution to the ML equations for the global maximum is well behaved can be checked by plotting the profile log-likelihood in one parameter after maximizing over other parameters. The profile log-likelihood for φ, when no nugget is present, is given by (5.20) with $\tau^2 = 0$. In other cases, the profile log-likelihood must be calculated numerically.

Using the ideas from Section A.12.5, confidence intervals can also be computed from profile log-likelihood plots. Plots of profile log-likelihoods for an exponential model with no nugget term are given in Figure 5.3. The parameters are $\log\varphi$ for the range and either $\log\sigma^2$ or $\log\sigma_e^2$ for the scale parameter. In all cases, the likelihood-based intervals are slightly right-skewed. In Section 5.6, it was argued that when the range parameter is large, the modified scale parameter σ_e^2 in (5.30) can be estimated more accurately than σ^2. However, in this example with $\nu = 0.5$ and no nugget, the estimated range parameter $\hat{\varphi} = \exp(0.24) = 1.3$ is not large relative to the domain of the data (a square of size 6×6). Indeed, it turns out that for $\nu = 0.5$, $\mathrm{corr}(\hat{\sigma}^2, \hat{\varphi}) = 0.66$ is not too large, and $\log\hat{\sigma}_e^2$ has a slightly larger estimated standard error than $\log\hat{\sigma}^2$. The bottom two rows of Table 5.1 indicate what happens with other choices for ν. The main change is that the nugget effect increases. The reason is that the semivariogram changes from a linear function to a quadratic function at $|h| = 0$, and a larger nugget term is needed to accommodate the gap. Figure 5.4 gives the fitted semivariograms for the three models with nugget effects.

In Example 5.8, a quadratic drift is included in the model to fit the basin-shaped pattern in the data. □

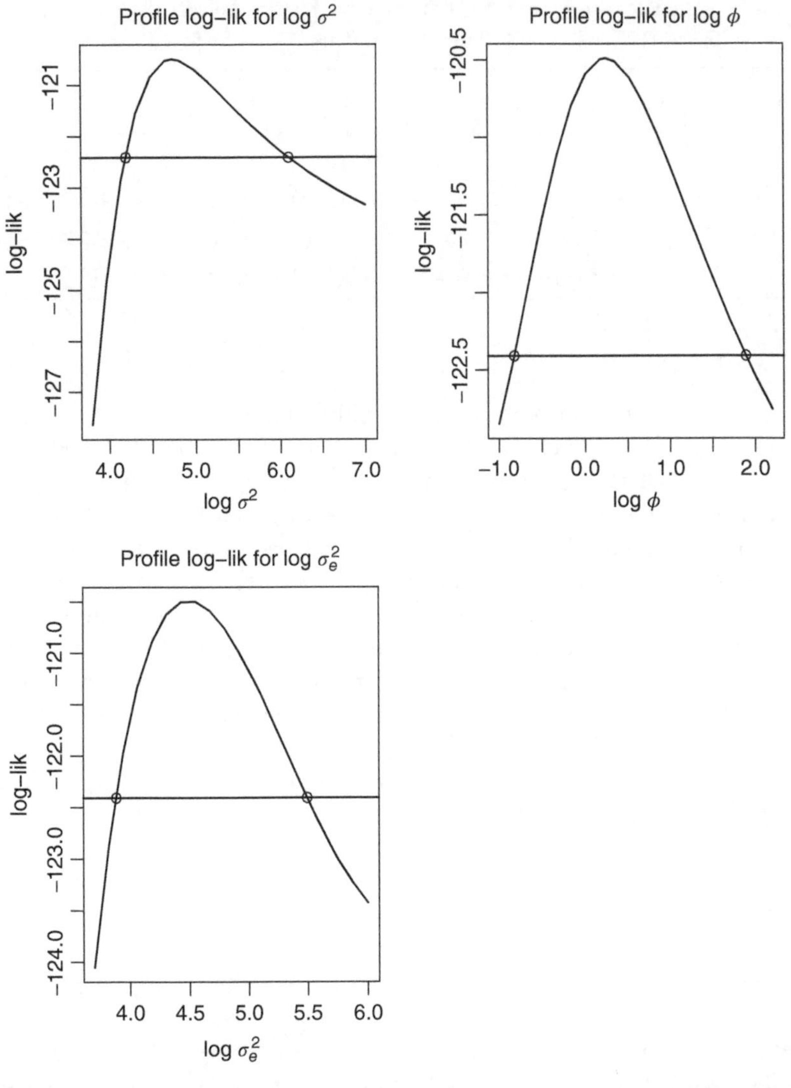

Figure 5.3 Bauxite data: Profile log-likelihoods together with 95% confidence intervals. Exponential model, no nugget effect.

Example 5.6 ***ML estimation for the elevation data, stationary Matérn models***

Consider the elevation data described earlier. Davis (1973) analyzed the data, and it was reanalyzed subsequently by various authors (e.g., Mardia, 1990; Ripley, 1988; Warnes and Ripley, 1987; Diggle and Ribeiro, 2007). Here, we fit several isotropic stationary models to illustrate some of the issues in maximum likelihood estimation.

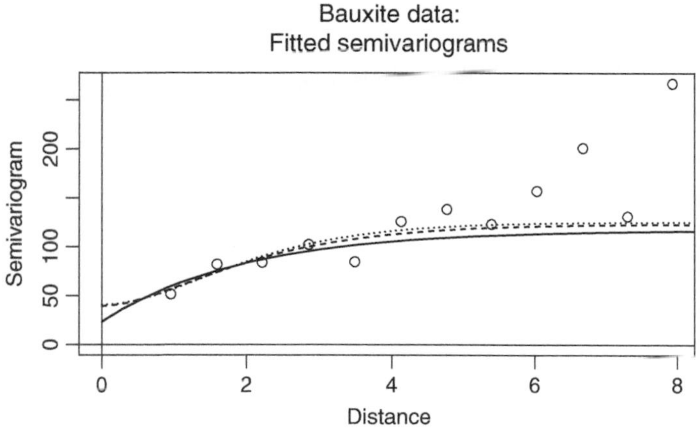

Figure 5.4 Bauxite data: sample isotropic semivariogram values and fitted Matérn semivariograms with a nugget effect, for $\nu = 0.5$ (solid), $\nu = 1.5$ (dashed), and $\nu = 2.5$ (dotted).

Consider various Matérn models with different levels of smoothness, $\nu = 0.5, 1.5, 2.5$. Table 5.2 gives the MLEs and standard errors. As in Example, 5.5, we also include the parameterization involving σ_e^2 instead of σ^2.

To begin with, look at the exponential model ($\nu = 0.5$). Note that the estimated range parameter $\hat{\varphi} = \exp(1.8) = 6.1$ is nearly the same as the width of the data set. That is, the fitted model has a large range parameter, which causes confounding between the estimates of φ and σ^2. Indeed, it turns out that the estimated correlation between their estimates is $\text{corr}(\hat{\sigma}^2, \hat{\varphi}) = \text{corr}(\log \hat{\sigma}^2, \log \hat{\varphi}) = 0.97$. This confounding can be partly removed by using the parameters σ_e^2 and φ; the correlation between the estimates is now reduced to -0.31. Also note that the standard error of $\log \hat{\sigma}_e^2$ (0.2) is much smaller than the standard error of $\log \hat{\sigma}^2$ (0.8). However, the standard error of $\log \hat{\varphi}$ is very wide (0.8) (yielding a very wide 95% confidence interval for φ of (1.3, 30)), reflecting the fact that there is insufficient independent information in the data to estimate this quantity accurately.

Table 5.2 Parameter estimates (and standard errors) for the elevation data using the Matérn model with a constant mean and with various choices for the index ν.

Index ν	Nugget	$\hat{\mu}$	$\log \hat{\sigma}^2$	$\log \hat{\sigma}_e^2$	$\log \hat{\varphi}$	$\hat{\tau}^2$	$\log L$
0.5	No	864 (45)	8.32 (0.77)	6.50 (0.21)	1.81 (0.81)	0 (0)	−244.6
0.5	Yes	864 (45)	8.32 (0.77)	6.50 (0.36)	1.81 (0.86)	0	−244.6
1.5	Yes	848 (28)	8.2 (0.5)	7.6 (0.4)	0.18 (0.25)	48 (46)	−242.1
2.5	Yes	845 (24)	8.1 (0.43)	9.6 (0.7)	−0.30 (0.18)	71 (44)	−242.3

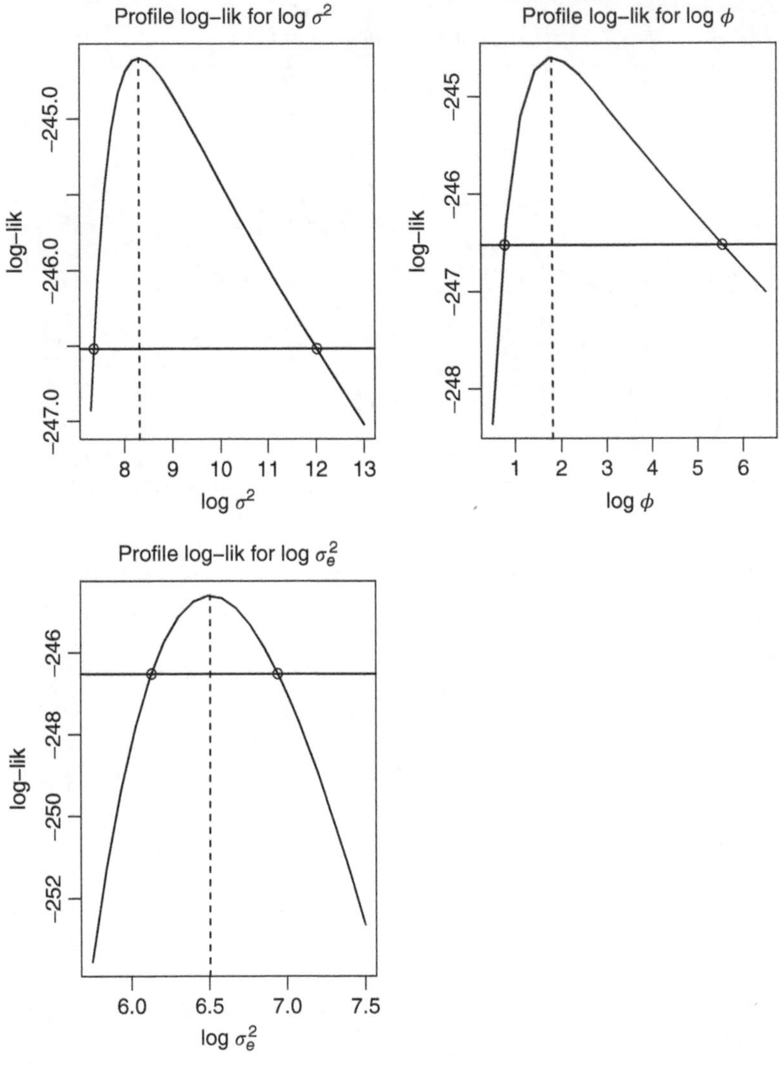

Figure 5.5 Elevation data: Profile log-likelihoods together with 95% confidence intervals. Exponential model, no nugget effect.

Confidence intervals for the parameters can also be produced from the profile log-likelihood for each parameter. Plots of profile log-likelihoods for the exponential model with no nugget term are given in Figure 5.5. The likelihood-based interval for $\log \varphi$ is slightly right-skewed, perhaps reflecting the problems described in the last paragraph. However, the likelihood-based interval for $\log \hat{\sigma}_e^2$ is fairly symmetric.

Next, consider the index values $\nu = 1.5, 2.5$. Note that there is no evidence of a nugget effect when $\nu = 0.5$, but the nugget effect becomes more prominent as ν increases. The reason is that the Matérn covariance function becomes smoother near $h = 0$ as ν increases. The result is a poorer fit between the fitted model (without the nugget term) and the observed semivariograms for small h, which is partly cured by adding a nugget term. Overall, in terms of likelihoods, there is little to choose between $\nu = 0.5$ without a nugget effect and a variety of choices for ν with a nugget effect. □

5.8 Restricted Maximum Likelihood (REML)

As in Section 5.5.2, let $\boldsymbol{y}$ be an observation from $N_n(\mu \mathbf{1}_n, \Sigma)$, where μ is a constant mean and where Σ depends on a set of unknown parameters $\boldsymbol{\theta}$, as in Section 5.5.2. The log-likelihood for μ and $\boldsymbol{\theta}$ is given by (5.16). The "restricted maximum likelihood principle" (REML) states that estimation of these two sets of parameters should be separated. That is, rather than estimating μ and $\boldsymbol{\theta}$ jointly, $\boldsymbol{\theta}$ should be estimated first using a restricted likelihood, after which μ can be estimated by (5.18).

The REML principle was set out in the spatial context by Harville (1977). There are two objectives: (i) to lessen or remove the bias of the covariance parameters and (ii) to simplify the likelihood equations.

The restricted likelihood is constructed using increments of the data. As in Section 5.5.2, let G be an $n \times (n-1)$ column orthonormal matrix, not depending on the data, such that each column is orthogonal to the constant vector of ones, so that $G^T\mathbf{1} = \mathbf{0}$ and $G^TG = I_{n-1}$ and set $\boldsymbol{z} = G^T\boldsymbol{y}$ to be an $(n-1)$ vector of contrasts in the data, from an $N_{n-1}(\mathbf{0}, \Sigma_{GG})$ distribution, where $\Sigma_{GG} = G^T\Sigma G$. The log-likelihood based on $\boldsymbol{z}$ for the covariance parameters $\boldsymbol{\theta}$ is given by

$$\log L(\boldsymbol{\theta}; \boldsymbol{z}) = -\frac{1}{2}\{\boldsymbol{z}^T\Sigma_{GG}^{-1}\boldsymbol{z} + \log|\Sigma_{GG}| + n'\log(2\pi)\}, \tag{5.31}$$

where $n' = n - 1$. Note that the log-likelihood no longer depends on μ.

As before, let $\boldsymbol{\theta}$ be partitioned into an overall scale parameter and the remaining parameters, $\boldsymbol{\theta} = (\boldsymbol{\theta}_c^T, \sigma^2)^T$, and write $\Sigma_{GG} = \sigma^2 P_{GG}$, where $P_{GG} = G^TPG$ is the transformed correlation matrix for $\boldsymbol{x}$. Note P_{GG} itself is not a correlation matrix since its diagonal elements are not equal to 1. However, (5.31) can still be maximized first over σ^2 to give a profile log-likelihood

$$\log L_{\text{profile,REML}}(\boldsymbol{\theta}_c; \boldsymbol{z}) = -\frac{1}{2}\left\{n' + n'\log(\boldsymbol{z}^TP_{GG}^{-1}\boldsymbol{z}/n') + \log|P_{GG}| + n'\log(2\pi)\right\}. \tag{5.32}$$

It is interesting to contrast the REML likelihood with the full maximum likelihood profile log-likelihood (5.20). The two log-likelihoods have the same quadratic form but different multipliers, n and n'. Further, the determinants are related by

$$\log|P| = \log|P_{GG}| - \log(\mathbf{1}^T P^{-1}\mathbf{1}_n/n). \tag{5.33}$$

See Exercise 5.3 for more details.

In the setting of n i.i.d. $N(\mu, \sigma^2)$ observations, the difference between the full likelihood approach and the REML approach is the choice of either n or $n-1$ as the divisor in the estimate of σ^2, respectively. The latter choice is generally preferred. The REML approach can also be used for intrinsic processes, and is indeed the only approach available. Suppose $\boldsymbol{y}$ contains n observations from an intrinsic process of intrinsic order 0 with a semivariogram $\gamma(h)$ and set $\sigma(h) = -\gamma(h)$ to be the intrinsic covariance function, defined up to a constant term. Then, $\boldsymbol{z} = G^T\boldsymbol{y} \sim N_{n-1}(\mathbf{0}, G^T\Sigma G)$ and the log-likelihood continues to take the form (5.31). When the drift just consists of a constant mean, REML and ML are generally similar. However, in more complicated settings (i.e. nontrivial drift as in Section 5.11), REML can be more sensitive to misspecification than MLE. See, e.g., Diggle and Ribeiro (2007, p. 117).

5.9 Vecchia's Composite Likelihood

If n is large, computation of the MLEs can be computationally prohibitive because of the need to invert and multiply $n \times n$ matrices; each of these operations involves $O(n^3)$ calculations. In this section, we develop an approximate likelihood due to Vecchia (1988) that is much quicker to compute.

In principle, Vecchia's method described here works with any n-dimensional random vector with a multivariate normal distribution, $\boldsymbol{x} \sim N_n(\boldsymbol{\mu}, \Sigma)$, where $\boldsymbol{\mu}$ is an n-dimensional vector and where $\Sigma = \Sigma(\theta)$ with elements $\sigma_{ij} = \sigma_{ij}(\boldsymbol{\theta})$ depends on a p-dimensional set of parameters $\boldsymbol{\theta}$.

First, suppose the components of $\boldsymbol{x}$, written $x_1, \ldots, x_n$, have been listed in a specific order. From the properties of conditional likelihood, the joint probability density of $\boldsymbol{x}$ can be split into a product of conditional densities

$$f(\boldsymbol{x}) = \prod_{i=1}^{n} f(x_i | x_j : j = 1, \ldots, i-1), \tag{5.34}$$

where in the first factor $f(x_1)$, the conditioning set is empty. For a Gaussian random field, each of these conditional densities is a one-dimensional normal density, which can be written in closed form. So far there has been no simplification in the computation. However, Vecchia suggested replacing the full conditioning set by a smaller set, which we call the "partial past." Let $P_i \subset \{1, \ldots, i-1\}$ be a subset of the indices preceding i, for $i = 1, \ldots, n$ and consider the product of these conditional densities

$$L(\boldsymbol{\theta};\boldsymbol{x}) = \prod_{i=1}^{n} f(x_i|x_j : j \in P_i; \boldsymbol{\theta}), \tag{5.35}$$

where the dependence of the conditional densities on a parameter vector $\boldsymbol{\theta}$ is now highlighted. In general, the phrase "composite likelihood" is used to describe a product of selected marginal and conditional densities. Hence, Vecchia's likelihood (5.35) is an example of a composite likelihood, which can be viewed as an approximation to the full likelihood. If the sets P_i are not large, then the composite likelihood will be quick to compute. The preceding discussion is valid for any multivariate vector $\boldsymbol{x}$. Next, we consider how to construct the subsets P_i in a spatial setting. If the data lie on a rectangular lattice, and if the ordering of the sites is determined by a half-space $\mathcal{H}$ as in Section 4.8.1, then a natural choice for the partial past is to take a shifted version of a single set $D_0 \subset \mathcal{H}$

$$P_i = \{j : \; t_j \in t_i - D_0\}. \tag{5.36}$$

For example, using the lexicographic half-space $\mathcal{H} = \mathcal{L}$ in $\mathbb{Z}^2$, recall that a site $t = (t[1], t[2])$ lies in $\mathcal{L}$ if either $t[1] > 0$ or $t[1] = 0$ and $t[2] > 0$. Hence, a possible choice for D_0 is

$$D_0 = \{(0,1), (1,-1), (1,0), (1,1)\}. \tag{5.37}$$

In this example, D_0 consists of a set of sites close to the origin, so the composite likelihood will be good at estimating features of the covariance function depending on short-range dependence.

A number of suggestions have been made in the literature about implementing Vecchia's ideas in practice. Stein et al. (2004) argue that it is also a good idea to include a few sites more distant from the origin to help estimate features of the covariance function depending on longer range dependence. Pardo-Igúzquiza and Dowd (1997) suggest taking the sites in a random order with a random selection of k sites from the past, where k typically lies between 10 and 20. Mardia (2011), in a discussion of work by Pardo-Igúzquiza and Dowd (1997), reports that it is better to take the sites in lexicographic order, again with a random selection of sites from the past. Caragea and Smith (2007) argue for organizing the sites into blocks and using a multivariate version of Vecchia's approach. Thus, the use of Vecchia-type approximations continues to be an active area of current research, and it remains to be seen what turns out to be the best approach in practice.

The density in (5.35) can be expressed as

$$f(\boldsymbol{x}) = \prod (2\pi\sigma_i^2)^{-1/2} \exp\left\{ -\frac{1}{2\sigma_i^2}(\boldsymbol{\ell}_i^T(\boldsymbol{x} - \mu\mathbf{1}_n))^2 \right\}, \tag{5.38}$$

where $\boldsymbol{\ell}_i$ is the vector of coefficients for the residual of the regression of x_i on the previous components, with variance $\sigma_i^2 = \boldsymbol{\ell}_i^T \Sigma \boldsymbol{\ell}_i$. The vector $\boldsymbol{\ell}_i$ takes the form

$$\begin{aligned} \ell_{ij} &= 1, \quad j = i, \\ (\ell_{ij}, j \in P_i) &= -\Sigma_i^{-1} \boldsymbol{\alpha}_i, \\ \ell_{ij} &= 0, \text{otherwise.} \end{aligned} \tag{5.39}$$

Here, Σ_i is the submatrix of Σ and $\boldsymbol{\alpha}_i$ is the subvector of the ith column of Σ corresponding to the sites in P_i. If $\Sigma = (\sigma_{ij}(\boldsymbol{\theta}))$ depends on a parameter vector $\boldsymbol{\theta}$, then it is straightforward to write down the derivatives of $\boldsymbol{\ell}_i$ and σ_i^2 with respect to $\boldsymbol{\theta}$ in terms of the derivatives $d\sigma(h; \boldsymbol{\theta})/d\boldsymbol{\theta}$. When $P_i = \{1, \dots, i-1\}$ consists of the full past, this representation reexpresses the joint density of dependent observations as a product of densities of independent random variables. In the $d = 1$-dimensional setting, this factorization approximates the description of a stationary time series as a (possibly infinite) autoregression representation on its past plus an innovation error. The evidence so far suggests that Vecchia's method can have high statistical efficiency using only a moderately sized sets P_i. For an illustration, see Example 5.7.

5.10 REML Revisited with Composite Likelihood

In Section 5.9, the linear combination $e_i = \boldsymbol{\ell}_i^T(\mathbf{x} - \mu \mathbf{1}_n), i \geq 1$ in (5.38) is the residual after predicting x_i from some of the previous components using the best linear predictor (BLP). However, $\boldsymbol{\ell}_i^T \mathbf{1}_n \neq 0$ so that the resulting likelihood still involves μ as well as the vector of parameters $\boldsymbol{\theta}$ in $\sigma(h; \boldsymbol{\theta})$. Another way to proceed, which removes any dependence of the likelihood on μ, is to use REML, which is constructed from the distribution of residuals from the best linear unbiased predictors (BLUPs). This approach is due to Kitandis (1991) and Stein et al. (2004). To describe the BLUP methodology, it is helpful to set up the notation in a general multivariate setting. Let $\boldsymbol{u}$ be a p-dimensional random vector and let v be a random variable. Suppose the means are constant, $E(u_i) = \mu, i = 1, \dots, p$, and $E(v) = \mu$, with covariance structure

$$\text{var}(\boldsymbol{u}) = A \ (p \times p), \quad \text{cov}(\boldsymbol{u}, v) = \mathbf{b} \ (p \times 1), \quad \text{var}(v) = c \ (\text{scalar}).$$

A linear predictor of v based on $\boldsymbol{u}$ takes the form $\hat{v} = f + \mathbf{g}^T \boldsymbol{u}$ for some scalar f and a p-vector of coefficients $\mathbf{g}$. The predictor is called unbiased if the residual $e = v - \hat{v}$ has mean $E(e) = 0$ for all μ, which implies that $f = 0$ and $\mathbf{1}_n^T \mathbf{g} = \sum g_i = 1$. The linear unbiased predictor is called *best* if $\text{var}(v - \hat{v})$ is minimized. It is straightforward to show that the BLUP is uniquely determined with $f = 0$ and

$$\mathbf{g} = A^{-1} \left(\mathbf{b} + \frac{1 - q_{1b}}{q_{11}} \mathbf{1}_p \right) \tag{5.40}$$

with residual variance

$$\tau^2 = \text{var}(e) = c - q_{bb} + (q_{1b} - 1)^2/q_{11}$$

in terms of the quadratic forms

$$q_{11} = \mathbf{1}_p^T A^{-1}\mathbf{1}_p, \quad q_{1b} = \mathbf{1}_p^T A^{-1}\mathbf{b}, \quad q_{bb} = \mathbf{b}^T A^{-1}\mathbf{b}.$$

More details can be found in Section 7.4 in the context of kriging.

Now let us return to the spatial setting, with data $\mathbf{y} = (y_1, \ldots, y_n)^T$ at sites $t_1, \ldots, t_n$, where we assume a constant mean $E(y_i) = \mu, i = 1, \ldots, n$ and a covariance matrix

$$\text{var}(\mathbf{y}) = \Sigma = (\sigma_{jj'} : 1 \le j, j' \le n).$$

The REML likelihood for $\mathbf{y}$ can be constructed using the ideas from Section 5.8, and it can be recast in terms of BLUPs, where for each $i = 2, \ldots, n$ we predict $v = y_i$ using $\boldsymbol{u} = (y_1, \ldots, y_{i-1})^T$. Define the residual e_i^U, say, as the linear combination $e_i^U = (\boldsymbol{\ell}_i^U)^T\mathbf{y}$, where $\boldsymbol{\ell}_i^U$ is the unique n-vector $\boldsymbol{\ell}_i$ satisfying the BLUP constraints at site i; that is,

(i) $\ell_{i,i+1} = \cdots = \ell_{i,n} = 0$.
(ii) $\ell_{ii} = 1$.
(iii) $\boldsymbol{\ell}_i$ is an increment, that is, $\mathbf{1}^T\boldsymbol{\ell}_i = 0$.
(iv) $\text{var}(\boldsymbol{\ell}_i^T\mathbf{y})$ is minimal among all linear combinations satisfying (i)–(iii).

The superscript "U" is used to indicate the BLUP predictor. The subvector $(-\ell_{i1}^U, \ldots, -\ell_{i,i-1}^U)^T$ is the vector of predictor coefficients labeled $\mathbf{g}$ in (5.40). The matrix A in (5.40) is the covariance matrix of $y_1, \ldots, y_{i-1}$ and the vector $\boldsymbol{b}$ contains the covariances between y_i and $y_1, \ldots, y_{i-1}$. Properties (i) and (ii) ensure that $e_i^U = (\boldsymbol{\ell}_i^U)^T\mathbf{y}$ is a difference between y_i and its predicted value based on $y_1, \ldots, y_{i-1}$; property (iii) ensures that the predictor is unbiased; and property (iv) ensures that this predictor is "best." See Exercise 5.5 for more details.

It is also the case that the random variables e_i^U are uncorrelated. To verify this claim, suppose, if possible, for some $j < i$ that $\text{cor}(e_i^U, e_j^U) = \rho \neq 0$. Then, for any real value c, $\boldsymbol{\ell}_i = \boldsymbol{\ell}_i^U - c\boldsymbol{\ell}_j^U$ is a coefficient vector satisfying the first three BLUP constraints (i)–(iii) at site i. However, for $c = \rho\tau_i/\tau_j$, $\text{var}\{\boldsymbol{\ell}_i^T\mathbf{y}\} = (1-\rho^2)\tau_i^2 < \tau_i^2$, which would contradict the optimality of $\boldsymbol{\ell}_i^U$. Hence, $\rho = 0$. Here $\tau_i^2 = \text{var}\{(\boldsymbol{\ell}_i^U)^T\mathbf{y}\} = (\boldsymbol{\ell}_i^U)^T\Sigma\boldsymbol{\ell}_i^U$.

Now, additionally suppose that the data are normally distributed, and collect the vectors $\boldsymbol{\ell}_i$ together as the rows of an $(n-1)\times n$ matrix L. The preceding paragraph ensures that $L\Sigma L^T$ is diagonal, with elements τ_i^2. Then the REML log-likelihood can be written in the following form based on the BLUP residuals:

$$\log L_{\text{profile,REML},1} = -\frac{1}{2}\{\mathbf{y}^T L^T (L\Sigma L^T)^{-1} L\mathbf{y} + \log|L\Sigma L^T| + (n-1)\log(2\pi)\} \quad (5.41)$$

$$= -\frac{1}{2}\left[\sum_{i=2}^{n}\{((\boldsymbol{\ell}_i^U)^T\mathbf{y})^2/\tau_i^2 + \log\tau_i^2\} + (n-1)\log(2\pi)\right], \quad (5.42)$$

where the ℓ_i^U and τ_i^2 depend on the vector of parameters θ through the underlying covariance function as shown in (5.40). It is confirmed in Exercise 5.7 that $\log L_{\text{profile,REML},1}$ is the same as the REML log-likelihood in (5.31), up to a constant additive term not depending on the data or the parameters.

The vector ℓ_i^U can be written more explicitly using the notation $\ell(i; \{1, \dots, i-1\})$ to indicate that $\ell(i; \{1, \dots, i-1\})^T \mathbf{y}$ is a residual after predicting y_i using the full set of past values at sites $\{1, \dots, i-1\}$. More generally, we can consider coefficient vectors $\ell(i; P_i)$, where $P_i \subset \{1, \dots, i-1\}$. This approach gives a BLUP variant of Vecchia's composite likelihood

$$\log L_{\text{profile,REML,Vecchia}} = -\frac{1}{2}\left\{\sum_{i=2}^{n} (\ell(i; P_i)^T \mathbf{y})^2 / \tau^2(i; P_i) + \log \tau^2(i; P_i) + (n-1)\log(2\pi)\right\}, \tag{5.43}$$

where $\tau^2(i; P_i) = \operatorname{var}(\ell_i(1; P_i)^T \mathbf{y})$. However, for general subsets P_i of the past, the linear combinations $\ell(i; P_i)^T \mathbf{y}$ are no longer uncorrelated for different i.

Example 5.7 ***Vecchia ML for the synthetic Landsat data***
In this example, we use the 200×200 synthetic Landsat data set from Example 1.6. Recall that the data have been simulated from a stationary Gaussian process with an exponential covariance function, with parameters $\mu = 0, \sigma^2 = 1$, $\varphi = 8, \tau^2 = 0$. The purpose is to investigate the stability and accuracy of the Vecchia ML estimates, using different-sized full unilateral lexicographic

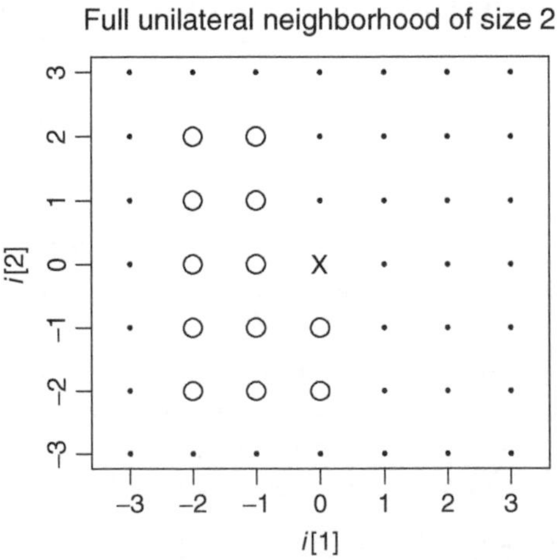

Figure 5.6 Unilateral lexicographic neighborhood of full size $k = 2$ for lattice data; current site marked by $\times$; neighborhood sites in the lexicographic past marked by $\circ$. Other sites are marked by a dot.

Table 5.3 Parameter estimates (with standard errors in parentheses) for Vecchia's composite likelihood in Example 5.7 for the synthetic Landsat data, using different sizes of neighborhood k

k	$\hat{\sigma}^2$	$\hat{\varphi}$	$\hat{\tau}^2$
1	0.984 (0.067)	8.095 (0.678)	0.039 (0.003)
2	1.054 (0.074)	8.952 (0.700)	0.041 (0.002)
3	1.020 (0.069)	8.674 (0.636)	0.042 (0.002)
4	1.027 (0.072)	8.714 (0.654)	0.042 (0.002)
5	1.011 (0.071)	8.564 (0.646)	0.041 (0.002)

The true parameter values are $\sigma^2 = 1, \varphi = 8, \tau^2 = 0$.

neighborhoods of size k. The full unilateral lexicographic neighborhood of size k is defined by

$$D = \{h = (h[1], h[2]) \,:\, \max(|h[1], |h[2]|) \leq k, \text{ and either } h[1] > 0 \text{ or } (h[1] = 0, h[2] > 0)\}$$

and is shown in Figure 5.6 for $k = 2$.

For simplicity, we assume the mean is known and focus on the covariance parameters. The parameter estimates are given in Table 5.3, together with estimated standard errors obtained from the Hessian matrix. For this example, note that the estimates are quite stable as k varies. The estimates for σ^2 and φ are well within two standard errors of the true values. However, the estimate for τ^2, though small, is significantly above 0, perhaps because the true value of τ^2 lies at the boundary of the parameter region. □

5.11 Spatial Linear Model

The preceding sections assume that the mean of the stochastic process $\{X(t)\}$ is constant. In this section, we allow the mean to vary systematically.

Definition 5.11.1 Let $\{X(t) : t \in \mathbb{R}^d\}$ be a random field, which can be decomposed as

$$X(t) = \mu(t) + Z(t), \quad t \in \mathbb{R}^d, \tag{5.44}$$

where $Z(t)$ is a zero mean stationary random field with a covariance function $\sigma(h; \boldsymbol{\theta})$, depending on a parameter vector $\boldsymbol{\theta}$, and where the mean function takes the form

$$\mu(t) = \sum_{j=1}^{q} \beta_j f_j(t) \tag{5.45}$$

in terms of known specified functions $f_j(t)$ and unknown coefficients β_j. Then, $\{X(t)\}$ is said to follow the *spatial linear model*.

Usually, it is assumed that $f_1(t) \equiv 1$, so that if $q = 1$, the random field is stationary with unknown mean β_1. In the stationary case, the name *ordinary spatial linear model* can be used to match the terminology of ordinary kriging in Chapter 7.

More generally, it is common to let the drift space be given by $\mathcal{F}_k$, the space of polynomials of degree k introduced in Section 3.4. Thus, the functions $f_j(t)$ represent powers of t. For example, if $d = 2$ and $k = 2$, then $q = 6$, where the constant, linear, and quadratic functions of t are given by $(1, t[1], t[2], t[1]^2, t[2]^2, t[1]t[2])$. When the drift space is given by $\mathcal{F}_k$, the name *universal spatial linear model* can be used to match the terminology of universal kriging in Chapter 7.

The drift functions at a set of data sites $D = \{t_1, \dots, t_n\}$ can be combined into an $n \times q$ matrix F with entries

$$f_{ij} = f_j(t_i), \quad i = 1, \dots, n, \quad j = 1, \dots, q. \tag{5.46}$$

The only condition made on the drift functions and the data sites is that F should have full column rank q.

5.11.1 MLEs

A general treatment of maximum likelihood estimation for a multivariate normal distribution with drift is given in Section A.12. Here are the key details in the current setting. Assume the same covariance structure as in the stationary case in Section 5.5. In particular, suppose that a vector of observations $\boldsymbol{y} = (y_1, \dots, y_n)^T$ at the set of sites $D = (t_1, \dots, t_n)$ in $\mathbb{R}^d$ is available from the x-process, possibly including a nugget effect. As before, let $\Sigma = (\sigma_{ij})$ where $\sigma_{ij} = \sigma(t_i - t_j)$ denote the covariance matrix of the data, depending on a p-dimensional parameter vector $\boldsymbol{\theta} = (\sigma^2, \boldsymbol{\theta}_c^T)^T$ partitioned into an overall variance σ^2 and a set of correlation parameters $\boldsymbol{\theta}_c$), and write $\Sigma = \sigma^2 P$, where P is the correlation matrix of the data, as in (5.6). The formulas for the stationary case carry over with little change. The log-likelihood is now

$$\log L = -\frac{1}{2}\{(\boldsymbol{y} - F\boldsymbol{\beta})^T \Sigma^{-1} (\boldsymbol{y} - F\boldsymbol{\beta}) + \log|\Sigma| + n\log(2\pi)\}, \tag{5.47}$$

and formula (5.18) for the estimated mean becomes the *generalized least squares (GLS)* estimator

$$\hat{\boldsymbol{\beta}} = (F^T \hat{\Sigma}^{-1} F)^{-1} F^T \hat{\Sigma}^{-1} \boldsymbol{y}, \tag{5.48}$$

for the estimated vector of regression coefficients.

Using Section A.3.7, the profile log-likelihood simplifies in the same way as (5.24) to give

$$\log L_{\text{profile,ML}}(\boldsymbol{\theta}_c; \boldsymbol{y}) = -\frac{1}{2}\{n + n\log(\boldsymbol{z}^T P_{GG} \boldsymbol{z}/n) + \log|P| + n\log(2\pi)\}, \tag{5.49}$$

where G is now an $n \times (n-q)$ column orthonormal matrix, whose columns are orthogonal to the drift matrix, $G^T G = I_{n-q}, G^T F = 0$. Also, $\boldsymbol{z} = G^T \boldsymbol{y}$ and $P_{GG} = G^T P G$. The profile log-likelihood needs to be maximized numerically over the correlation parameters $\boldsymbol{\theta}_c$. The asymptotic distribution of $(\boldsymbol{\beta}, \boldsymbol{\theta})$ takes a form similar to the stationary case in Eq. (5.26). For the spatial linear model, the Fisher information matrix is given by

$$\mathcal{I} = \begin{bmatrix} \mathcal{I}^{(\beta)} & 0 \\ 0 & \mathcal{I}^{(\theta)} \end{bmatrix} = \begin{bmatrix} F^T \Sigma^{-1} F & 0 \\ 0 & A \end{bmatrix}, \tag{5.50}$$

where A is defined in (5.27).

Example 5.8 ***Bauxite data with quadratic drift***
In Example 5.5 for the bauxite data, a stationary model using the exponential covariance function with no nugget term was selected. But since the data exhibit a basin shape (see Example 1.4), it is also natural to consider fitting a quadratic drift, with spatial dependence still modeled by an exponential covariance function with no nugget term.

For this quadratic drift model, it turns out that the estimated range parameter is $\hat{\varphi} = 0$, i.e. there is no spatial dependence in the data and the covariance matrix is proportional to the identity. The log-likelihood is $\log L = -113.4$.

As noted earlier, one way to choose between statistical models is to minimize the AIC. For the stationary model with no nugget in Table 5.1, and for the quadratic drift model with independent errors (i.e., $\phi = 0$), these values are

$$\text{AIC}_{\text{stationary}} = 2(3 - (-120.2)) = 246.4 > \text{AIC}_{\text{quadratic}} = 2(7 - (-113.4)) = 240.8.$$

Hence, there is some evidence to prefer the quadratic model. Note that the stationary model has three parameters (a scale parameter, a range parameter, and a mean). The quadratic drift model includes five more drift parameters (two linear terms and three quadratic terms), but does not include the range parameter, for a total of seven parameters. □

For the bauxite data, it seems clear-cut to prefer the quadratic drift model over the stationary model. However, in general, it can be challenging to model spatial data for which nearby sites tend to have similar values. This pattern can be modeled through spatial dependence or drift, or some combination of both. Of course, the bauxite data set is small. If it had been observed over a larger region, then more local extreme points in the data might have appeared and the quadratic drift (or even higher order polynomial drift) would be less appealing. Similarly, if the model is extrapolated beyond the domain of observation, then the quadratic drift will typically deviate more quickly from the observed data values. Knowledge of the application area can offer guidance about whether a drift is sensible or not.

5.11.2 Outfill Asymptotics for the Spatial Linear Model

Regularity conditions under outfill asymptotics were discussed for the stationary case in Section 5.5.3. In the setting of the spatial linear model, it is further necessary to assume that there is enough growth of the domain in all dimensions to ensure that the regression parameters can be consistently estimated. Then, it can be shown (Mardia and Marshall, 1984) that the eigenvalues of $\mathcal{I}$ increase at a rate of at least n, where

$$\mathcal{I} = \begin{bmatrix} (F^T\Sigma^{-1}F)^{-1} & 0 \\ 0 & A^{-1} \end{bmatrix}, \tag{5.51}$$

where A is given in (5.27), and that the maximum likelihood estimates are asymptotically normally distributed

$$\begin{bmatrix} \hat{\boldsymbol{\beta}} \\ \hat{\boldsymbol{\theta}} \end{bmatrix} \sim N_{p+1}\left(\begin{bmatrix} \boldsymbol{\beta} \\ \boldsymbol{\theta} \end{bmatrix}, \mathcal{I} \right). \tag{5.52}$$

In particular, $\hat{\boldsymbol{\beta}}$ and $\hat{\boldsymbol{\theta}}$ are asymptotically independent.

Next, consider the asymptotic bias. From the Mardia–Watkins Theorem in Section A.13.2, the bias in $\boldsymbol{\beta}$ is negligible ($E(\hat{\boldsymbol{\beta}} - \boldsymbol{\beta}) = o(n^{-1})$). The bias in $\boldsymbol{\theta}$ is typically of order $1/n$ and takes the form

$$E(\hat{\boldsymbol{\theta}}) - \boldsymbol{\theta} = (\mathcal{I}^{(\theta)})^{-1}\boldsymbol{\delta} + o(n^{-1}), \tag{5.53}$$

where $\boldsymbol{\delta}$ is a vector of length p, with components

$$\delta_i = \frac{1}{2}\text{tr}\left\{(\mathcal{I}^{(\beta)})^{-1}\partial\mathcal{I}^{(\beta)}/\partial\theta_i\right\} + \frac{1}{2}\text{tr}\left\{(\mathcal{I}^{(\theta)})^{-1}M_i\right\}, \tag{5.54}$$

where M_i is a $p \times p$ matrix with elements

$$(M_i)_{jk} = \frac{1}{2}\text{tr}(\Sigma_{ij}\Sigma^{-1}\Sigma_k\Sigma^{-1} - \Sigma_{ik}\Sigma^{-1}\Sigma_j\Sigma^{-1} - \Sigma_{jk}\Sigma^{-1}\Sigma_i\Sigma^{-1}), \tag{5.55}$$

and $\Sigma_i = \partial\Sigma/\partial\theta_i$ and $\Sigma_{ij} = \partial^2\Sigma/\partial\theta_i\partial\theta_j$ for $i, j, k = 1, \ldots, p$.

5.12 REML for the Spatial Linear Model

In the setting of the spatial linear model, the REML approach involves projecting the data onto the orthogonal complement of the drift space, represented by the $p \times q$ matrix F from Eq. (5.46). Let G $(n \times (n-q))$ be a column orthonormal matrix orthogonal to the columns of F; that is,

$$G^TG = I_{n-q}, \quad G^TF = 0. \tag{5.56}$$

The matrix G is not uniquely determined; it is determined only up to multiplication on the right by an orthogonal $(n-q) \times (n-q)$ orthogonal matrix R, say.

However, the span of the columns of G is unique, which is all that matters for REML estimation. Then, the REML log-likelihood is

$$\log L = -\frac{1}{2}\{\mathbf{y}^T G(G^T \Sigma G)^{-1} G^T \mathbf{y} + \log |G^T \Sigma G| + n \log(2\pi)\} \tag{5.57}$$

(Kitandis, 1983). It can be easily checked that the log-likelihood is unchanged if G is rotated on the right to GR.

As described in Section A.12.4, it is helpful to separate an overall scale parameter from the remaining parameters in $\boldsymbol{\theta}$, i.e. $\Sigma = \sigma^2 P = \sigma^2 P(\boldsymbol{\theta}_c)$. Then the log-likelihood can be maximized analytically over σ^2, to give a profile log-likelihood to be maximized numerically over $\boldsymbol{\theta}_c$.

5.13 Intrinsic Random Fields

Suppose the data $\{x(t_i) : i = 1, \ldots, n\}$ come from a pth order intrinsic process, $p \geq 0$. For an intrinsic process, the covariance function is defined only up to an equivalence class of functions. Thus, the covariance matrix Σ is not well defined.

To study intrinsic processes, it is first necessary to consider the "drift matrix" F $(n \times r)$ whose columns contain the r, say, monomials in t up to order p, evaluated at n data sites.

The most important case is $p = 0$ for which F has one column, given by

$$F^T = [1, \ldots, 1].$$

The cases $p = 1$ and $p = 2$ are also important. In $d = 2$ dimensions, the ith row of F is given by

$$(p = 1) \quad f_i^T = \begin{bmatrix} 1 & t_i[1] & t_i[2] \end{bmatrix},$$
$$(p = 2) \quad f_i^T = \begin{bmatrix} 1 & t_i[1] & t_i[2] & t_i[1]^2 & t_i[1]t_i[2] & t_i[2]^2 \end{bmatrix},$$

where $t_i = (t_i[1], t_i[2])$ denote the two components of the site t_i. Thus, the columns of F contain the constant function, linear functions, and (for $p = 2$) quadratic functions of t.

Given F, the next step is to define an "increment matrix" $G(n \times (n - r)$, which has the properties that G has full column rank and

$$G^T F = 0.$$

In practice, it is convenient to assume that G is column orthonormal, $G^T G = I_{n-r}$.

Then for an intrinsic process, the covariance matrix of the increments is a matrix Ω, say, given by

$$\Omega = \text{var}(G^T \mathbf{y}) = \text{var}(\mathbf{y}) = G^T \Sigma G \tag{5.58}$$

is well defined where $\boldsymbol{y} = G^T\boldsymbol{x}$. Further, the increments have a log-likelihood given by

$$\log L = -\frac{1}{2}\left\{\boldsymbol{y}^T\Omega^{-1}\boldsymbol{y} + \log|\Omega| + (n-r)\log(2\pi)\right\}. \tag{5.59}$$

This function can then be optimized with respect to any parameters in the covariance function. Cressie and Lahiri (1993) have considered the asymptotic distribution of the REML estimator, following Mardia and Marshall (1984).

The log-likelihood of the increments for a pth-order intrinsic processes turns out to be identical to the REML log-likelihood for a spatial linear model with a stationary covariance function when the drift functions consist of polynomials in t up to degree p. Of course, for the spatial linear model, there is also the option of using the full likelihood; this option is not available for intrinsic processes.

The REML log-likelihood was discussed in detail earlier for a stationary process with unknown mean. This is a special case of a spatial linear model where the drift functions consist of polynomials of degree 0, i.e. the constant function. In the REML section, we used the sub-Helmert matrix to construct the increments. The matrix H in that section is an example of the increment matrix G defined above.

Example 5.9 ***Gravimetric data***

Consider the gravimetric data introduced in Section 1.5 and notice two features in Figure 1.13. First, the semivariogram in panel (b) appears to be unbounded in several directions as the lag increases. Hence, we consider fitting a self-similar intrinsic model, both without and with a nugget term, to the data. Second, there appears to be an approximate linear trend from the lower right to the upper left in the bubble plot in panel (a). Hence, we focus on fitting IRF(1) models; for these models, the likelihood removes the effects of a linear trend.

The self-similar models were introduced in Section 3.10 and depend on an index α. For the models with intrinsic order $k = 1$, the index is restricted to the range $0 < \alpha < 2$, with the corresponding intrinsic covariance function

$$\sigma_I(h) = \begin{cases} -\sigma^2|h|^{2\alpha}, & 0 < \alpha < 1, \\ \sigma^2|h|^{2\alpha}\log|h|, & \alpha = 1, \\ \sigma^2|h|^{2\alpha}, & 1 < \alpha < 2, \end{cases} \tag{5.60}$$

where the scale parameter σ^2 absorbs the proportionality constant $c_{\alpha,d}$ in (3.46). The model has two parameters, the index α and the scale parameter σ^2. The interpretation of an intrinsic covariance function needs care, as described in Section 3.10. The intrinsic covariance function (5.60) can be used to define Σ in (5.58). If a nugget term is present in the model, then Σ should be replaced by $\Sigma + \tau^2 I$.

The log-likelihood is given by (5.59). The profile log-likelihood, that is, the log-likelihood after maximizing over σ^2 (and τ^2, if present), is shown in Figure 5.7, as a function of the index α. Several conclusions can be drawn from this figure.

(a) If no nugget is included, the MLE for α is $\hat{\alpha} = 0.59$, with a 95% likelihood-based confidence interval (0.42, 0.80). The maximized log-likelihood is 93.43.
(b) If a nugget is included, the MLE for α is $\hat{\alpha} = 1.13$, with a 95% likelihood-based confidence interval (0.54, 1.70). The maximized log-likelihood is 95.23.
(c) Hence, when a nugget is included, the estimated α is larger than without a nugget. If the models are compared by a likelihood ratio test, twice the difference between the log-likelihoods is

$$2(95.23 - 93.43) = 3.60,$$

which is not quite significant at the upper 5% critical value 3.84 of the χ_1^2 distribution, and suggesting a slight preference for the model with no nugget. If the two models are compared in terms of their AIC values, the two AIC values are

$$\text{AIC}_{\text{no-nugget}} = 2(2 - 93.43) = -182.86, \quad \text{AIC}_{\text{nugget}} = 2(3 - 95.23) = -184.46.$$

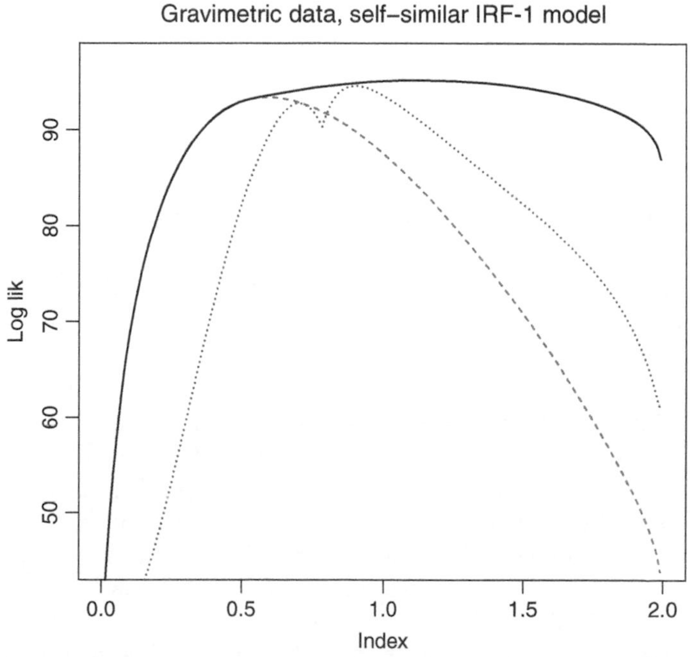

Figure 5.7 Profile log-likelihoods for self-similar models of intrinsic order $p = 1$, as a function of the index α, $0 < \alpha < 2$, both without a nugget effect (dashed line) and with a nugget effect (solid line). In addition, the log-likelihood for each α, with the parameters estimated by MINQUE (described In Section 5.13), is shown (dotted line).

Since the AIC value for the model with a nugget is smaller, there is a slight preference for the model with a nugget.

(d) In summary, the data can be described either by a self-similar model without nugget and α about 0.5 or by a self-similar model with nugget and α about 1.0.

The theory of minimum norm quadratic unbiased estimation (MINQUE) was developed by Rao (see, e.g. 1973, pp. 302–305). Its application was originally to the problem of heteroscedasticity and the estimation of variance components in random effects models. It was extended to spatial analysis by Marshall and Mardia (1985). Details are given in Exercise 5.6.

Figure 5.7 also includes the log-likelihood for the model with a nugget when σ^2 and τ^2 are estimated by the MINQUE method, again indexed by α. The maximizing value of the index is $\hat{\alpha}_{\text{MINQUE}} = 0.90$, which is compatible with the MLE. A strange feature in this plot is the cusp. For indices below the cusp, the estimated nugget term is 0, whereas for indices above the cusp, the estimated nugget term is nonzero. □

Part of the motivation behind REML and MINQUE is to reduce bias in the estimation of covariance parameters. REML is a likelihood-based method designed to reduce the bias arising from the presence of drift terms in the model. MINQUE is a moment-based method that is applicable when the covariance function can be written as a linear combination of parameter-free building blocks.

5.14 Infill Asymptotics and Fractal Dimension

This setting deals with increasing numbers of observations within a fixed bounded region. For convenience, we limit the mathematical treatment to the regularly spaced case of n^d data sites in the unit hyper-cube, with spacing $\Delta = 1/n$ between the sites.

In this setting, the data converge to a single continuous realization of the process on a bounded region as $n \to \infty$. Because the region is bounded, the observations cannot be far enough apart to be asymptotically independent, and, in general, it is not possible to estimate parameters, such as range parameters, consistently. However, provided the covariance function or semivariogram follows a power law behavior at the origin, it is possible to estimate the "micro" parameters associated with the infinitesimal behavior of the process.

The regularity conditions are more delicate in this setting than in the outfill setting, and much work focuses on a specific parametric model, the Matérn covariance function,

$$\sigma(h) = \sigma^2 \frac{2}{\Gamma(\nu)} \left(\frac{1}{2} |h|/\varphi \right)^{\nu} K_\nu(|h|/\varphi), \tag{5.61}$$

given in Section 2.6, where σ^2 is a scale parameter, $\varphi > 0$ is a range parameter, and the index $\nu > 0$ governs the smoothness of the process. The modified variance parameter

$$\sigma_e^2 = \sigma^2/\varphi^{2\nu}$$

was defined in (5.30). This reparameterization was suggested by Zhang (2004); see also Zhang and Zimmerman (2005) and Kaufman and Shaby (2013).

In regular statistical problems, the order of asymptotic variance of the maximum likelihood estimator is $O(1/n^d)$ as $n \to \infty$. For the purposes of this section, say, an estimator has *efficient order* if the order of its asymptotic variance is $O(1/n^d)$, and, say, it has *subefficient order* if the order of its asymptotic variance is greater than $O(1/n^d)$.

It turns out that the infill asymptotics setting is not regular enough for all the parameters to be estimated with efficient order by maximum likelihood. There is enough information in the short-lag increments of the process to estimate $\sigma^2/\varphi^{2\nu}$ with efficient order, but there is not enough information to estimate the range parameter φ with efficient order. One consequence is that the correlation between $\hat{\sigma}^2$ and $\hat{\varphi}$ tends to 1 as the sample size gets large. Stein (1999, pp. 188–199) gives a detailed treatment of infill asymptotics, under a slightly simplified model based on a circulant approximation. Then, the following results can be established under various assumptions about the parameters.

(a) All three parameters $(\sigma_e^2, \varphi, \nu)$ are unknown.
   (i) $\hat{\sigma}_e^2$ has slightly subefficient order with variance of order $O(\log^2 n/n^d)$.
   (ii) $\hat{\varphi}$ has very subefficient order with variance of order $O(1)$ in dimensions $d = 1,2,3$, $O(1/\log n)$ in dimension $d = 4$, and $O(1/n^{d-4})$ in dimensions $d \geq 5$. In particular, $\hat{\varphi}$ is not even consistent in dimensions $d \leq 3$.
   (iii) $\hat{\nu}$ has efficient order with variance of order $O(1/n^d)$.

(b) ν is known. This is a common assumption in spatial applications since ν is a difficult parameter to estimate unless the data are very finely spaced. The asymptotic variance of $\hat{\varphi}$ is the same as in (a)(ii), but $\hat{\sigma}_e^2$ can now be estimated efficiently.

To help understand the preference for the modified variance parameter σ_e^2 rather than σ^2, consider the case where the index parameter lies in the interval $0 < \nu < 1$. Then, it can be shown that the behavior of the semivariogram near the origin takes the power law form

$$\gamma(h) = \sigma^2 - \sigma(h) = C_\nu \sigma_e^2 |h|^{2\nu} + O(|h|^2), \tag{5.62}$$

where

$$C_\nu = \frac{\Gamma(1-\nu)}{\Gamma(1+\nu)} 2^{2\nu} \tag{5.63}$$

is a positive constant depending on ν (see Exercise 5.2). A similar but more complicated result holds for larger values of ν (e.g. Kent, 1989, p. 1439).

For any $\nu > 0$, the parameter ν in (5.62) describes the local smoothness of the process, the process being rougher for ν near 0. For $0 < \nu < 1$, this smoothness can also be characterized in terms of the fractal dimension of the process

$$D = d + 1 - \nu;$$

see also Section 2.9.

Estimation of ν, especially for $0 < \nu < 1$, has also received considerable attention in the literature in a nonparametric setting. That is, the semivariogram is assumed to satisfy (5.62), but not necessarily to be the Matérn model. When $0 < \nu < 1$ and the remainder term vanishes, this semivariogram defines a zeroth-order intrinsic process known as fractional Brownian motion (Mandelbrot and van Ness, 1968); see Section 3.10. It can also be viewed as a limiting case of the Matérn model as $\varphi \to \infty$.

In the nonparametric setting of (5.62), here is a simple method to estimate the smoothness index ν and the scale parameter $\delta = C_\nu \sigma_e^2$ given in Kent and Wood (1997); see also Istas and Lang (1997). The starting point is a fixed lag vector $h \in \mathbb{Z}^d$ and an integer "dilation" parameter $u \geq 1$. Define a "re-scaled dilated mean of squared increments"

$$\overline{Z}_n^u = \sum n^{2\nu} \{X((i + uh)/n) - X(i/n)\}^2,$$

where the sum is over all sites $i \in \mathbb{Z}^d$ such that i/n and $(i + uh)/n$ lie in the unit cube $D = \{x \in \mathbb{R}^d : 0 \leq x[\ell] \leq 1 \text{ for } \ell = 1, \dots, d\}$. Up to boundary effects, there are approximately n^d terms in this sum. It is straightforward to verify that $E\{\overline{Z}_n^u\} = \delta u^{\nu/2} + o(1)$ as $n \to \infty$. Further, it can be shown that $\overline{Z}_n^u$ converges in probability to this limiting average value. Hence, given two distinct dilation parameters u and υ (e.g. $u = 1, \upsilon = 2$), natural estimators of ν and δ are

$$\hat{\nu} = 2\frac{\log \overline{Z}_n^\upsilon - \log \overline{Z}_n^u}{\upsilon - u}, \quad \log \hat{\delta} = \frac{\upsilon \log \overline{Z}_n^u - u \log \overline{Z}_n^\upsilon}{\upsilon - u}.$$

However, some comments are needed.

(a) When more than two dilations are used, regression methods can be used for estimation; $\hat{\nu}$ is the estimated slope of the regression line.

(b) Describing the accuracy of $\hat{\nu}$ is rather complicated. If $0 < \nu < 3/4$ and $d = 1$, $\hat{\nu}$ has an asymptotic variance of order $O(1/n)$ and it is asymptotically normal. However, for $3/4 < \nu < 1$, its variance is of lower order and its asymptotic distribution is nonnormal. The reason for this nonstandard behavior is that for ν near 1, the process becomes smoother and the increments are more correlated. However, by using higher order increments, it is possible to construct an estimator for $3/4 < \nu < 1$, which is asymptotically normally distributed with variance of order $O(1/n)$. Similar issues arise in $d \geq 2$ dimensions.

(c) It is also possible to define the concept of regularity of a covariance function for an index $\nu \geq 1$ using ideas from Section 3.4. Further, by using increments of high enough order, it is possible to construct asymptotically normal estimates of ν with variance of order $O(n^{-d})$.

Davies and Hall (1999) have given a good example of estimating surface roughness in $d = 2$ dimensions. Kent and Wood (1995) have provided some technical details.

Exercises

5.1 The spherical scheme is defined by the covariance function

$$\sigma(h; \sigma^2, a) = \begin{cases} 1 - \frac{3}{2}|h/a| + \frac{1}{2}|h/a|^3, & |h| \leq a, \\ 0, & |h| > a, \end{cases}$$

which is positive definite in dimensions $d = 1, 2, 3$. The parameters are the scale parameter σ^2 and the range parameter a. Clearly, $\sigma(h)$ is smooth in σ^2, but its dependence on a is more delicate. Show that for fixed $h \neq 0$,
(a) $\sigma(h; \sigma^2, a)$ is continuous in a for all $a > 0$.
(b) $\partial\sigma(h; \sigma^2, a)/\partial a$ is continuous in a for all $a > 0$.
(c) $\partial^2\sigma(h; \sigma^2, a)/\partial a^2$ is not continuous in a at $a = |h|$.

5.2 This exercise looks at the regularity of the Matérn covariance function for small lags. This behavior is important for the study of infill asymptotics in Section 5.14. Suppose the real index ν is not a negative integer and let z be a positive number. The modified Bessel function of the first kind $I_\nu(z)$ has the limiting behavior

$$I_\nu(z) = \frac{(z/2)^\nu}{\Gamma(1+\nu)} \left\{1 + O(z^2)\right\}$$

as $z \to 0$ for fixed ν (e.g. Abramowitz and Stegun, 1964, p. 375). The modified Bessel function of the second kind $K_\nu(z)$ is defined by

$$K_\nu(z) = \frac{\pi/2}{\sin \nu\pi} \left\{I_{-\nu}(z) - I_\nu(z)\right\}.$$

For $0 < \nu < 1$, deduce the limiting behavior

$$\frac{2}{\Gamma(\nu)}(z/2)^\nu K_\nu(z) = 1 - \frac{\Gamma(1-\nu)}{\Gamma(1+\nu)}(z/2)^{2\nu} + O(z^2)$$

as $z \to 0$. Hence, for $0 < \nu < 1$, confirm that the Matérn covariance function in (5.61) has the limiting behavior in (5.62). Hint: Use the reflection formula for the gamma function

$$\Gamma(\nu)\Gamma(1-\nu) \sin \pi\nu = \pi.$$

5.3

(a) Let $G = \begin{bmatrix} G_1 & G_2 \end{bmatrix}$ be an $n \times n$ orthogonal matrix partitioned into two blocks, with n_1 and n_2 columns, respectively, $n_1 + n_2 = n$, so that $G_1^T G_1 = I, G_2^T G_2 = I, G_1^T G_2 = 0$. Let B be an $n \times n$ positive definite matrix, and set

$$B_{ij} = G_i^T B G_j, \quad B^{ij} = G_i^T B^{-1} G_j, \quad i, j = 1, 2.$$

Using the results from Section A.3.4 on partitioned matrices, show that

$$\begin{aligned} |B| &= |B_{11}|\ |B_{22.1}| = |B_{11}|/|B^{22}| \\ &= |B_{22}|\ |B_{11.2}| = |B_{22}|/|B^{11}|. \end{aligned}$$

(b) Let $F(n \times n_1)$ be a matrix of full column rank. Show that the matrix $P_F = F(F^T F)^{-1/2}$ is column orthonormal, i.e. $P_F^T\, P_F = I_{n_1}$.

(c) If G_1 in part (a) is related to F in part (b) by $G_1 = F(F^T F)^{-1/2}$, show that

$$|B_{11}| = |F^T B F|/|F^T F|, \quad |B^{11}| = |F^T B^{-1} F|/|F^T F|.$$

Hence, show that the determinantal identities in part (a) can be rewritten as

$$|B| = |F^T B F|/(|F^T F|\ |B^{22}|) = |B_{22}|\ |F^T F|/|F^T B^{-1} F|.$$

(d) If $F = \mathbf{1}_n$ has one column, confirm that the identity in part (c) reduces to Eq. (5.33).

5.4 The $n \times n$ Helmert matrix H, $n \geq 2$, is an orthogonal matrix whose rows are defined as follows:

- For $j = 1, \ldots, n-1$, the j row is given by

$$(1, \ldots, 1, -j, 0, \ldots, 0)/\sqrt{j(j+1)},$$

where 1 is repeated j times and 0 is repeated $n - j - 1$ times.

- The nth row is given by $(1, \ldots, 1)/\sqrt{n}$.

That is, for $n = 3$

$$H = \begin{bmatrix} 1/\sqrt{2} & -1/\sqrt{2} & 0 \\ 1/\sqrt{6} & 1/\sqrt{6} & -2/\sqrt{6} \\ 1/\sqrt{3} & 1/\sqrt{3} & 1/\sqrt{3} \end{bmatrix}$$

(a) Show that the rows of H are orthonormal; that is, they have unit norm and they are orthogonal to each other. Show that in matrix form, these statements take the form $HH^T = I_n$, that is, H is an orthogonal matrix.

(b) The *sub-Helmert* matrix, denoted $H_{\backslash n}$, say, is defined as the first $n - 1$ rows of H. Show that $H_{\backslash n}$ is "row-orthonormal," that is, $H_{\backslash n} H_{\backslash n}^T = I_{n-1}$. Show that each row of $H_{\backslash n}$ is orthogonal to the constant vector $\mathbf{1}_n$, $H_{\backslash n} \mathbf{1}_n = \mathbf{0}_{n-1}$. Thus, if $\boldsymbol{y}$ is an n-dimensional vector and $\boldsymbol{z} = H_{\backslash n} \boldsymbol{y}$, then each element of $\boldsymbol{z}$ is a contrast of the elements of $\boldsymbol{y}$.

(c) Show that the matrix $H_{\backslash n}^T$, of size $(n-1) \times n$, is the same as the matrix G, which was used in Sections 5.5 and 5.8 to construct a vector of contrasts, $\boldsymbol{z} = G^T\boldsymbol{y}$, for a data vector $\boldsymbol{y}$.

5.5 Consider the setting of Section 5.9 with $n = 3$. Let $\boldsymbol{X} \sim N_3(0, \Sigma)$ and suppose the covariance matrix Σ satisfies $\sigma_{11} = \sigma_{22} = \sigma_{33} = 1$. In the notation of this section, show that the coefficient vectors $\boldsymbol{\ell}_i^T = (\ell_{i1}, \ell_{i2}, \ell_{i3})$ are given by

$$\ell_{11} = 1, \quad \ell_{12} = 0, \quad \ell_{13} = 0,$$
$$\ell_{21} = -\sigma_{12}, \quad \ell_{22} = 1, \quad \ell_{23} = 0,$$
$$\ell_{31} = \frac{\sigma_{23}\sigma_{12} - \sigma_{31}}{1 - \sigma_{12}^2}, \quad \ell_{32} = \frac{\sigma_{13}\sigma_{12} - \sigma_{23}}{1 - \sigma_{12}^2}, \quad \ell_{33} = 1.$$

Set $e_j = \boldsymbol{\ell}_j^T\boldsymbol{x}, j = 1, 2, 3$. Show from first principles that e_1, e_2, and e_3 are uncorrelated.

In the REML setting of Section 5.10, the second and third coefficient vectors are now required to be orthogonal to the constant vector, $(\boldsymbol{\ell}_2^U)^T\mathbf{1} = 0$ and $(\boldsymbol{\ell}_3^U)^T\mathbf{1} = 0$. Show that

$$\ell_{21}^U = -1, \quad \ell_{22}^U = 1, \quad \ell_{23}^U = 0,$$
$$\ell_{31}^U = -\frac{1}{2}\left(1 + \frac{\sigma_{31} - \sigma_{32}}{1 - \sigma_{12}}\right), \quad \ell_{32}^U = -\frac{1}{2}\left(1 - \frac{\sigma_{31} - \sigma_{32}}{1 - \sigma_{12}}\right), \quad \ell_{33}^U = 1.$$

Note that $\ell_{31}^U + \ell_{32}^U = -1$. Further, show that $e_2^U = (\boldsymbol{\ell}_2^U)^T\boldsymbol{x}$ and $e_3^U = (\boldsymbol{\ell}_3^U)^T\boldsymbol{x}$ are uncorrelated.

5.6 (Marshall and Mardia, 1985) This exercise develops the principle of MINQUE for certain spatial processes for which the mean vanishes and the covariance function is linear in the unknown parameters.

(a) let $\boldsymbol{x} \sim N_p(\mathbf{0}, \Psi)$ where $\Psi = \sigma^2 A + \tau^2 I$. Here, A is a known positive-definite symmetric matrix from a parametric model, and the second term represents a nugget effect. The objective is to estimate the scaling parameters $\sigma^2 \geq 0$ and $\tau^2 \geq 0$ from a single realization of the vector $\boldsymbol{x}$. See Mardia and Marshall (1984) for more details.

From the data, construct two statistics, $u = \boldsymbol{x}^T\boldsymbol{x}$, and $v = \boldsymbol{x}^T A\boldsymbol{x}$. Also set

$$m_0 = p, \quad m_1 = \mathrm{tr}(A), \quad m_2 = \mathrm{tr}(A^2).$$

Recall for a general symmetric matrix B that $E(\boldsymbol{x}^T B\boldsymbol{x}) = E\{\mathrm{tr}(B\boldsymbol{x}\boldsymbol{x}^T)\} = \mathrm{tr}(B\Psi)$. Hence, deduce that

$$E(u) = \sigma^2 m_1 + \tau^2 m_0, \quad E(v) = \sigma^2 m_2 + \tau^2 m_1.$$

(b) The MINQUE estimator is given by matching the observed values of u and v to their expected values. Show that the resulting estimator takes the form

$$\begin{bmatrix} \hat{\sigma}^2 \\ \hat{\tau}^2 \end{bmatrix} = \begin{bmatrix} m_1 & m_0 \\ m_2 & m_1 \end{bmatrix}^{-1} \begin{bmatrix} u \\ v \end{bmatrix}.$$

It is possible for this procedure to produce negative estimates for either σ^2 or τ^2 (but not both). If the solution to the matrix equation yields a negative estimate for τ^2, then set $\hat{\tau}^2 = 0$ and $\hat{\sigma}^2 = u/m_1$. Similarly, if the solution to the matrix equation yields a negative estimate for σ^2, then set $\hat{\sigma}^2 = 0$ and $\hat{\tau}^2 = u/m_0$.

(c) Show how MINQUE can be applied to a self-similar process with a nugget effect, where the index of self-similarity is known. In particular, in the notation of (5.58), set $\sigma^2 A = \Omega$ and note that G is a column orthonormal matrix so that $G^T G = I_{n-r}$.

5.7 The purpose of this exercise is to show that the two forms of the REML log-likelihood (5.31) and (5.41) are the same, up to an additive constant not depending on the data or the parameters. Let $\boldsymbol{y}$ be an n-vector assumed to come from $N_n(\mu \mathbf{1}_n, \Sigma)$, where Σ is a positive definite matrix, where $n \geq 2$ is fixed. If A is an $((n-1) \times n$ contrast matrix (i.e. $A\mathbf{1}_n = 0$), the REML log-likelihood is given up to a constant term by

$$\log L = -\frac{1}{2}\{(\boldsymbol{y} - \mu\mathbf{1})^T A^T (A\Sigma A^T)^{-1} A(\boldsymbol{y} - \mu\mathbf{1}) + \log |A\Sigma A^T|\}.$$

In (5.31), the rows of A are assumed to be orthonormal, and the choice of A is irrelevant. In (5.41), A satisfies the assumptions of part (c) below.

(a) Let A and B be $(n-1) \times n$ matrices, both of which are full rank $n-1$ and contrast matrices – so that $A\mathbf{1}_n = B\mathbf{1}_n = 0_{n-1}$. Explain why there exists a nonsingular $(n-1) \times (n-1)$ matrix C such that $B = CA$. Hence, show that the following two quadratic forms are equal

$$(\boldsymbol{y} - \mu\mathbf{1})^T A^T (A\Sigma A^T)^{-1} A(\boldsymbol{y} - \mu\mathbf{1}) = (\boldsymbol{y} - \mu\mathbf{1})^T B^T (B\Sigma B^T)^{-1} B(\boldsymbol{y} - \mu\mathbf{1})$$

and that the quadratic forms do not depend on μ.

(b) For $2 \leq k \leq n$, define the vector subspace

$$S_k = \{\boldsymbol{a} \in \mathbb{R}^n : a_{k+1} = \cdots = a_n = 0 \text{ and } \mathbf{1}_n^T \boldsymbol{a} = 0\}$$

of n-vectors whose final $n-k$ elements vanish and which are contrast vectors (i.e. $\mathbf{1}_n^T \boldsymbol{a} = 0$). Show that S_k has dimension $k-1$.

(c) Let A be an $(n-1) \times n$ matrix partitioned in terms of its rows as

$$A = \begin{bmatrix} \boldsymbol{a}_1^T \\ \vdots \\ \boldsymbol{a}_{n-1}^T \end{bmatrix},$$

where $\boldsymbol{a}_k^T = (a_{k1}, \ldots, a_{kn})$. Since the convention in this book is that all vectors are column vectors, there is a need for a transpose when they are treated as row vectors. Suppose the rows of A have the following properties for all $k = 1, \ldots, n-1$:

(i) $\boldsymbol{a}_k \in \mathcal{S}_{k+1}$,

(ii) $a_{k,k+1} = 1$,

Show that the vectors $\boldsymbol{a}_1, \ldots, \boldsymbol{a}_k$ form a basis for $\mathcal{S}_{k+1}$.

(d) Let B be another $(n-1) \times n$ matrix satisfying (i)–(ii). Show that there is a lower triangular matrix $C((n-1) \times (n-1))$ with $c_{kk} = 1, k = 1, \ldots, n-1$ such that $B = CA$. Hint: Explain why the kth row of B can be written as a linear combination of the first k rows of A, i.e. $\boldsymbol{b}_k = \sum_{j=1}^{k} c_{kj}\boldsymbol{a}_j$ for suitable constants c_{kj} with $c_{kk} = 1$.

(e) Note that $|C| = 1$ since for a lower triangular square matrix, the determinant is the product of the diagonal elements. Hence, deduce that $|A\Sigma A^T| = |B\Sigma B^T|$.

(f) The sub-Helmert matrix $H_{\backslash n}$ of Exercise 5.4 has the property that row k lies in $\mathcal{S}_{k+1}, k = 1, \ldots n-1$. If row k is divided by $-\sqrt{(k+1)/k}$, then the resulting matrix satisfies the conditions of part (c) above. Hence, deduce that $n|H_{\backslash n}\Sigma H_{\backslash n}^T| = |A\Sigma A^T|$ is the same for all matrices A satisfying the conditions of part (b).

(g) In (5.41), the coefficient matrix L depends on the underlying parameters, and at first sight the REML log-likelihood depends on the covariance parameters $\boldsymbol{\theta}$ both through the covariance matrix Σ and through L. However, as shown in this exercise, the REML log-likelihood can be written in a form that depends just on Σ.

6

Estimation for Lattice Models

6.1 Introduction

This chapter investigates the problem of fitting a stationary autoregression Gaussian model to a set of spatial data on a rectangular grid in $\mathbb{Z}^d$. The autoregression models include unilateral autoregressions (UARs), conditional autoregressions (CARs), and simultaneous autoregressions (SARs). In principle, the methods of Chapter 5 can be used, but there are specialized methods that can be much faster.

Recall that for any stationary Gaussian process, the covariance function $\{\sigma_h\}$ has a spectral representation. When a spectral density exists, this representation takes the form

$$\sigma_h = \int_{(-\pi,\pi)^d} f(\omega) \cos(h^T \omega) d\omega, \quad h \in \mathbb{Z}^d,$$

and conversely the spectral density can be written in terms of the covariance function

$$f(\omega) = \frac{1}{(2\pi)^d} \sum_{h \in \mathbb{Z}^d} \sigma_h \cos(h^T \omega).$$

For simplicity, suppose the spectral density is bounded away from 0,

$$f(\omega) \geq c > 0.$$

Then the *inverse spectral density*, defined by

$$g(\omega) = 1/\{(2\pi)^{2d} f(\omega)\}, \tag{6.1}$$

is a bounded function. It can be used to define the *inverse covariance function*

$$\psi_h = \int_{(-\pi,\pi)^d} g(\omega) \cos(h^T \omega) \, d\omega, \quad h \in \mathbb{Z}^d, \tag{6.2}$$

Spatial Analysis, First Edition. John T. Kent and Kanti V. Mardia.

and conversely the inverse spectral density can be written in terms of the inverse covariance function

$$g(\omega) = \frac{1}{(2\pi)^d} \sum_{h \in \mathbb{Z}^d} \psi_h \cos(h^T \omega). \tag{6.3}$$

For all the autoregression models based on finite neighborhoods (i.e. UARs, CARs, and SARs), the inverse spectral density is assumed to have a *finite* trigonometric expansion

$$g(\omega) = \frac{1}{(2\pi)^d} \sum_{h \in \mathcal{N}_0} \psi_h \cos(h^T \omega),$$

where $\mathcal{N}_0 = \mathcal{N} \cup \{0\}$ and $\mathcal{N}$ is a finite symmetric neighborhood of the origin (Section 4.4). The simplest model is discrete white noise, for which $\sigma_h = \sigma^2 \delta_h$, a multiple of the Kronecker delta function, and the spectral density is constant, $f(\omega) = \sigma^2/(2\pi)^d$. Then $\psi_h = \sigma^{-2}\delta_h$ and $g(\omega) = 1/\{(2\pi)^d \sigma^2\}$.

The covariance function defines an infinite-dimensional covariance matrix Σ_∞ on $\mathbb{Z}^d$,

$$\Sigma_\infty = (\sigma_{st}), \quad \sigma_{st} = \sigma_{s-t}, \quad s, t \in \mathbb{Z}^d,$$

with the elements listed in lexicographic order, say. Similarly, the inverse covariance function defines an infinite-dimensional covariance matrix Ψ_∞ on $\mathbb{Z}^d$

$$\Psi_\infty = (\psi_{st}), \quad \psi_{st} = \psi_{s-t}, \quad s, t \in \mathbb{Z}^d.$$

For an autoregression model, the precision matrix is *sparse*; that is, within each row and column of the matrix, there are only finitely many nonzero entries.

The infinite covariance and precision matrices are inverses of each other

$$\Sigma_\infty \Psi_\infty = I_\infty.$$

Throughout the chapter the data sites are assumed to lie on a regular rectangular grid

$$D = \{t \in \mathbb{Z}^d : 1 \le t[\ell] \le n[\ell], \ \ell = 1, \dots, d\} \tag{6.4}$$

of dimensions $N = (n[1], \dots, n[d])$, with corresponding data values given by the vector $\boldsymbol{x} = \{x_t, \ t \in D\}$, with the sites listed in lexicographic order. The size of D can be denoted in two ways

$$|N| = |D| = n[1] \times \cdots \times n[d], \tag{6.5}$$

depending on whether the region itself or just its size is of interest.

Let Σ_D and Ψ_D denote the restrictions of Σ_∞ and Ψ_∞ to the sites in D. Then Σ_D denotes the covariance matrix of the process on D. However, Ψ_D is no longer the precision matrix on D

$$\Sigma_D^{-1} \ne \Psi_D. \tag{6.6}$$

Minus twice the log-likelihood is

$$-2\log L = (\boldsymbol{x} - \mu\mathbf{1}_N)^T \Sigma_D^{-1} (\boldsymbol{x} - \mu\mathbf{1}_N) + \log|\Sigma_D|. \tag{6.7}$$

There are three main computational challenges in the calculation and maximization of the log-likelihood for autoregression models.

(a) For many autoregression models, the covariance function does not have a simple explicit form. Hence, it is not straightforward to compute the elements of Σ_D.
(b) The inversion of the $|N| \times |N|$ matrix Σ_D requires $O(|N|^3)$ calculations in general.
(c) Similarly, the evaluation of $|\Sigma_D|$ requires $O(|N|^3)$ calculations in general.

For autoregression models, the inverse covariance function is more tractable than the covariance function. Hence, it is worth considering a variety of other estimators in this situation. These methods fall into two broad classes.

- *Moment estimators.* These are based on identities satisfied by the covariance functions. They can be developed most easily for UARs in Section 6.4.1, and for CARs in Section 6.4.2. For UAR models, these are essentially the same as maximum likelihood estimators, but for CAR models, they are less efficient (though simpler to compute) than maximum likelihood estimators. Moment estimators for both UARs and CARs involve solving a set of linear equations where the coefficients depend on the data only through a small set of sample moments.
- *Approximations to the maximum likelihood estimator.* These are based on tractable approximations for Σ_D^{-1} and $\log|\Sigma_D|$ in (6.7). For a stationary model on $\mathbb{Z}^d$, there are a variety of approximations based on Toeplitz, circulant, and folded circulant approximations. These are explored in Section 6.5.

All the estimators depend on the data through estimates of the population covariance function. There are several ways to define the sample covariance function, and the most important choices are given in Section 6.2. The chapter looks at the AR(1) process in one dimension (Section 6.3) in detail. This is an excellent motivating example for the methods of this chapter because it can be viewed as both a UAR and a CAR and the various approximations for the log-likelihood can be evaluated explicitly.

6.2 Sample Moments

Given data $\{x_t : t \in D\}$ from a stationary random field with mean μ, the objective is to estimate the parameters of the model. For simplicity, let μ be estimated by the sample mean $\hat{\mu} = \bar{x}$.

It is important to note that all the estimators in this chapter for the parameters of the covariance function are based on sample estimates of the covariance function for various lags $h \in \mathbb{Z}^d$. There are several ways to define the sample covariance function based on second-order moments, depending on the choice of divisor and boundary conditions. Section 4.6.5 gives a description of different boundary conditions.

The first choice is the *unbiased* sample covariance function

$$s_h^{(u)} = \frac{1}{|D_h|} \sum_{t \in D_h} (x_t - \bar{x})(x_{t+h} - \bar{x}). \tag{6.8}$$

Here, $D_h = \{s \in D : \ t + h \in D\}$ denotes the sites t such that t and $t + h$ both lie in D, of size

$$|D_h| = (n[1] - |h[1]|) \times \cdots \times (n[d] - |h[d]|). \tag{6.9}$$

The divisor $|D_h|$ is the number of terms in the sum, so that $s_h^{(u)}$ is an unbiased estimate of σ_h.

The second choice is the *biased* sample covariance function, which uses a common divisor for all lags

$$s_h^{(b)} = \frac{1}{|D|} \sum_{t \in D_h} (x_t - \bar{x})(x_{t+h} - \bar{x}). \tag{6.10}$$

The difference between the two versions is a divisor of $|D_h|$ or $|D|$, respectively. An advantage of the biased estimator is that the sequence of sample covariances is guaranteed to be positive semidefinite (Exercise 6.1). Further, the presence of bias is sometimes offset by smaller mean squared errors.

A third choice is the *interior* sample covariance function. Its construction requires the choice of a finite symmetric neighborhood $\mathcal{N}^*$ of the origin, at least as big as the symmetric neighborhood $\mathcal{N}$ in Section 6.1 for the autoregression model of interest. Define $D_{\text{int}} = \{t \in D : t + h \in D \text{ for all } h \in \mathcal{N}^*\}$ to be the "interior" of D, and set

$$s_h^{(i)} = \frac{1}{|D_{\text{int}}|} \sum_{t \in D_{\text{int}}} (x_t - \bar{x})(x_{t+h} - \bar{x}). \tag{6.11}$$

A fourth choice uses a periodic boundary condition, so that data sites on opposite sides of D are treated as adjacent to one another

$$s_h^{(p)} = \frac{1}{|D|} \sum_{s \in D} (x_s - \bar{x})(x_{(s-h) \text{Mod } N} - \bar{x}). \tag{6.12}$$

The Mod operator is defined by Eq. (A.2).

The fifth, and final, choice based on a reflecting boundary condition will be discussed later. All the choices except (6.8) produce a sample covariance function that is guaranteed to be positive semidefinite.

6.3 The AR(1) Process on $\mathbb{Z}$

The simplest process to motivate the ideas of this chapter is given by the one-dimensional AR(1) process. This process $\{X_t : t \in \mathbb{Z}\}$, say, has two natural representations (Sections 4.8 and 4.6). For notational convenience, suppose the mean vanishes, $\mu = 0$ for this section.

The first representation is a UAR, obtained by regressing X_t on its past values. The conditional expectation takes the form

$$E\left(X_t | \{X_s,\ s < t\}\right) = \lambda X_{t-1}.$$

The *one-sided innovation* terms, defined by

$$\varepsilon_t = X_t - \lambda X_{t-1},$$

are i.i.d. $N(0, \sigma_\varepsilon^2)$, say. For each t, the current innovation term ε_t is independent of $x_s,\ s < t$. The spectral density is given by

$$f(\omega) = \frac{1}{2\pi} \frac{\sigma_\varepsilon^2}{|1 - \lambda e^{i\omega}|^2} = \frac{1}{2\pi} \frac{\sigma_\varepsilon^2}{1 + \lambda^2 - 2\lambda \cos \omega}. \tag{6.13}$$

The Fourier series for the functions $f,\ 1/f$, and $\log f$ can be computed explicitly

$$2\pi f(\omega) = \frac{\sigma_\varepsilon^2}{1 - \lambda^2} \left\{ 1 + 2\sum_{j=1}^{\infty} \lambda^j \cos j\omega \right\}, \tag{6.14}$$

$$1/\{2\pi f(\omega)\} = \frac{1}{\sigma_\varepsilon^2}\{1 + \lambda^2 - 2\lambda \cos \omega\}, \tag{6.15}$$

$$\log\{2\pi f(\omega)\} = \log \sigma_\varepsilon^2 + 2\sum_{j=1}^{\infty} \frac{\lambda^j}{j} \cos j\omega \tag{6.16}$$

(Exercise 6.3).

The second representation is a CAR, obtained by regressing X_t on all the values of the process except itself. The conditional expectation takes the form

$$E\left(X_t | \{X_s,\ s \neq t\}\right) = \beta(X_{t-1} + X_{t+1}).$$

The *two-sided innovation* terms η_t, defined by the residuals

$$\eta_t = X_t - \beta(X_{t-1} + X_{t+1}),$$

are normally distributed $N(0, \sigma_\eta^2)$, say, but are not i.i.d. in this case. For each t, the current innovation term η_t is independent of $\{X_s,\ s \neq t\}$. The spectral density can be written in two forms, from the CAR representation and from (6.13), as

$$f(\omega) = \frac{1}{2\pi} \frac{\sigma_\eta^2}{1 - 2\beta \cos \omega} = \frac{1}{2\pi} \frac{\sigma_\varepsilon^2}{1 + \lambda^2 - 2\lambda \cos \omega}. \tag{6.17}$$

The two forms are the same, provided we set

$$\beta = \frac{\lambda}{1+\lambda^2}, \quad \sigma_\eta^2 = \frac{\sigma_\varepsilon^2}{1+\lambda^2}. \tag{6.18}$$

Note that as λ ranges between -1 and 1, β ranges between $-1/2$ and $1/2$.

The covariance function of the process takes the form

$$\sigma_h = \sigma^2 \lambda^{|h|}, \quad h \in \mathbb{Z}. \tag{6.19}$$

In particular, $\sigma_0 = \sigma^2$, $\sigma_1 = \lambda\sigma^2$. The marginal variance σ^2 is related to the innovation variances by

$$\sigma^2 = \frac{\sigma_\varepsilon^2}{1-\lambda^2} = \sigma_\eta^2 \frac{1+\lambda^2}{1-\lambda^2}.$$

Two simple identities connect the values of the autocovariance function to the autoregression parameters λ and β

$$\lambda = \sigma_1/\sigma_0, \quad \beta = \sigma_1/(\sigma_0+\sigma_2) \tag{6.20}$$

(Exercise 6.3). The first identity forms the basis of the UAR moment estimator and the second forms the basis for the CAR moment estimator.

Next consider n observations $x_1, \ldots, x_n$ from an AR(1) process. An attractive feature of this process is that it is straightforward to write down the covariance matrix $\Sigma_n = \Sigma$, its exact inverse $\Psi^{(\text{exact})} = \Sigma^{-1}$, and the various approximations. In particular,

$$\Sigma = \sigma^2 \begin{bmatrix} 1 & \lambda & \lambda^2 & \ldots & \lambda^{n-1} \\ \lambda & 1 & \lambda & \ldots & \lambda^{n-2} \\ \lambda^2 & \lambda & 1 & \ldots & \lambda^{n-3} \\ \vdots & \vdots & \vdots & \vdots & \vdots \\ \lambda^{n-1} & \lambda^{n-2} & \lambda^{n-3} & \ldots & 1 \end{bmatrix}, \tag{6.21}$$

is a Toeplitz matrix (Section A.3.8), with the exact inverse

$$\Psi^{(\text{exact})} = \frac{1}{(1-\lambda^2)\sigma^2} \begin{bmatrix} 1 & -\lambda & 0 & \ldots & 0 & 0 \\ -\lambda & 1+\lambda^2 & -\lambda & \ldots & 0 & 0 \\ 0 & -\lambda & 1+\lambda^2 & \ldots & 0 & 0 \\ \vdots & \vdots & \vdots & \vdots & \vdots & \vdots \\ 0 & 0 & 0 & \ldots & 1+\lambda^2 & -\lambda \\ 0 & 0 & 0 & \ldots & -\lambda & 1 \end{bmatrix}. \tag{6.22}$$

This matrix is almost Toeplitz, in the sense that the entries in positions $(1,1)$ and (n,n) differ from the other diagonal elements. Modifying the exact inverse covariance matrix to be Toeplitz yields the approximation

$$\Psi^{(\text{Toep})} = \frac{1}{(1-\lambda^2)\sigma^2}\begin{bmatrix} 1+\lambda^2 & -\lambda & 0 & \dots & 0 & 0 \\ -\lambda & 1+\lambda^2 & -\lambda & \dots & 0 & 0 \\ 0 & -\lambda & 1+\lambda^2 & \dots & 0 & 0 \\ \vdots & \vdots & \vdots & \vdots & \vdots & \vdots \\ 0 & 0 & 0 & \dots & 1+\lambda^2 & -\lambda \\ 0 & 0 & 0 & \dots & -\lambda & 1+\lambda^2 \end{bmatrix}. \quad (6.23)$$

Note that the upper-left and lower-right elements of $\Psi^{(\text{Toep})}$ are bigger than the corresponding values of $\Psi^{(\text{exact})}$. This means that $\Psi^{(\text{Toep})}$ tends to underestimate the variability of the data at sites 1 and n. The effect is particularly pronounced when $|\lambda|$ is near 1.

Two other approximations of $\Psi^{(\text{exact})}$ are also of interest. The first is a circulant approximation

$$\Psi^{(\text{circ})} = \frac{1}{(1-\lambda^2)\sigma^2}\begin{bmatrix} 1+\lambda^2 & -\lambda & 0 & \dots & 0 & -\lambda \\ -\lambda & 1+\lambda^2 & -\lambda & \dots & 0 & 0 \\ 0 & -\lambda & 1+\lambda^2 & \dots & 0 & 0 \\ \vdots & \vdots & \vdots & \vdots & \vdots & \vdots \\ 0 & 0 & 0 & \dots & 1+\lambda^2 & -\lambda \\ -\lambda & 0 & 0 & \dots & -\lambda & 1+\lambda^2 \end{bmatrix}, \quad (6.24)$$

for which sites 1 and n are treated as neighbors.

The second is the *folded circulant* approximation described in Section A.11. Basically, the data are reflected to give a doubled data set $x_1, \dots, x_n, x_n, \dots, x_1$ whose inverse matrix is approximated by a circulant matrix of dimension $2n$. Then the quadratic form is written in terms of the elements of $x_1, \dots, x_n$ as

$$\Psi^{(\text{fold})} = \frac{1}{(1-\lambda^2)\sigma^2}\begin{bmatrix} 1-\lambda+\lambda^2 & -\lambda & 0 & \dots & 0 & 0 \\ -\lambda & 1+\lambda^2 & -\lambda & \dots & 0 & 0 \\ 0 & -\lambda & 1+\lambda^2 & \dots & 0 & 0 \\ \vdots & \vdots & \vdots & \vdots & \vdots & \vdots \\ 0 & 0 & 0 & \dots & 1+\lambda^2 & -\lambda \\ 0 & 0 & 0 & \dots & -\lambda & 1-\lambda+\lambda^2 \end{bmatrix}. \quad (6.25)$$

This approximation deals more effectively with the variances at the end sites than the Toeplitz approximation, especially for $|\lambda|$ near 1.

Note that for both the circulant and folded approximations, all the row and column sums are the same. Hence, these approximations also work in the intrinsic case, $\lambda = 1$, when all the rows and columns sum to 0. In this case, $\Psi^{(\text{circ})}$ and $\Psi^{(\text{fold})}$ are both singular, $\Psi^{(\text{circ})}\mathbf{1}_n = \mathbf{0}_n$ and $\Psi^{(\text{fold})}\mathbf{1}_n = \mathbf{0}_n$. Hence, only linear combinations $\boldsymbol{a}^T\boldsymbol{x}$, where $\boldsymbol{a}^T\mathbf{1} = 0$, have well-defined distributions.

It is also possible to evaluate the normalizing constant $\log|\Sigma|$ for these approximations as a sum of log eigenvalues

$$\log\left|\Sigma^{(\text{Toep})}\right| = n\log\sigma_\eta^2 - \sum_{k=1}^{n}\log\{1 - 2\beta\cos(\pi k/(n+1))\}, \tag{6.26}$$

$$\log\left|\Sigma^{(\text{circ})}\right| = n\log\sigma_\eta^2 - \sum_{k=0}^{n-1}\log\{1 - 2\beta\cos(2\pi k/n)\}, \tag{6.27}$$

$$\log\left|\Sigma^{(\text{fold})}\right| = n\log\sigma_\eta^2 - \sum_{k=0}^{n-1}\log\{1 - 2\beta\cos(\pi k/n)\}. \tag{6.28}$$

The expressions for (6.27) and (6.28) also make sense in the intrinsic case $\beta = 1/2$ (i.e. $\lambda = 1$), when appropriately interpreted. If $\log|\Sigma^{(\text{circ})}|$ and $\log|\Sigma^{(\text{fold})}|$ are replaced by the log of the product of the nonzero eigenvalues (rather than the product of all the eigenvalues), then (6.27) and (6.28) continue to be valid, provided the sums start at $k = 1$ rather than $k = 0$.

The circulant approximation extends easily to any CAR in any dimension. The folded approximation extends easily to any CAR in any dimension possessing full symmetry (see Section A.11.4). However, the Toeplitz expansion (6.26) is only valid for the AR(1) case. For large n, all the approximations are very similar, forming in each case a discrete approximation to the integral

$$\frac{n}{2\pi}\int_0^{2\pi}\log f(\omega)\,d\omega \tag{6.29}$$

which appears in the Whittle approximation to the log-likelihood in (6.43). The circulant version given here and another possible circulant version of the AR(1) process are explored further in Exercises 6.7 and 6.10.

6.4 Moment Methods for Lattice Data

In Chapter 4, several types of autoregression models for lattice random fields were investigated, including unilateral autoregressions (UARs) and conditional autoregressions (CARs). In this section, we see how moment equations can be derived for UARs and CARs, enabling explicit estimates of the parameters to be obtained in terms of the sample covariance function.

For UARs, these moment estimates are essentially the same as maximum likelihood estimates (MLEs) up to boundary conditions; hence, the estimates are consistent and efficient. For CARs, the estimates are consistent, but they are not as efficient as MLEs especially when strong dependence is present.

6.4.1 Moment Methods for Unilateral Autoregressions (UARs)

UARs were studied in Section 4.8. They are artificial in a spatial context because they are based on an arbitrary notion of spatial ordering. However, they are still important theoretically as a generalization of time series in one dimension. Further, a version of this idea forms the underlying justification for Vecchia's approximation in Section 5.9.

Let $\{X_t,\ t \in \mathbb{Z}^d\}$ be modeled by a stationary *UAR*

$$X_t - \mu = \sum_{s \in K} d_s (X_{t-s} - \mu) + \varepsilon_t, \tag{6.30}$$

where K is a finite subset of some half-space $\mathcal{H} \subset \mathbb{Z}^d$. The noise terms $\varepsilon_t \sim N(0, \sigma_\varepsilon^2)$ are independent of each other and ε_t is independent of $\{X_{t-s} : s \in \mathcal{H}\}$. The covariance function $\{\sigma_h\}$ satisfies some moment equations. Multiply (6.30) by $(X_{t-h} - \mu)$, $h \neq 0$, and take expectations to get

$$\sigma_h = \sum_{s \in K} d_s \sigma_{h-s}, \quad h \in \mathcal{H}, \tag{6.31}$$

$$\sigma_0 = \sum_{s \in K} d_s \sigma_s + \sigma_\varepsilon^2. \tag{6.32}$$

Define a matrix $A(|K| \times |K|)$ with entries

$$a_{st} = \sigma_{s-t}, \quad s, t \in K \tag{6.33}$$

and a vector $\mathbf{g}(|K| \times 1)$ with entries

$$g_s = \sigma_s, \quad s \in K, \tag{6.34}$$

where the elements of K are listed in lexicographic order, say. Then (6.31), restricted to $h \in K$, yields an equation for $\boldsymbol{d} = \{d_s,\ s \in K\}$

$$\boldsymbol{d} = A^{-1}\mathbf{g} \tag{6.35}$$

in terms of the covariance function. Also, σ_ε^2 can be determined from (6.32)

$$\sigma_\varepsilon^2 = \sigma_0 - \mathbf{g}^T A^{-1} \mathbf{g}. \tag{6.36}$$

Equations (6.35)–(6.36) are the analogs of the Yule–Walker equations in time series.

Replacing σ_h by $s_h^{(b)}$ in the definitions of A and $\mathbf{g}$ yields the moment estimates, denoted $\hat{\boldsymbol{d}}_{\text{mom}}$ and $\hat{\sigma}^2_{\varepsilon,\text{mom}}$.

In finite samples, there is no guarantee that $\hat{\boldsymbol{d}}_{\text{mom}}$ will define a stationary UAR. For example, in a one-dimensional AR(1) process, there is no guarantee that $\hat{d}_{\text{mom}} \in \mathbb{R}$ will satisfy $|\hat{d}_{\text{mom}}| < 1$. However, these problems disappear in large samples when the data come from the assumed model.

6.4.2 Moment Estimators for Conditional Autoregression (CAR) Models

Consider the CAR model on $\mathbb{Z}^d$

$$E[X_t|\{X_s, s \neq t\}] = \mu + \sum_{s\in\mathcal{N}} \beta_s(X_{t-s} - \mu), \quad \text{var}[X_t|\{X_s, s \neq t\}] = \sigma_\eta^2, \tag{6.37}$$

where $\mathcal{N}$ is a finite symmetric neighborhood of the origin, as in Section 4.4, and where σ_η^2 is the conditional noise variance. This model was studied in Section 4.6. In particular, Eqs. (4.35) and (4.36) in Chapter 4 describe the relationship between the covariance function and the parameters of the model:

$$\sigma_h = \sum_{s\in\mathcal{N}} \beta_s \sigma_{h-s}, \quad h \neq 0, \tag{6.38}$$

$$\sigma_0 = \sum_{s\in\mathcal{N}} \beta_s \sigma_s + \sigma_\eta^2. \tag{6.39}$$

We can invert these equations to solve for β_s, $s \in \mathcal{N}$ and σ_η^2 in terms of $\{\sigma_h\}$. Let $\mathcal{N}^\dagger$ be a half-neighborhood of $\mathcal{N}$, so that $0 \notin \mathcal{N}^\dagger$ and only one of each pair $\pm h \in \mathcal{N}$ lies in $\mathcal{N}^\dagger$. Next, let $\boldsymbol{\beta}$ denote the vector of $\{\beta_s : s \in \mathcal{N}^\dagger\}$, with the elements arranged in lexicographic order. Similarly, let $\mathbf{g}$ denote the vector $\{2\sigma_h : h \in \mathcal{N}^\dagger\}$, and define an $|\mathcal{N}^\dagger| \times |\mathcal{N}^\dagger|$ matrix A with entries

$$a_{st} = 2(\sigma_{s-t} + \sigma_{s+t}). \tag{6.40}$$

Then (6.38) and (6.39) can be rewritten as

$$\sigma_0 = \boldsymbol{\beta}^T\mathbf{g} + \sigma_\eta^2, \quad \mathbf{g} = A\boldsymbol{\beta} \tag{6.41}$$

so that

$$\boldsymbol{\beta} = A^{-1}\mathbf{g}, \quad \sigma_\eta^2 = \sigma_0 - \mathbf{g}^T A^{-1}\mathbf{g}. \tag{6.42}$$

Replacing σ_h by $s_h^{(b)}$ in the definitions of A and $\mathbf{g}$ yields the moment estimates, denoted $\hat{\boldsymbol{\beta}}_{\text{mom}}$ and $\hat{\sigma}^2_{\eta,\text{mom}}$.

However, just as for UAR models, there is no guarantee in finite samples that the resulting moment estimate $\hat{\boldsymbol{\beta}}_{\text{mom}}$ will define a valid CAR model. Further, if the unbiased covariances $s_h^{(u)}$ were used to estimate A, there would be no guarantee that the estimated matrix is positive definite.

Moment estimators for CAR models can also be viewed as maximum composite likelihood estimators. Details are in Section 6.7.

Example 6.1 ***Mercer–Hall data***

The Mercer–Hall data was studied in Example 1.14. The data take the form of a 20×25 matrix of grain yields from a uniformity trial. Ideally, in an agricultural setting, there would be no systematic effects of spatial location on the yield.

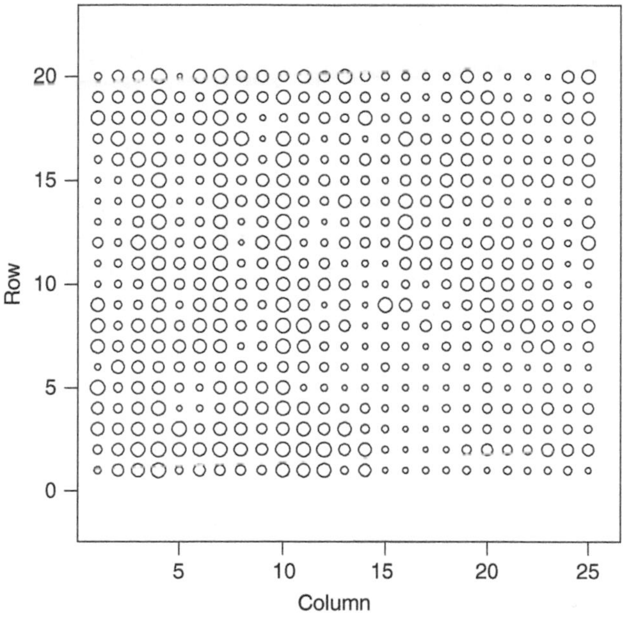

Figure 6.1 Mercer–Hall data: bubble plot. See Example 6.1 for an interpretation.

Such a situation would justify the use of standard experimental designs in other experiments with an assumption of i.i.d. errors. However, for this data set, a plot of the data in Figure 6.1 shows noticeable oscillations between the columns, especially on the left-hand side of the data. This effect is apparently due to an ancient ridge and furrow plowing system that has left its mark on the fertility of the modern field.

For this exercise, we fit a CAR model using moment estimation and maximum likelihood estimation to the 20×13 matrix containing left-hand half of the data (columns 1–13), where the column effects are most pronounced. We fit the CAR model using a full second-order neighborhood $\mathcal{N} = \{h \in \mathbb{Z}^2 : \max(|h[1]|, |h[2]|) \leq 2, h \neq 0\}$, so that $\boldsymbol{\beta}$ is a 12-dimensional vector. A plot of the sample and fitted covariance functions is given in Figure 6.2. The fitted covariances have been computed by numerical integration from the spectral density. The moment and maximum likelihood fitted covariance functions are very similar to each other. For both methods, the fitted covariances decay monotonically in the 0^0 direction (i.e. the vertical direction) and exhibit an oscillating decay in the other three directions. This behavior provides an approximate match to the sample covariance function in the four directions.

The CAR model has been presented here as an illustrative example to show that such models can have oscillating covariance functions analogously to the AR(2)

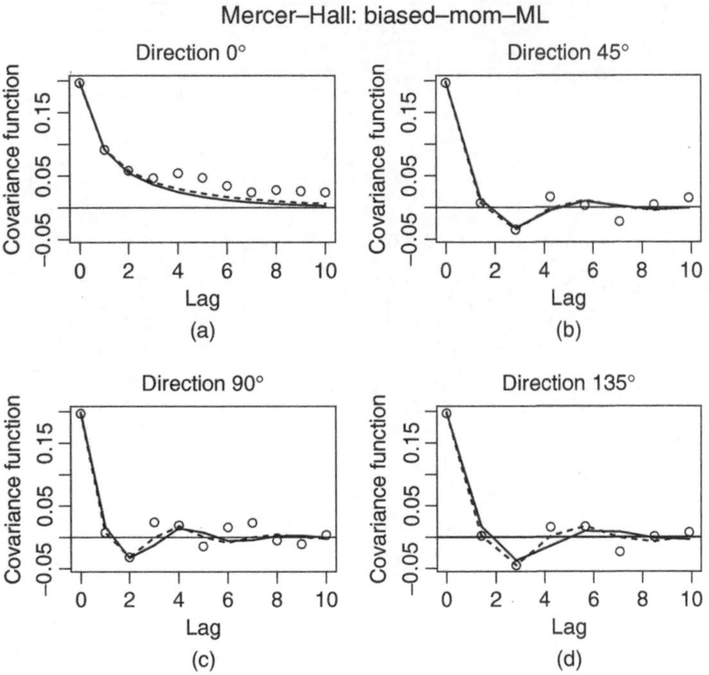

Figure 6.2 A plot of the sample and two fitted covariance functions ("biased-mom-ML") for a CAR model fitted to the leftmost 13 columns of the Mercer–Hall data (Example 6.1). The data have been summarized by the biased sample covariance function. The four panels show the covariance function in the four principal directions with the sample covariances (open circles) together with the fitted covariances using moment estimation (solid lines) and maximum likelihood estimation (dashed lines).

model in time series. More serious attempts to model the Mercer–Hall data have been made by numerous authors, notably Whittle (1954, 1986) and Gaetan and Guyon (2010). See also Cressie (1993, pp. 248–259, 446–447, 454–458) for a detailed analysis and a critique of earlier attempts.

An attempt was also made to construct the moment estimator for this CAR model using the unbiased sample covariance function. Unfortunately, the resulting spectral density was invalid – it was negative for some frequencies. □

6.5 Approximate Likelihoods for Lattice Data

For data $\boldsymbol{x}$ from a stationary Gaussian process, the exact log-likelihood is given by (6.7), where $\Psi_D^{(\text{exact})} = \Sigma_D^{-1}$ appears in the quadratic form and $\log|\Psi_D^{(\text{exact})}| = -\log|\Sigma_D|$ appears in the normalizing constant. For the purposes of this section, suppose that the mean vanishes, $\mu = 0$ and that the interest is in estimating the covariance parameters. Working with $\Psi_D^{(\text{exact})}$ can be problematic for the reasons suggested in the Introduction to this chapter. Hence, in this section, several approximations for the log-likelihood of autoregression models are considered, partly based on Kent and Mardia (1996). These are motivated by the possible treatment of boundary conditions suggested in Section 4.6.5.

(a) The starting point is the approximation to the log-likelihood proposed by Whittle (1954). In the notation here, it takes the form

$$-2\log L^{(W)} = Q^{(W)} + \frac{|D|}{(2\pi)^d}\int_{(-\pi,\pi)^d} \log f(\omega)\, d\omega. \tag{6.43}$$

The quadratic form $Q^{(W)}$ can be written in two equivalent forms. The first representation is in terms of the biased sample covariance function and the matrix Ψ_D described above (6.6),

$$Q^{(W)} = \boldsymbol{x}^T\Psi_D\boldsymbol{x} = |D| \sum_{h\in\mathcal{N}_0} s_h^{(b)}\psi_h = |D|\left\{ s_0^{(b)} + 2\sum_{h\in\mathcal{N}^\dagger} s_h^{(b)}\psi_h \right\}. \tag{6.44}$$

The second representation is in terms of the "biased" periodogram

$$Q^{(W)} = |D|\frac{1}{(2\pi)^d}\int_{(-\pi,\pi)^d} I(\omega)/f(\omega)\, d\omega, \tag{6.45}$$

where

$$I^{(b)}(\omega) = \sum_{h\in\mathbb{Z}^d} e^{i\omega h} s_h^{(b)}, \quad \omega\in\mathbb{R} \tag{6.46}$$

(Exercise 6.2). Recall that $s^{(b)} = 0$ unless $|h[\ell]| \le n[\ell]-1$, $\ell = 1,\dots,d$, so that the number of terms in the sum is finite. The second form brings out the role of the spectral density more clearly. It can be shown that the biased periodogram is always nonnegative (Exercise 6.1).

(b) As noticed by Guyon (1982), a drawback in Whittle's approximate log-likelihood is that the asymptotic bias in $s^{(b)}$ for large domains is nontrivial in dimensions $d \ge 2$. See Exercise 6.6. Hence, he proposed replacing the $Q^{(W)}$ by an unbiased version $Q^{(G)}$, say, using the unbiased sample covariance function from (6.8), to give

$$Q^{(G)} = \sum_{h\in\mathcal{N}_0} |D_h|\psi_h s_h^{(u)}. \tag{6.47}$$

(c) However, the unbiased analog of the periodogram is not guaranteed to be nonnegative in finite samples, which means its use in maximum likelihood can lead to invalid solutions. Therefore, Künsch (1987) proposed an intermediate solution based on tapering. Let w_t, $t\in Z^2$, be a tapering window for D such that $w_t = 1$ well inside D, $w_t = 0$ outside D, and w_t varies smoothly near the boundary of D. Then set

$$s_h^{(K)} = \frac{1}{|D|}\sum_{t\in D_h} w_t(x_t - \overline{x})\, w_{t+h}(x_t - \overline{x}),$$

$$s_h^{(K)} = 0 \text{ if } D_h = \emptyset.$$

Note that the effect of the taper is to downweight sites near the boundary. A judicious choice of taper, depending on the size of the domain D, will ensure that $s_h^{(K)}$ is asymptotically unbiased for σ_h. Further, since $\{s_h^{(K)}\}$ can be expressed as a convolution of two sequences, the periodogram $I^{(K)}(\omega)$ is always nonnegative. However, a problem in practice is how to choose the taper.

(d) Mathematically, a simple approximation is based on the block circulant approximation $\Psi_D^{(\mathrm{circ})}$ with elements

$$\{\Psi_D^{(\mathrm{circ})}\}_{st} = \psi_{(s-t)\mathrm{Mod}\ N}, \quad s, t \in D,$$

where the block Mod operation is described in Eq. (A.2). From this point of view, each edge of D is regarded as adjacent to its opposite edge. The quadratic form can be written as a simple linear combination of terms involving the periodic sample covariances from (6.12)

$$\boldsymbol{x}^T \Psi_D^{(\mathrm{circ})} \boldsymbol{x} = |D| \sum_{h \in \mathcal{N}} \psi_h s_h^{(p)}. \tag{6.48}$$

Further, the normalizing constant becomes a Riemann sum

$$\log |\Psi_D^{(\mathrm{circ})}| = \sum_{j \in \mathbb{Z}_N^d} \log f(\omega_j), \tag{6.49}$$

where ω_j is a d-vector with components $\omega_j[\ell] = 2\pi j[\ell]/n[\ell], \ell = 1, \ldots, d$. The main problem with this approximation is that observations on opposite sides of D are treated as neighbors of one another.

(e) Another choice is the folded approximation $\Psi_D^{(\mathrm{fold})}$. This approximation, based on a reflection boundary condition, is only defined for models with full symmetry. First, "double" the data by reflecting D about each side to get a new domain of size $2^d|D|$. Call the doubled data $\boldsymbol{y}$, treat it as stationary, and use the block circulant approximation (d) for its precision matrix. Finally, write the quadratic form in $\boldsymbol{y}$ in terms of the original data $\boldsymbol{x}$. Details are given in Section A.11.

The folded approximation is particularly appealing under high autocorrelation and especially in the limiting case of IRF-0 processes (Besag and Mondal, 2005; Mondal, 2018).

Example 6.2 ***Mercer–Hall data***

We revisit the Mercer–Hall data of Example 6.1, this time summarizing the data using the folded sample covariance function. The fit is similar to the earlier one; see Figure 6.3. □

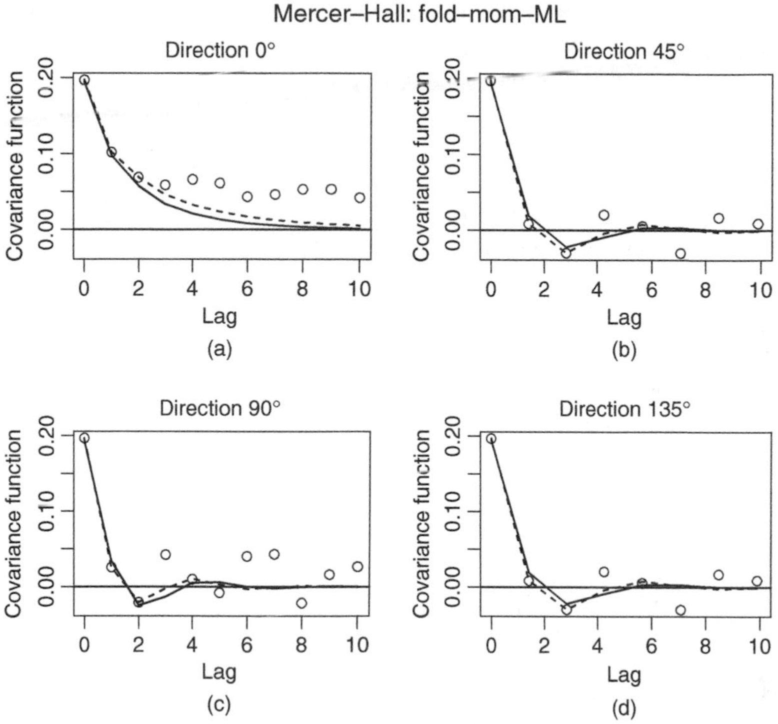

Figure 6.3 A plot of the sample and two fitted covariance functions ("fold-mom-ML") for a CAR model fitted to the leftmost 13 columns of the Mercer–Hall data (Example 6.2). The data have been summarized by the folded sample covariance function. The four panels show the covariance function in the four principal directions with the sample covariances (open circles) together with the fitted covariances using moment estimation (solid lines) and maximum likelihood estimation (dashed lines).

6.6 Accuracy of the Maximum Likelihood Estimator

Consider data $\{x_t,\ t \in D\}$, where D is a rectangular domain D of size $|D| = |N| = n[1] \times \cdots \times n[d]$, represented as a vector $\boldsymbol{x}$, with the elements in lexicographic order. Assume the data come from a stationary process on $\mathbb{Z}^d$ with mean 0. Let $\Sigma = \Sigma(\boldsymbol{\theta})$ denote the covariance matrix at the data sites and let $f(\omega; \boldsymbol{\theta})$ be the corresponding spectral density of the covariance function. Here, $\boldsymbol{\theta}$ is a p-dimensional vector of parameters to estimate. The purpose of this section is to investigate the asymptotic variance of the MLE as each of the dimensions of D gets large, $n[\ell] \to \infty$, $\ell = 1, \ldots, d$.

Thus, $\boldsymbol{x}$ is a realization from $N_{|N|}(\mathbf{0}, \Sigma)$. Asymptotically, under mild regularity conditions, the MLE $\hat{\boldsymbol{\theta}}$ is normally distributed about the true value

$$\hat{\boldsymbol{\theta}} \sim N_p(\boldsymbol{\theta}, \mathcal{I}^{-1}),$$

where $\mathcal{I}$ is the Fisher information matrix. For the covariance parameters in the multivariate normal distribution, the elements of $\mathcal{I}$ take the form

$$(\mathcal{I})_{ij} = \frac{1}{2}\text{tr}\{\Sigma^{-1}\Sigma_i\Sigma^{-1}\Sigma_j\}, \tag{6.50}$$

where $\Sigma_i = \partial\Sigma/\partial\theta_i$; see Section A.12.2.

To understand this formula more clearly, it is helpful to switch to the Fourier or spectral domain and to use a circulant approximation to the covariance matrix. Let

$$\boldsymbol{y} = G^T\boldsymbol{x}, \tag{6.51}$$

where $G = G_N^{(\text{DFT,rea})}$ is defined in Eqs. (A.44) and (A.45). Under the circulant approximation, the elements of $\boldsymbol{y}$, indexed by the cyclic integers $k \in \mathbb{Z}_N^d$, are independent with variances $f(\omega_k; \boldsymbol{\theta})$, where $\omega_k = (2\pi k[1]/n[1], \ldots, 2\pi k[d]/n[d])$. Hence, the log-likelihood is

$$\log L = -\frac{1}{2}\sum_{k\in\mathbb{Z}_N^d}\{y_k^2/f(\omega_k;\boldsymbol{\theta}) + \log f(\omega_k;\boldsymbol{\theta})\}. \tag{6.52}$$

Using again the results in Section A.12.2, the elements of the Fisher information become

$$\begin{aligned}(\mathcal{I})_{ij} &= \frac{1}{2}\sum_{k\in\mathbb{Z}_N^d}\frac{f_i(\omega_k)f_j(\omega_k)}{f^2(\omega_k)} \\ &\approx \frac{1}{2}\frac{|N|}{(2\pi)^d}\int\frac{f_i(\omega)f_j(\omega)}{f^2(\omega)}\,d\omega \\ &= \frac{1}{2}\frac{|N|}{(2\pi)^d}\int\frac{g_i(\omega)g_j(\omega)}{g^2(\omega)}\,d\omega,\end{aligned} \tag{6.53}$$

where $f_i(\omega) = f_i(\omega;\boldsymbol{\theta}) = \partial f(\omega)/\partial\theta_i$. In the second line, the sum has been replaced by an integral. In the final line, the information has been represented in terms of the inverse spectral density, $g(\omega)$, defined in (6.1), with derivatives $g_i(\omega) = g_i(\omega;\boldsymbol{\theta}) = \partial g(\omega)/\partial\theta_i$.

Example 6.3 ***Outfill asymptotics for the Matérn model***
The Matérn process $X(t)$, $t \in \mathbb{R}^d$, from Section 2.6 has spectral density

$$f(\omega) = \frac{1}{(A + B|\omega|^2)^{\nu+d/2}}, \quad \omega \in \mathbb{R}^d.$$

Let $X_t = X(t),\ t \in \mathbb{Z}^d$, denote the Matérn process restricted to the integer lattice. The spectral density of the restricted process is obtained by wrapping the original spectral density onto the torus

$$f_T(\omega) = \sum_{j \in \mathbb{Z}^d} f(\omega + 2\pi j), \quad \omega \in (-\pi, \pi)^d. \tag{6.54}$$

For data on an n^d integer lattice, the asymptotic variance matrix of the maximum likelihood estimates for the parameters $\boldsymbol{\theta} = (A, B, \nu)$ is given by the inverse of the asymptotic 3×3 Fisher information matrix $\mathcal{I}$ with elements

$$(\mathcal{I})_{ij} = \frac{n^d}{2(2\pi)^d} \int_{(-\pi,\pi)^d} \frac{f_{T,i} f_{T,j}}{f_T^2}\, d\omega, \quad i,j = 1,2,3, \tag{6.55}$$

where

$$f_{T,i}(\omega) = \partial f_T / \partial \theta_i = \sum_{j \in \mathbb{Z}^d} \partial f(\omega + 2\pi j)/\partial \theta_i, \quad i = 1,2,3, \tag{6.56}$$

and where the partial derivatives of the Matérn spectral density $f(\omega)$ with respect to the parameters are given by

$$\begin{aligned} \partial f/\partial \theta_1 &= -(\nu + d/2) \frac{1}{(A + B|\omega|^2)^{1+\nu+d/2}}, \\ \partial f/\partial \theta_2 &= -(\nu + d/2) \frac{|\omega|^2}{(A + B|\omega|^2)^{1+\nu+d/2}}, \\ \partial f/\partial \theta_3 &= -\log(A + B|\omega|^2) f. \end{aligned} \tag{6.57}$$

It can be checked that $f_T(\omega)$ is bounded away from 0 and ∞ and that the first two derivatives of $f_T(\omega; h)$ are bounded in absolute value. Hence, the integrals in the Fisher information matrix in (6.55) are all finite; see Exercise 6.11. □

An important class of models is given by the CAR models, with spectral density

$$f(\omega) = (2\pi)^{-d} \sigma_\eta^2 / \tilde{b}(\omega), \tag{6.58}$$

where

$$\tilde{b}(\omega) = 1 - \sum_{s \in \mathcal{N}} \beta_s \cos(s^T \omega) = 1 - 2 \sum_{s \in \mathcal{N}^\dagger} \beta_s \cos(s^T \omega). \tag{6.59}$$

The parameters are the coefficients $\{\beta_s,\ s \in \mathcal{N}^\dagger\}$ and σ_η^2. For algebraic calculations, it is more convenient to work with

$$\tau = 1/\sigma_\eta^2$$

instead of σ_η^2.

For $k \in \mathbb{Z}_N^d$, the elements y_k are realizations from independent normal distributions, $N(0, (\tau \tilde{b}_k)^{-1})$, where

$$\tilde{b}_k = \tilde{b}(\omega_k), \quad \omega_k = 2\pi(k[1]/n[1], \ldots, k[d]/n[d]). \tag{6.60}$$

For later use, recall the basic fact about fourth moments of the normal distribution. If $X \sim N(0, \sigma^2)$,

$$\mathrm{var}(X) = \sigma^2, \quad \mathrm{var}(X^2) = 2\sigma^2. \tag{6.61}$$

Up to a constant term, the log-likelihood from (6.52) simplifies to

$$\log L = -\frac{1}{2} \sum_{k \in \mathbb{Z}_N^d} \{\tau \tilde{b}_k y_k^2 - \log \tau - \log \tilde{b}_k\}. \tag{6.62}$$

The parameters of the model are $\theta_0 = \tau$ and $\theta_h = \beta_h$, $h \in \mathcal{N}^\dagger$. From Section A.12, the Fisher information is the matrix $\mathcal{I}$ with entries

$$\begin{aligned}(\mathcal{I})_{ij} &= \sum_{k \in \mathbb{Z}_N^d} \frac{g_i(\omega_k) g_j(\omega_k)}{g^2(\omega_k)} \\ &\approx \frac{|N|}{2\pi} \int_0^{2\pi} \frac{g_i(\omega) g_j(\omega)}{g^2(\omega)}\, d\omega,\end{aligned} \tag{6.63}$$

where $g(\omega) = (2\pi)^{-d} \tau \tilde{b}(\omega)$ is the inverse spectral density, with derivatives with respect to the parameters given by

$$g_0(\omega) = \partial g(\omega)/\partial \theta_0 = (2\pi)^{-d} \left\{ 1 - 2 \sum_{h \in \mathcal{N}^\dagger} \beta_h \cos(h^T \omega) \right\},$$
$$g_h(\omega) = \partial g(\omega)/\partial \theta_h = -2\tau(2\pi)^{-d} \cos(h^T \omega), \quad h \in \mathcal{N}^\dagger.$$

The integral form is the asymptotic form of the Fisher information matrix for both the exact likelihood and all the approximate versions considered in this chapter.

In particular, the MLE is asymptotically normally distributed,

$$\hat{\theta}_{\mathrm{ML}} \sim N(\boldsymbol{\theta}, \mathcal{I}^{-1}).$$

6.7 The Moment Estimator for a CAR Model

Another estimator for CAR models is the moment estimator of Section 6.4.2. To study this estimator, it is helpful first to give it an interpretation as a composite maximum likelihood estimator based on the full conditional densities (e.g. Mardia et al. 2010). This composite likelihood is different from the composite likelihood used in Vecchia's estimation procedure in Section 5.9.

This composite likelihood here is given by the product of the full conditional densities. For a CAR model with mean 0, this likelihood is obtained by writing down the product of the densities for $x_t | \boldsymbol{x}_{\setminus t}$. Effectively, the residuals $\eta_t = x_t - \sum_{s \in \mathcal{N}} \beta_s x_{t-s}$, after regressing x_t on its neighboring values, are falsely treated as independent of one another. In convolution notation, we can write $\boldsymbol{\eta} = \boldsymbol{b} * \boldsymbol{x}$, where $\boldsymbol{b}$ is a set of coefficients with $b_0 = 1$ and $b_s = -\beta_s$, $s \in \mathcal{N}$. The composite log-likelihood is obtained by treating the convolved data $\boldsymbol{b} * \boldsymbol{x}$ as i.i.d $N(0, \sigma_\eta^2)$ random variables.

Let the notation cMLE be used for the composite maximum likelihood estimator. For the CAR model, it can be shown that the cMLE is the same as a version of the moment estimator in Section 6.4.2. See Exercise 6.5 for details. In general, composite maximum likelihood estimators will be consistent, but less efficient than the full maximum likelihood estimator, and this is the situation for the CAR model.

To study the efficiency of the cMLE, it is easiest to work in the spectral domain with a circulant approximation, using (6.51) to define $\boldsymbol{y}$. Since convolution in the spatial domain corresponds to multiplication in the Fourier domain, the convolved data in the Fourier domain have elements $\tilde{b}_k y_k$, which are treated as i.i.d. $N(0, 1/\tau)$ random variables, using the notation from (6.60), with $\tau = 1/\sigma_\eta^2$. Hence, the composite log-likelihood becomes

$$\log L_c = -\frac{1}{2}\sum_k \{\tau \tilde{b}_k^2 y_k^2 - \log \tau\}. \tag{6.64}$$

The *composite likelihood information matrix* is defined by

$$\mathcal{I}_{\text{CL}} = \mathcal{H}\mathcal{J}^{-1}\mathcal{H},$$

where $\mathcal{J} = E\{(\partial l_c/\partial \boldsymbol{\theta})(\partial l_c/\partial \boldsymbol{\theta})^T\}$ is the "expected squared score" and $\mathcal{H} = -E\{\partial^2 l_c/\partial \boldsymbol{\theta}\partial \boldsymbol{\theta}^T\}$ is the matrix of "expected score derivatives," after changing the sign, where $l_c = \log L_c$. The composite MLE is asymptotically normally distributed

$$\hat{\boldsymbol{\theta}}_{\text{cML}} \sim N(\boldsymbol{\theta}, \mathcal{H}^{-1}\mathcal{J}\mathcal{H}^{-1}).$$

The relative efficiency of the cMLE compared to the MLE can be summarized by the eigenvalues of $A = \mathcal{I}_{\text{CL}}^{-1}\mathcal{I}$, which always lie between 0 and 1, and which, in turn, can be summarized by the scalar quantity $|A|^{1/p}$, where p is the dimension of $\boldsymbol{\theta}$. The example of the AR(1) process is examined in detail in Exercise 6.8 and the relative efficiency, as a function of β is plotted in Figure 6.4.

For the basic first-order CAR model in $\mathbb{Z}^2$ (Example 4.4), the cMLE was termed a "coding estimator" by Besag and Moran (1975). In particular, if a planar grid is divided into black and white squares as in a chess board, then the white squares are conditionally independent given the black squares and vice versa. In that case, the composite likelihood becomes the product of two conditional likelihoods (black given white times white given black).

Exercises

6.1 Given observations x_t, $t \in D$, where D is a rectangular region in $\mathbb{Z}^d$ as in (6.4), let $\bar{x}$ denote the sample mean of the data and define the centered

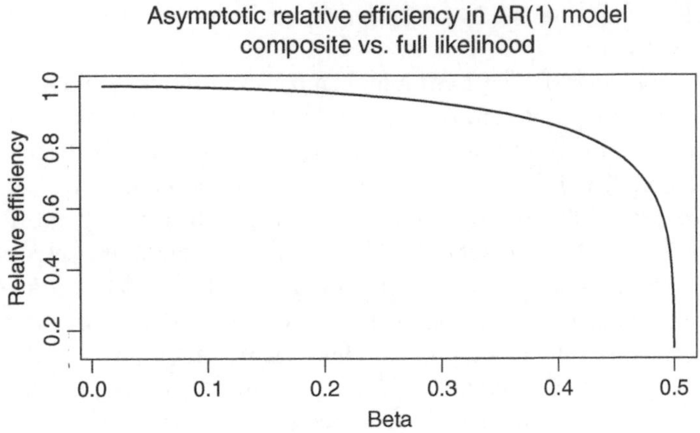

Figure 6.4 Relative efficiency of the composite likelihood estimator in AR(1) model relative to the ML estimator.

data $\{y_t,\ t \in \mathbb{Z}^d\}$ by

$$y_t = \begin{cases} x_t - \bar{x}, & t \in D, \\ 0, & t \notin D. \end{cases}$$

(a) Show that the biased sample covariance function can be defined by the convolution

$$s_h^{(b)} = \frac{1}{|D|}(y * \check{y})_h = \frac{1}{|D|}\sum_{t\in\mathbb{Z}^d} y_t y_{t+h}.$$

(b) For any set of complex coefficients $\{\alpha_t\}$, show that

$$\sum_{s,t\in\mathbb{Z}^d} \alpha_s \overline{\alpha}_t s_{s-t}^{(b)} = \left|\sum_{t\in\mathbb{Z}^d} \alpha_t y_t\right|^2 \geq 0.$$

Hence, deduce that the sequence $\{s_h^{(b)}\}$ is positive semidefinite.

(c) Show that setting $\alpha_t = \exp(-i\omega t)$ yields the periodogram $I(\omega)$ in (6.46). Hence, deduce that the periodogram is nonnegative for all ω.

6.2 The quadratic form $Q^{(W)}$ in the Whittle log-likelihood for a CAR model is specified as a linear combination of terms involving the biased sample covariance function. Show that it can also be expressed in terms of the periodogram as in (6.45).
Hint: Recall or prove the one-dimensional result for $k \in \mathbb{Z}$, $\int_0^{2\pi} \exp(ik\omega)\, d\omega = 2\pi$ if $k = 0$, and 0 for $k \neq 0$. Similarly, prove the d-dimensional result for $k \in \mathbb{Z}^d$,

$$\int_{(-\pi,\pi)^d} \exp(ik^T\omega)\, d\omega = \begin{cases} (2\pi)^d, & k = 0, \\ 0, & k \neq 0. \end{cases}$$

Then expand out the integral $\int_{(-\pi,\pi)^d} I(\omega)/f(\omega)\, d\omega$ term by term.

6.3 (a) Using the ideas from Sections 4.8.2 and 4.6.1, show that the AR(1) and CAR(1) models in Section 6.3 have the spectral densities given in (6.13) and (6.17).

(b) Show that the spectral densities are the same as one another under the conditions in (6.18). Further show that λ and β in (6.20) are related by

$$\beta = \lambda / \left(1 + \lambda^2\right).$$

(c) For $z = \exp(i\omega)$, show that

$$z + \bar{z} = 2\cos\omega,$$
$$(1 - \lambda z)(1 - \lambda \bar{z}) = 1 + \lambda^2 - 2\lambda\cos\omega,$$
$$\frac{1}{1-\lambda^2}\left\{\frac{1}{1-\lambda z} + \frac{1}{1-\lambda\bar{z}} - 1\right\} = \frac{1}{(1-\lambda z)(1-\lambda\bar{z})},$$

and use the geometric and logarithmic series expansions

$$\frac{1}{1-\lambda z} = \sum_{j=0}^{\infty} \lambda^j z^j, \quad -\log(1-\lambda z) = \sum_{j=1}^{\infty} \frac{1}{j}\lambda^j z^j.$$

to verify the Fourier expansions (6.14)–(6.16).

(d) Prove the identities (6.20) to express λ and β in terms of the covariance function (6.19).

6.4 Consider n equally spaced data sites in one dimension from the exponential covariance function. The covariance matrix Σ was given in (6.21). Show that the inverse of Σ is given by $\Psi^{(\text{exact})}$ in (6.22) by confirming that the matrix product reduces to the identity matrix, $\Sigma\,\Psi^{(\text{exact})} = I_n$.

6.5 *A regression interpretation of the moment estimator for a CAR model.* Let D be a finite domain in $\mathbb{Z}^d$ and let $\mathcal{N}$ be a finite symmetric neighborhood of the origin, with half-neighborhood $\mathcal{N}^\dagger$. Let $D_{\mathcal{N}} = \{t \in D : t + s \in D$ for all $s \in \mathcal{N}\}$, and let $\boldsymbol{y}$ denote the vector $\{x_t : t \in D_{\mathcal{N}}\}$ with elements arranged in lexicographic order.

Next define a "design matrix" X of size $|D_{\mathcal{N}}| \times |\mathcal{N}^\dagger|$ with entries

$$x_{ts} = x_{t-s} + x_{t+s}, \quad t \in D_{\mathcal{N}},\ s \in \mathcal{N}^\dagger.$$

A regression of $\boldsymbol{y}$ on X yields estimates for regression coefficient vector $\boldsymbol{\beta}$ and the residual variance σ_η^2 given by a sample version of the moment identity (6.42), where the elements of the matrix A and the vector $\mathbf{g}$ and $\boldsymbol{\sigma}_0$ are estimated by

$$a_{\text{reg};h_1h_2} = \frac{1}{|D_{\mathcal{N}}|}\sum_{t\in D_{\mathcal{N}}} (x_{t-h_1} + x_{t+h_1})(x_{t-h_2} + x_{t+h_2}),$$
$$g_{\text{reg};h} = \frac{1}{|D_{\mathcal{N}}|}\sum_{t\in D_{\mathcal{N}}} x_t(x_{t+h} + x_{t-h}),$$

$$\sigma_{\text{reg};0} = \frac{1}{|D_{\mathcal{N}}|} \sum_{t \in D_{\mathcal{N}}} x_t^2.$$

6.6 Guyon (1982). This exercise looks in more detail at the unbiased sample covariance function used in Section 6.5. Suppose D is a rectangular lattice of length n in each direction so that it contains $|D| = n^d$ data sites. For simplicity, consider the lag $h = \begin{bmatrix} 1 & 0 & \dots & 0 \end{bmatrix}$ representing one step along the first coordinate axis. Assume the data x_t, $t \in D$, come from a stationary random field with the known mean 0 and with the covariance function $\{\sigma_h : h \in \mathbb{Z}^d\}$. Consider the unbiased and biased sample covariance functions

$$s_h^{(u)} = \frac{1}{|D_h|} \sum_{t \in D_h} x_t x_{t+h}, \quad s_h^{(b)} = \frac{1}{|D|} \sum_{t \in D_h} x_t x_{t+h}.$$

These definitions are almost the same as in Eqs. (6.8)–(6.10), except they have been centered at 0 rather than the sample mean to make the calculations simpler.

(a) Show that the size of D_h is $|D_h| = (n-1)n^{d-1}$.

(b) Show that the unbiased covariance function $s_h^{(u)}$ in (6.8) is unbiased, i.e. $E\{s_h^{(u)}\} = \sigma_h$.

(c) Hence, show that the biased sample covariance $s_h^{(b)}$ in (6.10) is biased,

$$E\{s_h^{(b)}\} = \frac{|D_h|}{|D|}\sigma_h = \left(1 - \frac{1}{n}\right)\sigma_h.$$

(d) Under mild regularity conditions (e.g. Section 5.5.3) it can be shown that $s_h^{(u)}$ is asymptotically normally distributed for large n,

$$n^{d/2}(s_h^{(u)} - \sigma_h) \sim N(0, \kappa_h^2)$$

for some variance κ_h^2, say, not depending on n. Hence, deduce that asymptotic distribution of (6.10) is

$$n^{d/2}(s_h^{(b)} - \sigma_h) \sim N(-n^{d/2-1}\sigma_h, \kappa_h^2).$$

(e) In dimensions $d = 1, 2, 3$, deduce that for this distribution

(i) if $d = 1$, the bias $n^{-1/2}\sigma_h$ is negligible in comparison to the standard deviation κ_h.

(ii) if $d = 2$, the bias $n^0\sigma_h = \sigma_h$ has the same order as the standard deviation κ_h.

(iii) if $d = 3$, the bias $n^{1/2}\sigma_h$ dominates the standard deviation.

That is, if $d \geq 2$, $s_h^{(b)}$ is not asymptotically unbiased for σ_h. As a consequence, it can be shown that maximizing the Whittle approximation (6.43) does not produce an estimator asymptotically equivalent to the MLE. Instead, it is necessary to replace $s_h^{(b)}$ by $s_h^{(u)}$ as in (6.47) or to use tapering.

(f) At the same time, deduce that $E\{(s_h^{(b)} - \sigma_h)^2\} \to 0$ as $n \to \infty$ so that $s_h^{(b)}$ is a consistent estimator of σ_h.

6.7 *Information matrix for MLE from circular CAR(1) process.* The simplest nontrivial stationary Gaussian process on the line is the AR(1) model or equivalently the CAR(1) model. The spectral density of the covariance function has two equivalent representations, given in (6.17), i.e.

$$f(\omega) = \frac{1}{2\pi}\frac{\sigma_\eta^2}{1 - 2\beta\cos\omega} = \frac{1}{2\pi}\frac{\sigma_\varepsilon^2}{1 + \lambda^2 - 2\lambda\cos\omega}. \tag{6.65}$$

The easiest setting in which to investigate the issues related to maximum likelihood estimation is to consider the circulant version of this process for which the circulant covariance matrix has eigenvalues

$$f_k = (2\pi)f(\omega_k), \quad \omega_k = 2\pi k/n, \quad k = 0, \dots, n-1.$$

All circulant matrices have the same eigenvectors as shown in Section A.7. The eigenvectors are given in complex coordinates by the columns of the matrix $G_n^{(\mathrm{DFT,com})}$ in (A.41) or equivalently in real coordinates by the columns of $G_n^{(\mathrm{DFT,rea})}$ in (A.44) and (A.45).

(a) Let Σ denote the covariance matrix for this circulant model and let $\boldsymbol{x}$ denote an n-vector of data from this model, with log-likelihood

$$\log L = -\frac{1}{2}\left\{\boldsymbol{x}^T\Sigma^{-1}\boldsymbol{x} + \log|\Sigma| + n\log(2\pi)\right\}.$$

By rotating the data to $\boldsymbol{y} = G_n^{(\mathrm{DFT,rea})}\boldsymbol{x}$, show that the log-likelihood can also be written as

$$\log L = -\frac{1}{2}\left\{\sum y_k^2/f_k + \sum \log f_k + n\log(2\pi)\right\}.$$

(b) For the purposes of this exercise, parameterize the model by $\theta = (\theta_1, \theta_2)^T = (\sigma_\eta^2, \beta)$. Show that the information matrix $\mathcal{I}$ has elements

$$(\mathcal{I})_{ij} = \frac{1}{2}\sum_{k=0}^{n-1}(\partial \log f_k/\partial\theta_i)(\partial \log f_k/\partial\theta_j).$$

Hence, for this model show that

$$\mathcal{I} = \frac{1}{2}\begin{bmatrix} n/\sigma_\eta^4 & (2/\sigma_\eta^2)\sum c_k/d_k \\ (2/\sigma_\eta^2)\sum c_k/d_k & 4\sum c_k^2/d_k^2 \end{bmatrix},$$

where $c_k = \cos 2\pi k/n$, $d_k = 1 - 2\beta c_k$.

(c) For large n, the sums in the above matrices can be approximated by integrals. Some of the integrals are simpler to state by expressing $\beta = \lambda/(1+\lambda^2)$ in terms of the parameter λ from (6.18), with inverse $\lambda = (1-\sqrt{1-4\beta^2})/(2\beta)$. Also, it is convenient to set

$$P = 1+\lambda^2, \quad M = 1-\lambda^2,$$

so that

$$1+\lambda^2 - 2\lambda\cos\omega = P(1-2\beta\cos\omega),$$

and

$$\sqrt{1-4\beta^2} = \sqrt{1-\frac{4\lambda^2}{P^2}} = \sqrt{M^2/P^2} = M/P.$$

Then from Gradshteyn and Ryzhik (1980, p. 394, Formula 3.616.7), for $k \geq 0$ the following integrals can be defined and evaluated:

$$R_k = \frac{1}{2\pi}\int_0^{2\pi} \frac{\cos k\omega}{1-2\beta\cos\omega}\, d\omega = (P/M)\lambda^k \tag{6.66}$$

and

$$\begin{aligned} Q_k &= \frac{1}{2\pi}\int_0^{2\pi} \frac{\cos k\omega}{(1-2\beta\cos\omega)^2}\, d\omega \\ &= \frac{\lambda^k P^2}{M^3}(2\lambda^2 + (k+1)M) \\ &= \begin{cases} P^3/M^3, & k=0, \\ 2\lambda P^2/M^3, & k=1, \\ (3\lambda^2-\lambda^4)P^2/M^3, & k=2. \end{cases} \end{aligned} \tag{6.67}$$

Since $2\cos^2\omega = 1+\cos(2\omega)$, it is also convenient to define a quantity

$$\tilde{Q}_2 = (Q_0+Q_2)/2 = (1+4\lambda^2-\lambda^4)P^2/M^3.$$

Hence, deduce that

$$\mathcal{I} \approx n\begin{bmatrix} 1/(2\sigma_\eta^4) & R_1 \\ R_1 & 2\tilde{Q}_2 \end{bmatrix}.$$

6.8 *Information matrix for composite MLE from circular CAR(1) process.* Consider again the circulant CAR(1) model of the previous exercise. The purpose of this exercise is to investigate the accuracy of the composite estimator of Section 6.7.

The composite log-likelihood function is given by (6.64), i.e.

$$\log L_c = -\frac{1}{2}\sum_k \left\{\tilde{b}_k^2 y_k^2/\sigma_\eta^2 + \log\sigma_\eta^2 + \log(2\pi)\right\},$$

where $y_k \sim N(0, \sigma_\eta^2/\tilde{b}_k)$ independently for $i = 1, \ldots, n$. Recall the formulas for $\text{var}(y_k)$ and $\text{var}(y_k^2)$ from Eq. (6.61).
The composite score function is obtained by differentiating composite log-likelihood function with respect to $\theta_1 = \sigma_\eta^2$ and $\theta_2 = \beta$. Show that it has components

$$U_1^{(c)} = -\frac{1}{2}\sum_k \left\{-\tilde{b}_k^2 y_k^2/\sigma_\eta^4 + 1/\sigma_\eta^2\right\}, \quad U_2^{(c)} = -\frac{1}{2}\sum_k \{-4c_k\tilde{b}_k y_k^2/\sigma_\eta^2\}.$$

A basic property of trigonometric functions implies that $\sum c_k = 0$. Hence, check that the expected composite scores are 0.
Next, let $\mathcal{H}$ denote the matrix of minus the expected second derivatives of the composite log-likelihood. Show that it has elements

$$(\mathcal{H})_{11} = \frac{1}{2}\sum_k 1/\sigma_\eta^4 = \frac{n}{2\sigma_\eta^4}, \quad (\mathcal{H})_{12} = \frac{1}{2}\sum_k \{4c_k\tilde{b}_k/\sigma_\eta^2\} = 0,$$

$$(\mathcal{H})_{22} = \frac{1}{2}\sum_k \{8c_k^2/\tilde{b}_k\} \approx 4nR_2.$$

Expressions for R_k and Q_k are explained and derived in Exercise 6.7.
In addition, let $\mathcal{J} = E\{\boldsymbol{U}^{(c)}\boldsymbol{U}^{(c)T}\}$ denote the matrix of expected "squared" scores, which, since the Fourier coefficients y_k are independent, can be computed termwise. Show that the elements are

$$(\mathcal{J})_{11} = \frac{1}{4}\sum_k\left\{\tilde{b}_k^4/\sigma_\eta^8\right\}\text{var}(y_k^2) = \frac{1}{2\sigma_\eta^4}\sum_k\{1 - 4\beta c_k + 4\beta^2 c_k^2\} = \frac{n}{2\sigma_\eta^4}\{1 + 2\beta^2\},$$

$$(\mathcal{J})_{12} = \frac{1}{4}\sum_k \left\{4c_k\tilde{b}_k^3/\sigma_\eta^6\right\}\text{var}(y_k^2) = \frac{2}{\sigma_\eta^2}\sum_k \{c_k\tilde{b}_k\} = \frac{2}{\sigma_\eta^2}\sum_k \{c_k - 2\beta c_k^2\}$$

$$= -\frac{2\beta n}{\sigma_\eta^2},$$

$$(\mathcal{J})_{22} = \frac{1}{4}\sum_k \left\{16c_k^2\tilde{b}_k^2/\sigma_\eta^4\right\}\text{var}(y_k^2) = 8\sum_k c_k^2 = 4n.$$

A plot of the efficiency of the composite estimator is given in Figure 6.4.

6.9 *Information matrix for MLE from circular CAR(1) process with nugget effect.* Following on from Exercise 6.7, include a nugget effect in the CAR(1) model. Recall that a CAR(1) process with regression parameter β and conditional variance σ_η^2 can also be viewed as an AR(1) process with regression parameter λ and residual variance σ_ε^2. The parameters are related by

$$\beta = \lambda/P, \quad P = 1 + \lambda^2,$$

$$\lambda = (1 - \Delta)/(2\beta), \quad \Delta = \sqrt{1 - 4\beta^2},$$

$$\sigma_\eta^2 = \sigma_\varepsilon^2/P.$$

The covariance matrix of a circulant version of the process, observed at sites $t = 1, \dots, n$, has eigenvalues

$$\begin{aligned} f_k &= \frac{\sigma_\eta^2}{1 - 2\beta \cos \omega_k} + \tau^2, \quad \omega_k = 2\pi k/n, \quad k = 1, \dots, n, \\ &= \delta_1 \frac{1 - 2\delta_3 \cos \omega_k}{1 - 2\delta_2 \cos \omega_k}, \end{aligned}$$

where $\Delta = \sqrt{1 - 4\beta^2}$, $\delta_1 = \sigma_\eta^2 \Delta + \tau^2$, $\delta_2 = \beta$, and $\delta_3 = \beta\tau^2/\delta_1$. The model can be parameterized in several ways. For the purposes of this exercise, three natural parameterizations are

$$\begin{aligned} \boldsymbol{\delta} &= (\delta_1, \delta_2, \delta_3)^T = (\sigma_\eta^2 \Delta + \tau^2, \beta, \beta\tau^2/(\sigma_\eta^2 \Delta + \tau^2))^T, \\ \boldsymbol{\theta} &= (\theta_1, \theta_2, \theta_3)^T = (\sigma_\eta^2, \beta, \tau^2)^T, \\ \boldsymbol{\chi} &= (\chi_1, \chi_2, \chi_3)^T = (\sigma_\varepsilon^2, \lambda, \tau^2)^T. \end{aligned}$$

For a model with nontrivial dependence, the parameters satisfy the constraints, $\delta_1, \sigma_\eta^2, \sigma_\varepsilon^2 > 0$, $\tau^2 \geq 0$, $0 < |\delta_2|$, $|\delta_3| < 1/2, 0 < |\lambda| < 1$.
The first representation is easiest to differentiate, and from a time- series perspective represents a circulant version of an ARMA(1,1) process. The second and third representations emphasize the CAR(1) and AR(1) interpretations of the parameters.
Define an $n \times 3$ matrix of coefficients $A = (a_{kj})$, $k = 1, \dots, n$, $j = 1, 2, 3$ by $a_{kj} = \partial \log f_k / \partial \delta_j$. Show that

$$\begin{aligned} a_{k1} &= 1/\delta_1, \quad a_{k2} = 2\cos\omega_k/(1 - 2\delta_2 \cos\omega_k), \\ a_{k3} &= -2\cos\omega_k/(1 - 2\delta_3 \cos\omega_k). \end{aligned}$$

Show that the Fisher information matrix for $\boldsymbol{\delta}$ is given by

$$\mathcal{I}^{(\delta)} = \frac{1}{2}\sum_{k=1}^{n} (\partial \log f_k/\partial\boldsymbol{\delta})(\partial \log f_k/\partial\boldsymbol{\delta})^T = \frac{1}{2} A^T A.$$

Using the notation introduced in Exercise 6.7, show that the elements of $\mathcal{I}^{(\delta)}$ can be approximated by integrals with values

$$\mathcal{I}^{(\delta)} = n \begin{bmatrix} 1/(2\delta_1^2) & R_1/\delta_1 & -R_1^*/\delta_1 \\ R_1/\delta_1 & 2\tilde{Q}_2 & (*) \\ -R_1^*/\delta_1 & (*) & 2\tilde{Q}_2^* \end{bmatrix},$$

where

$$(*) = -\{\beta(R_0 + R_2) - \beta^*(R_0^* + R_2^*)\}/(\beta - \beta^*).$$

Here, the unstarred quantities R_0 and R_2 are based on $\delta_2 = \beta$ and the starred quantities R_0^* and R_2^* are based on $\delta_3 = \beta^*$, say.

Show that the mapping from $\boldsymbol{\theta}$ to $\boldsymbol{\delta}$ has the Jacobian matrix

$$J^{(1)} = \partial\boldsymbol{\delta}/\partial\boldsymbol{\theta}^T = \begin{bmatrix} 1 & 0 & 1 \\ 0 & 1 & 0 \\ -\beta\tau^2/\delta_1^2 & \tau^2/\delta_1 & \beta\sigma_\eta^2/\delta_1^2 \end{bmatrix}.$$

Show that the mapping from $\boldsymbol{\chi}$ to $\boldsymbol{\theta}$ has the Jacobian matrix

$$J^{(2)} = \partial\boldsymbol{\theta}/\partial\boldsymbol{\chi}^T = \begin{bmatrix} 1/P & -2\lambda\sigma_\varepsilon^2/P^2 & 0 \\ 0 & M/P^2 & 0 \\ 0 & 0 & 1 \end{bmatrix}.$$

Hence deduce that the information matrix for $\boldsymbol{\chi}$ is given by

$$\mathcal{I}^{(\chi)} = J^{(2)T} J^{(1)T} \mathcal{I}^{(\delta)} J^{(1)} J^{(2)}.$$

In particular show that if $\tau^2 = 0$, then the estimates of σ_ε^2 and λ are asymptotically uncorrelated. (On the other hand, this last result can be obtained more simply by writing the eigenvalues f_k directly in terms of the parameters σ_ε^2 and λ.)

6.10 Another circulant approximation to the covariance matrix for a circular CAR(1) model at n sites is given by

$$\Sigma_{\text{circ}} = \sigma^2 \text{circ}(1, \lambda, \lambda^2, \ldots, \lambda^m, \lambda^m, \ldots, \lambda^2, \lambda), \quad n \text{ odd}, m = (n-1)/2,$$
$$\Sigma_{\text{circ}} = \sigma^2 \text{circ}(1, \lambda, \lambda^2, \ldots, \lambda^{m-1}, \lambda^m, \lambda^{m-1}, \ldots, \lambda^2, \lambda), \quad n \text{ even}, m = n/2,$$

Show that the eigenvalues are given by

$$\alpha_j = \frac{1 - \lambda^2 - 2\lambda^{m+1}\{a_j - \lambda b_j\}}{1 + \lambda^2 - 2\lambda c_j}, \quad n \text{ odd}, m = (n-1)/2,$$
$$\alpha_j = \frac{1 - \lambda^2 - \lambda^m\{b_j + 2\lambda a_j - 2\lambda c_j b_j - \lambda^2 b_j\}}{1 + \lambda^2 - 2\lambda c_j}, \quad n \text{ even}, m = n/2,$$

where $a_j = \cos(m+1)\omega_j$, $b_j = \cos m\omega_j$, $c_j = \cos\omega_j$ in terms of $\omega_j = 2\pi j/n$, and $j = 0, 1, \ldots, n-1$. In particular note the special cases

$$(n=2) \quad \alpha_0 = 1 + \lambda, \ \alpha_1 = 1 - \lambda,$$
$$(n=3) \quad \alpha_0 = 1 + 2\lambda, \ \alpha_1 = \alpha_2 = 1 - \lambda,$$
$$(n=4) \quad \alpha_0 = 1 + 2\lambda + \lambda^2, \ \alpha_1 = \alpha_3 = 1 - \lambda^2, \ \alpha_2 = 1 - 2\lambda + \lambda^2.$$

Hint: From Fuller (1996, p. 151, eqn (4.2.8)), we have

$$\alpha_j = \begin{cases} \sum_{h=-m}^{m} \lambda^{|h|} \cos(2\pi h/n), & n \text{ odd}, m = (n-1)/2, \\ \sum_{h=-m+1}^{m} \lambda^{|h|} \cos(2\pi h/n), & n \text{ even}, m = n/2. \end{cases}$$

Further, from Gradshteyn and Ryzhik (1980, p. 31, eqn (1.353.3))

$$\sum_{h=0}^{n-1} \lambda^{|h|} \cos hx = \frac{1 - \lambda \cos x - \lambda^n \cos nx + \lambda^{n+1} \cos(n-1)x}{1 + \lambda^2 - 2\lambda \cos x}.$$

6.11 Consider the Matérn process restricted to the integer lattice $t \in \mathbb{Z}^d$, and suppose data are observed on a rectangular region of size $D \subset \mathbb{Z}^d$ of size $n_1 \times \cdots \times n_d$. The purpose of this exercise is look at the outfill asymptotics problem and in particular to show that the elements of the limiting Fisher information matrix in (6.55) are finite. There are several steps.

(a) The first step is to develop a general criterion to ensure the convergence of an infinite sum, by bounding the terms of the sum by a monotone decreasing function whose integral is finite. Thus let f_k, $k \in \mathbb{Z}^d$ be a sequence of finite real numbers, and suppose the sequence can be bounded by

$$|f_k| \le \varphi(|k|) \tag{6.68}$$

for $|k| \ge r_0$ where $\varphi(r)$, a continuous function of a real argument r, is monotone decreasing for $r \ge r_0$ for some $r_0 > 0$, and where $\int_{r_0}^{\infty} \varphi(r) r^{d-1}\, dr < \infty$. Then $\sum_{k \in \mathbb{Z}^d} |f_k| < \infty$. One way to prove this result is as follows. Divide $\mathbb{R}^d$ into rectangles

$$R_k = \{\omega \in \mathbb{R}^d : |\omega[\ell] - \omega_k[\ell]| \le \pi,\ \ell = 1, \ldots, d\},$$

in terms of center points $\omega_k = 2\pi k$, $k \in \mathbb{Z}^d$. Then $\mathbb{R}^d = \cup_k R_k$ and the rectangles are disjoint apart from adjoining boundaries. Note that for $\omega \in R_k$, $|\omega - \omega_k| \le \pi\sqrt{d}$.

Suppose $|k| \ge (3/2)\sqrt{d}$ (so that $\omega_k \ge 3\pi\sqrt{d}$) and let $\omega \in R_k$. Using two versions of the triangle inequality

$$|\omega_k| \le |\omega_k - \omega| + |\omega|, \quad |\omega| \le |\omega_k - \omega| + |\omega_k|,$$

deduce from the first formula that

$$|\omega| \ge |\omega_k| - |\omega_k - \omega| \ge |\omega_k| - \pi\sqrt{d} \ge 3\pi\sqrt{d} - \pi\sqrt{d} = 2\pi\sqrt{d},$$

and hence from the second formula that

$$|\omega_k| \ge |\omega| - |\omega_k - \omega| \ge |\omega| - \pi\sqrt{d} \ge |\omega|/2.$$

If $|k| \ge (3/2)\sqrt{d}$ and $|k| \ge r_0$, then integrating over R_k yields $(2\pi)^d |f_k| \le \int_{R_k} \varphi(\omega/2)\, d\omega$. Hence, deduce that $\sum_{k \in \mathbb{Z}^d} |f_k|$ can be bounded by a sum of a finite sum of terms (for $|k| < \max\{(3/2)\sqrt{d}, r_0\}$) plus an infinite sum, which is bounded by a convergent integral.

(b) Next adapt this result to the partial derivatives $\partial f(\omega)/\theta_j$, $j = 1, 2, 3$, in (6.57). In particular, verify that the three functions $\varphi(r) = 1/(A + Br^2)^{1+\nu+d/2}$, $\varphi(r) = r^2/(A + Br^2)^{1+\nu+d/2}$ and $\varphi(r) = \log(A + B|r|^2)/(A + Br^2)^{\nu+d/2}$ satisfy the monotonicity and integrability conditions of part (a).

(c) Lastly, write

$$\partial \log f_T(\omega)/\partial\theta_j = \frac{\partial f_T(\omega)/\partial\theta_j}{f_T(\omega)}.$$

Note that since the Matérn spectral density is strictly positive for all ω, the wrapped version can be bounded below by a positive constant c on the torus, $f_T(\omega) \geq c > 0$, $|\omega[\ell]| \leq \pi$, $\ell = 1, \ldots, d$. Use part (b) to deduce that $\partial f_T(\omega)/\partial\theta_j$ can be written as a convergent infinite sum, where the bound is uniform over $\omega \in (-\pi, \pi)^d$.

(d) Hence the elements of the Fisher information matrix can be expressed as the integrals over $(-\pi, \pi)^d$ of the product of two bounded continuous functions, and hence the integrals are finite.

7

Kriging

7.1 Introduction

In this chapter, we consider the question of prediction for random fields. For spatial data, prediction is mainly concerned with interpolation or smoothing between the existing data sites rather than with extrapolation beyond the data sites. In contrast, in time series analysis, prediction is usually concerned with forecasts into the future.

Prediction in a spatial context is known as "kriging" in the geostatistics literature, and we shall use this term throughout the book. D. G. Krige, a South African mining engineer, developed these ideas for identifying gold reserves in the mining areas in South Africa. Cressie (1990) gives a historical review of the ideas behind kriging. Some further comments are given in Appendix B.

There are several ways to predict the values of an unknown function given observations at a finite set of sites. Different approaches give different perspectives about the problem, but from this book's perspective, they can all be unified under the umbrella of kriging.

(a) Best linear unbiased prediction for a Gaussian process (Sections 7.2–7.8)
(b) Bayesian prediction for a Gaussian process (Section 7.12)
(c) Nonparametric regression in machine learning (Section 7.13)
(d) Splines (Section 7.14)
(e) Reproducing kernel Hilbert spaces (RKHSs) (Section 7.15)

For statistical purposes, the two most important approaches are (a) and (b). Both start from the same model. There is an underlying "signal" following a Gaussian process. Given noisy observations on the signal, the objective is to predict the signal at a new set of sites. When the mean function is completely known, the two approaches are essentially identical. However, if the mean function contains unknown parameters, the approaches will differ.

Spatial Analysis, First Edition. John T. Kent and Kanti V. Mardia.

The derivation of the kriging equations in approach (a) can quickly become rather complicated. Therefore, we shall develop the subject in stages, starting with the simplest situations, and focus on a notation that highlights the underlying simplicity of the method. The covariance part of the model is either a known fully-specified covariance function (which specifies the variance of all linear combinations) or a known intrinsic covariance function (which only specifies the variance of certain increments). We look at each of the following cases in turn.

- Random field with known covariance function and known mean 0 (Section 7.3). This case is called *simple kriging*.
- Random field with known covariance function and unknown mean function (Sections 7.4–7.5). The case of a constant unknown mean is called *ordinary kriging*, and the case of a general parametric drift with unknown coefficients is called *universal kriging*.
- Random field with known intrinsic covariance function (Section 7.8). This case is also covered by the equations of universal kriging.

The kriging equations in approach (a) are initially derived under the assumption of a fully specified covariance function. However, the equations extend with little or no change to intrinsic covariance functions (which only partially specify the covariances of the random field) (Section 7.8). For certain intrinsic models, this kriging surface is closely related to the fitted surface obtained from thin-plate splines and related splines (Section 7.14).

If the model of spatial dependence is smooth enough, the kriging surface can be differentiated with respect to the site. In addition, the data can take the form of specified derivatives of the random field at specified sites as well as the values of the random field at the sites. Questions related to differentiability are discussed in Section 7.11.

In conventional kriging, the drift parameters are treated as unknown constants. The drift parameters can also treated by Bayesian methods in approach (b) by giving them a prior distribution. Bayesian kriging is discussed in Section 7.12. The importance of Bayesian kriging in machine learning is discussed in Section 7.13.

Certain kriging predictors also arise in other settings, notably in splines, in approach (c), and RKHSs, in approach (d), which are studied in Sections 7.14 and 7.15.

Finally, a collection of d splines can be used to construct deformations of $\mathbb{R}^d$. This topic is discussed in Section 7.16.

Examples are given throughout the chapter to illustrate the calculations and to visualize the kriging predictor (especially Sections 7.9 and 7.10). The examples use both small artificial data sets and larger real data sets.

The discussion in this chapter assumes that there are no additional explanatory variables to help predict the random process, but in many practical examples, such information is available. For example, Rathbun (1998) developed kriging methods for estuarine data using environmental and geographic information about an estuary including its irregular shape.

7.2 The Prediction Problem

Let $\{U(t)\}$ be a Gaussian random field, or Gaussian process,

$$U(\cdot) \sim GP(\mu(\cdot), \sigma(\cdot, \cdot)), \tag{7.1}$$

with a mean function $\mu(\cdot)$ and a covariance function $\sigma(\cdot, \cdot)$, considered as a *signal.* Suppose that noisy measurements are made at a set of sites $\{t_i,\ i = 1, \ldots, n\}$

$$X_i = U(t_i) + \varepsilon_i, \quad i = 1, \ldots, n, \tag{7.2}$$

where it is generally assumed that the error terms ε_i are i.i.d. $N(0, \tau^2)$ random variables, independent of the signal. The vector $\boldsymbol{X} = [X(t_1), \ldots, X(t_n)]^T$ is called the *observation vector.* The presence of the error term can also be described as a nugget effect. This framework also covers the noiseless situation when $\tau^2 = 0$; in this case, $\varepsilon_i = 0$ and $X_i = U(t_i),\ i = 1, \ldots, n.$

The data sites $t_1, \ldots, t_n$ are sometimes known as *training sites.* Given observations at the training sites, the objective is to predict the value of the signal at a new site t_0, say. The new site is sometimes known as a *test site.*

The Gaussian assumption is made here to simplify the presentation. However, except for Section 7.12 on Bayesian kriging, most of the results just depend on the second moments of the process and continue to hold for non-Gaussian random fields.

Throughout the chapter, the following notation is used. Let

$$\Sigma = (\sigma_{ij}), \quad \sigma_{ij} = \sigma(t_i, t_j), \quad i,j = 1, \ldots, n, \tag{7.3}$$

denote the $n \times n$ covariance matrix of the signal at the data sites. Similarly, let

$$\sigma(t) = \begin{bmatrix} \sigma(t, t_1) \\ \vdots \\ \sigma(t, t_n) \end{bmatrix} \tag{7.4}$$

denote the $n \times 1$ column vector of covariances for the signal at the data sites and a site t. It is convenient to abbreviate the notation for the covariance vector at the test site as

$$\sigma(t_0) = \sigma_0. \tag{7.5}$$

Finally, let

$$\Omega = \Sigma + \tau^2 I_n \tag{7.6}$$

denote the covariance matrix of the observation vector $\boldsymbol{X}$. That is, Ω has elements

$$\omega_{ij} = \begin{cases} \sigma_{ij} + \tau^2, & i = j, \\ \sigma_{ij}, & i \neq j, \end{cases}$$

for $i, j = 1, \ldots, n$.

In addition, suppose the mean of the signal lies in a p-dimensional vector space of functions $\mathcal{F}$. (Note that p, the dimension of the drift space, was denote by q in Section 5.11.) Let

$$\boldsymbol{f}(t) = [f_1(t), \ldots, f_p(t)]^T \tag{7.7}$$

be a p-dimensional vector of known basis functions in $\mathcal{F}$. The basis functions can also be thought of as drift functions. The expected signal as a function of t takes the form

$$E\{U(t)\} = \boldsymbol{\beta}^T \boldsymbol{f}(t) \tag{7.8}$$

in terms of an unknown coefficient vector $\boldsymbol{\beta}$. The values of the drift functions at the data sites and at a new site t_0, say, can be collected together as an $n \times p$ matrix F and a p-dimensional column vector $\boldsymbol{f}_0$ given by

$$\begin{aligned} F &= (f_{ij}), \quad f_{ij} = f_j(t_i), \quad i = 1, \ldots, n, \ j = 1, \ldots, p, \\ \boldsymbol{f}_0 &= [f_1(t_0), \ldots, f_p(t_0)]^T, \end{aligned} \tag{7.9}$$

and they can be combined together as an $(n+1) \times p$ matrix

$$F^{(0)} = \begin{bmatrix} \boldsymbol{f}_0^T \\ F \end{bmatrix} = (f_{ij}), \quad f_{ij} = f_j(t_i), \quad i = 0, \ldots, n, \ j = 1, \ldots, p. \tag{7.10}$$

A list of the notation is given in Table 7.1. The objective is to predict the value of the signal $U(t_0)$ at a new site t_0 using a linear predictor

$$\hat{U}(t_0) = c + \boldsymbol{\gamma}^T \boldsymbol{X}, \ \text{say.} \tag{7.11}$$

The coefficients $c \in \mathbb{R}$ and $\boldsymbol{\gamma} \in \mathbb{R}^n$ depend on t_0 and are chosen to minimize the expected prediction mean squared error (PMSE)

$$\text{PMSE}(t_0; \boldsymbol{\gamma}, c) = E\{[U(t_0) - \hat{U}(t_0)]^2\}, \tag{7.12}$$

subject to suitable constraints. The optimal predictor in (7.11) is called the *kriging predictor* and equations for the optimal coefficients c and $\boldsymbol{\gamma}$ are called the *kriging equations*.

The PMSE for the optimal values of c and $\boldsymbol{\gamma}$ is denoted $\text{PMSE}(t_0)$. It can also be called the *kriging variance* $\sigma_K^2(t_0)$ so that

$$\sigma_K^2(t_0) = \text{PMSE}(t_0) = E\{[U(t_0) - \hat{U}(t_0)]^2\}, \tag{7.13}$$

when the optimal predictor $\hat{U}(t_0)$ is used. Its square root is called the *kriging standard error*.

Regularity Conditions. Most of the chapter deals with random fields with a fully specified covariance function $\sigma(s, t)$ (for which stationary random fields are an

Table 7.1 Notation used for kriging at the data sites $t_1, \ldots, t_n$, and at the prediction and data sites $t_0, \ldots, t_n$.

Object	Training sites	Test site and training sites
Sites	$t_1, \ldots, t_n$	$t_0, \ldots, t_n$
Signal	$U(t_1), \ldots, U(t_n)$	$U(t_0), \ldots, U(t_n)$
Data	$\boldsymbol{X} = [X(t_1), \ldots, X(t_n)]^T$	—
Kriging coefficient vector	$\gamma = [\gamma_1, \ldots, \gamma_n]^T$	$\delta = [\delta_0, \ldots, \delta_n]^T$ $= [-1, \gamma_1, \ldots, \gamma_n]^T$
Drift matrix	$F = (f_{ij})\ (n \times p)$ $f_{ij} = f_j(t_i)$ $i = 1, \ldots, n,\ j = 1, \ldots, p$	$F^{(0)} = \begin{bmatrix} \boldsymbol{f}_0^T \\ F \end{bmatrix}$ $\boldsymbol{f}_0\ (p \times 1)$ has elements $f_j(t_0),\ j = 1, \ldots, p$
Projection matrix	$P = F(F^TF)^{-1}F^T$	$P^{(0)} = F^{(0)}(F^{(0)T}F^{(0)})^{-1}F^{(0)T}$
Covariance matrices	$\Sigma = \text{var}\{[U(t_1), \ldots, U(t_n)]^T\}$ $= (\sigma_{ij}),\ \sigma_{ij} = \sigma(t_i, t_j)$ $1 \le i,j \le n$ $\Omega = \text{var}\{[X_1, \ldots, X_n]^T\}$ $= \Sigma + \tau^2 I$	$\Phi = \text{var}\{[U(t_0), X_1, \ldots, X_n]^T\}$ $= \begin{bmatrix} \sigma(t_0, t_0) & \boldsymbol{\sigma}_0^T \\ \boldsymbol{\sigma}_0 & \Omega \end{bmatrix}$ where $\boldsymbol{\sigma}_0$ has elements $\sigma(t_0, t_i),\ i = 1, \ldots, n$

The first column is used in Sections 7.2–7.6.4. The second column is used in Section 7.6.5.

important special case) and two regularity conditions are assumed at the data sites $t_1, \ldots, t_n$.

(a) Ω is positive definite, and
(b) F has full column rank p.

Thus, there are no linear dependencies between the values of the observations at the data sites or between the drift functions at the data sites. The assumption that Ω is invertible will be true if either $\tau^2 > 0$ or Σ is invertible (or both). Theorem 2.3.3 guarantees the invertibility of Σ for a stationary process under mild regularity conditions.

Parts of the chapter deal with intrinsic random fields and Condition (a) can be relaxed to

(a') Ω is conditionally positive definite of order k, for some known value of $k \ge 0$, and $\mathcal{F}$ contains $\mathcal{F}_k$, the space of polynomials in t, of degree $\le k$.

The development of the kriging equations is given in Sections 7.3–7.5. In each setting, the notation PMSE(t_0; $\boldsymbol{\gamma}$, c) is used for the PMSE of a general predictor, PMSE(t_0; $\boldsymbol{\gamma}$) for the case when $c = 0$, and $\sigma_K^2(t_0) = \text{PMSE}(t_0)$ for the kriging variance, i.e. the PMSE of the optimal predictor. In each setting, the optimal c is $c = 0$. The formulas for the optimal $\boldsymbol{\gamma}$ in the three settings are given in Eqs. (7.14), (7.21), and (7.30), respectively.

It should be emphasized that the expectations in equations such as (7.12) are unconditional. That is, although we are interested in predicting the value of the random field once $\boldsymbol{X}$ has been observed, for assessing the quality of prediction, we average over the joint distribution of $U(t_0)$ and $\boldsymbol{X}$. The reason for using unconditional expectations is to deal properly with the case when drift parameters are unknown.

Sections 7.3–7.5 emphasize the dependence of the kriging predictor on the observation vector $\boldsymbol{X}$ for a fixed new site t_0. However, once the kriging predictor $\hat{U}(t_0)$ is obtained for a single site t_0, the site can be varied to define a *kriging surface* for $t_0 \in \mathbb{R}^d$. An alternative approach to the kriging equations, which also emphasizes the dependence on a new site t_0, is given in Section 7.6.

Comment on Uppercase and Lowercase Notation. An observation vector treated as random is written as $\boldsymbol{X}$. Similarly, an observation vector for a specific set of data is written as $\boldsymbol{x}$. In particular, the formulas for unbiasedness and PMSE involve expectations over $\boldsymbol{X}$. On the other hand, formulas for a predictor for a specific set of data, such as (7.61), will generally use $\boldsymbol{x}$. Similarly, the notations $\hat{U}(t_0)$ and $\hat{u}(t_0)$ are used for the predictor. In particular, Bayesian kriging looks at the posterior distribution of $U(t_0)$ given the data; hence the notation $\hat{u}(t_0)$ for the kriging predictor is appropriate in this case.

7.3 Simple Kriging

The simplest situation for prediction occurs when the random field has a known mean 0. Hence, the vector of drift functions $\boldsymbol{f}(t)$ is not present in this case. A predictor of the form (7.11) is called the *best linear predictor (BLP)* if it minimizes the PMSE (7.12).

Theorem 7.3.1 (Simple kriging) *Consider the problem of predicting a signal $U(t_0)$ at a new site t_0 given observations $X_1, \ldots, X_n$ at sites $t_1, \ldots, t_n$ as set out in Section 7.2. Suppose the mean function vanishes, $\mu(t) = 0$ for all t. The BLP predictor in this setting is called the "simple kriging predictor" and is given by $\hat{U}(t_0) = \boldsymbol{\gamma}^T\boldsymbol{X}$ where*

$$\boldsymbol{\gamma} = \Omega^{-1}\boldsymbol{\sigma}_0. \tag{7.14}$$

Further, the kriging variance is

$$\sigma_K^2(t_0) = \sigma(t_0, t_0) - \boldsymbol{\sigma}_0^T \Omega^{-1} \boldsymbol{\sigma}_0. \tag{7.15}$$

Proof: The PMSE for the predictor (7.11) is given from (7.12) by

$$\begin{aligned}
\text{PMSE}(t_0; \boldsymbol{\gamma}, c) &= E\left\{ \left[U(t_0) - \hat{U}(t_0)\right]^2 \right\} \\
&= E\left\{ \left[U(t_0) - c - \boldsymbol{\gamma}^T \mathbf{X}\right]^2 \right\} \\
&= E\left\{ \left[U(t_0) - \sum_{i=1}^{n} \gamma_i X(t_i) - c\right]^2 \right\} \\
&= \sigma(t_0, t_0) - 2\boldsymbol{\sigma}_0^T \boldsymbol{\gamma} + \boldsymbol{\gamma}^T \Omega \boldsymbol{\gamma} + c^2,
\end{aligned} \tag{7.16}$$

where $\sigma(t_0, t_0)$ is the variance of the signal $U(t_0)$ at the new site t_0, and $\boldsymbol{\sigma}_0$ and Ω are defined in (7.3)–(7.6).

To minimize (7.16), first minimize over c to get $c = 0$. Then, substitute $c = 0$ in (7.16), differentiate with respect to $\boldsymbol{\gamma}$, and set the derivative to $\mathbf{0}$

$$2\Omega\boldsymbol{\gamma} - 2\boldsymbol{\sigma}_0 = \mathbf{0}, \tag{7.17}$$

to obtain the predictor (7.14). Plugging the solution for $\boldsymbol{\gamma}$ into the PMSE (7.16) yields minimum value (7.15). □

Note that for a Gaussian random field, $\hat{U}(t_0)$ can be interpreted as the conditional expectation of $X(t_0)$ given $\mathbf{X}$ (Eq. (A.18)).

This derivation can be easily extended to the case where the random field has a general, but known, mean function $E\{U(t)\} = \mu(t)$. Thus, let $\boldsymbol{\mu} = [\mu(t_1), \ldots, \mu(t_n)]^T$ denote the n-vector of means at the data sites and let $\mu(t_0)$ denote the mean at the new site. By applying the previous derivation to the centered process $\{U(t) - \mu(t)\}$ and then adding in the means, the predictor takes the form

$$\hat{U}(t_0) = \mu(t_0) + \boldsymbol{\gamma}^T(\mathbf{X} - \boldsymbol{\mu}), \quad \boldsymbol{\gamma} = \Omega^{-1}\boldsymbol{\sigma}_0. \tag{7.18}$$

The kriging variance remains unchanged.

The kriging covariance at two new test sites s and t can be derived by a similar argument to give

$$\sigma_K(s, t) = \sigma(s, t) - \boldsymbol{\sigma}(s)^T \Omega^{-1} \boldsymbol{\sigma}(t). \tag{7.19}$$

The simple kriging predictor can be interpreted as a posterior mean in a Bayesian analysis. Similarly, the kriging variance can be interpreted as a posterior variance. See Section 7.12.2 for more details.

7.4 Ordinary Kriging

When the drift function includes unknown parameters, the prediction problem becomes more complicated. To bring out some of the key ideas, we start with the simple case of a constant unknown mean (known as *ordinary kriging*) in this section and cover the general case (known as *universal kriging*) in Section 7.5. The names "ordinary kriging" and "universal kriging" come from the geosciences literature. Note the term "ordinary" in the kriging context is unrelated to its use when describing an "ordinary random field" in Chapter 3.

Thus, for this section, suppose that the mean of the signal is constant, $E\{U(t)\} = \mu$, but μ is unknown. In the notation of (7.7), the vector drift function $\boldsymbol{f}(t)$ is now the scalar function $f(t) = 1$ and we write the coefficient β as μ here. An "unbiasedness" constraint

$$E\{U(t_0) - \hat{U}(t_0)\} = 0, \quad \text{for all } \mu \tag{7.20}$$

is imposed to get a unique solution for the predictor. As before, this expectation is unconditional. A predictor of the form (7.11) is called the *best linear unbiased predictor (BLUP)* if it minimizes the PMSE (7.12) subject to the unbiasedness constraint (7.20).

Theorem 7.4.1 (Ordinary kriging) *Consider the problem of predicting a signal $U(t_0)$ at a new site t_0 given observations $X_1, \ldots, X_n$ at sites $t_1, \ldots, t_n$ as set out in Section 7.2. Suppose the mean function is constant, $\mu(t) = \mu$ for all t, where μ is unknown. The BLUP in this setting is called the "ordinary kriging predictor" and is given by $\hat{U}(t_0) = \boldsymbol{\gamma}^T\mathbf{X}$ where*

$$\boldsymbol{\gamma} = \Omega^{-1}\boldsymbol{\sigma}_0 + \left(\frac{1 - \mathbf{1}^T\Omega^{-1}\boldsymbol{\sigma}_0}{\mathbf{1}^T\Omega^{-1}\mathbf{1}}\right)\Omega^{-1}\mathbf{1}. \tag{7.21}$$

Further, the kriging variance is

$$\sigma_K^2(t_0) = \sigma(t_0, t_0) - \boldsymbol{\sigma}_0^T\Omega^{-1}\boldsymbol{\sigma}_0 + \frac{(1 - \mathbf{1}^T\Omega^{-1}\boldsymbol{\sigma}_0)^2}{\mathbf{1}^T\Omega^{-1}\mathbf{1}}. \tag{7.22}$$

Proof: For a linear predictor $\hat{U}(t_0) = c + \boldsymbol{\gamma}^T\mathbf{X}$, the expected prediction error becomes

$$E\left\{U(t_0) - c - \boldsymbol{\gamma}^T\mathbf{X}\right\} = \mu\left(1 - \sum \gamma_i\right) - c. \tag{7.23}$$

If (7.23) equals 0 for all μ, it follows that

$$c = 0, \quad \sum \gamma_i = 1. \tag{7.24}$$

Next, modify the formula (7.12) for the PMSE (taking $\mu = 0$ without loss of generality) incorporating a Lagrange multiplier to accommodate the constraint $\sum \gamma_i = \gamma^T \mathbf{1} = 1$. The result is

$$E\left\{ \left[U(t_0) - \gamma^T \mathbf{X}\right]^2 \right\} + 2\lambda\left(1 - \sum \gamma_i\right) = \sigma(t_0, t_0) - 2\sigma_0^T\gamma + \gamma^T\Omega\gamma + 2\lambda(1 - \mathbf{1}^T\gamma). \tag{7.25}$$

Differentiating (7.25) with respect to γ and setting the derivative to $\mathbf{0}$ yields

$$2\Omega\gamma - 2\sigma_0 - 2\lambda\mathbf{1} = \mathbf{0},$$

which can be solved to give $\gamma = \Omega^{-1}(\sigma_0 + \lambda\mathbf{1})$. The constraint $\gamma^T\mathbf{1} = 1$ implies the Lagrange multiplier λ is given by

$$\lambda = \frac{1 - \mathbf{1}^T\Omega^{-1}\sigma_0}{\mathbf{1}^T\Omega^{-1}\mathbf{1}}.$$

Inserting this value for λ in the formula for γ yields the optimal coefficient vector (7.21). The corresponding linear combination $\hat{U}(t_0) = \gamma^T\mathbf{X}$ is called the *ordinary kriging predictor*. It is easily checked that $\mathbf{1}^T\gamma = 1$ as required for unbiasedness. Substituting (7.21) in (7.25) yields the optimal PMSE (7.22). Note that (7.22) can be expanded as (7.15) (the PMSE with known μ) plus an extra positive term to allow for the estimation of μ. □

Formula (7.21) for γ is rather cumbersome. Therefore, it is helpful to rewrite it in a more intuitive form. Start from formula (7.18) for $\hat{U}(t_0)$, which in the case of a constant known mean μ takes the form

$$\hat{U}(t_0) = \mu + \sigma_0^T\Omega^{-1}(\mathbf{X} - \mu\mathbf{1}_n). \tag{7.26}$$

But since μ is not known in the current setting, it is natural to replace it by an estimate. The log-likelihood for μ, given the observation $\mathbf{X} = \mathbf{x}$, say, takes the form

$$\log L(\mu) = -\frac{1}{2}(\mathbf{x} - \mu\mathbf{1})^T\Omega^{-1}(\mathbf{x} - \mu\mathbf{1}) + \text{const.}$$

Differentiating L with respect to μ and setting the derivative to 0 yields $\mathbf{1}^T\Omega^{-1}(\mathbf{x} - \mu\mathbf{1}) = 0$, which can be solved to give the maximum likelihood estimator

$$\hat{\mu} = \frac{\mathbf{1}^T\Omega^{-1}\mathbf{x}}{\mathbf{1}^T\Omega^{-1}\mathbf{1}}. \tag{7.27}$$

This estimator is also known as the generalized least squares (GLS) estimator of μ; see Eq (5.18). It differs from the ordinary least squares (OLS) estimator because the covariance matrix Ω of $\mathbf{X}$ is taken into account. Substituting $\hat{\mu}$ into (7.26) yields the predictor

$$\hat{U}(t_0) = \hat{\mu} + \sigma_0^T\Omega^{-1}(\mathbf{X} - \hat{\mu}\mathbf{1}). \tag{7.28}$$

Somewhat fortuitously, the "plug-in" predictor (7.28) is identical to the optimal predictor $\hat{U}(t_0) = \gamma^T\mathbf{X}$, with γ from (7.21) (Exercise 7.1). Thus, (7.28) provides a more intuitive representation for the ordinary kriging predictor than (7.21).

7.5 Universal Kriging

Next, suppose the mean function $\mu(t)$ takes the parametric form

$$\mu(t) = \sum_{j=1}^{p} \beta_j f_j(t), \tag{7.29}$$

where the $f_j(t)$, $j = 1, \dots, p$ are known functions of t, but where the coefficients β_j are unknown and need to be estimated. Generally, the first drift function is $f_1(t) \equiv 1$ to accommodate a constant term.

In many applications, the drift functions represent polynomials in t. For example, to accommodate polynomials up to a quadratic degree in $d = 2$ dimensions, we can use $p = 6$ monomials,

$$f_1(t) = 1,\ f_2(t) = t[1],\ f_3(t) = t[2],\ f_4(t) = t[1]^2,\ f_5(t) = t[1]t[2],\ f_6(t) = t[2]^2.$$

If $\mathcal{F}_k$ denotes the polynomials in d dimensions of degree $\leq k$, then this vector space has dimension

$$p = p(k) = \binom{d+k}{k}$$

(Equation (3.16) and Exercise 3.9).

Theorem 7.5.1 (Universal kriging) *Consider the problem of predicting a signal $U(t_0)$ at a new site t_0 given observations $X_1, \dots, X_n$ at sites $t_1, \dots, t_n$ as set out in Section 7.2. Suppose the mean function is given by (7.29), where the coefficients $\beta_1, \dots, \beta_p$ are unknown. The BLUP in this setting is called the "universal kriging predictor" and is given by $\hat{U}(t_0) = \boldsymbol{\gamma}^T \boldsymbol{X}$ where*

$$\boldsymbol{\gamma} = \Omega^{-1} \left\{ \boldsymbol{\sigma}_0 + F\left(F^T \Omega^{-1} F\right)^{-1} (\boldsymbol{f}_0 - F^T \Omega^{-1} \boldsymbol{\sigma}_0) \right\}. \tag{7.30}$$

Further, the kriging variance is

$$\sigma_K^2(t_0) = \sigma(t_0, t_0) - \boldsymbol{\sigma}_0^T \Omega^{-1} \boldsymbol{\sigma}_0 + (\boldsymbol{f}_0 - F^T \Omega^{-1} \boldsymbol{\sigma}_0)^T (F^T \Omega^{-1} F)^{-1} (\boldsymbol{f}_0 - F^T \Omega^{-1} \boldsymbol{\sigma}_0). \tag{7.31}$$

Proof: The drift matrix F and the drift vector $\boldsymbol{f}_0$ are defined in (7.9). The unbiasedness constraint

$$E\{U(t_0) - \hat{U}(t_0)\} = 0, \quad \text{for all } \boldsymbol{\beta},$$

can be written in matrix form as $c + \boldsymbol{\beta}^T F^T \boldsymbol{\gamma} = \boldsymbol{\beta}^T \boldsymbol{f}_0$ for all $\boldsymbol{\beta}$; that is,

$$c = 0,\ F^T \boldsymbol{\gamma} = \boldsymbol{f}_0. \tag{7.32}$$

Minimizing the PMSE with respect to vectors $\boldsymbol{\gamma}$ satisfying (7.32), using a *vector* Lagrange multiplier this time, yields (7.30). Substituting this result into (7.12) yields the PMSE (7.31). □

As in the last section, the formula for γ is rather cumbersome. It can be written in a more intuitively comprehensible form using the GLS estimator of $\boldsymbol{\beta}$, which takes the form $\hat{\boldsymbol{\beta}} = (F^T\Omega^{-1}F)^{-1}F^T\Omega^{-1}\mathbf{X}$; see Eq. (5.48). Then

$$\hat{U}(t_0) = \boldsymbol{f}_0^T\hat{\boldsymbol{\beta}} + \boldsymbol{\sigma}_0^T\Omega^{-1}(\mathbf{X} - F\hat{\boldsymbol{\beta}}) \tag{7.33}$$

(Exercise 7.1).

The first two terms of (7.31) represent the kriging variance if $\boldsymbol{\beta}$ is known; the last term is positive and represents an effect due to the estimation of $\boldsymbol{\beta}$.

The kriging covariance at two new test sites s and t can be derived by a similar argument to give

$$\begin{aligned}\sigma_K(s,t) &= \sigma(s,t) - \boldsymbol{\sigma}(s)^T\Omega^{-1}\boldsymbol{\sigma}(t)\\ &\quad + (\boldsymbol{f}(s) - F^T\Omega^{-1}\boldsymbol{\sigma}(s))^T(F^T\Omega^{-1}F)^{-1}(\boldsymbol{f}(t) - F^T\Omega^{-1}\boldsymbol{\sigma}(t)).\end{aligned} \tag{7.34}$$

7.6 Further Details for the Universal Kriging Predictor

The form of the coefficient vector γ in (7.30) for the universal kriging predictor is rather uninformative and nonintuitive. In this section, we give two alternative representations that will be useful later in the chapter. The first representation is in terms of "transfer matrices" A and B, which can be found either directly (Sections 7.6.1–7.6.3) or in terms of a set of linear equations involving a "bordered" covariance matrix (Section 7.6.4). Transfer matrices were introduced in Mardia et al. (1991) and Kent and Mardia (1994). The second representation involves an "augmented" covariance matrix (Section 7.6.5).

7.6.1 Transfer Matrices

Write the universal kriging predictor (7.33) in the form

$$\begin{aligned}\hat{U}(t_0) &= \gamma^T\mathbf{X} = \mathbf{X}^T\gamma\\ &= \mathbf{X}^TA\boldsymbol{f}_0 + \mathbf{X}^TB\boldsymbol{\sigma}_0,\end{aligned} \tag{7.35}$$

where

$$B = \Omega^{-1} - \Omega^{-1}F(F^T\Omega^{-1}F)^{-1}F^T\Omega^{-1}, \tag{7.36}$$

$$A = \Omega^{-1}F(F^T\Omega^{-1}F)^{-1}, \tag{7.37}$$

so that

$$\gamma = A\boldsymbol{f}_0 + B\boldsymbol{\sigma}_0. \tag{7.38}$$

It is convenient to call $B(n \times n)$ the *transfer covariance matrix* and $A(n \times p)$ the *transfer drift matrix*. Both matrices depend only on the data sites and not on the new site t_0.

Equation (7.35) demonstrates how the predictor depends on the observation vector $\boldsymbol{X}$ (namely, by multiplying A and B on the left by $\boldsymbol{X}^T$). It also demonstrates how the predictor depends on the new data site t_0 (namely, by multiplying A and B on the right by $\boldsymbol{f}_0$ and $\boldsymbol{\sigma}_0$, respectively) in terms of $\boldsymbol{f}_0$, the p-vector of the drift functions at t_0, and $\boldsymbol{\sigma}_0$, the n-vector of the covariances for the random field between the new site t_0 and the data sites. In particular, the predictor $\hat{U}(t_0)$ will be r-times differentiable, $r \geq 0$, as a function of t_0, if the drift and covariance functions are r-times differentiable.

7.6.2 Projection Representation of the Transfer Matrices

It is also possible to recast A and B in more intuitive forms using projection matrices and generalized inverses. First, we need to recall some facts from matrix algebra. For any $(n \times p)$ matrix G whose columns are linearly independent (so G^TG is nonsingular), let $P_G = G(G^TG)^{-1}G^T$ denote the $n \times n$ projection matrix onto the column space of G. Then $P_G = P_G^T$, $P_G G = G$, $P_G^2 = P_G$, and $P_G\boldsymbol{y} = \boldsymbol{0}$ if $G^T\boldsymbol{y} = \boldsymbol{0}$. See Section A.3.3. The Moore–Penrose generalized inverse of a symmetric matrix will also be needed (Section A.3.2).

Using this notation B and A can be reexpressed as

$$B = [(I - P_F)\Omega(I - P_F)]^- \tag{7.39}$$

$$A = (I - B\Omega)F(F^TF)^{-1}. \tag{7.40}$$

Here, B is given as a Moore–Penrose generalized inverse, which can be interpreted as the Moore–Penrose generalized inverse of Ω restricted to the orthogonal complement of the column space of F.

To prove (7.39) and (7.40), it is convenient to rotate the coordinate system. Let $\Xi = \begin{bmatrix}\Xi_1 & \Xi_2\end{bmatrix}$ be an orthogonal matrix such that the columns of Ξ_1 span the column space of F. A convenient choice for Ξ_1 is $\Xi_1 = F(F^TF)^{-1/2}$ (Section A.3.3). Using the partitioning of Ξ, define partitioned matrices

$$\Omega^* = \Xi^T\Omega\Xi = \begin{bmatrix}\Omega_{11}^* & \Omega_{12}^* \\ \Omega_{21}^* & \Omega_{22}^*\end{bmatrix}, \quad \Omega^{*-1} = \begin{bmatrix}\Omega^{*11} & \Omega^{*12} \\ \Omega^{*21} & \Omega^{*22}\end{bmatrix}, \quad F^T = \Xi^T F = \begin{bmatrix}F_1^* \\ 0\end{bmatrix}.$$

Note that Ω^* is invertible because Ω is. Also, F_1^* is a $p \times p$ matrix, which is invertible because F has full rank. Then Eqs. (7.36) and (7.37) for $B^* = \Xi^T B\Xi$ and $A^* = \Xi^T A$ simplify to

$$\begin{aligned} B^* &= \Omega^{*-1} - \Omega^{*-1}\begin{bmatrix}F_1^* \\ 0\end{bmatrix}\left\{F_1^{*T}\Omega^{*11}F_1^*\right\}^{-1}\begin{bmatrix}F_1^{*T} & 0\end{bmatrix}\Omega^{*-1} \\ &= \begin{bmatrix}\Omega^{*11} & \Omega^{*12} \\ \Omega^{*21} & \Omega^{*22}\end{bmatrix} - \begin{bmatrix}\Omega^{*11} & \Omega^{*12} \\ \Omega^{*21} & \Omega^{*21}\{\Omega^{*11}\}^{-1}\Omega^{*12}\end{bmatrix} \\ &= \begin{bmatrix}0 & 0 \\ 0 & (\Omega_{22}^*)^{-1}\end{bmatrix}, \end{aligned} \tag{7.41}$$

and

$$
\begin{aligned}
A^* &= \Omega^{*-1}\begin{bmatrix} F_1^* \\ 0 \end{bmatrix}\{F_1^{*T}\Omega^{*11}F_1^*\}^{-1} \\
&= \begin{bmatrix} I \\ \Omega^{*21}(\Omega^{*11})^{-1} \end{bmatrix}(F_1^{*T})^{-1} \\
&= \begin{bmatrix} I \\ -(\Omega_{22}^*)^{-1}\Omega_{21}^* \end{bmatrix}(F_1^{*T})^{-1}.
\end{aligned} \tag{7.42}
$$

using two results about the inverse of a partitioned matrix

$$
(\Omega_{22}^*)^{-1} = \Omega^{*22} - \Omega^{*21}\{\Omega^{*11}\}^{-1}\Omega^{*12}, \quad \Omega^{*21}\{\Omega^{*11}\}^{-1} = -(\Omega_{22}^*)^{-1}\Omega_{21}^* \tag{7.43}
$$

(Section A.3.4).

Since $P_{F^*} = \begin{bmatrix} I & 0 \\ 0 & 0 \end{bmatrix}$ represents a projection onto the first p coordinate axes of $\mathbb{R}^n$, the starred version of (7.39) reduces to

$$
[(I - P_F^*)\Omega^*(I - P_F^*)]^- = \begin{bmatrix} 0 & 0 \\ 0 & \Omega_{22}^* \end{bmatrix}^- = \begin{bmatrix} 0 & 0 \\ 0 & \Omega_{22}^{*-1} \end{bmatrix},
$$

which is the same as (7.41). Similarly, the starred version of (7.40) reduces to

$$
\begin{aligned}
(I - B^*\Omega^*)F^*(F^{*T}F^*)^{-1} &= \begin{bmatrix} I & 0 \\ -\Omega_{22}^{*-1}\Omega_{21}^* & 0 \end{bmatrix}\begin{bmatrix} (F_1^{*T})^{-1} \\ 0 \end{bmatrix} \\
&= \begin{bmatrix} I \\ -\Omega_{22}^{*-1}\Omega_{21}^* \end{bmatrix}(F_1^{*T})^{-1},
\end{aligned}
$$

which is the same as (7.42).

The kriging variance (7.31) can also be written directly in terms of A and B. In particular,

$$
\begin{aligned}
\sigma_K^2(t_0) &= E\left\{\left(U(t_0) - \boldsymbol{\gamma}^T\mathbf{X}\right)^2\right\} \\
&= \sigma(t_0, t_0) - 2\boldsymbol{\gamma}^T\boldsymbol{\sigma}_0 + \boldsymbol{\gamma}^T\Omega\boldsymbol{\gamma} \\
&= \sigma(t_0, t_0) - 2\boldsymbol{f}_0^TA^T\boldsymbol{\sigma}_0 - 2\boldsymbol{\sigma}_0^TB\boldsymbol{\sigma}_0 + (A\boldsymbol{f}_0 + B\boldsymbol{\sigma}_0)^T\Omega(A\boldsymbol{f}_0 + B\boldsymbol{\sigma}_0) \\
&= \sigma(t_0, t_0) - 2\boldsymbol{f}_0^TA^T\boldsymbol{\sigma}_0 - \boldsymbol{\sigma}_0^TB\boldsymbol{\sigma}_0 + \boldsymbol{f}_0^TA^T\Omega A\boldsymbol{f}_0,
\end{aligned} \tag{7.44}
$$

where we have used the identities $B\Omega B = B$ and $A^T\Omega B = 0$ to derive the last line. These identities are most easily established using the starred forms (7.41) and (7.42). Note that if formulas (7.39) and (7.40) are used to compute B and A, then formula (7.44) for the kriging variance does not involve the evaluation of Ω^{-1}. This property is important in Section 7.8, where it is shown that (7.44) remains valid for intrinsic random fields defined in terms of intrinsic covariance functions.

7.6.3 Second Derivation of the Universal Kriging Predictor

In this section, we give another proof of the optimality of $\boldsymbol{\gamma}$, using the representation (7.38) in terms of transfer matrices, where the modified formulas (7.39) and (7.40) are used for A and B. Although a second proof is not needed here, it will be needed for intrinsic random fields.

The strategy of the proof is to show that any value of $\boldsymbol{\gamma}'$ different from the optimal $\boldsymbol{\gamma}$ leads to a higher value of the objective function.

The first step is to confirm that $\boldsymbol{\gamma}$ in (7.38) satisfies the constraint (7.32). Since $F^T A = I_p$ and $F^T B = 0$ (exercise for the reader), we get

$$F^T\boldsymbol{\gamma} = F^T A\boldsymbol{f}_0 + F^T B\boldsymbol{\sigma}_0 = \boldsymbol{f}_0,$$

as required.

Next consider a general value of $\boldsymbol{\gamma}'$ satisfying the constraint; $\boldsymbol{\gamma}'$ can be written as

$$\boldsymbol{\gamma}' = \boldsymbol{\gamma} + \boldsymbol{\eta},$$

where $\boldsymbol{\eta}$ is a nonzero n-vector satisfying $F^T\boldsymbol{\eta} = \mathbf{0}$. Then the formula for the PMSE based on $\boldsymbol{\gamma}'$ becomes

$$\begin{aligned}\text{PMSE}(t_0; \boldsymbol{\gamma}') &= \sigma(t_0, t_0) - 2\boldsymbol{\gamma}'^T\boldsymbol{\sigma}_0 + \boldsymbol{\gamma}'^T\Omega\boldsymbol{\gamma}' \\ &= \sigma(t_0, t_0) - 2\boldsymbol{\gamma}^T\boldsymbol{\sigma}_0 + \boldsymbol{\gamma}^T\Omega\boldsymbol{\gamma} + \boldsymbol{\eta}^T\Omega\boldsymbol{\eta} + 2\boldsymbol{\eta}^T\left\{-\boldsymbol{\sigma}_0 + \Omega\boldsymbol{\gamma}\right\},\end{aligned}$$

which can be viewed as the optimal value of the PMSE, plus a quadratic form, $\boldsymbol{\eta}^T\Omega\boldsymbol{\eta}$ (which is strictly positive provided $\boldsymbol{\eta} \neq \mathbf{0}$), and a cross-product term, $2\boldsymbol{\eta}^T\left\{-\boldsymbol{\sigma}_0 + \Omega\boldsymbol{\gamma}\right\}$. The optimality of (7.30) will be proved if we can show the cross-product term vanishes. To do this, we first set out some preliminary results.

(a) $\boldsymbol{\eta}^T = \boldsymbol{\eta}^T(I - P_F)$ (since $F^T\boldsymbol{\eta} = \mathbf{0}$ and $P_F = F(F^TF)^{-1}F^T$, it follows that $P_F\boldsymbol{\eta} = \mathbf{0}$).
(b) $\boldsymbol{\eta}^T\Omega = \boldsymbol{\eta}^T(I - P_F)\Omega$ (from (a)).
(c) $(I - P_F)\Omega B = I - P_F$; this is most easily proved in starred form, where it states

$$\begin{bmatrix}0 & 0\\ 0 & I\end{bmatrix}\begin{bmatrix}\Omega^*_{11} & \Omega^*_{12}\\ \Omega^*_{21} & \Omega^*_{22}\end{bmatrix}\begin{bmatrix}0 & 0\\ 0 & (\Omega^*_{22})^{-1}\end{bmatrix} = \begin{bmatrix}0 & 0\\ 0 & I\end{bmatrix}.$$

(d) $\boldsymbol{\eta}^T\Omega B = \boldsymbol{\eta}^T(I - P_F)$ (using (b) to get the intermediate form $\boldsymbol{\eta}(I - P_F)\Omega B$ and (c) to get the final form).
(e) $\boldsymbol{\eta}^T(\Omega - \Omega B\Omega) = \mathbf{0}^T$ (using (b) and (c) get the intermediate forms $\boldsymbol{\eta}^T[(I - P_F)\Omega - (I - P_F)\Omega B\Omega] = \boldsymbol{\eta}^T[(I - P_F)\Omega - (I - P_F)\Omega] = \mathbf{0}^T$).

Hence, the cross-product term becomes

$$-2\boldsymbol{\eta}^T\boldsymbol{\sigma}_0 + 2\boldsymbol{\eta}^T\Omega(I - B\Omega)F(F^TF)^{-1}\boldsymbol{f}_0 + 2\boldsymbol{\eta}^T\Omega B\boldsymbol{\sigma}_0.$$

The first term becomes $-2\boldsymbol{\eta}^T(I - P_F)\boldsymbol{\sigma}_0$ by (a). In the second term $\boldsymbol{\eta}^T(\Omega - \Omega B\Omega) = \mathbf{0}^T$ by (e) and so this term vanishes. The third term simplifies to $2\boldsymbol{\eta}^T\Omega B\boldsymbol{\sigma}_0 = 2\boldsymbol{\eta}^T(I - P_F)\Omega B\boldsymbol{\sigma}_0 = 2\boldsymbol{\eta}^T(I - P_F)\boldsymbol{\sigma}_0$ using (d). Putting the pieces together, we see that the cross-product term vanishes, as required.

7.6.4 A Bordered Matrix Equation for the Transfer Matrices

The matrices B and A in (7.36) and (7.37) can also be derived by inverting a certain $(n+p) \times (n+p)$-dimensional matrix. This approach can be found in the spline literature, e.g., Wahba (1990, pp. 12–13). Define the *bordered covariance matrix*

$$M = \begin{bmatrix} \Omega & F \\ F^T & 0 \end{bmatrix} \tag{7.45}$$

by adding rows and columns for the drift matrix F. Then M^{-1} exists and can be partitioned as

$$M^{-1} = \begin{bmatrix} B & A \\ A^T & C \end{bmatrix}, \tag{7.46}$$

where

$$C = -(F^T F)^{-1} F^T (\Omega - \Omega B \Omega) F (F^T F)^{-1}. \tag{7.47}$$

A proof of the decomposition (7.46) is given in Exercise 7.2.

One reason for introducing M is that it provides an elegant way of computing the optimal coefficient vector $\boldsymbol{\gamma}$ for universal kriging. Consider the equation

$$M \begin{bmatrix} \boldsymbol{\gamma} \\ \boldsymbol{\lambda} \end{bmatrix} = \begin{bmatrix} \boldsymbol{\sigma}_0 \\ \boldsymbol{f}_0 \end{bmatrix}$$

for two vectors $\boldsymbol{\gamma}(n \times 1)$ and $\boldsymbol{\lambda}(p \times 1)$, with solution

$$\begin{bmatrix} \boldsymbol{\gamma} \\ \boldsymbol{\lambda} \end{bmatrix} = M^{-1} \begin{bmatrix} \boldsymbol{\sigma}_0 \\ \boldsymbol{f}_0 \end{bmatrix}.$$

In particular, the first block of rows yields $\boldsymbol{\gamma} = B\boldsymbol{\sigma}_0 + A\boldsymbol{f}_0$, which is identical to the universal kriging coefficient vector $\boldsymbol{\gamma}$ in (7.38.)

7.6.5 The Augmented Matrix Representation of the Universal Kriging Predictor

There is yet another representation of the optimal coefficient vector $\boldsymbol{\gamma}$ that is useful for certain purposes. To develop this representation, we use $(n+1)$-dimensional matrices and vectors instead of n-dimensional quantities.

The site t_0 can be included with the other data sites to give an augmented $(n+1) \times (n+1)$ covariance matrix

$$\Phi = \begin{bmatrix} \sigma(t_0, t_0) & \boldsymbol{\sigma}_0^T \\ \boldsymbol{\sigma}_0 & \Omega \end{bmatrix}, \tag{7.48}$$

where the rows and columns of Φ are indexed by $i, j = 0, \ldots, n$. Note that the upper-left entry $\sigma(t_0, t_0)$ is the variance of the signal $U(t_0)$; the remaining diagonal elements contain the variances of the observation vector $\boldsymbol{X}$. Similarly, t_0 can be included in the augmented drift matrix $F^{(0)}$ defined in (7.10).

The PMSE for a general n-vector of coefficients γ is given in (7.16), and when $c = 0$ it can also be written as a quadratic form in an $(n+1)$-vector $\boldsymbol{\delta}$,

$$\begin{aligned} \text{PMSE}(t_0; \gamma) &= \sigma(t_0, t_0) - 2\gamma^T \sigma_0 + \gamma^T \Omega \gamma \\ &= \boldsymbol{\delta}^T \Phi \boldsymbol{\delta}, \end{aligned} \tag{7.49}$$

where

$$\boldsymbol{\delta} = \begin{bmatrix} 1 \\ -\gamma \end{bmatrix}, \tag{7.50}$$

with the elements of $\boldsymbol{\delta}$ indexed $i = 0, \ldots, n$. Further, the constraint (7.32) can be written as

$$F^{(0)T} \boldsymbol{\delta} = \mathbf{0}. \tag{7.51}$$

Thus, the prediction problem can be restated as minimizing (7.49) over $\boldsymbol{\delta}$ subject to $\delta_0 = 1$ and to the constraint (7.51). In fact, it is easier to find the solution in this notation. The solution is given by

$$\boldsymbol{\delta} = \mathbf{d}_{(0)}/d_{00}, \quad \text{where } D = [(I - P^{(0)})\Phi(I - P^{(0)})]^-. \tag{7.52}$$

Here, D^- is the Moore–Penrose generalized inverse of D and

$$P^{(0)} = F^{(0)}(F^{(0)T}F^{(0)})^{-1}F^{(0)T} \tag{7.53}$$

is the projection matrix onto the column space of $F^{(0)}$ (Sections A.3.2 and A.3.3). The notation here indicates that $\boldsymbol{\delta}$ is given by the initial column of D, scaled by its initial element, so that $\delta_0 = 1$. Then $-\gamma$ is given by the remaining elements of $\boldsymbol{\delta}$.

To show that the optimal $\boldsymbol{\delta}$ can be written in the form (7.52), another transformation is helpful. This time we rotate to an orthogonal basis in R^{n+1} where the first p basis elements lie in the space spanned by the columns of $F^{(0)}$. Let $\Xi^{(0)} = \left[\Xi_1^{(0)}\ \Xi_2^{(0)}\right]$ be an $(n+1) \times (n+1)$ orthogonal matrix with these properties. A convenient choice for the $(n+1) \times p$ matrix $\Xi_1^{(0)}$ is $\Xi_1^{(0)} = F^{(0)}(F^{(0)T}F^{(0)})^{-1/2}$. Note that $\Xi_1^{(0)T}\Xi_1^{(0)} = I_p$ so that $\Xi_1^{(0)}$ has orthonormal columns. Further $\Xi_1^{(0)}\Xi_1^{(0)T} = F^{(0)}(F^{(0)T}F^{(0)})^{-1}F^{(0)T} = P^{(0)}$ in (7.53). Using an asterisk to denote rotated matrices, set

$$\Phi^* = \Xi^{(0)T}\Phi\Xi^{(0)} = \begin{bmatrix} \Phi_{11}^* & \Phi_{12}^* \\ \Phi_{21}^* & \Phi_{22}^* \end{bmatrix}, \quad P^{(0)*} = \Xi^{(0)T}P^{(0)}\Xi^{(0)} = \begin{bmatrix} I & 0 \\ 0 & 0 \end{bmatrix},$$

$$F^{(0)*} = \Xi^{(0)T}F^{(0)} = \begin{bmatrix} F_1^{(0)*} \\ 0 \end{bmatrix},$$

where $F_1^{(0)*}$ is a $p \times p$ nonsingular matrix. Also set

$$\boldsymbol{\delta}^* = \Xi^{(0)T}\boldsymbol{\delta} = \begin{bmatrix} \boldsymbol{\delta}_1^* \\ \boldsymbol{\delta}_2^* \end{bmatrix}, \quad \mathbf{e}^* = \Xi^{(0)T}\mathbf{e} = \begin{bmatrix} \mathbf{e}_1^* \\ \mathbf{e}_2^* \end{bmatrix},$$

where $\boldsymbol{e} = [1,0,\ldots,0]^T$ is the unit $(n+1)$-vector in the direction of the first coordinate axis. Then, the constraint $\delta_0 = 1$ on $\boldsymbol{\delta}$ is equivalent to the constraint $\boldsymbol{e}^{*T}\boldsymbol{\delta}^* - 1$ on $\boldsymbol{\delta}^*$. The PMSE (7.49) can be recast as $\boldsymbol{\delta}^{*T}\Phi^*\boldsymbol{\delta}^*$ and the constraints $\delta_0 = 1$ and (7.51) become $\boldsymbol{e}^{*T}\boldsymbol{\delta}^* = 1$ and $F^{(0)*T}\boldsymbol{\delta}^* = \mathbf{0}$. Since $F^{(0)*T}\boldsymbol{\delta}^* = F_1^{(0)*T}\boldsymbol{\delta}_1^*$ and $F_1^{(0)*}$ is nonsingular, the second constraint is equivalent to $\boldsymbol{\delta}_1^* = \mathbf{0}$.

Thus, the optimization problem reduces to minimizing

$$\boldsymbol{\delta}_2^{*T}\Phi_{22}^*\boldsymbol{\delta}_2^*$$

over $\boldsymbol{\delta}_2^*$ such that $\boldsymbol{e}_2^{*T}\boldsymbol{\delta}_2^* = 1$. By standard matrix theory (Exercise 7.4), the optimal choice of $\boldsymbol{\delta}_2^*$ is given by

$$\Phi_{22}^{*-1}\boldsymbol{e}_2^*/(\boldsymbol{e}_2^{*T}\Phi_{22}^{*-1}\boldsymbol{e}_2^*).$$

Note that $\Phi_{22}^{*-1} = \{[(I - P^{(0)*})\Phi^*(I - P^{(0)*})]^-\}_{22}$. Hence, rotating back to the original coordinates yields (7.52).

Finally, a small cautionary note is needed about the representation (7.52) that is not needed for the other methods of computing $\boldsymbol{\gamma}$. The derivation assumes that the matrix $\Phi^{(0)}$ is positive definite, not just that Φ is positive definite. It means that in practice t_0 cannot equal any of the data sites when the nugget variance vanishes, $\tau^2 = 0$ (of course, prediction is not interesting in this case anyway).

7.6.6 Summary

This section has described several methods of determining the kriging predictor, sometimes directly in terms of the kriging coefficient vector $\boldsymbol{\gamma}$, in terms of the transfer matrices A, B, and in terms of the augmented kriging coefficient vector $\boldsymbol{\delta}$. In addition, some of the derivations have made different assumptions on the drift functions, though simple and ordinary kriging can be viewed as special cases of universal kriging. Table 7.2 summarizes the different methods of representing $\boldsymbol{\gamma}$.

Table 7.2 Various methods of determining the kriging predictor $\hat{U}(t_0) = \boldsymbol{\gamma}^T\mathbf{X}$, where $\boldsymbol{\gamma}$ can be defined in terms of transfer matrices by $\boldsymbol{\gamma} = A\boldsymbol{f}_0 + B\boldsymbol{\sigma}_0$ or in terms of $\boldsymbol{\delta}$ by $\boldsymbol{\delta} = [1, -\boldsymbol{\gamma}^T]^T$.

Quantity	Method	Location
$\boldsymbol{\gamma}$	Direct simple	(7.14)
	Direct ordinary	(7.21)
	Direct universal	(7.30)
A, B	Direct	(7.36) and (7.37)
	Projection matrix	(7.39) and (7.40)
	Bordered matrix	(7.45) and (7.46)
$\boldsymbol{\delta}$	Augmented matrix	(7.52)

The different representations all have their strengths. The direct representation for γ and the transfer matrices A and B is the most straightforward to derive, but requires that Ω be invertible. However, the projection representation for the transfer matrices makes it clear that there is no need to invert Ω. Instead Ω only needs to be conditionally positive definite with respect to F, a fact that is important when the same formula is used for the intrinsic case. The bordered matrix representation is the simplest to present algebraically. Finally, the augmented representation brings out the relationship to simple kriging most clearly.

7.7 Stationary Examples

This section contains some artificial examples in one dimension to illustrate different aspects of kriging in elementary situations.

Example 7.1 ***Stationary model, $n = 5$, $\mu = 0$ known, squared exponential covariance function (7.55), no nugget***
Suppose there is a "true" function

$$U(t) = 2 + \sin t, \tag{7.54}$$

where t is in radians. Thus, $U(t)$ is a shifted sine function. Suppose the function is unknown but that a limited number of (noiseless) observations are available at $n = 5$ sites, $t = -4, -2, 0, 1, 4.5$. We would like to "predict" or reconstruct $U(t)$ from these observations. For the purpose of this example, it is assumed that the unknown true function is a realization from a Gaussian process with mean 0 and the squared exponential covariance function

$$\sigma(h) = \sigma^2 \exp(-|h|^2/\varphi). \tag{7.55}$$

The objective is to construct the kriging predictor and to see how it varies, depending on the choice of the range parameter φ. Figure 7.1 shows the kriging predictor for three choices of the range parameter, $\varphi = 0.1, 10, 100$, respectively, when there is no nugget effect. The true function is the shifted sine function in (7.54) and is shown by the solid curve in each panel.

For a small range parameter, e.g., $\phi = 0.1$, the covariance model provides little information about $U(t)$ between the data sites, and the kriging predictor is pulled toward the assumed mean 0. This behavior is prominent in Panel (a). Panel (b), with a larger range parameter, provides a smooth interpolator between the data points that tracks the true function $U(t)$ remarkably closely in this example. In particular, the presence of the shift term does not have a strong effect on the predictor. In Panel (c), the range parameter seems to be too large. The increase in the data

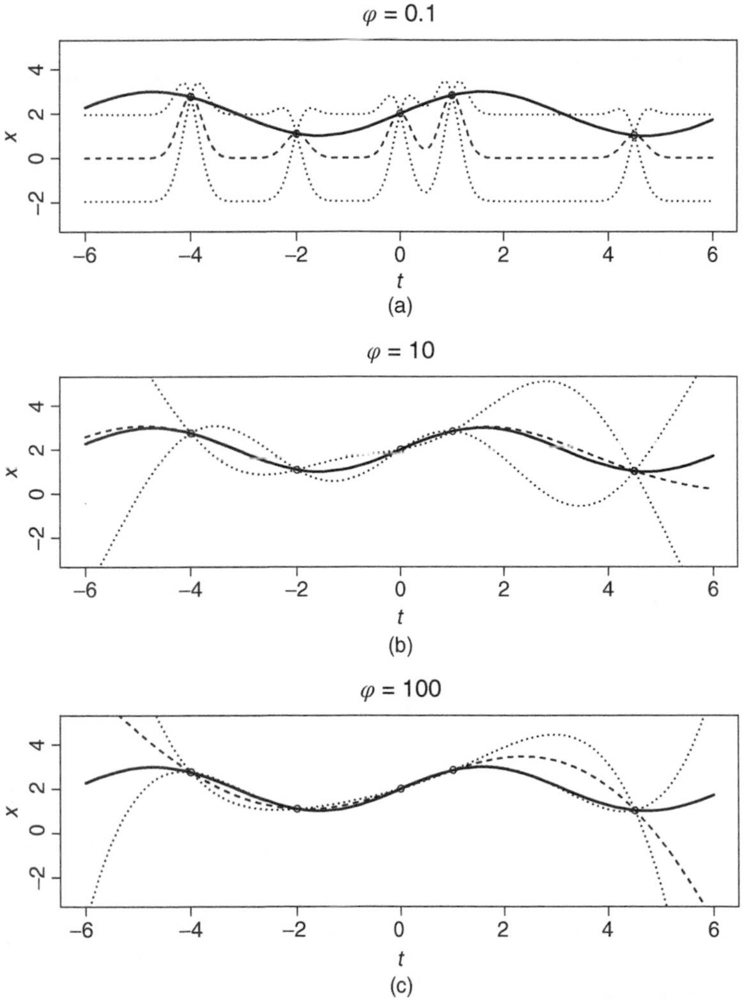

Figure 7.1 Kriging predictor for $n = 5$ data points assumed to come from a stationary random field with a squared exponential covariance function (7.55), without a nugget effect, with mean 0. Panels (a)–(c) show the kriging predictor for three choices of the range parameter, $\varphi = 0.1,\ 10,\ 100$, respectively. Each panel shows the true unknown shifted sine function (solid), together with the fitted kriging curve (dashed), plus/minus twice the kriging standard errors (dotted).

between $t = -2$ and $t = 2$ leads to a kriging predictor between $t = 2$ and $t = 4.5$ that overshoots the true function given by (7.54).

As might be expected, none of the predictors is guaranteed to be very accurate when predicting beyond the range of the data. All the predictors will be pulled

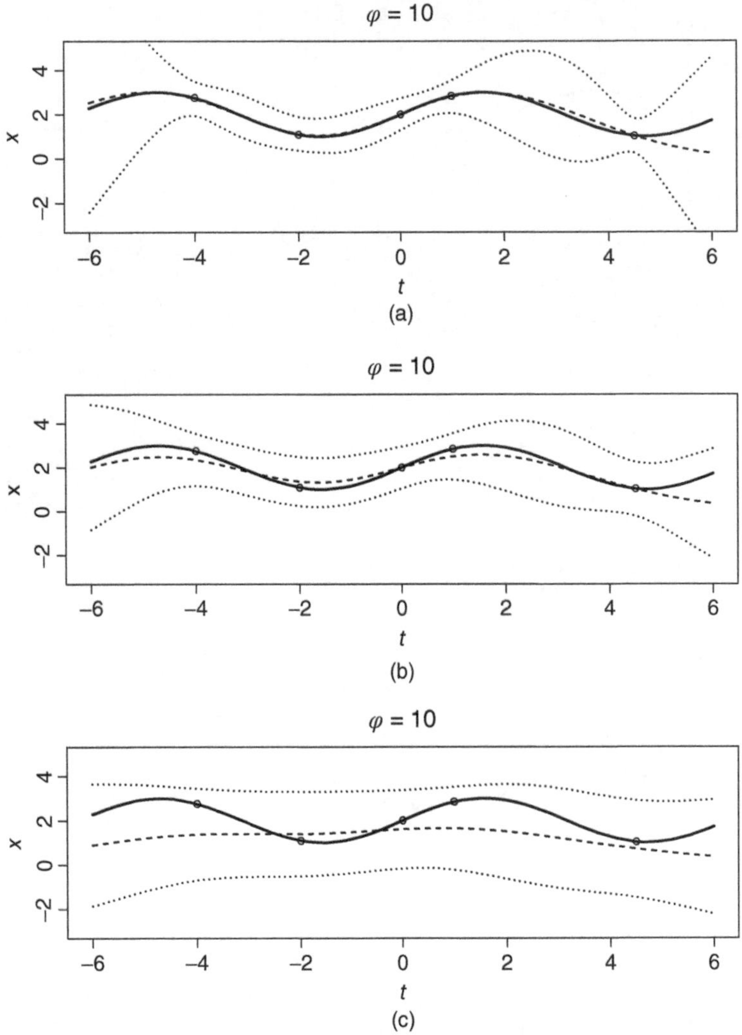

Figure 7.2 Kriging predictor for $n = 5$ data points assumed to come from a stationary random field with a squared exponential covariance function (7.55), plus a nugget effect, with mean 0. The size of the relative nugget effect in Panels (a)–(c) is given by $\psi = \tau^2/(\sigma^2 + \tau^2) = 0.01, 0.1, 0.5$, respectively. Each panel shows the true unknown shifted sine function (solid), together with the fitted kriging curve (dashed), plus/minus twice the kriging standard errors (dotted).

back to the assumed true mean for values of t far enough from the data sites, a feature most visible in Panel (a).

Figure 7.2 illustrates a kriging predictor when a nugget effect is included. The three panels of Figure 7.2 use the same range parameter that was used for the middle panel of Figure 7.1, namely $\varphi = 10$, and include relative nugget effects of sizes $\psi = \tau^2/(\sigma^2 + \tau^2) = 0.01, 0.1, 0.5$, respectively. For the small nugget effect in Panel (a), the kriging predictor is very similar to the interpolating predictor in Panel (b) of Figure 7.1, but note the wider standard error bands at the data sites. As the relative nugget effect increases in Panels (b) and (c), the predictor damps down peaks and troughs of the sine function, and the whole curve is pulled toward the assumed mean 0.

In this example, the variance σ^2 has been chosen visually to provide a clearly visible standard error bands, and it has been chosen separately in each panel in Figures 7.1 and 7.2. The main feature to notice in each panel is how the standard error bands vary with t. □

Next, two examples based on the elevation data and the bauxite data are given to illustrate two-dimensional kriging in practice.

Example 7.2 In Chapter 5, a stationary model was fitted to the elevation data using an exponential covariance function with no nugget. From Table 5.2, the estimated marginal variance and range parameter are $\hat{\sigma}^2 = \exp(8.32) = 1105$ and $\hat{\varphi} = \exp(1.81) = 6.11$. Using this model, the data have been interpolated using ordinary kriging to give a fitted surface. Figure 7.3a gives a contour plot. The minimum value of the surface lies near the top middle of the plot; the maximum occurs

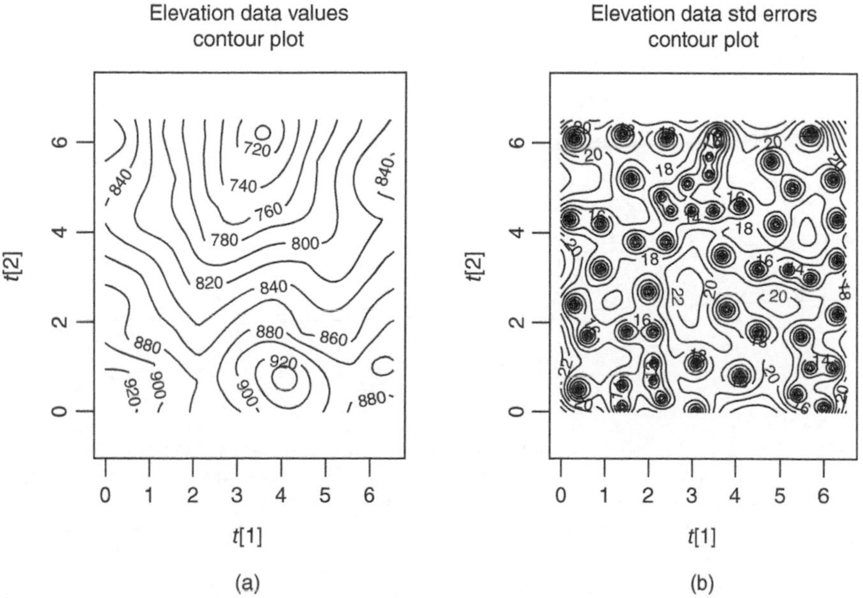

Figure 7.3 Panel (a) shows the interpolated kriging surface for the elevation data, as a contour map. Panel (b) shows a contour map of the corresponding kriging standard errors. This figure is also included in Figure 1.3.

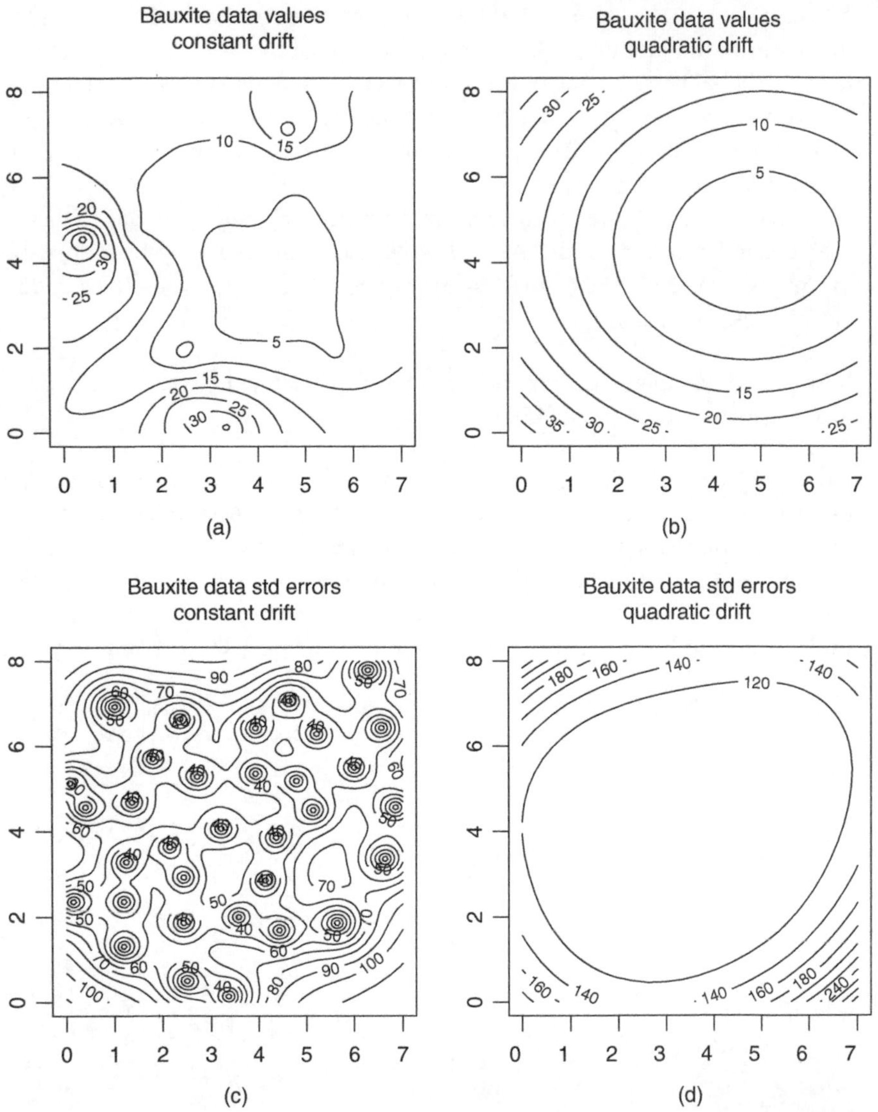

Figure 7.4 Panel (a) shows a contour plot for the kriged surface fitted to the bauxite data assuming a constant mean and an exponential covariance function for the error terms. Panel (b) shows the same plot assuming a quadratic trend and independent errors. Panels (c) and (d) show the kriging standard errors for the models in (a) and (b), respectively.

at a couple of peaks along the bottom of the plot. Since there is no nugget effect, the surface is predicted exactly with no error at the data sites; the standard error increases away from the data sites. Figure 7.3b gives a contour plot showing how the standard errors vary across the region. Some other ways of visualizing the kriging surface were given in Figure 1.3 using a perspective plot and an image plot. □

Example 7.3 In Chapter 5, various models were fitted to the bauxite data, including a stationary exponential model (i.e. a Matérn model of index $\nu = 0.5$) with range $\hat{\varphi} = \exp(0.24) = 1.27$ (Table 5.1), and a model with quadratic drift and independent errors, i.e. no spatial autocorrelation (Example 5.8). Contour maps for the two kriging surfaces and their standard errors are plotted in Figure 7.4. Both contour plots have a basin in the middle of the region. However, Panel (a) is much more sensitive to the fluctuations in the data with strong local peaks at the bottom and left and a smaller local peak at the upper right. Panel (b) is much more regular, being an exact quadratic surface. In Example 5.8, it was found that the second model gave a better fit to the data in terms of the Akaike information criterion (AIC), but it should be remembered that this is only a small data set ($n = 33$ sites).

The kriging standard errors for the two models are plotted in Panels (c) and (d), respectively. In Panel (c), note how the kriging standard error drops to 0 at each data point since there is no nugget effect. In Panel (d), the kriging standard error is reasonably constant in the middle of the plot, but increases toward the boundaries due to the estimation error for the quadratic drift coefficients. □

7.8 Intrinsic Random Fields

7.8.1 Formulas for the Kriging Predictor and Kriging Variance

In this section, kriging is extended from ordinary random fields to intrinsic random fields. An intrinsic random field of order k is defined in terms of an intrinsic covariance function $\sigma_I(h)$ with respect to the vector space $\mathcal{F}_k$ of polynomials in t of degree $\leq k$. The space $\mathcal{F}_k$ was introduced in Section 3.4, with dimension $p(k)$ in (3.16).

Consider values of the random field at sites $t_0, t_1, \ldots, t_n$. Recall that a linear combination $\sum_{i=0}^{n} \delta_i X(t_i)$ is an increment if and only if the $(n+1)$-dimensional coefficient vector $\boldsymbol{\delta}$ satisfies $\sum_{i=1}^{n} \delta_i f(t_i) = 0$ for all functions $f \in \mathcal{F}_k$. For an intrinsic random field, only increments of the process are assumed to have a well-defined distribution.

Thus, assume a signal $U(t)$ follows an intrinsic Gaussian random random field with an intrinsic mean $\mu_I(t)$ lying in a finite-dimensional vector space of drift functions $\mathcal{F}$ and with an intrinsic covariance function $\sigma_I(h)$. Without loss of generality, assume the drift space $\mathcal{F}$ includes the space of polynomials $\mathcal{F}_k$, since such functions are killed by increments. The problem is the same as in Section 7.2. Given observations, $x_1, \ldots, x_n$ which are (possibly noisy) versions of the signal at sites $t_1, \ldots, t_n$, the objective is to predict the signal at a new site t_0.

It is sometimes helpful to split the drift space $\mathcal{F}$ of dimension p, say, into two parts: the *intrinsic* drift space $\mathcal{F}_k$ and an *extrinsic* drift space $\mathcal{F}_{\text{ext}}$, say, so that

$$\mathcal{F} = \mathcal{F}_k + \mathcal{F}_{\text{ext}}.$$

Construct the $(n+1) \times p$ drift matrix $F^{(0)}$ as in Section 7.2 and partition it as

$$F^{(0)} = \begin{bmatrix} F_k^{(0)} & F_{\text{ext}}^{(0)} \end{bmatrix},$$

where $F_k^{(0)}$ has $p(k)$ columns corresponding to functions in $\mathcal{F}_k$. Of course, if the extrinsic drift is not present, then there is some simplification; namely $\mathcal{F} = \mathcal{F}_k$, $F^{(0)} = F_k^{(0)}$ and $p = p(k)$.

If $\boldsymbol{\delta}$ is a coefficient vector, suppose $\delta_0 = 1$ and partition $\boldsymbol{\delta} = [1, -\boldsymbol{\gamma}^T]^T$ as in Section 7.6.5. Then $\sum_{i=0}^n \delta_i X(t_i) = X(t_0) - \sum_{i=1}^n \gamma_i X(t_i)$ can be regarded as a residual between the true value of $X(t_0)$ and a linear predictor in terms of $X(t_1), \ldots, X(t_n)$. The algebraic notation needed to minimize the PMSE in this intrinsic setting is exactly the same as that developed for ordinary random fields, especially as developed in Section 7.6.5. More specifically, we wish to minimize $\boldsymbol{\delta}^T \Phi \boldsymbol{\delta}$ over coefficient vectors $\boldsymbol{\delta}$ such that $\delta_0 = 1$ and

$$F^{(0)T} \boldsymbol{\delta} = \mathbf{0}. \tag{7.56}$$

The first $p(k)$ components of the constraint correspond to the condition on $\boldsymbol{\delta}$ for it to define a valid increment; the remaining components are an unbiasedness constraint for the extrinsic drift.

Hence, the predictor for an intrinsic random field takes exactly the same form as given by various formulas in Section 7.6, especially the representation (7.52) in terms of the augmented covariance and drift matrices. Another valid representation is (7.35) in terms of transfer matrices, with the transfer matrices either specified directly by (7.39) and (7.40) or found by inverting the bordered set of linear equations (7.46). However, the direct representation for the transfer matrices (7.36) and (7.37) is not valid, nor is the direct representation of the predictor in (7.33). The reason is that the $n \times n$ covariance matrix Ω need not be nonsingular in the intrinsic setting. More details of this issue are given in Section 7.8.2.

7.8.2 Conditionally Positive Definite Matrices

In this section, we look in more detail at conditional positive definiteness; see also Section 3.4. For this discussion, it is notationally simplest to assume there is no extrinsic drift, so that $F^{(0)} = F_k^{(0)}$; then the notation in Section 7.2 can be used without any changes.

Conditional positive definiteness means that for any $(n+1)$-dimensional coefficient vector $\boldsymbol{\delta}$ satisfying $F^{(0)T}\boldsymbol{\delta} = \mathbf{0},\ \boldsymbol{\delta} \neq \mathbf{0}$, the quadratic form

$$\boldsymbol{\delta}^T \Phi \boldsymbol{\delta} > 0$$

must be positive.

Recall that one way to view a conditionally positive definite covariance function $\sigma_I(h)$ is as an equivalence class of functions, not a single function. In terms of the matrix Φ, this means that the only part of Φ relevant to increments is the part that is orthogonal to $F^{(0)}$. That is, if Φ is replaced by

$$\Phi' = \Phi + F^{(0)}V^{(0)T} + V^{(0)}F^{(0)T} = \begin{bmatrix} \sigma^2 + 2\boldsymbol{f}_0^T\boldsymbol{v}_0 & \boldsymbol{\sigma}_0^T + \boldsymbol{f}_0^T V^T \\ \boldsymbol{\sigma}_0 + V\boldsymbol{f}_0 & \Omega + FV^T + VF^T \end{bmatrix}$$

for any $(n+1) \times p$ matrix of coefficients $V^{(0)}$, where

$$V^{(0)} = \begin{bmatrix} \boldsymbol{v}_0^T \\ V \end{bmatrix},$$

has been partitioned into its initial row $\boldsymbol{v}_0^T$ and the remaining rows V, then the formula for the kriging predictor and the kriging variance remain unchanged. To understand why this claim is true, we consider two representations of the kriging predictor, one in terms of transfer matrices from (7.39) and (7.40), and one in terms of the augmented covariance matrix from Section 7.6.5. In each case we demonstrate invariance.

First consider the transfer matrices. The kriging coefficient vector is given in (7.38) by $\boldsymbol{\gamma} = A\boldsymbol{f}_0 + B\boldsymbol{\sigma}_0$. Under the change by V, B remains unchanged since the new version of B is

$$B' = [(I - P_F)(\Omega + FV^T + VF^T)(I - P_F)]^- = [(I - P_F)\Omega(I - P_F)]^- = B.$$

On the other hand, A changes to

$$A' = [I - B(\Omega + FV^T + VF^T)]F(F^TF)^{-1} = [I - B(\Omega + VF^T)]F(F^TF)^{-1} = A - BV,$$

and $\boldsymbol{\sigma}_0$ changes to

$$\boldsymbol{\sigma}_0' = \boldsymbol{\sigma}_0 + V\boldsymbol{f}_0.$$

Thus, the kriging vector (7.38) is unchanged,

$$\boldsymbol{\gamma}' = A'\boldsymbol{f}_0 + B\boldsymbol{\sigma}_0' = A\boldsymbol{f}_0 - BV\boldsymbol{f}_0 + B\boldsymbol{\sigma}_0 + BV\boldsymbol{f}_0 = \boldsymbol{\gamma}.$$

In terms of the starred matrices in Section 7.6.2, the property of conditional positive definiteness means that Ω_{22}^* is positive definite, but no claims are made about the remaining submatrices, Ω_{11}^*, Ω_{12}^*, Ω_{21}^*, other than that Ω^* should be symmetric. In particular, Ω_{11}^* need not be positive definite and might even be singular.

A second justification of invariance can be given using the representation (7.52) based on the augmented covariance matrix. The formula for $(I - P^{(0)})\Phi(I - P^{(0)})$ is invariant under modification by $V^{(0)}$; hence (7.52) is unchanged.

7.9 Intrinsic Examples

Example 7.4 ***Linear semivariogram, small n***
Using the notation from Eq. (3.50), consider the self-similar $\text{IRF}_1(1,0)$ random field $\{U(t)\}$ in $d = 1$ dimension with an intrinsic covariance function

$$\sigma_I(h) = -\sigma^2|h|$$

and with an implicit constant drift term. Assume there is no nugget effect and let the data be denoted $x_1, \ldots, x_n$ at sites $t_1, \ldots, t_n$.

(a) $n = 1$. In this case, the predictor takes a simple constant form $\hat{u}(t_0) = x_1$.
(b) $n \geq 2$. Here the kriging predictor reduces to a piecewise linear interpolator. That is, for each $1 \leq i \leq n-1$, $\hat{u}(t)$ is linear between t_i and t_{i+1} with $\hat{u}(t_i) = x_i$ and $\hat{u}(t_{i+1}) = x_{i+1}$. Further, $\hat{u}(t)$ is constant below t_1 and is constant above t_n. A proof is given in Exercise 7.6.

It is also possible to evaluate the kriging variance explicitly. It is a quadratic function of t between each pair of data points and is linear for $t \leq t_1$ and for $t \geq t_n$.

As a particular case, consider an artificial data set with $n = 3$ and observations $x_1 = 1,\ x_2 = 4,\ x_3 = 3$ at sites $t_1 = 1,\ t_2 = 2,\ t_3 = 3$. Each panel of Figure 7.5 shows the kriging predictor (solid curve) and error bands for the kriging standard errors (dashed curves). In Panel (a), the model only includes the constant intrinsic drift term. Note that the predictor is constant outside the range of the data (with different values to the left and to the right).

In Panel (b), the model also includes a linear extrinsic drift term. The predictor is still piecewise linear between the data points, but is now linear outside the range of the data (with the same slope but with different intercepts at each end). □

Example 7.5 In Example 5.9, various self-similar intrinsic models with linear drift were fitted to the gravimetric data. The best-fitting model without a nugget effect had a self-similarity index $\alpha = 0.59$. The corresponding kriging surface and standard errors are plotted in Figure 7.6. A linear regression fitted to the data would generate a kriging surface with equally spaced contour lines. It can be seen from Panel (a) that the fitted surface is nearly linear except in the upper-left corner. The standard errors are shown in Panel (b). As expected, the standard errors drop to 0 at the data points. In addition, the standard error increases for sites near the boundary of the plotting region. □

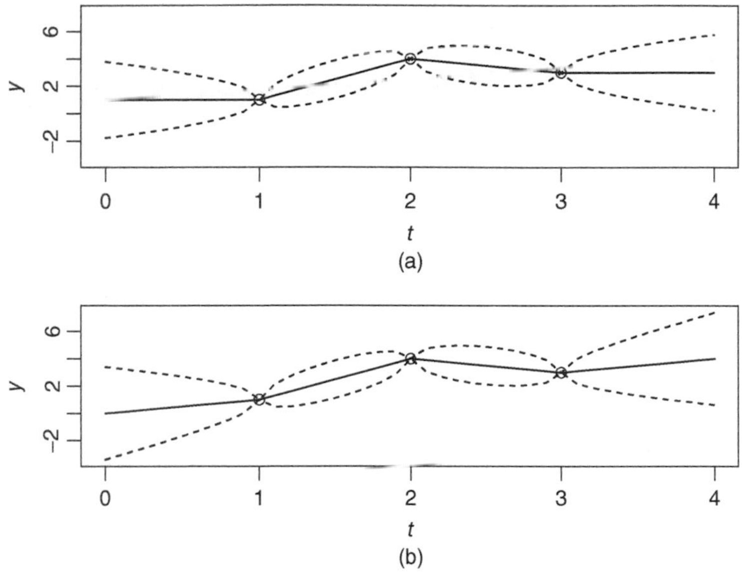

Figure 7.5 Kriging predictor and kriging standard errors for $n = 3$ data points assumed to come from an intrinsic random field, $\sigma_I(h) = -\frac{1}{2}\sigma^2|h|$, no nugget effect. The intrinsic drift is constant. Panel (a): no extrinsic drift; Panel (b): linear extrinsic drift. Each panel shows the fitted kriging curve (solid), plus/minus twice the kriging standard errors (dashed).

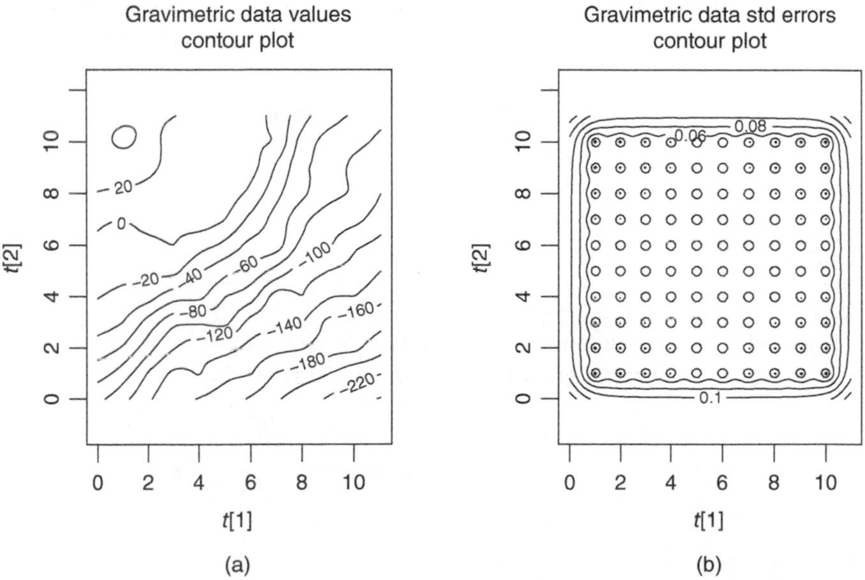

Figure 7.6 Panel (a) shows the interpolated kriging surface for the gravimetric data, as a contour map. Panel (b) shows a contour map of the corresponding kriging standard errors.

7.10 Square Example

The next example is motivated by an application to deformations in Section 7.16. Consider $n = 4$ sites in the plane $\mathbb{R}^2$ lying on a square with coordinates

$$\begin{bmatrix} t_1^T \\ t_2^T \\ t_3^T \\ t_4^T \end{bmatrix} = \begin{bmatrix} 0 & 1 \\ -1 & 0 \\ 0 & -1 \\ 1 & 0 \end{bmatrix},$$

and with associated data values $\boldsymbol{x} = [x_1\ x_2\ x_3\ x_4]^T$.

Consider a $p = 3$-dimensional drift function

$$\boldsymbol{f}(t) = [1/2,\ t[1]/\sqrt{2},\ t[2]/\sqrt{2}]^T.$$

The drift matrix F at the four sites is

$$F = \begin{bmatrix} 1/2 & 0 & 1/\sqrt{2} \\ 1/2 & -1/\sqrt{2} & 0 \\ 1/2 & 0 & -1/\sqrt{2} \\ 1/2 & 1/\sqrt{2} & 0 \end{bmatrix}.$$

The scalings in the drift functions have been chosen so that the columns of F are unit vectors. Further, it can be seen that the columns of F are orthogonal to one another so that $F^T F = I_3$.

Next let $\sigma(h)$ denote a stationary covariance function. The covariance vector of the process between the data sites and a new site t_0 and the covariance matrix for the four data sites can be written as

$$\sigma_0 = \begin{bmatrix} \sigma(t_0 - t_1) \\ \sigma(t_0 - t_2) \\ \sigma(t_0 - t_3) \\ \sigma(t_0 - t_4) \end{bmatrix}, \quad \Omega = \Sigma = \begin{bmatrix} c & a & b & a \\ a & c & a & b \\ b & a & c & a \\ a & b & a & c \end{bmatrix},$$

where

$$a = \sigma(\sqrt{2}),\ b = \sigma(2),\ c = \sigma(0).$$

Note that Ω is a circulant matrix and that three of its standardized eigenvectors, $\xi_{(1)}, \xi_{(2)}, \xi_{(3)}$, say, are given by the columns of F with eigenvalues

$$\lambda_1 = c + 2a + b, \quad \lambda_2 = \lambda_3 = c - b.$$

It is easily checked that the remaining standardized eigenvector is given by

$$\xi_{(4)} = \begin{bmatrix} 1 & -1 & 1 & -1 \end{bmatrix}^T/2$$

with eigenvalue

$$\lambda_4 = c - 2a + b.$$

The fact that the columns of F are eigenvectors of Ω simplifies the calculation of the transfer matrices B and A from (7.39) and (7.40)

$$B = \lambda_4^{-1} \xi_{(4)} \xi_{(4)}^T, \quad A = F. \tag{7.57}$$

From (7.35), the kriging predictor at a new site t_0 takes the form

$$\begin{aligned} \hat{u}(t_0) &= \boldsymbol{x}^T A \boldsymbol{f}_0 + \boldsymbol{x}^T B \boldsymbol{\sigma}_0 \\ &= \bar{x} + \frac{(x_4 - x_2)}{2} t_0[1] + \frac{(x_1 - x_3)}{2} t_0[2] \\ &\quad + \{4(c + b - 2a)\}^{-1} \left(\sum_{i=1}^{4} (-1)^i x_i \right) \left(\sum_{j=1}^{4} (-1)^j \sigma(t_0 - t_j) \right), \end{aligned} \tag{7.58}$$

where $\bar{x} = (x_1 + x_2 + x_3 + x_4)/4$. From (7.44), the kriging variance reduces to

$$\begin{aligned} \sigma_K^2(t_0) &= c - 2\boldsymbol{f}_0^T A^T \boldsymbol{\sigma}_0 - \boldsymbol{\sigma}_0^T B \boldsymbol{\sigma}_0 + \boldsymbol{f}_0^T A^T \Omega A \boldsymbol{f}_0 \\ &= c - 2 \sum_{j=1}^{3} f_{0j} (\xi_{(j)}^T \boldsymbol{\sigma}_0) - \frac{1}{\lambda_4} (\xi_{(4)}^T \boldsymbol{\sigma}_0)^2 + \sum_{j=1}^{3} \lambda_j f_{0j}^2. \end{aligned} \tag{7.59}$$

In particular, if $t_0 = 0$ lies at the center of the square, then

$$\hat{u}(t_0) = \bar{x}, \quad \sigma_K^2(t_0) = \frac{5}{4}\sigma(0) - 2\sigma(1) + \frac{1}{2}\sigma(\sqrt{2}) + \frac{1}{4}\sigma(2). \tag{7.60}$$

Equation (7.60) remains valid if the stationary covariance function is replaced by an intrinsic covariance function of order 0 or 1. In particular, note that the predictor and kriging variance are unchanged if $\sigma(h)$ is replaced by $\sigma(h) + \alpha + \beta h^2$ for arbitrary constants α and β.

7.11 Kriging with Derivative Information

In some examples, the information about a process consists of not only values of the process at specific locations but also values of various derivatives of the process (e.g. Mardia et al., 1996). We work in the context of a covariance function $\sigma(s, t)$, possibly intrinsic, plus a drift space $\mathcal{F}$, which includes the intrinsic drift in the intrinsic case.

These various types of information can be unified through the concept of a linear functional L, say, of the process $\{U(t)\}$. Examples of linear functionals include

$$U(L) = U(t_0), \text{ or } U(L) = \partial U(t_0)/\partial t[j]$$

for the value of the process itself, or the value of a partial derivative of the process (for some $j \in \{1, \dots, d\}$), at a specific site t_0.

The machinery of kriging can be easily extended to linear functionals, both in terms of the data provided and in terms of the prediction made. The main technical constraint is that the linear functionals must be "continuous." A unified notation for linear functionals involving derivative constraints is

$$U(L_0) = \frac{\partial^{|\kappa|} U(t_0)}{\partial t_0^\kappa} = \frac{\partial^{|\kappa|} U(t_0)}{\partial t_0[1]^{\kappa[1]} \cdots \partial t_0[d]^{\kappa[d]}},$$

where $\kappa = (\kappa[1], \ldots, \kappa[d])$, is a multi-index with nonnegative integer components $\kappa[j] \geq 0$, $j = 1, \ldots, d$ and with size $|\kappa| = \kappa[1] + \cdots + \kappa[d]$. The choice $\kappa = 0$ corresponds to a value of the process and $\kappa \neq 0$ corresponds to a particular partial derivative of the process.

Then $U(L_0)$ exists as a bounded linear functional of the process if

(a) $\partial^{|\kappa|} g(t)/\partial t^\kappa$ exists and is a continuous function of t for all $g \in \mathcal{F}$.
(b) $\partial^{2|\kappa|} \sigma(s,t)/\partial s^\kappa \partial t^\kappa$ exists and is a continuous function of s and t.

That is, the covariance function needs to be twice as differentiable as the the order of the derivative constraint.

We are now ready to describe the kriging problem in this context. Suppose the observations are given by values of the $U(\cdot)$ process for n linear functionals, $U(L_i) = u(L_i)$, say, where for simplicity it is assumed there is no noise so the data are $x_i = u(L_i)$, $i = 1, \ldots, n$. It is desired to predict the process at a new linear functional $U(L_0)$ by a linear predictor

$$\hat{U}(L_0) = \sum_{i=1}^{n} \gamma_i X_i$$

in such a way as to minimize the prediction error $E[\{U(L_0) - \hat{U}(L_0)\}^2]$ (treating the predictor as random) under the restriction that $L_0 - \sum \gamma_i L_i$ is an increment with respect to $\mathcal{F}$.

It turns out that the algebra of previous sections proceeds without change provided we interpret

$$\Omega = \Sigma = (\sigma_{ij}), \quad \sigma_{ij} = \sigma(L_i, L_j), \quad i,j = 1, \ldots, n,$$

$$\sigma_0 = \sigma_{0,i}, \quad \sigma_{0,i} = \sigma(L_0, L_i), \quad i = 1, \ldots, n,$$

$$F = (f_{ij}), \quad f_{ij} = f_j(L_i), \quad i = 1, \ldots, n, \quad j = 1, \ldots, p.$$

Plugging these values into (7.39) and (7.40) yields the kriging coefficients γ in (7.35).

It is also possible to introduce a nugget effect by setting $\Omega = \Sigma + D$, where D is diagonal, perhaps with different values for different sorts of linear functionals.

To illustrate the calculations in a bit more detail, consider a one-dimensional example with data values $u(t_i) = x_i$, $i = 1, \ldots, n$, at sites $t_1, \ldots, t_n$ and first-derivative constraints $u'(t_i) = y_i$ at the same sites, where a prime "′"

denote a derivative in t. Let $\sigma(h)$ be the covariance function, an even function of $h \in \mathbb{R}$. Then the $2n \times 2n$ covariance matrix Ω can be split into $n \times n$ blocks

$$\Omega = \begin{bmatrix} \Omega^{(1)} & \Omega^{(2)} \\ \Omega^{(3)} & \Omega^{(4)} \end{bmatrix},$$

with

$$\omega_{ij}^{(1)} = \sigma(t_i - t_j), \quad \omega_{ij}^{(2)} = -\sigma'(t_i - t_j)$$
$$\omega_{ij}^{(3)} = \sigma'(t_i - t_j), \quad \omega_{ij}^{(4)} = -\sigma''(t_i - t_j)$$

for $i, j = 1, \ldots, n$. Similarly, if $f_1(t), \ldots, f_p(t)$ denote a basis for the drift functions, then the $2n \times p$ drift matrix can be partitioned as

$$F = \begin{bmatrix} F^{(1)} \\ F^{(2)} \end{bmatrix},$$

with elements

$$f_{ij}^{(1)} = f_j(t_i), \quad f_{ij}^{(2)} = f_j'(t_i).$$

When predicting the process itself at a new site t_0, the vector $\boldsymbol{\sigma}_0$ has entries $\sigma(t_0 - t_i)$, $i = 1, \ldots, n$ and entries $-\sigma'(t_0 - t_{i-n})$, $i = n+1, \ldots, 2n$ and the vector $\boldsymbol{f}_0$ has entries $f_j(t_0)$, $j = 1, \ldots, p$. When predicting the first derivative of the process at a new site t_0, the vector $\boldsymbol{\sigma}_0$ has entries $\sigma'(t_0 - t_i)$, $i = 1, \ldots, n$ and entries $-\sigma''(t_0 - t_{i-n})$, $i = n+1, \ldots, 2n$ and the vector $\boldsymbol{f}_0$ has entries $f_j'(t_0)$, $j = 1, \ldots, p$.

Example 7.6 Consider the same toy data set that was used earlier in Example 7.4, with data values $x = 1, 4, 3$ at sites $t = 1, 2, 3$. Consider fitting an intrinsic covariance function $\sigma_I(h) = |h|^3$. Since this process has intrinsic order $k = 1$, there is also an implicit linear drift, with drift functions $f_1(t) = 1$, $f_2(t) = t$. The kriging predictor for this process is identical to the interpolating cubic spline; see Section 7.14.

The kriging predictor based on data constraints alone is plotted in Figure 7.7a as a function of $t = t_0$. Note that the fitted response is piecewise cubic and is twice differentiable at the knots $t = 1, 2, 3$.

Since this covariance function is twice differentiable at $h = 0$, it is possible to include first-order derivative constraints. The covariance function has derivatives $\sigma'(h) = 3\,\mathrm{sign}(h)|h|^2$ and $\sigma''(h) = 6|h|$.

Suppose the first derivative of the process has values $(y_1, y_2, y_3) = (-1, 0, 2)$ at the sites $t = 1, 2, 3$. The kriging predictor is plotted in Figure 7.7b. Note how the kriging predictor has been twisted at the knots to match the specified first derivative values.

Linear functionals can also be used to describe integration as well as differentiation. The prediction of the integral of the random field over a bounded region is known as block kriging. Some basics on block integrals and the related idea of dispersion variance are investigated in Section 3.12. □

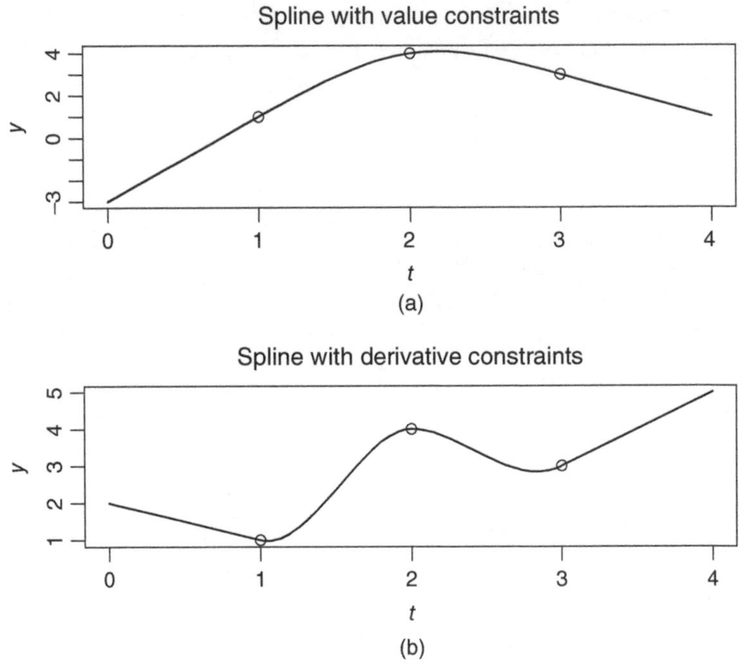

Figure 7.7 Kriging predictors for Example 7.6. For Panel (a), the kriging predictor is based on value constraints at sites 1,2,3. For Panel (b), the kriging predictor is additionally based on derivative constraints at the same sites.

7.12 Bayesian Kriging

7.12.1 Overview

Kriging can also be approached from a Bayesian perspective, starting with Kitandis (1986). See also Le and Zidek (1992), Handcock and Stein (1993), and Schabenberger and Gotway (2005, p. 391). It is helpful to partition the observations and parameters into a hierarchy of different "levels" following Gelfand and Ghosh (2013, p. 41); see also Section 8.4.

- *Level 1.* The observations $X_i = x_i$ at sites t_i, $i = 1, \ldots, n$.
- *Level 2.* The unknown signal $U(t)$ representing a functional parameter following a Gaussian process, whose mean depends on a parameter of regression or drift parameters $\boldsymbol{\beta}$ through (7.8).
- *Level 3a.* The drift parameters $\boldsymbol{\beta}$.
- *Level 3b.* The covariance parameters, typically a scale parameter σ^2 and a range parameter φ for the covariance function, and possibly a nugget variance τ^2. Let $\boldsymbol{\theta}$ be a vector containing the parameters σ^2, φ, and τ^2. For simplicity of presentation, suppose $\tau^2 > 0$.

The Bayesian model specifies the joint distribution for all the random quantities as

- hyper-priors for $\boldsymbol{\theta}$ and $\boldsymbol{\beta}$;
- a conditional GP prior for $U(\cdot)$ given $\boldsymbol{\theta}$ and $\boldsymbol{\beta}$;
- a conditional distribution for $\boldsymbol{X}$ given $U(\cdot), \boldsymbol{\theta}, \boldsymbol{\beta}$

$$X_i \mid \{U(\cdot), \boldsymbol{\beta}, \boldsymbol{\theta}\} \sim N(U(t_i), \tau^2).$$

The signal $U(\cdot)$ follows a GP whose mean depends on $\boldsymbol{\beta}$ and whose covariance function depends on $\boldsymbol{\theta}$.

The purpose of kriging is to predict the signal given the observations, and the form of the kriging predictor depends on what is assumed about higher level parameters. It is helpful to highlight three possible scenarios.

(a) *Simple Bayesian kriging.* If $\boldsymbol{\beta}$ and $\boldsymbol{\theta}$ are known, then the Gaussian random field model for $\{U(t)\}$ can be viewed as a prior distribution. Given the data $\boldsymbol{x}$, the posterior distribution for $\{U(t)\}$ is another Gaussian random field. Its mean and covariance function are identical to the those obtained from simple kriging in Section 7.3. See Section 7.12.2.

(b) *Empirical Bayesian kriging.* If $\boldsymbol{\beta}$ and $\boldsymbol{\theta}$ are not known, then they can be estimated from $\boldsymbol{x}$ e.g. by maximum likelihood, as in Chapter 5. Once they are estimated, their values can be plugged into the method of the previous paragraph. This procedure can be viewed as an example of empirical Bayes inference. Watkins and Mardia (1992) showed that this approach works well under certain asymptotic assumptions.

   In a simpler setting where $\boldsymbol{\theta}$ is known, let $\hat{\boldsymbol{\beta}}$ denote the maximum likelihood estimator (MLE) of $\boldsymbol{\beta}$. Then $\hat{\boldsymbol{\beta}}$ has an explicit form as a GLS estimator. The resulting predictor, with $\boldsymbol{\theta}$ replaced by $\hat{\boldsymbol{\theta}}$, is identical to the universal kriging predictor of Section 7.5. However, the formulas for the kriging variance will be different. The empirical Bayes predictor uses the value in (7.15), which is smaller than the correct value (7.31) used in the universal kriging predictor.

(c) *Bayesian kriging with drift.* Suppose $\boldsymbol{\theta}$ is known (or estimated) but that $\boldsymbol{\beta}$ has a prior multivariate Gaussian distribution. Averaging over the prior distribution of $\boldsymbol{\beta}$ gives a marginalized GP prior for the signal $U(\cdot)$. Then as in simple Bayesian kriging, the posterior distribution for $U(\cdot)$ is another Gaussian random field, and the posterior mean function and covariance function for $\{U(t)\}$ can be computed explicitly. The resulting predictor is different from the universal kriging predictor (unless the prior for $\boldsymbol{\beta}$ is improper). Details are given in Section 7.12.3.

(d) *Full Bayesian kriging.* Now assume a prior is also introduced for $\boldsymbol{\theta}$ as well. In general, simulation-based inference such as Markov chain Monte Carlo (MCMC) is needed to carry out the inference in this case. For some further discussion, see Section 8.4.

7.12.2 Details for Simple Bayesian Kriging

Here is a summary of simple kriging from a Bayesian perspective. Suppose the signal $\{U(t),\ t \in \mathbb{R}^d\}$ follows a GP prior with known a mean function $\mu(t)$ and a known covariance function $\sigma(s,t)$. Noisy observations take the form

$$X_i = U(t_i) + \varepsilon_i, \quad \varepsilon_i \sim N(0, \tau^2),$$

at specified sites $t_1, \ldots, t_n$, where the error terms ε_i are independent of one another and the signal.

Suppose data $x_1, \ldots, x_n$ are observed. Then the conditional distribution of the signal given that the data is again a Gaussian process. The posterior mean function is given by the kriging predictor

$$\mu_{\text{post}}(t) = \hat{u}(t) = \mu(t) + \boldsymbol{\sigma}(t)^T \Omega^{-1}(\boldsymbol{x} - \boldsymbol{\mu}), \tag{7.61}$$

where

$$\boldsymbol{x} = \begin{bmatrix} x_1 \\ \vdots \\ x_n \end{bmatrix}, \quad \boldsymbol{\mu} = \begin{bmatrix} \mu(t_1) \\ \vdots \\ \mu(t_n) \end{bmatrix}, \quad \boldsymbol{\sigma}(t) = \begin{bmatrix} \sigma(t - t_1) \\ \vdots \\ \sigma(t - t_n) \end{bmatrix}$$

and

$$\Omega = \Sigma + \tau^2 I, \quad \Sigma = (\sigma_{ij}), \quad \sigma_{ij} = \sigma(t_i - t_j).$$

The posterior covariance function is given by

$$\sigma_{\text{post}}(s,t) = \sigma(s,t) - \boldsymbol{\sigma}(s)^T \Omega^{-1} \boldsymbol{\sigma}(t). \tag{7.62}$$

The proof is a simple consequence of a conditioning argument for a multivariate normal distribution (Eqs. (A.17) and (A.18)) for $[U(s),\ U(t),\ X_1, \ldots, X_n]^T$ with the mean $[\mu(s),\ \mu(t),\ \mu(t_1), \ldots, \mu(t_n)]^T$ and the covariance matrix

$$\begin{bmatrix} \sigma(s,s) & \sigma(s,t) & \boldsymbol{\sigma}(s)^T \\ \sigma(t,s) & \sigma(t,t) & \boldsymbol{\sigma}(t)^T \\ \boldsymbol{\sigma}(s) & \boldsymbol{\sigma}(s) & \Omega \end{bmatrix}.$$

Note that the kriging variance in (7.15) is the same as the posterior variance

$$\sigma_K^2(t) = \sigma_{\text{post}}(t,t).$$

7.12.3 Details for Bayesian Kriging with Drift

As in Section 7.2 suppose that, given $\boldsymbol{\beta}$, the signal $U(t)$ follows a Gaussian process with a mean function and covariance function

$$E\{U(t) \mid \boldsymbol{\beta}\} = \boldsymbol{\beta}^T \boldsymbol{f}(t), \quad \text{cov}\{U(s), U(t) \mid \boldsymbol{\beta}\} = \sigma(s,t), \tag{7.63}$$

in terms of a p-vector of known drift functions and a known covariance function. In earlier sections, $\boldsymbol{\beta}$ was viewed as an unknown coefficient vector. But now suppose $\boldsymbol{\beta}$ is also treated as random and is given a prior normal distribution

$$\boldsymbol{\beta} \sim N_p(\boldsymbol{\nu}_0, \Delta), \tag{7.64}$$

where $\boldsymbol{\nu}_0$ is a known p-vector and Δ is a known positive definite matrix.

Averaging over $\boldsymbol{\beta}$ gives a marginal distribution of $U(\cdot)$ as a Gaussian process with a marginal mean function $\mu_M(t)$ and marginal covariance function $\sigma_M(s,t)$, say, given by

$$\mu_M(t) = \boldsymbol{f}(t)^T \boldsymbol{v}_0, \quad \sigma_M(s,t) = \sigma(s,t) + \boldsymbol{f}(s)^T \Delta \boldsymbol{f}(t). \tag{7.65}$$

Note that even if the original covariance function $\sigma(s,t)$ is stationary, so $\sigma(s,t)$ depends only on the lag $s-t$, the marginal covariance function is not stationary.

Consider data $X_i = x_i$ at sites t_i, $i = 1, \dots, n$, and suppose it is desired to estimate the parameters of the process and to predict the latent process $U(t_0)$ at a new site t_0. The posterior distribution for $U(t_0)$ is easy to work out using conditional moments for a multivariate normal distribution. The result is identical to the simple kriging predictor (7.18) in Section 7.3, but working here with the marginal moments (7.65) of the latent process. It is also straightforward to compute the posterior distribution of $\boldsymbol{\beta}$.

Given $\boldsymbol{\beta}$, and allowing for a nugget effect in the data measurements, the joint distribution of $U(t_0)$ and $\boldsymbol{X}$ is multivariate normal with the mean and covariance matrix

$$E\left\{\begin{bmatrix} U(t_0) \\ \boldsymbol{X} \end{bmatrix} \mid \boldsymbol{\beta}\right\} = \begin{bmatrix} \boldsymbol{f}_0^T \boldsymbol{\beta} \\ F\boldsymbol{\beta} \end{bmatrix}, \quad \text{var}\left\{\begin{bmatrix} U(t_0) \\ \boldsymbol{X} \end{bmatrix} \mid \boldsymbol{\beta}\right\} = \begin{bmatrix} \sigma(t_0, t_0) & \boldsymbol{\sigma}_0^T \\ \boldsymbol{\sigma}_0 & \Omega \end{bmatrix}, \tag{7.66}$$

using the same notation as in Section 7.5. Averaging over the prior distribution of $\boldsymbol{\beta}$ yields the marginal mean and variance of $U(t_0)$, $\boldsymbol{X}$, and $\boldsymbol{\beta}$

$$E\left\{\begin{bmatrix} U(t_0) \\ \boldsymbol{X} \\ \boldsymbol{\beta} \end{bmatrix}\right\} = \begin{bmatrix} \boldsymbol{f}_0^T \boldsymbol{v}_0 \\ F\boldsymbol{v}_0 \\ \boldsymbol{v}_0 \end{bmatrix} \tag{7.67}$$

$$\begin{aligned} \text{var}\left\{\begin{bmatrix} U(t_0) \\ \boldsymbol{X} \\ \boldsymbol{\beta} \end{bmatrix}\right\} &= \begin{bmatrix} \sigma(t_0,t_0) + \boldsymbol{f}_0^T \Delta \boldsymbol{f}_0 & \boldsymbol{\sigma}_0^T + \boldsymbol{f}_0^T \Delta F^T & \boldsymbol{f}_0^T \Delta \\ \boldsymbol{\sigma}_0 + F\Delta \boldsymbol{f}_0 & \Omega + F\Delta F^T & F\Delta \\ \Delta \boldsymbol{f}_0 & \Delta F^T & \Delta \end{bmatrix} \\ &= \begin{bmatrix} \Psi_{11}^{(\Delta)} & \Psi_{12}^{(\Delta)} & \Psi_{13}^{(\Delta)} \\ \Psi_{21}^{(\Delta)} & \Psi_{22}^{(\Delta)} & \Psi_{23}^{(\Delta)} \\ \Psi_{31}^{(\Delta)} & \Psi_{32}^{(\Delta)} & \Psi_{33}^{(\Delta)} \end{bmatrix}, \text{ say.} \end{aligned} \tag{7.68}$$

The posterior distribution of $\boldsymbol{\beta}$ is given by conditioning on $\boldsymbol{X}$ in (7.67) and (7.68). This distribution is normal with a conditional mean and variance

$$\boldsymbol{v}_0 + \Psi_{32}^{(\Delta)} \{\Psi_{22}^{(\Delta)}\}^{-1} (\boldsymbol{x} - F\boldsymbol{v}_0) = \boldsymbol{v}_0 + \Delta F^T (\Omega + F\Delta F^T)^{-1} (\boldsymbol{x} - F\boldsymbol{v}_0)$$

and

$$\Psi_{33}^{(\Delta)} - \Psi_{32}^{(\Delta)} \{\Psi_{22}^{(\Delta)}\}^{-1} \Psi_{23}^{(\Delta)} = \Delta - \Delta F^T (\Omega + F\Delta F^T)^{-1} F\Delta.$$

Similarly, the predictive distribution of $U(t_0)$ is given by conditioning on $\boldsymbol{x}$ in (7.67) and (7.68), yielding a normal distribution with a conditional mean

$$E(U(t_0) \mid \boldsymbol{x}) = \boldsymbol{f}_0^T \boldsymbol{v}_0 + \boldsymbol{\gamma}_\Delta^T (\boldsymbol{x} - F\boldsymbol{v}_0), \tag{7.69}$$

where

$$\gamma_\Delta = \{\Psi_{22}^{(\Delta)}\}^{-1}\Psi_{21}^{(\Delta)}. \tag{7.70}$$

Setting

$$D = \Delta^{-1} + F^T\Omega^{-1}F, \tag{7.71}$$

it can be shown that the formula for γ_Δ can be simplified to

$$\gamma_\Delta = \Omega^{-1}\sigma_0 + \Omega^{-1}FD^{-1}\left(\boldsymbol{f}_0 - F^T\Omega^{-1}\sigma_0\right). \tag{7.72}$$

The conditional variance of $U(t_0)$ given $\boldsymbol{x}$ can be termed the *Bayesian kriging variance* and simplifies to

$$\begin{aligned} K_B(t_0) &= \mathrm{var}\{\hat{U}(t_0) \mid \boldsymbol{x}\} = \Psi_{11}^{(\Delta)} - \Psi_{12}^{(\Delta)}\{\Psi_{22}^{(\Delta)}\}^{-1}\Psi_{21}^{(\Delta)} \\ &= \sigma(t_0, t_0) - \sigma_0^T\Omega^{-1}\sigma_0 + (\boldsymbol{f}_0 - F^T\Omega^{-1}\sigma_0)^T D^{-1}(\boldsymbol{f}_0 - F^T\Omega^{-1}\sigma_0). \end{aligned} \tag{7.73}$$

It is interesting to look at what happens when the prior variance matrix Δ gets large. Write $\Delta = \lambda\Delta_0$ where Δ_0 is a fixed matrix and $\lambda > 0$ is a scale parameter. If $\lambda \to \infty$ (an improper prior distribution for $\boldsymbol{\beta}$), then Δ^{-1} tends to the zero matrix and $D \to F^T\Omega^{-1}F$. Remembering that $F^T\gamma = \boldsymbol{f}_0$ in (7.32) we see that the Bayesian kriging predictor (7.69) converges to the universal kriging predictor $\hat{u}(t_0) = \gamma^T\boldsymbol{x}$ with γ given in (7.30). Similarly, the Bayesian kriging variance (7.73) converges to the universal kriging variance in (7.31). Further details are explored in Exercise 7.7. See also Section B.4.

7.13 Kriging and Machine Learning

Bayesian prediction has become a popular tool in machine learning. For more details, see, e.g., Schölkopf and Smola (2002), Rasmussen and Williams (2006), and Sambasivan et al. (2020). Although Bayesian prediction in machine learning is essentially the same as kriging, the emphasis is somewhat different from the statistical perspective and Table 7.3 gives a comparison of the terminology and notation used in the two disciplines. The discussion here is based on Kalaitzis and Lawrence (2011).

Consider the problem of regressing a real-valued variable x on an explanatory variable t using a response function $g(t)$. Given data (t_i, x_i), $i = 1, \ldots, n$, the regression model takes the form

$$x_i = g(t_i) + \varepsilon_i,$$

where the ε_i are i.i.d. $N(0, \tau^2)$ random variables. In general, the response function contains unknown parameters that need to be estimated. To keep the discussion simple, suppose t is one dimensional. The simplest and most important regression model is linear regression with a response function $g(t) = a_0 + a_1 t$. This response function is sometimes augmented by extra terms to accommodate nonlinearity. An example is quadratic regression with $g(t) = a_0 + a_1 t + a_2 t^2$. In any

Table 7.3 Comparison between the terminology and notation of this book for simple kriging and Rasmussen and Williams (2006, pp. 13–17) for simple Bayesian kriging. The posterior mean and covariance function take the same form in both formulations, given by (7.61) and (7.62).

This book	Rasmussen and Williams
Simple kriging	Bayesian prediction
Sites $t \in \mathbb{R}^d$	Locations $\boldsymbol{x} \in \mathbb{R}^D$
Signal $U(t)$	Underlying function $f(\boldsymbol{x})$
Prior mean and covariance function of $U(t)$ $\mu(t),\ \sigma(s,t)$	Prior mean and covariance function of $f(\boldsymbol{x})$ $m(\boldsymbol{x}),\ k(\boldsymbol{x},\boldsymbol{x}')$
Data sites $t_1, \ldots, t_n$	Training data locations $\boldsymbol{x}_1, \ldots, \boldsymbol{x}_n$
Covariance matrix at data sites $\Sigma\ (n \times n)$	Covariance matrix at training locations $K\ (n \times n)$
Noisy observations $x_1, \ldots, x_n$	Noisy observations $y_1, \ldots, y_n$
Variance matrix of observations $\Omega = \Sigma + \tau^2 I$	Variance matrix of observations $K + \sigma^2 I$
New site t	Test location x_*
Posterior mean of $U(t)$ = kriging predictor $\mu_{\text{post}}(t) = \hat{u}(t)$	Posterior mean of $f(x_*)$ = predictive mean $\hat{f}_* = \hat{f}(x_*)$
Posterior covariance function $\sigma_{\text{post}}(s,t)$	Posterior covariance function $\text{cov}\{(f(x_*), f(x'_*)\}$

case, the number of unknown parameters is generally small relative to the sample size n.

Another approach is nonparametric regression, which allows a very flexible response function. Let $\varphi(t)$ be a specified function of t, assumed to be square-integrable. Typically, $\varphi(t)$ is a bounded continuous even function of t, which has a mode at $t = 0$ and dies away quickly in the tails. One example is given in (7.75) below. The function $\varphi(t)$ is often called a "kernel function" or (especially in higher dimensions for a function depending only on $|t|$) a "radial basis function." Also, let t_j^*, $j = 1, \ldots, M$, denote a large number M, say, prespecified sites, sometimes known as "knots." Then, the response function is assumed to take the form of a linear combination of shifted versions of the kernel. That is,

$$g(t) = \sum b_j \varphi(t - t_j^*), \tag{7.74}$$

where the b_j are coefficients to be estimated from data. Note the knot sites are not the same as the data sites. The knots t_j^* are often chosen to be equally spaced on a fine grid so that the response function can describe a wide range of possible behaviors.

Further, M is allowed to be much larger than n. Thus, there will be far too many parameters to estimate from the data. One way to regularize the problem is to impose prior distributions on the parameters. To simplify the discussion here, give the coefficients b_j i.i.d. $N(0, \sigma_b^2)$ distributions.

This is exactly the framework described Section 7.12.3 for Bayesian kriging with drift. The response function $g(t)$ can be viewed as a Gaussian process with the covariance function

$$\sigma(s, t) = \sigma_b^2 \sum_{j=1}^{M} \varphi(s - t_j^*)\varphi(t - t_j^*).$$

So far, this covariance function depends heavily on the choice of knots. To get a simpler result, consider the limit as the knots become densely spaced on a wide interval. For example, consider $M = 2m^2 + 1$ sites located at $t_j^* = j/m$ for $j = -m^2, \ldots, m^2$. That is, the sites have a spacing of $1/m$ and span the interval $[-m, m]$. Letting $m \to \infty$ and scaling the problem appropriately yields the following limiting covariance function:

$$\begin{aligned}\sigma(s, t) &\propto \int_{-\infty}^{\infty} \varphi(s - h)\varphi(t - h)\, dh \\ &= \int \varphi(s - h)\check{\varphi}(h - t)\, dh \\ &= \int \varphi((s - t) - h')\check{\varphi}(h')\, dh',\end{aligned}$$

where $\check{\varphi}(t) = -\varphi(t)$ and the last line involves the substitution $h' = h - t$.

As a specific example, let

$$\varphi(t) = \check{\varphi}(t) \propto \exp\{-t^2/(2\kappa^2)\} \tag{7.75}$$

denote the density in t of the $N(0, \kappa^2)$ distribution and remember that the convolution of two normal densities with variance κ^2 is again normal, with variance $2\kappa^2$. In other words,

$$\sigma(s, t) \propto \exp\{-(s - t)^2/(4\kappa^2)\} \tag{7.76}$$

is equal to the squared exponential covariance function. Example 7.1 used this covariance function to illustrate the behavior of the fitted nonparametric regression function when the true response function is a sine wave. This example is often used in the machine learning literature to motivate the use of Gaussian process regression.

Similar constructions to (7.74) were used in Section 2.14.3 to simulate a Gaussian process with an arbitrary spectral density as a sum of random cosine waves. The procedure there, for $t \in \mathbb{R}^d$, involved the kernel function $\varphi(t) = \cos(t^T\omega)$. The use of knots here corresponds there to a phase shift in the cosine function, and, in addition, the construction there included a random distribution for ω.

Another difference between statistics and machine learning can be highlighted. Machine learning tends to emphasize simple kriging rather than more complicated ordinary and universal kriging formulations. Practitioners often either pretend (or preprocess the data to assume) the mean is 0, or they have so much data; the effect of allowing a mean or higher order drift terms is negligible.

One topic of major interest in machine learning and statistics is computational efficiency. In terms of computational effort, kriging is a $O(n^3)$ operation, which can be a major bottleneck for large data sets. Various approximations can be used to improve computation times. These are not studied here but are very important.

7.14 The Link Between Kriging and Splines

The kriging theory developed for spatial prediction turns out to have very close links to the theory of smoothing splines in nonparametric regression. The models used for spatial dependence in the kriging setting can be recast as smoothness constraints on an underlying regression function. In this section, we set out these links in detail. For more detail about splines, see, e.g., Wahba (1990), Green and Silverman (1994), Gu (2002), and Berlinet and Thomas-Agnan (2004).

7.14.1 Nonparametric Regression

This section gives a different perspective on nonparametric regression from the description in Section 7.13. Consider a function $g(t)$ whose values x_i are known at a finite set of sites t_i, $i = 1, \dots, n$ in $\mathbb{R}^d$, are either exactly,

$$x_i = g(t_i),$$

or subject to noise

$$x_i = g(t_i) + \varepsilon_i, \quad \varepsilon_i \sim N(0, \tau^2).$$

The objective in nonparametric regression is to estimate the function g for all $t \in \mathbb{R}^d$. Of course, without some further constraints, this task is too general to have a solution.

In classic regression, g is assumed to have a parametric form, which is linear in a set of parameters. For example, g may be assumed to be a linear or quadratic function in t. Then, the classic methods of multiple regression analysis can be used to fit the parameters.

In nonparametric regression, a constraint is placed on the roughness of g. For the purposes of this book, we limit our attention to the roughness penalty

$$J^d_{r+1}(g) = \int |\nabla^{\{r+1\}} g(t)|^2 \, dt, \tag{7.77}$$

where the integral is over $\mathbb{R}^d$. Here, $\nabla^{\{r+1\}}$ denotes the $(r+1)$-fold iterated gradient of g, an $(r+1)\,d$-dimensional vector containing all the rth-order partial derivatives of g, with replication. For example, if $d = 2, r = 1$, then $\nabla^{\{2\}} g = (g_{11}, g_{12}, g_{21}, g_{22})$ with $g_{ij}(t)$ denoting $\partial^2 g/\partial t[i]\partial t[j]$. Combining replications, we can rewrite J^d_{r+1} as

$$J^d_{r+1}(g) = \sum_{|m|=r+1} \binom{r+1}{m} \int \left\{ \frac{\partial^{r+1} g(t)}{\partial t[1]^{m[1]} \ldots \partial t[d]^{m[d]}} \right\}^2 dt, \tag{7.78}$$

where $m = (m[1], \ldots, m[d])$ is a multi-index and

$$\binom{r+1}{m} = \frac{(r+1)!}{m[1]! \cdots m[d]!}$$

is a multinomial coefficient. As in (3.35) write $D^m g = \partial^{|m|} g/\partial t^m$ as a concise notation for the corresponding partial derivative.

One of the reasons this particular penalty is chosen is that it is invariant under rotations and translations of t. If t is replaced by $t^{(\text{new})} = Et + \delta$, say, where $E(d \times d)$ is orthogonal, and we set $g^{(\text{new})}(t^{(\text{new})}) = g(t) = g(E^T(t^{(\text{new})} - \delta))$, then $\nabla^{\{r+1\}}(\text{new})(t^{(\text{new})}) = (\otimes E)\nabla^{\{r+1\}} g(t)$, where $\otimes E$, the $(r+1)$-fold tensor product of E with itself is an $(r+1)d \times (r+1)d$ orthogonal matrix. Hence, $J^d_{r+1}(g) = J^d_{r+1}(g^{(\text{new})})$.

Then the smoothing spline problem can be phrased as follows: find a function g to minimize

$$\sum_{i=1}^{n} |x_i - g(t_i)|^2 + \lambda J^d_{r+1}(g) \tag{7.79}$$

over all functions g, which are smooth enough for the required derivatives to exist and such that the integral is finite. The value of the smoothing parameter $\lambda > 0$ controls the trade-off between data fidelity and smoothness. As $\lambda \to \infty$, more emphasis is placed on smoothness and the solution converges to the classical OLS solution based on a multiple regression on polynomials up to degree r. (Such functions have penalty 0 under J^d_{r+1}.)

On the other hand, for small λ, the fitted function is closer to data values. The limiting case $\lambda \to 0$ is known as the interpolating spline and can be phrased as a solution to the following optimization problem: find a function g to

$$\text{minimize } J^d_{r+1}(g) \text{ such that } g(t_i) = x_i, \quad i = 1, \ldots, n. \tag{7.80}$$

It turns out that for a solution to exist, the order r of derivatives cannot be too small, depending on the dimension d. Further, and perhaps more surprisingly, the solution can be identified with a kriging predictor for a certain intrinsic random field model. The next two sections describe this identification in detail. It is simpler to treat the interpolating case first.

7.14.2 Interpolating Splines

Let $\text{IRF}_d(\alpha, r)$ denote the self-similar intrinsic random field of index $\alpha > 0$ in d dimensions with a drift space given by $\mathcal{F}_r$, the polynomials in t up to degree r where $r \geq 0$ is an integer. As shown in Section 3.10, the intrinsic order of this process is $[\alpha]$, the integer part of α, so that necessarily $r \geq [\alpha]$. If $r = [\alpha]$, then the drift space consists solely of intrinsic drift; if $r > [\alpha]$, there is also extrinsic drift. This section uses the self-similar intrinsic covariance functions from Section 3.10 given by

$$\sigma_\alpha(h) = \begin{cases} c_{\alpha,d}|h|^{2\alpha}, & \alpha > 0 \text{ not an integer,} \\ c'_{\alpha,d}|h|^{2\alpha} \log|h|, & \alpha > 0 \text{ an integer,} \end{cases}$$

with corresponding spectral density

$$f(\omega) = (2\pi)^{-d}|\omega|^{-d-2\alpha}.$$

The normalizing constants from (3.46) and (3.47) have the property that $\text{sign}(c_{\alpha,d}) = (-1)^{[\alpha]-1}$ for $\alpha > 0$ not an integer, and $\text{sign}(c'_{\alpha,d}) = (-1)^{\alpha-1}$ for $\alpha > 0$ an integer, Here $[\alpha]$ denotes the integer part of α.

The following theorem looks at the interpolating problem and identifies the interpolating spline with a kriging predictor. Consider data (x_i, t_i), $x_i \in \mathbb{R}$ and $t_i \in \mathbb{R}^d$, $i = 1, \ldots, n$.

Theorem 7.14.1 *Given an integer r satisfying $r + 1 > d/2$, set the self-similar index parameter to $\alpha = r + 1 - d/2 > 0$. Then the problem of interpolating the data $(x_i, t_i), i = 1, \ldots, n$ subject to minimizing the roughness criterion (7.77) has a solution $g^*(t)$, say, given for any particular site $t = t_0$ by*

$$g^*(t_0) = \mathbf{x}^T A \boldsymbol{f}_0 + \mathbf{x}^T B \boldsymbol{\sigma}_0, \tag{7.81}$$

where $A, B, \boldsymbol{f}_0$, and $\boldsymbol{\sigma}_0$ were determined earlier (see especially Section 7.8 and Table 7.1) to express the kriging solution for the self-similar $\text{IRF}_d(\alpha, r)$ model when there is no nugget effect.

Proof: Dropping the subscript 0 on t, the solution in (7.81) can be written in the form

$$g^*(t) = \sum_{|m| \leq r} a_m t^m + \sum_{j=1}^{n} b_j \sigma_\alpha(t - t_j), \tag{7.82}$$

where the coefficients, written as vectors, are given by $\boldsymbol{a} = A^T\boldsymbol{x}$ and $\boldsymbol{b} = B\boldsymbol{x}$, respectively, both of which are linear functions of the data $\boldsymbol{x}$. Thus, $g^*(t)$ is a polynomial in t of order r plus a linear combination of n copies of the covariance function centered at the data sites $\{t_j\}$. With hindsight of an explicit form for the solution, it is easy to construct a proof.

Given any function $g_1(t)$, which interpolates the data, and for which the elements of $\nabla^{\{r+1\}}g_1(t)$ are square-integrable, write $g_1(t) = g^*(t) + g_2(t)$, where $g_2(t) = g_1(t) - g^*(t)$ satisfies $g_2(t_i) = 0, i = 1, \ldots, n$. The roughness penalty can be expanded as

$$J^d_{r+1}(g_1) = J^d_{r+1}(g^*) + J^d_{r+1}(g_2) + 2\int (\nabla^{\{r+1\}}g^*)^T(\nabla^{\{r+1\}}g_2)\, dt. \tag{7.83}$$

If we can show that the last term vanishes, then we will have shown that $J^d_{r+1}(g_1) \geq J^d_{r+1}(g^*)$ with equality if and only if $g_1 = g^*(t)$. The theorem will then follow.

To prove that the last term vanishes, use integration by parts $r+1$ times to write it as

$$\begin{aligned} 2(-1)^{r+1}\int \{\Delta^{\{r+1\}}g^*(t)\}g_2(t)\, dt &= 2(2\pi)^d \int \left\{\sum_{j=1}^n b_j\delta_0(t-t_j)\right\} g_2(t)\, dt \\ &= 2\sum_{j=1}^n b_j g_2(t_j) = 0. \end{aligned} \tag{7.84}$$

Here, we have used the fact that $-\Delta\sigma_\alpha(t) = \sigma_{\alpha-1}(t)$, where Δ is the Laplacian operator, and that $\sigma_{-d/2}(t) = \delta_0(t)$ is the Dirac delta function (Section 3.10).

The integration by parts in (7.84) can also be formulated in the Fourier domain. Let $\tilde{g}_2(\omega)$ denote the Fourier transform of $g_2(t)$, and let $\psi(\omega) = \sum b_j \exp(-i\omega^T t_j)$, so that $g^*(t)$ is the Fourier transform of $\psi(\omega)f_\alpha(\omega)$. Note that $\psi(\omega) = O(|\omega|^{r+1})$ as $\omega \to 0$, and that

$$\sum_{|m|=r+1} \binom{r+1}{m} \omega^{2m} = |\omega|^{2r+2}, \tag{7.85}$$

where $\omega^m = \omega[1]^{m[1]} \cdots \omega[d]^{m[d]}$. Since $d + 2\alpha = 2r+2$, the last integral in (7.83) can be rewritten by the Parseval relation for L^2 functions (Section A.5) as

$$\begin{aligned} &2\int \nabla^{\{r+1\}}g^*(t)^T\nabla^{\{r+1\}}g_2(t)\, dt \\ &= 2\sum_{|m|=r+1}\binom{r+1}{m}\int [\psi(\omega)(i\omega)^m|\omega|^{-d-2\alpha}][(-i\omega)^m\tilde{g}_2(\omega)]\, d\omega \\ &= 2\int 1\cdot[\psi(\omega)\tilde{g}_2(\omega)]\, d\omega \\ &= 2(2\pi)^d \sum b_j g_2(t_j). \end{aligned} \tag{7.86}$$

To gain more insight into (7.86), we note how the singularity of various functions about $\omega = 0$ affects the interpretation of their Fourier transforms, as a function of t, similarly to Section 3.8. Recall the spectral density function $f_\alpha(\omega)$ has a singularity of order $2r+2$, so its Fourier transform is $\sigma_\alpha(t)$ plus an arbitrary polynomial of degree $2r+2$. The function $\psi(\omega)f_\alpha(\omega)$ has a singularity of order $r+1$. Its Fourier transform is $g^*(\omega)$ plus an arbitrary polynomial of degree $r+1$. The function $(i\omega)^m\psi(\omega)f_\alpha(\omega)$ has no singularity so its Fourier transform is $\nabla^m g^*(\omega)$, which is well defined with no ambiguity.

The function $\tilde{g}_2(\omega)$ has a singularity of order $r+1$, so its Fourier transform is $(2\pi)^d g_2(-t)$ plus an arbitrary polynomial of degree $r+1$. The function $(-i\omega)^m\tilde{g}_2(\omega)$ has no singularity and its Fourier transform $(2\pi)^d\nabla^m g_2(-t)$, or equivalently its inverse Fourier transform $\nabla^m g_2(t)$, is well defined. Similarly, the function $\psi(\omega)\tilde{g}_2(\omega)$ has no singularity and its Fourier transform $(2\pi)^d\sum b_j g_2(t_j - t)\tilde{g}_2(\omega)$ is well defined. The last line of (7.86) is twice this Fourier transform evaluated at $t = 0$. □

7.14.3 Comments on Interpolating Splines

1. Using the same argument as in (7.84), the value of the roughness penalty for $g^*(t)$ can be calculated as

$$\begin{aligned} J^d_{r+1}(g^*) &= (2\pi)^d \sum_{j=1}^{n} b_j g^*(t_j) \\ &= (2\pi)^d \sum_{j,k=1}^{n} b_j b_k \sigma(t_j - t_k) + \sum_{|m|\le r} a_m \sum_{j=1}^{n} b_j t_j^m \\ &= (2\pi)^d \boldsymbol{b}^T\Sigma\boldsymbol{b} \\ &= (2\pi)^d \boldsymbol{x}^T B\boldsymbol{x}, \end{aligned} \tag{7.87}$$

since $\boldsymbol{b} = B\boldsymbol{x}$ and $B\Sigma B = B$. Here, we have used the fact that $\boldsymbol{b}$ defines an rth-order increment so that

$$\sum_{|m|\le r} a_m \sum_{j=1}^{n} b_j t_j^m = 0. \tag{7.88}$$

2. The most popular choices of r in dimensions $d = 1,2$ are as follows. For $d = 1$, the choice $r = 0$ ($\alpha = 1/2$) yields a piecewise linear path, and $r = 1$ ($\alpha = 3/2$) yields the usual cubic spline. For $d = 2$, the choice $r = 1$ ($\alpha = 1$) yields the usual thin-plate spline; see Section 7.16. Note that as d increases, the smallest feasible value of r increases.
3. In dimensions $d = 1, 2$, note that $[\alpha] = r$ so that the order r of the polynomial drift is the minimum order compatible with the order $[\alpha]$ of the intrinsic random field. However, in dimensions $d \ge 3, [\alpha] < r$, so that an extra

polynomial drift term needs to be included in the specification of the intrinsic random field in order for the kriging and thin-plate spline approaches to match. Thus, kriging offers, in principle, a more flexible modeling strategy than splines by allowing the order of the polynomial drift to vary (both higher in any dimension and lower in dimensions $d \geq 3$).

4. Note how kriging and thin-plate splines arrive at (7.81) from very different directions. In kriging, (7.81) is viewed as a linear combination of $\boldsymbol{x}$ to be optimized, with the site t_0 held fixed. In contrast, in splines, (7.81) is viewed as a function of t_0 to be optimized, with the data $\boldsymbol{x}$ held fixed.
5. In Section 7.16, we will consider a pair of thin-plate splines to model a deformation of the plane.

7.14.4 Smoothing Splines

To solve the smoothing spline problem, first note that if the values of $g(t_i)$ are fixed, $g(t_i) = z_i$ say, $i = 1, \dots, n$, then the optimal choice of g is the interpolating spline passing through $\{z_i\}$. Thus, setting $\boldsymbol{z} = [z_1, \dots, z_n]^T$ and $\kappa = (2\pi)^d \lambda$, the smoothing problem can be rephrased as

$$\begin{aligned} \min_g L(g, \lambda) &= \min_{\boldsymbol{z} \in \mathbb{R}^n} \min_{g: g(t_i) = z_i} L(g, \lambda) \\ &= \min_{\boldsymbol{z}} |\boldsymbol{x} - \boldsymbol{z}|^2 + \kappa \boldsymbol{z}^T B \boldsymbol{z}. \end{aligned} \tag{7.89}$$

Minimizing (7.89) with respect to $\boldsymbol{z}$ is straightforward and yields $\boldsymbol{z} = (I + \kappa B)^{-1} \boldsymbol{x}$. Hence, the overall optimal choice of g is given by

$$\begin{aligned} g(t_0) &= \boldsymbol{z}^T A \boldsymbol{f}_0 + \boldsymbol{z}^T B \boldsymbol{\sigma}_0 \\ &= \boldsymbol{x}^T (I + \kappa B)^{-1} A \boldsymbol{f}_0 + \boldsymbol{x}^T (I + \kappa B)^{-1} B \boldsymbol{\sigma}_0. \end{aligned} \tag{7.90}$$

After a bit of algebra, this formula can be rewritten as

$$g(t_0) = \boldsymbol{x}^T A_\kappa \boldsymbol{f}_0 + \boldsymbol{x}^T B_\kappa \boldsymbol{\sigma}_0, \tag{7.91}$$

where A and B are the transfer matrices when there is no nugget effect and A_κ and B_κ are the transfer matrices when there is a nugget effect of size $\tau^2 = \kappa$.

7.15 Reproducing Kernel Hilbert Spaces

The theory of RKHSs is based on the following idea. Let E be an index set; typically $E = \mathbb{R}^d$. Every positive definite covariance function on E can be identified with a certain Hilbert space of functions on E. Further, the problem of kriging for a Gaussian random field can be identified with a problem of optimization on this function space. The details are beyond the scope of this book.

See, e.g., Wahba (1990), Berlinet and Thomas-Agnan (2004), and Schölkopf and Smola (2002) for more information.

There is a simple setting where the idea of an RKHS is easy to understand. If E is finite of size n, then a covariance function is equivalent to an $n \times n$ covariance matrix Σ, say. A "function" on E is just an n-vector $\boldsymbol{f}$, say. In this case, the Hilbert space squared norm is given by $\boldsymbol{f}^T \Sigma^{-1} \boldsymbol{f}$.

The theory of RKHSs is usually developed for fully specified covariance functions. However, a closely related theory can also be developed for intrinsic covariance functions, where an intrinsic covariance function can be identified with a *semi-norm* on a *semi-RKHS*.

The classic use of RKHSs is in spline theory, where the problem of finding a spline can be identified with an optimization problem in an RKHS. However, since the penalty (7.77) has a nonzero nullspace, it only defines a seminorm in function space and it is necessary to impose some arbitrary boundary conditions to get a norm. A more natural approach is to formulate the problem of finding a spline as an optimization problem in a semi-RKHS rather than an RKHS (e.g. Mosamam and Kent, 2010).

7.16 Deformations

A deformation is a mapping from a region S to a region T in $\mathbb{R}^d$, where typically $d = 2$ or 3. The main use of deformations in shape analysis is to measure shape change, whereas in image analysis the aim is to bring a set of objects to a common registration system so that they can be compared (or averaged) pixel by pixel, in order to look for fine-scale differences.

Suppose that the source object S has landmarks

$$t_i = (t_i[1], \ldots, t_i[d]), \quad i = 1, \ldots, n,$$

and that the target shape T has corresponding landmarks

$$x_i = (x_i[1], \ldots, x_i[d]), \quad i = 1, \ldots, n.$$

The aim is to fit a deformation, that is, to find a smooth transformation $x = \varphi(t) = \left[\varphi_{[1]}(t) \quad \ldots \quad \varphi_{[d]}(t))\right]^T : \mathbb{R}^d \to \mathbb{R}^d$ that interpolates the data

$$x_i = \varphi(t_i), \quad i = 1, \ldots, n. \tag{7.92}$$

Thus, the deformation takes the source landmarks in S to the target landmarks in T. Let $x(t) = (x_{[1]}(t), \ldots, x_{[d]}(t))$ represent a typical site in the target space and denote the $n \times 1$ data vectors for each component $\ell = 1, \ldots, d$ by $\boldsymbol{x}[\ell]$. Following our convention, vectors with d components have not been written in boldface.

One solution is to fit an interpolating random field model to each component $\varphi_{[\ell]}(t)$. In $d = 2$ dimensions, a common choice is a self-similar IRF$_2(\alpha, 1)$ model with $\alpha = 1$ (Section 3.10). This model is intrinsic of order 1 with an intrinsic covariance function

$$\sigma_I(h) = |h|^2 \log |h|$$

and linear drift (Bookstein, 1989). This solution is also known as an interpolating thin-plate spline, motivated by the penalty (7.77) with $d = 2$, $r = 1$. The two interpretations can be identified with each other by Theorem 7.1. The self-similar assumption is made because it is thought that the variability in biological deformations can exhibit self-similar behavior over a certain range of scales (e.g. Mardia, Bookstein, Kent and Meyer, 2006b; Mardia, Angulo and Goitía, 2006a).

The presence of linear drift ensures that if the target landmarks can be written as an affine transformation of the source landmarks (i.e. $x = A\,t + b$ where $A(d \times d)$ is nonsingular and b is a d-vector), then the roughness penalty (7.77) will equal 0 for each component $\ell = 1, 2$. Thus, the roughness penalty measures the extent to which the deformation is nonaffine. In practice, deformations will often involve small departures from an affine map, and in this setting, the methodology is most valuable.

It has only been possible to give a very brief introduction to the theory of deformations here. For applications of the theory here to problems in shape analysis, see, e.g., the books by Bookstein (1992), Dryden and Mardia (2016), and for a more general view of deformations, see, e.g., Joshi and Miller (2000) and Srivastava and Klassen (2016).

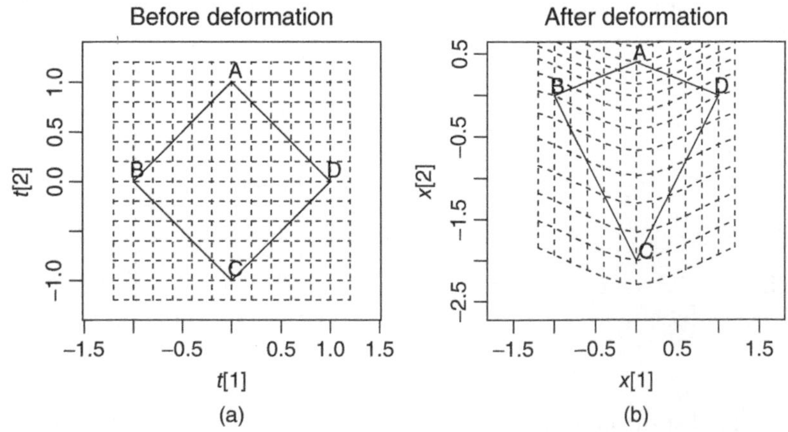

Figure 7.8 Deformation of a square (a) into a kite (b) using a thin-plate spline. The effect of the deformation on $\mathbb{R}^2$ can also be visualized: it maps a grid of parallel lines to a bi-orthogonal grid.

Example 7.7 ***"Square"-to-"kite" deformation***
Consider a square and a kite in the plane with landmarks A,B,C,D given by the rows of

$$\begin{bmatrix} 0 & 1 \\ -1 & 0 \\ 0 & -1 \\ 1 & 0 \end{bmatrix} \quad \text{and} \quad \begin{bmatrix} 0 & 0.4 \\ -1 & 0 \\ 0 & -2 \\ 1 & 0 \end{bmatrix},$$

respectively. A pair of thin-plate splines has been constructed taking the left-hand landmarks to each column, respectively, of the right-hand landmarks. The corresponding deformation is plotted in Figure 7.8. In particular, the effect of the deformation on an orthogonal grid of lines is plotted. The deformation includes vertical stretching near landmark C and vertical compression near landmark A. □

Exercises

7.1 In the setting of ordinary kriging, show that the maximum likelihood estimator of μ based on data $\boldsymbol{x} = [x_1, \ldots, x_n]^T$, also known as the generalized least squares estimator, is given by $\hat{\mu}$ in (7.27). Hence, show that the formula for the predictor $\hat{u}(t_0)$ in (7.28) is the same as $\boldsymbol{\gamma}^T\boldsymbol{x}$ with $\boldsymbol{\gamma}$ given in (7.21).

Similarly, in the setting of universal kriging, show that the maximum likelihood estimator of $\boldsymbol{\beta}$ is given by the GLS estimator $\hat{\boldsymbol{\beta}}$ in the line above (7.33). Hence, show that the formula predictor $\hat{u}(t_0)$ in (7.33) is the same as $\boldsymbol{\gamma}^T\boldsymbol{x}$ with $\boldsymbol{\gamma}$ given in (7.30).

7.2 The easiest way to prove that M^{-1} has the form in (7.46) is by rotating to the starred coordinates in Section 7.6.2. Show that M^* and the stated form for $(M^*)^{-1}$ reduce to

$$M^* = \begin{bmatrix} \Omega^*_{11} & \Omega^*_{12} & F^*_1 \\ \Omega^*_{21} & \Omega^*_{22} & 0 \\ (F^*_1)^T & 0 & 0 \end{bmatrix},$$

$$(M^*)^{-1} = \begin{bmatrix} 0 & 0 & (F^*_1)^{-T} \\ 0 & (\Omega^*_{22})^{-1} & -(\Omega^*_{22})^{-1}\Omega^*_{21}(F^*_1)^{-T} \\ (F^*_1)^{-1} & -(F^*_1)^{-1}\Omega^*_{12}(\Omega^*_{22})^{-1} & -(F^*_1)^{-1}\Omega^*_{11.2}(F^*_1)^{-T} \end{bmatrix},$$

where $\Omega^*_{11.2} = \Omega^*_{11} - \Omega^*_{12}(\Omega^*_{22})^{-1}\Omega^*_{21}$. Verify that $M^*(M^*)^{-1} = I_{n+p}$.

7.3 (a) Let $U(t)$ be a one-dimensional stationary process with an unknown mean μ and with a covariance function $\sigma(h) = \sigma^2\rho(h)$, where $\rho(h)$ is a specified correlation function. Consider predicting the value of the

process at a new site t_0 given noise-free observations $x_1 = u_1, x_2 = u_2$ at sites t_1, t_2.
Set $\alpha_j = \rho(|t_j - t_0|)$, $j = 1, 2$, and $\rho = \rho(|t_1 - t_2|)$, and suppose $|\rho| < 1$. Show that the predictor becomes

$$\hat{u}(t_0) = \left(1 - \frac{\alpha_1 + \alpha_2}{1+\rho}\right)\bar{x} + \frac{(\alpha_1 - \rho\alpha_2)x_1 + (\alpha_2 - \rho\alpha_1)x_2}{1-\rho^2}$$
$$= \bar{x} + \frac{\alpha_1 - \alpha_2}{1-\rho} d_x, \quad \bar{x} = (x_1 + x_2)/2, \; d_x = (x_1 - x_2)/2,$$

with prediction error

$$\sigma_K^2(t_0) = \sigma^2 \left\{1 - \frac{\alpha_1^2 + \alpha_2^2 - 2\rho\alpha_1\alpha_2}{1-\rho^2} + \frac{[1 + \rho - (\alpha_1 + \alpha_2)]^2}{2(1+\rho)}\right\}.$$

(b) Next include a nugget effect τ^2 in the model. Suppose the data x_1, x_2 are now noisy measurements of the underlying signal at the sites t_1, t_2. Show that the kriging predictor of the underlying signal becomes

$$\hat{u}(t_0) = \bar{x} + \frac{\alpha_1 - \alpha_2}{1 + \psi - \rho} d_x, \quad \bar{x} = (x_1 + x_2)/2, \; d_x = (x_1 - x_2)/2,$$

where $\psi = \tau^2/(\sigma^2 + \tau^2)$ the relative nugget effect.

(c) Suppose $\rho(h) \to 0$ as $h \to \infty$. As $t_0 \to \infty$ for fixed t_1 and t_2, show that under (a) and (b), $\hat{u}(t_0) \to \bar{x}$.

(d) Let $t_0 = t_1$ (one of the data sites), so that $\alpha_1 = 1$, $\alpha_2 = \rho$. Show that under (a)

$$\hat{u}(t_0) = u_1,$$

and under (b)

$$\hat{u}(t_0) = x_1 - \frac{\psi}{1 + \psi - \rho} d_x \neq x_1.$$

In the second case, note that the predictor at the data site t_1 is not equal to the observation x_1 if $\psi > 0$.

7.4 Let A be a symmetric positive definite $n \times n$ matrix and let $\boldsymbol{b}$ be an n-vector. Consider the minimization problem

$$\text{minimize } \boldsymbol{x}^T A \boldsymbol{x} \text{ such that } \boldsymbol{b}^T \boldsymbol{x} = 1,$$

over $\boldsymbol{x} \in \mathbb{R}^n$. Show that the solution is given by

$$\boldsymbol{x} = A^{-1}\boldsymbol{b}/(\boldsymbol{b}^T A^{-1} \boldsymbol{b}).$$

Hint: Using a Lagrange multiplier λ, minimize the unconstrained objective function $\boldsymbol{x}^T A \boldsymbol{x} + 2\lambda(1 - \boldsymbol{b}^T \boldsymbol{x})$ over $\boldsymbol{x} \in \mathbb{R}^n$ to get $\boldsymbol{x} = \lambda A^{-1}\boldsymbol{b}$, and show that the constraint is satisfied if $\lambda = 1/(\boldsymbol{b}^T A^{-1} \boldsymbol{b})$.

7.5 Consider the linear intrinsic covariance function $\sigma_I(h) = -|h|$, $h \in \mathbb{R}$ in one dimension. Consider equally spaced sites $t_i = i$, $i = 1, \ldots, n$. Hence, Σ_I

is an $n \times n$ matrix with elements $\sigma_{I;ij} = -|i - j|$ and the elements of $F(n \times 1)$ are all equal to 1.

Show that the matrices B and A in (7.39)–(7.40) take the following forms:

- B is an $n \times n$ matrix; the first two elements of the first row are 0.5, –0.5; the last two elements of the last row are –0.5, 0.5; all other rows contain the elements –0.5, 1, –0.5, straddling the main diagonal; all the remaining elements are 0.
- A is an $n \times 1$ matrix; the first and last elements are 0.5; the remaining elements are 0.

For example if $n = 4$,

$$B = \frac{1}{2}\begin{bmatrix} 1 & -1 & 0 & 0 \\ -1 & 2 & -1 & 0 \\ 0 & -1 & 2 & -1 \\ 0 & 0 & -1 & 1 \end{bmatrix}, \quad A = \frac{1}{2}\begin{bmatrix} 1 \\ 0 \\ 0 \\ 1 \end{bmatrix}.$$

Further, show that the kriging predictor is given by the piecewise linear interpolator

$$\hat{u}(t_0) = \begin{cases} x_1, & t_0 < 1, \\ x_i + (t_0 - t_i)(x_{i+1} - x_i), & i \le t_0 \le i + 1, \\ x_n, & t_0 \ge n. \end{cases}$$

and that the kriging variance is given by

$$\sigma_K^2(t_0) = \begin{cases} 2(i + 1 - t_0)(t_0 - i), & i \le t_0 \le i + 1, \\ 2(1 - t_0), & t_0 < 1, \\ 2(t_0 - n), & t_0 > n. \end{cases}$$

Note the kriging variance is quadratic in t_0 between the data sites and linear outside them.

Hint: For the first part, work with the bordered matrix M in Section 7.6.4. The representation for the M^{-1} in (7.46) depends on matrices A, B, C. Use the choices for B and A given here and let $C = \frac{n-1}{2}$. Confirm these choices are correct by showing that $MM^{-1} = I_{n+1}$. In passing note that B can be identified with the folded circulant approximation (6.25) to the inverse covariance matrix for an AR(1) process with autoregression parameter λ as $\lambda \to 1$.

For the second part, note that B can be written as a sum $B = \sum_{i=1}^{n-1} B_i$, where B_i has nonzero entries

$$\frac{1}{2}\begin{bmatrix} 1 & -1 \\ -1 & 1 \end{bmatrix} = \frac{1}{2}\begin{bmatrix} 1 \\ -1 \end{bmatrix}\begin{bmatrix} 1 & -1 \end{bmatrix}$$

for rows $i, i+1$ and columns $i, i+1$. Suppose $i_0 \le t_0 \le i_0 + 1$. Evaluate $\sum \boldsymbol{x}^T B_i \boldsymbol{\sigma}_0$ one term at a time, and note that the answer depends on whether $i < i_0$, $i = i_0$ or $i > i_0$. Also note that $\boldsymbol{x}^T A \boldsymbol{f}_0 = (x_1 + x_n)/2$.

For the kriging variance, expand out $\sum \boldsymbol{\sigma}_0^T B_i \boldsymbol{\sigma}_0$ one term at a time.

7.6 Exercise 7.5 can also be extended to unequally spaced time points $t_1 < \cdots < t_n$. Show that in this case B is a tri-diagonal matrix with diagonal elements

$$b_{ii} = \begin{cases} \frac{1}{2(t_2-t_1)}, & i = 1, \\ \frac{1}{2(t_i-t_{i-1})} + \frac{1}{2(t_{i+1}-t_i)}, & i = 2, \ldots, n-1, \\ \frac{1}{2(t_n-t_{n-1})}, & i = n, \end{cases}$$

and with super- and sub-diagonal elements

$$b_{i,i+1} = b_{i+1,i} = -\frac{1}{2(t_{i+1} - t_i)}, \quad i = 1, \ldots, n-1.$$

Further, $C = \frac{1}{2}(t_n - t_1)$.
Hence, deduce that the kriging predictor becomes the piecewise linear interpolator

$$\hat{u}(t_0) = \begin{cases} x_1, & t_0 < t_1, \\ x_i + \frac{t_0-t_i}{t_{i+1}-t_i}(x_{i+1} - x_i), & t_i \le t_0 \le t_{i+1}, \\ x_n, & t_0 \ge t_n. \end{cases}$$

Further, show that the kriging variance is given by

$$\sigma_K^2(t_0) = \begin{cases} 2(t_1 - t_0), & t_0 < t_1, \\ 2(t_{i+1} - t_0)(t_0 - t_i)/(t_{i+1} - t_i), & t_i \le t_0 \le t_{i+1}, \\ 2(t_0 - t_n), & t_0 > t_n. \end{cases}$$

Note the kriging variance is quadratic in t between the data sites and linear outside them.
Hint: Adapt the proof of Exercise 7.5.

7.7 Verify the formulas for Bayesian kriging prediction in Section 7.12. In particular, using the Woodbury formula for the inverse of a matrix

$$(\Omega + F\Delta F^T)^{-1} = \Omega^{-1} - \Omega^{-1}FD^{-1}F^T\Omega^{-1}, \quad \text{where } D = \Delta^{-1} + F^T\Omega^{-1}F$$

(see Section A.3.6), show that the formula for γ_Δ in the posterior mean (7.69) simplifies to (7.72). You may find it helpful to expand out both formulas and match the terms. In addition, rewrite the equation for D in (7.71) by multiplying on the left by D^{-1}, on the right by Δ, and rearranging the terms to get $\Delta = D^{-1} + D^{-1}F^T\Omega^{-1}F\Delta$. Similarly, show that the formula for the posterior variance in the first line of (7.73) simplifies to the second line.

Finally, confirm that as the prior variance matrix Δ gets large, the posterior mean and Bayesian kriging variance converge to the universal kriging predictor and its variance in Section 7.5.

Similarly, show that the posterior mean and variance of $\boldsymbol{\beta}$ converge to the generalized least squares estimate above (7.33) and its variance $(F^T\Omega^{-1}F)^{-1}$.

7.8 Let $U(t)$ be a stationary process, $t \in \mathbb{R}^2$ with a covariance function $\sigma(h)$ and an unknown mean μ. Consider $n = 3$ sites lying on an equilateral triangle with coordinates

$$\begin{bmatrix} t_1^T \\ t_2^T \\ t_3^T \end{bmatrix} = \begin{bmatrix} -\frac{\sqrt{3}}{2} & -\frac{1}{2} \\ 0 & 1 \\ \frac{\sqrt{3}}{2} & -\frac{1}{2} \end{bmatrix}.$$

The triangle is centered at the origin, and the three sides have length $\sqrt{3}$. Suppose noise-free measurements $\boldsymbol{x} = [x_1\ x_2\ x_3]^T$ of the process are available at these three sites (i.e. no nugget effect).

The covariance vector of the process between the data sites and a new site t_0, and the covariance matrix for the three data sites can be written as

$$\boldsymbol{\sigma}_0 = \begin{bmatrix} \sigma_{0,1} \\ \sigma_{0,2} \\ \sigma_{0,3} \end{bmatrix} = \begin{bmatrix} \sigma(t_0 - t_1) \\ \sigma(t_0 - t_2) \\ \sigma(t_0 - t_3) \end{bmatrix}, \quad \Omega = \Sigma = \begin{bmatrix} b & a & a \\ a & b & a \\ a & a & b \end{bmatrix},$$

where

$$a = \sigma(\sqrt{3}), \ b = \sigma(0).$$

Further,

$$\Sigma^{-1} = c \begin{bmatrix} a+b & -a & -a \\ -a & a+b & -a \\ -a & -a & a+b \end{bmatrix},$$

where $c = 1/\{(b-a)(b+2a)\}$.

Show that $\Sigma^{-1}\mathbf{1} = d\mathbf{1}, \quad d = 1/(b+2a)$, and thus

$$\mathbf{1}^T\Sigma^{-1}\mathbf{1} = 3d, \quad \mathbf{1}^T\Sigma^{-1}\boldsymbol{\sigma}_0 = d\mathbf{1}^T\boldsymbol{\sigma}_0, \quad \boldsymbol{x}^T\Sigma^{-1}\boldsymbol{\sigma}_0 = d\boldsymbol{x}^T\boldsymbol{\sigma}_0 - ac(\mathbf{1}^T\boldsymbol{x})(\mathbf{1}^T\boldsymbol{\sigma}_0).$$

Using the formulas for the kriging predictor and kriging variance in (7.21) and (7.22), show that

$$\hat{u}(t_0) = \bar{x} + \frac{\sum_{i=1}^3 \sigma_{0,i}x_i - (\sum_{i=1}^3 \sigma_{0,i})(\sum_{i=1}^3 x_i)/3}{b-a}$$

and

$$\sigma_K^2(t) = b - \frac{(2a+b)\sum_{i=1}^3 \sigma_{0,i}^2 - a(\sum_{i=1}^3 \sigma_{0,i})^2}{(2a+b)(b-a)} + \frac{(2a+b-\sum_{i=1}^3 \sigma_{0,i})^2}{3(2a+b)}.$$

Further, for the center of the triangle, $t_0 = (0,0)^T$, show that

$$\hat{u}(t_0) = \bar{x}, \quad \sigma_K^2(t) = \frac{4}{3}\sigma(0) + \frac{2}{3}\sigma(\sqrt{3}) - 2\sigma(1).$$

7.9 The ordinary kriging predictor and kriging variance for a stationary random field in Theorem 7.1 have been expressed in terms of the covariance function $\sigma(h)$. These formulas can also be expressed in terms of the semivariogram $\gamma(h) = \sigma^2 - \sigma(h)$, where $\sigma^2 = \sigma(0)$. For simplicity, suppose that the nugget variance vanishes, $\tau^2 = 0$, so that in the notation of the theorem, the matrices $\Omega = \Sigma$ are the same. In a natural notation, define the quantities

$$\Gamma = \sigma^2 \mathbf{1}\mathbf{1}^T - \Sigma, \quad \boldsymbol{\gamma}_0 = \sigma^2 \mathbf{1} - \boldsymbol{\sigma}_0. \tag{7.93}$$

It is helpful to use the following version of the Woodbury formula for a matrix inverse (see Section A.3.6),

$$(A - c\boldsymbol{u}\boldsymbol{v}^T)^{-1} = A^{-1} + c\frac{A^{-1}\boldsymbol{u}\boldsymbol{v}^T A^{-1}}{1 - c\boldsymbol{v}^T A^{-1}\boldsymbol{u}}.$$

Assuming the matrix Γ is nonsingular, show that the kriging predictor and kriging variance can be written in the following forms:

$$\hat{u}(t_0) = \boldsymbol{\gamma}_0^T \Gamma^{-1}\boldsymbol{x} + \left(\frac{1 - \mathbf{1}^T\Gamma^{-1}\boldsymbol{\gamma}_0}{\mathbf{1}^T\Gamma^{-1}\mathbf{1}}\right)\mathbf{1}^T\Gamma^{-1}\boldsymbol{x}, \tag{7.94}$$

$$\sigma_K^2(t_0) = \boldsymbol{\gamma}_0^T\Gamma^{-1}\boldsymbol{\gamma}_0 - \frac{(1 - \mathbf{1}^T\Gamma^{-1}\boldsymbol{\gamma}_0)^2}{\mathbf{1}^T\Gamma^{-1}\mathbf{1}}. \tag{7.95}$$

Hint. Using the shorthand notation

$$\boldsymbol{a} = \Sigma^{-1}\mathbf{1}, \quad \boldsymbol{b} = \Sigma^{-1}\boldsymbol{\sigma}_0,$$

and

$$\alpha = \mathbf{1}^T\Sigma^{-1}\mathbf{1}, \quad \beta = \mathbf{1}^T\Sigma^{-1}\boldsymbol{\sigma}_0,$$

show that the kriging predictor in (7.21) becomes

$$\hat{u}(t_0) = \left(\boldsymbol{b} + \frac{1-\beta}{\alpha}\boldsymbol{a}\right)^T \boldsymbol{x}.$$

Also show that the Woodbury formula for Γ^{-1} becomes

$$-\Gamma^{-1} = \Sigma^{-1} + \sigma^2\boldsymbol{a}\boldsymbol{a}^T/(1 - \sigma^2\alpha)$$

so that

$$\Gamma^{-1}\boldsymbol{\gamma}_0 = \boldsymbol{b} + \frac{\sigma^2(\beta - 1)}{1 - \sigma^2\alpha}\boldsymbol{a}, \quad -\Gamma^{-1}\mathbf{1} = \frac{1}{1 - \sigma^2\alpha}\boldsymbol{a}$$

and

$$\mathbf{1}^T\Gamma^{-1}\boldsymbol{\gamma}_0 = \frac{\beta - \sigma^2\alpha}{1 - \sigma^2\alpha}, \quad -\mathbf{1}^T\Gamma^{-1}\mathbf{1} = \frac{\alpha}{1 - \sigma^2\alpha}.$$

Substitute these results into the kriging predictor (7.94) and show that it reduces to (7.21). A similar calculation can be used to show that (7.95) is the same as (7.22).

Caution 1. The matrix $-\Gamma$ is not necessarily nonsingular. It is only guaranteed to be conditionally positive definite. If the semivariogram is to be used to construct the kriging predictor, it is better to use the transfer matrices than (7.94) and (7.95). Replace $\Sigma = \Omega$ by $-\Gamma$ in Eqs. (7.39) and (7.40), and replace $\boldsymbol{\sigma}_0$ by $-\boldsymbol{\gamma}_0$ in (7.38).

Caution 2. Do not confuse the notation for the vector of semivariogram values $\boldsymbol{\gamma}^0$ with the use of $\boldsymbol{\gamma}$ for the kriging coefficient vector in (7.21).

8

Additional Topics

8.1 Introduction

This chapter gives a brief discussion of various topics that it has not been possible to cover in detail in the book. The basic initial model in the book has been a stationary Gaussian random field involving a *response variable* $X(t)$ indexed by a *site* t in a *domain* D, where typically $D = \mathbb{R}^d$ or $D = \mathbb{Z}^d$. Observations $x_1, \ldots, x_n$ are available at sites $t_1, \ldots, t_n$ and are assumed to come from this model. The key statistical challenges include the estimation of any unknown parameters and the prediction of the random field at new sites (kriging).

This basic framework can be generalized in various directions.

- *Gaussianity.* The simplest way to construct a non-Gaussian random field is to transform a Gaussian random field. Section 8.2 looks at log-normal random fields.
- *Latent random fields.* Suppose noisy observations are made of a random field; i.e. the random field $X(t)$ is not directly observed. The introduction of a nugget effect in Chapter 5 is a step in this direction, where the observations are still normally distributed. It is also possible to construct generalized linear models where the response variable follows, e.g., a binomial or Poisson distribution (Section 8.3).
- *Bayesian inference.* Most of the book has emphasized the use of frequentist ideas, especially maximum likelihood, to estimate any unknown parameters. However, it can be argued that a better statistical approach is to treat any unknown parameters as coming from a prior distribution. The Bayesian approach can also easily accommodate latent random fields. However, the price is an increased numerical complexity. A brief discussion is given in Section 8.4.
- *Multivariate random fields.* The response variable $X(t)$ can be vector-valued instead of real-valued. Many of the standard ideas carry over with little change. In addition, there is a concept of *co-kriging* (Section 8.5).

Spatial Analysis, First Edition. John T. Kent and Kanti V. Mardia.

- *Spatial–temporal models.* Another change is to increase the dimension of the domain to include time as well as space. Of course, time should not usually be treated as just another spatial dimension, so new modeling strategies are needed. See Section 8.6. There is also a brief discussion in Section 4.5.2.
- *Constrained kriging.* As discussed in Chapter 7, kriging can be used in the construction of a deformation of a continuous domain D. A modified version of kriging, called clamped-plate kriging, can be used to describe a deformation that is constrained to be fixed on the boundary of D when D is the interior of a disk. See Section 8.7.
- *Irregular lattice domains.* The discussion in Chapters 4 and 6 emphasized lattice random fields defined on $\mathbb{Z}^d$. However, in many applications it is beneficial to extend the lattice methods of Chapters 4 and 6 to irregularly spaced sites. Lindgren et al. (2011) have developed a powerful methodology that extends much of the tractability of condiional autoregression (CAR) and simultaneous autoregression (SAR) models to this setting (Section 8.8).
- *Selection of new sites.* Throughout the book, it has been assumed that the location of the sites is fixed. Section 8.9 considers the situation where an experimenter can add a new site to an existing network. The objective is to maximize the "information" in some sense provided by the new site.

8.2 Log-normal Random Fields

In most part of the book, we have made the assumption of a Gaussian random field. This is a powerful modeling strategy, both for its theoretical and practical importance. However, there are also situations involving nonnormal data. In this section and Section 8.3, we discuss two modeling strategies to deal with nonnormal data by relating the data to observations from a Gaussian process. These strategies are based on transformations and latent processes, respectively.

In this section, we look at nonlinear pointwise transformations of a random field. For positive random fields ($X(t) > 0$ for all t), a popular class is given by the Box–Cox transformations

$$Y(t) = (X(t)^{\lambda} - 1)/\lambda,$$

where λ is a real-valued parameter. When $\lambda < 1$ the effect of the transformation is to compress the right-hand tail of the distribution of $X(t)$ relative to the left-hand tail. The hope is that for some value of λ, the process $\{Y(t)\}$ will be approximately a Gaussian random process. For $\lambda \neq 0$, $Y(t)$ can lie in only a restricted part of the real line so the transformation to normality can only be approximate. The Box–Cox transformation for general λ in spatial applications was studied in Mardia and Goodall (1993).

The most important special case is the limiting case $\lambda \to 0$, for which the Box–Cox transformation becomes

$$Y(t) = \log X(t).$$

In this case, $Y(t)$ ranges through all the real numbers if $X(t)$ ranges through all the positive numbers, and the transformation to normality can be exact. If the transformed process $\{Y(t)\}$ is a Gaussian random field, then the original process $\{X(t)\}$ is called a "log-normal random field."

To work out the mean and covariance function of the original process, let us first we recall some background about the log-normal distribution. If $Y \sim N(\mu, \sigma^2)$ is normally distributed, then $X = \exp(Y)$ is said to follow a log-normal distribution. The first two moments of X, which can be computed simply from the moment generating function of Y, are given by

$$E(X) = \exp\left(\mu + \frac{1}{2}\sigma^2\right), \quad E(X^2) = \exp(2\mu + 2\sigma^2),$$
$$\operatorname{var}(X) = \exp(2\mu + \sigma^2)\left\{\exp(\sigma^2) - 1\right\}. \tag{8.1}$$

Similarly, if Y_1 and Y_2 are bivariate normal with means μ_1, μ_2, variances σ_1^2, σ_2^2 and covariance σ_{12}, then

$$\begin{aligned} E(X_1X_2) &= \exp\left\{\mu_1 + \mu_2 + \frac{1}{2}(\sigma_1^2 + \sigma_2^2 + 2\sigma_{12})\right\}, \\ \operatorname{cov}(X_1, X_2) &= \exp\left\{\mu_1 + \mu_2 + \frac{1}{2}(\sigma_1^2 + \sigma_2^2)\right\}\left\{\exp\left(\sigma_{12}\right) - 1\right\}. \end{aligned} \tag{8.2}$$

See Exercise 8.1. This construction extends in a straightforward way to the multivariate log-normal distribution and to log-normal stochastic processes.

The log-normal process is popular in mining applications (Dowd, 1982), where $X(t)$ represents the ore concentration. One complication is that integrals of interest (e.g. total mineral concentration in a block of ore) take place in the X domain; however, sums and integrals of log-normal random variables no longer have log-normal distributions. Dowd worked out a variety of numerical approximations to facilitate inferences in this setting. Some basics of block kriging in the normal case were discussed in Sections 3.12 and 7.11.

8.3 Generalized Linear Spatial Mixed Models (GLSMMs)

Start with an underlying latent Gaussian spatial process $S(t)$. The adjective "latent" means that this process represents an underlying reality that cannot be observed directly. Instead, noisy measurements are made, which depend on the latent process. More specifically, we make the following assumptions:

(a) $\{S(t)\}$ is a stationary Gaussian process on $t \in \mathbb{R}^d$ with drift function $E(S(t)) = \mu(t)$ and covariance function $\sigma(h)$.
(b) At sites $t_1, \ldots, t_n$, observations Z_i are made from a generalized linear model, where the mean of Z_i depends on $S(t_i)$ through a link function. Further, conditional on the latent process $\{S(t)\}$, the observations at different sites are independent of one another.

To some extent, this model is already familiar. If $Z_i|\{S(t)\} \sim N(S(t_i), \tau^2)$, then τ^2 represents a nugget effect in the usual stationary Gaussian model with drift.

The drift function in $S(t)$ can be thought of as the "fixed effects" and the autocovariance function determines the "random effects" in terms of the observations. Hence, we call these models "generalized linear spatial mixed models (GLSMMs)." Other descriptions in the literature include generalized linear geostatistical models (Diggle and Ribeiro, 2007) and hierarchical models (Gelfand and Ghosh, 2013).

The most tractable example is the Poisson distribution with a log link function, $Z_i|\{S(t)\} \sim P(\lambda_i)$, where $\log \lambda_i = S(t_i)$, proposed in Diggle et al. (1998). In this case, we can work out the first two marginal moments of the observations explicitly using the results from Section 8.2; see Exercises 8.2–8.3.

However, inference for unknown parameters in this Poisson model cannot be carried out explicitly. A popular way to proceed is to use Bayesian methods based on Markov chain Monte Carlo (MCMC) calculations. Section 8.4 summarizes the details.

8.4 Bayesian Hierarchical Modeling and Inference

By giving prior distributions to the underlying parameters, the latent model of the last section can be described in a hierarchical manner as follows. We follow the description in Section 7.12 and in Gelfand and Ghosh (2013, p. 41).

Level 1 (Observation level). Observations Z_i are taken at sites t_i, $i = 1, \ldots, n$. Given an underlying latent process $S(t)$, the observations are assumed to be conditionally independent, with the conditional distribution of Z_i depending on $S(t_i)$ and perhaps an additional parameter τ^2. In the Poisson model, $Z_i \sim P(\exp\{S(t_i)\})$ and τ^2 is not present; in the normal theory example, $Z_i \sim N(S(t_i), \tau^2)$ and τ^2 represents a nugget effect.

Level 2 (Latent process level). The latent process $\{S(t)\}$ is assumed to follow a Gaussian process with mean $\mu(t) = \boldsymbol{\beta}^T \boldsymbol{f}(t)$ in terms of a known set of functions $f_1(t), \ldots, f_q(t)$ and with stationary covariance function $\sigma(h; \boldsymbol{\theta}) = \sigma^2 \rho(h, \boldsymbol{\theta}_c)$.

Here, $\boldsymbol{\theta} = (\sigma^2, \boldsymbol{\theta}_c)$ is a parameter vector for a stationary covariance function including a scale parameter σ^2 together with some further parameters for the covariance structure $\boldsymbol{\theta}_c$ ($\boldsymbol{\theta}_c$ typically includes a range parameter and possibly an index parameter for smoothness).

Level 3 (Hyperparameter level). A prior distribution is specified for the parameters $\boldsymbol{\beta}, \boldsymbol{\theta}, \tau^2$, typically with the three sets of parameters modeled independently.

Write $\boldsymbol{z} = [z_1, \ldots, z_n]^T$ for possible values of the response variables at the data sites, and $\boldsymbol{s} = [s_1, \ldots, s_n]^T$ for possible values of the latent process. Then the joint density of all the random variables and parameters can be written in the form

$$f(\boldsymbol{z}, \boldsymbol{s}, \boldsymbol{\beta}, \boldsymbol{\theta}, \tau^2) = f(\boldsymbol{z}|\boldsymbol{s}, \tau^2) f(\boldsymbol{s}|\boldsymbol{\beta}, \boldsymbol{\theta}) f(\boldsymbol{\beta}) f(\boldsymbol{\theta}) f(\tau^2).$$

The objective of Bayesian analysis is to find the posterior distribution of the parameters $\boldsymbol{\beta}, \boldsymbol{\theta}, \tau^2$ given the observations $\boldsymbol{z}$.

There is also the related problem of predictive Bayesian analysis, that is, to predict the value of the latent process $S(t_0)$ at a new site t_0.

This model can also be regarded as a "spatial mixed model." In the drift function $\mu(t) = \boldsymbol{\beta}^T \boldsymbol{f}(t)$, the parameter $\boldsymbol{\beta}$ represents the "fixed effects" in the model and $\varepsilon(t) = S(t) - \mu(t)$ represents the "random effects."

Unfortunately, it is not generally feasible to compute the posterior distribution for either $\boldsymbol{s}$ or $(\boldsymbol{\theta}, \boldsymbol{\beta}, \tau^2)$ analytically, and it is necessary to turn to numerical methods. One common method is to use MCMC methods to simulate a Markov chain whose equilibrium distribution is given by the desired posterior distribution. Similar comments apply to the predictive distribution of $S(t_0)$ given $\boldsymbol{z}$.

MCMC methods are not covered at all in this book; we refer to Banerjee et al. (2015) and Gelfand et al. (2010) for comprehensive treatments.

However, it is important to note that the choice of prior distribution for $\boldsymbol{\theta}$ can be delicate. We saw in Chapter 5 that it can be difficult to simultaneously estimate the scale parameter σ^2 and the range parameter φ in a Matérn model. Hence, it is important to ensure that the posterior distribution does not merely reproduce the prior in this case. Further, there are issues related to improper priors leading to improper posterior distributions in some cases. The question of appropriate priors for $\boldsymbol{\theta}$ is still an open topic of research (Berger et al., 2001; Gelfand and Ghosh, 2013; Handcock and Stein, 1993; Steel and Fuentes, 2010).

8.5 Co-kriging

So far we have limited our attention to a single real-valued measurement x_t at each site $t \in \mathbb{R}^d$. A natural extension is to allow a vector of measurements $\boldsymbol{x}(t) \in \mathbb{R}^q$, $q \geq 2$, with elements $x_u(t), u = 1, \ldots, q$ at each site. There is a rich literature on statistical methods to tackle such data, especially in environmental applications. References include Journel and Huijbregts (1978), Mardia and Goodall (1993), Diggle and Giorgi (2019), Huang et al. (2009), Wackernagel (2003), and Banerjee and Gelfand (2010). For this section, we limit our attention to some mathematical details.

Then more notation is needed to describe the drift and covariance functions. For the drift we adopt the same framework as before, except that the choice of drift functions may depend on u. Given a collection of p_u-dimensional set of basis functions $\{f_{1u}(t), \dots, f_{p_u,u}(t)\}$, we model the expected value of the uth element of a multivariate random field by a linear combination,

$$E(x_u(t)) = \sum_{j=1}^{p_u} \beta_{ju} f_{ju}(t) = \mu_u(t), \text{ say,}$$

where β_{ju} are parameters to estimate. Write $p^* = \sum p_u$.

The covariance function is now a matrix-valued function. Under stationarity, it takes the form

$$\text{cov}(x_u(t), x_v(t+h)) = \sigma_{uv}(h), \quad u, v = 1, \dots, q, \quad h \in \mathbb{R}^d. \tag{8.3}$$

The covariance function possesses the symmetry property

$$\sigma_{uv}(h) = \sigma_{vu}(-h),$$

but does not always possess the full symmetry property,

$$\sigma_{uv}(h) = \sigma_{vu}(h).$$

Formally, estimation and prediction are carried out similarly to the scalar random field case, though the notation and calculations become more cumbersome. The most convenient strategy is to combine a matrix of observations into a long vector. In a general framework, the sites at which observations are made may depend on the index u. Thus, suppose observations $x_u(t_{iu})$, $i = 1, \dots, n_u$ are available, for $u = 1, \dots, q$. Define

$$\boldsymbol{x}^* = \left[x_1(t_{11}), \dots, x_1(t_{n_1,1}), x_2(t_{12}), \dots, x_2(t_{n_2,2}), \dots, x_q(t_{1q}), \dots, x_q(t_{n_q,q})\right]^T$$

by stacking the observation vectors for the scalar random fields on top of one another to give a vector of length $n^* = \sum_{u=1}^q n_u$, say. Let Σ^{**} denote the corresponding $nq \times nq$ covariance matrix and let the expected value be denoted

$$\boldsymbol{\mu}^* = F^* \boldsymbol{\beta}^*,$$

where F^* is an $n^* \times p^*$ block diagonal matrix of the form

$$F^* = \begin{bmatrix} F_1 & & 0 \\ & \ddots & \\ 0 & & F_q \end{bmatrix},$$

where for each $u = 1, \ldots, q$, the $(n_u \times p_u)$ matrix F_u has entries $f_{ju}(t_{iu})$ in the ith row and jth column, and

$$\beta^* = [\beta_{11}, \ldots, \beta_{p_1,1}, \ldots, \beta_{p_q,q}]^T.$$

Then, up to a constant term, the log-likelihood can be written as

$$-\frac{1}{2}\mathrm{tr}\{(x^* - \mu^*)^T(\Sigma^{**})^{-1}(x^* - \mu^*)\} + \log|\Sigma^{**}|.$$

For prediction, we add a new site t_0 and denote the vector of observations at this new site by $x^\circ = [x_1(t_0), \ldots, x_q(t_0)]^T$ with mean value $\mu^\circ = [\mu_1(t_0), \ldots, \mu_q(t_0)]^T = F^\circ\beta^*$,

$$F^\circ = \left[f_{11}(t_0), \ldots, f_{p_1,1}(t_0), \ldots, f_{p_q,q}(t_0)\right].$$

Similarly, define the covariance matrices

$$\mathrm{cov}(x^*, x^\circ) = \Sigma^{*\circ}, \quad \mathrm{cov}(x^\circ, x^\circ) = \Sigma^{\circ\circ}.$$

Then the kriging predictor (with known coefficients (β_{ju})) takes the form

$$\hat{x}^\circ = \mu^\circ + \Sigma^{\circ *}(\Sigma^{**})^{-1}(x^* - \mu^*).$$

In the case of unknown (β_{ju}), they can be replaced by the generalized least squares (GLS) estimator

$$\hat{\beta}^* = \{F^{*T}(\Sigma^{**})^{-1}F^*\}^{-1}F^{*T}(\Sigma^{**})^{-1}x^*.$$

The GLS estimator also appears in estimation for the spatial linear model (5.48) and in universal kriging for a single random field (7.33). Some comments will be helpful to set this general result in context. For simplicity, we focus on the case $q = 2$.

(a) One variable ($u = 1$) may be the variable of interest and be difficult to measure; the other variable ($u = 2$) may be an auxiliary or concomitant variable and be easy to measure. In particular, the second variable may be measured much more densely than the first. It may even be the case that the new site t_0 coincides with one of the observation sites for the second variable. The purpose of co-kriging is to "borrow strength" from the measurements on the second variable to improve the accuracy when predicting the first variable. See Exercise 8.5.

(b) It may be the case that both variables are measured at the same sites with the same choice of drift functions. Although this framework is notationally more straightforward, there is no special simplification to the algebra and interpretation except in the particular case of a tensor product model.

(c) When both variables are measured at the same sites with the same choice of drift functions, the presentation of the data and model can be simplified.

Define a data matrix $X(n \times p)$ with $\boldsymbol{x}^* = \text{vec}(X)$, and write the drift matrix in the form

$$F^* = I \otimes F.$$

Make the additional assumption that the joint covariance function factorizes into the tensor product form

$$\text{cov}(x_u(t), x_v(t+h)) = a_{uv}\sigma(h), \quad u, v = 1, \ldots, q, \quad h \in \mathbb{R}^d, \tag{8.4}$$

where the $q \times q$ matrix A represents the covariance between variables, and the common spatial covariance function $\sigma(h)$ represents the covariance in space. Then the kriging estimator simplifies to what it would be if each variable were predicted on its own; see, for example, Mardia and Goodall (1993). That is, there is no strength to be borrowed from neighboring observations in this case.

(d) A simple way to construct a multiple covariance function that does not have the tensor product property is through linear combinations of independent processes. Here is an example for a bivariate process.
Let $x_1(t)$ and $x_2(t)$ be independent zero-mean stationary processes with different covariance functions $\sigma_1(h)$ and $\sigma_2(h)$, respectively. Set $y_1(t) = x_1(t) + x_2(t)$, $y_2(t) = x_1(t) - x_2(t)$. Then, both $y_1(t)$ and $y_2(t)$ have the same covariance function $\sigma_1(h) + \sigma_2(h)$, and the same cross-covariance functions

$$E\{y_1(t)y_2(t+h)\} = E\{y_2(t)y_1(t+h)\} = \sigma_1(h) - \sigma_2(h),$$

but the cross-covariance function $\sigma_1(h) - \sigma_2(h)$ is not a constant multiple of the marginal covariance functions $\sigma_1(h) + \sigma_2(h)$. Hence, the covariance structure for the $\{y_1(t), y_2(t)\}$ does not have a tensor product form.

(e) A more general version of this construction is given as follows. For $k \geq 2$, let $A^{(1)}, \ldots, A^{(k)}$ be $q \times q$ positive definite matrices and let $\sigma_1(h), \ldots, \sigma_k(h)$, $h \in \mathbb{R}^d$, be distinct stationary spatial covariance functions. Then

$$\sigma_{uv}(h) = \sum_{i=1}^{k} a_{uv}^{(i)} \sigma_i(h), \quad u, v = 1, \ldots, q,$$

defines a valid multivariate spatial covariance function, which is not separable (Journel and Huijbregts, 1978, p. 172).

(f) It is also possible to extend the theory to the case where one or both variables follow intrinsic processes rather than stationary processes. If both processes are IRF-0, a popular summary measure is the cross-variogram

$$E\{(Y_1(t+h) - Y_1(t))(Y_2(t+h) - Y_2(t))\}.$$

However, this measure captures only some of the dependence structure between the two processes. A richer set of expected products of increments is needed to fully capture the dependence (Huang et al., 2009).

8.6 Spatial–temporal Models

8.6.1 General Considerations

In some ways, a spatial–temporal model can be viewed as a multivariate spatial process where the number of variables equals the number of times at which the process is observed. However, there is often structure across time that can be modeled explicitly. Here, we describe several possible strategies, following Kent and Mardia (2002).

A key property of much spatial–temporal data is spatial–temporal continuity; that is, observations at nearby sites and times will tend to be similar to each other. This underlying smoothness of a process $X(s, t)$ at a spatial site s and a temporal site t can be captured in the following ways:

- *Parametrically*, using a finite-dimensional space of regression or drift functions or
- *Nonparametrically*, using autocorrelation to make nearby values similar.

Both of these approaches can be applied in space and/or time. Letting D and C stand for a parametric "drift" function and a nonparametric "correlation" approach, respectively, the following types of models can be considered:

(a) *D–D*. Tensor products of drift in space and drift in time. This approach is explored in Section 3. Such models are appropriate for highly structured data. For example, Kent et al. (2001) modeled the changing shapes of cross sections of rat skulls as they grow, using principal splines in space and in time. Principal splines can be considered as analogous to low-order polynomials, but their exact structure depends on the layout of the spatial (or temporal) sites.
(b) *D–C*. Drift in space and correlation in time. The kriged Kalman filter model (Fontanella et al. (2005); Mardia et al., 1998; Sahu and Mardia, 2005; Sahu et al. (2005); Wikle and Cressie, 1999) exemplifies this approach. The drift functions are built from principal splines in space, but the their coefficients evolve according to time-series models.
(c) *C–D*. Correlation in space and drift in time.
(d) *C–C*. Joint correlation in space and time. The simplest examples are separable (i.e. tensor products) in space and time, but these are generally not very realistic; there is a need for space and time to interact with each other. Models with a stochastic motivation include space–time autoregressive and related models (e.g. Cressie and Wikle, 2011) in discrete space–time and the "diffusion–injection" model (Whittle, 1986, pp. 430–433) in continuous space–time. There are also explicit covariance functions (Gneiting, 2002; Gneiting and Guttorp, 2010), though some of the models can exhibit non-intuitive "dimple" effects (Kent et al., 2011). The covariance-spectral model

of Stein (2005) is specified in terms of a covariance function in space and a spectral density in time; it is not fully symmetric. Gneiting et al. (2007) review many of the key issues. Mardia and Goodall (1993) and Goodall and Mardia (1994) developed methodology for multivariate spatial-temporal processes.

The density of data points in space and time can guide the choice of modeling strategy. When the data are sparse, there is often a preference for drift-style models, as there is not enough information to fit an autocorrelation structure. Of course, a disadvantage of regression models is that the class of fitted curves and surfaces can be rather inflexible, especially for prediction and extrapolation.

On the other hand, when the data are dense, covariance functions become more feasible and flexible. In particular, they allow for more adaptive prediction and extrapolation. As noted in Section 7.14, there is a close link between the use of covariance models and the use of splines to fit curves and surfaces to discrete data; see also, Kent and Mardia (1994), Wahba (1990, Ch. 3), and Cressie (1993, pp. 180–183). Thus, the use of covariance models has a nonparametric flavor to it.

Statistical modeling for spatial–temporal data continues to be a major research theme. Some recent books dedicated to this theme include Banerjee et al. (2015), Cressie and Wikle (2011), Finkenstädt et al. (2007), Gelfand et al. (2010), Lawson and Denison (2002), Sahu (2022) and Wikle et al. (2019).

8.6.2 Examples

(a) *Separable models.* In this case, the space–time covariance function factors into a product of a spatial covariance function $\sigma_S(h)$ and a temporal covariance function $\sigma_T(u)$,

$$\sigma(h, u) = \sigma_S(h)\sigma_T(u), \quad h \in \mathbb{R}^d, \ u \in \mathbb{R}.$$

(b) *Fully symmetric models.* All space–time covariance functions satisfy the simple symmetry property

$$\sigma(h, u) = \sigma(-h, -u).$$

That is, the covariance function is unchanged if both h and u change sign. Some covariance functions satisfy the more restrictive "full symmetry" property

$$\sigma(h, u) = \sigma(h, -u) = \sigma(-h, u) = \sigma(-h, -u).$$

That is, the covariance function is unchanged if either h or u changes sign. Next, we give some examples of nonseparable models.

(c) *Gneiting's model.* Gneiting (2002) developed a class of nonseparable fully symmetric covariance functions. One of the simplest examples in this class is

$$\sigma_G(h, u) = \frac{\sigma^2}{(1 + |h|^2)^{1/2}} \exp\left(-\frac{u^2}{1 + |h|^2)}\right), \quad (h, u) \in \mathbb{R}^d \times \mathbb{R} \tag{8.5}$$

(though he originally reversed the roles of space and time). Curiously, Gneiting's models can sometimes include a counterintuitive "dimple." That is, for a large enough choice of $u > 0$, a plot of σ_G vs. $|h| \geq 0$ can decrease, and then increase, before finally decreasing to 0 as $|h| \to \infty$ (Kent et al., 2011).

(d) *Phase-shift models.* In some sense, the opposite of full symmetry is a phase-shift model. Here are three possible versions.

(i) Consider a stationary time series $Z_T(t)$ with covariance $C_T(u)$. Let $v \in \mathbb{R}^d$ be a vector defining a "temporal wind," and define a spatial–temporal process with a corresponding covariance function by

$$Z(s,t) = Z_T(t + v^T s), \quad \sigma(h,u) = \sigma_T(u + v^T h). \tag{8.6}$$

Here (s,t) denotes a site in space-time, h is a spatial lag. and u is a temporal lag. Changing the spatial coordinate s for $Z(s,t)$ corresponds to a phase shift in time for Z_T. Such a process can be called "temporally phase-shifted."

(ii) Similarly, if $Z_S(s)$, $s \in \mathbb{R}^d$, is a stationary process in space with covariance function $\sigma_S(h)$, $h \in \mathbb{R}^d$, and if $\boldsymbol{\mu} \in \mathbb{R}^d$ is a vector defining a "spatial wind," define a spatial–temporal process with a corresponding covariance function by

$$Z(s,t) = Z_S(s + t\mu), \quad \sigma(h,u) = \sigma_S(h + u\mu). \tag{8.7}$$

Changing the temporal coordinate t for $Z(s,t)$ corresponds to a phase shift in space for Z_S. Such a process can be called "spatially phase-shifted."

(iii) More generally, if $Z_0(s,t)$ is a general stationary spatial–temporal process with covariance function $\sigma_0(h,u)$, then a new process with a phase-shift in space and time can be defined by (Ma, 2003)

$$Z(s,t) = Z_0(s + t\mu, u + v^T s), \quad \sigma(h,u) = \sigma_0(h + u\mu, u + v^T h). \tag{8.8}$$

(e) *Taylor's frozen field hypothesis.* A stationary spatiotemporal process satisfies Taylor's frozen field hypothesis (Taylor, 1938) if the purely temporal and purely spatial covariances are related by

$$\sigma(0,u) = \sigma(u\mu, 0), \quad u \in \mathbb{R}, \tag{8.9}$$

where $\mu \in \mathbb{R}^d$ is a fixed vector. This hypothesis arises in the theory of turbulence. The spatially frozen field always satisfies (8.9). The temporally and jointly phase shifted models, (8.6) and (8.8), satisfy (8.9) provided $v = \mu$ and $v^T v = 1$ is a unit vector. However, fully symmetric models can also satisfy (8.9). See, e.g., Gneiting et al. (2007) for details.

(f) *Stein's model.* Starting from Gneiting's model (8.5), incorporate a phase shift in time to get

$$\sigma_{\text{Stein}}(h,u) = \sigma_G(h, u + v^T h). \tag{8.10}$$

From this starting point, Stein (2005) develops a more flexible, and partly non-parametric, model.

8.7 Clamped Plate Splines

Kriging in Chapter 7 emphasized the use of stationary or intrinsic random fields. In particular, the use of self-similar intrinsic random fields allowed kriging to be reformulated in terms of splines. In this section, we discuss a version of the self-similar random fields (Section 3.10) modified to vanish on the boundary of the unit ball in $\mathbb{R}^d$. The result is an ordinary, but not stationary, random process on the interior of the unit disk. One of the main applications is the construction of deformations constrained to leave the boundary of the unit ball fixed.

Given two distinct spatial sites $s, t \in \mathbb{R}^d$ lying in the unit ball, $|s| \leq 1,\ |t| \leq 1$, $s \neq t$, define a function

$$A(s,t) = \sqrt{\frac{|s|^2|t|^2 - 2s^T t + 1}{|s|^2 - 2s^T t + |t|^2}}. \tag{8.11}$$

It can be shown that $A(s,t) \geq 1$ for all s, t, $A(s,t) = 1$ if $|s| = 1$ or $|t| = 1$, and that $|s - t|A(s,t)$ has a finite positive limit as $s \to t$ for a fixed t satisfying $|t| < 1$.

Given an integer parameter $m > d/2$, the "clamped-plate spline" is defined by Boggio (1905)

$$G_{m,d}(s,t) = k_{m,d}|s - t|^{2m-d} \int_1^{A(s,t)} \frac{(v^2 - 1)^{m-1}}{v^{d-1}} dv \tag{8.12}$$

for suitable constants $k_{m,d}$.

It can be shown that

(a) $G(s,t) = G_{m,d}(s,t)$ is a $(2m - d - 1)$-times differentiable function of s and t; it and its derivatives up to order $m - 1$ tend to 0 as s and t tend to the boundary of the unit ball.

(b) $G(s,t)$ satisfies the differential equation

$$(-\Delta_s)^m G(s,t) = \delta_t(s), \tag{8.13}$$

where Δ_s is the Laplacian operator in s and $\delta_t(s)$ is a Dirac delta function centered at t. That is, $G(s,t)$ is the Green's function for the iterated Laplacian operator on the interior of the unit ball, with vanishing boundary conditions on the boundary of the unit ball.

(c) $G(s,t)$ can be written as an irregular term ($c_{\alpha,d}|s - t|^{2m-d}$ for d odd and $c'_{\alpha,d}|s - t|^{2m-d} \log|s - t|$ for d even), plus an infinitely differentiable function of s and t on the interior of the unit ball. The normalizing constants $c_{\alpha,d}$

and $c'_{\alpha,d}$ are taken from (3.46) and (3.48) for the self-similar processes on $\mathbb{R}^d$, where $\alpha = m - d/2$.

(d) $G(s,t)$ is a positive definite function on the unit ball and hence defines a valid covariance function. The corresponding processes can be viewed as versions of the self-similar processes in Chapter 3 modified to vanish on the boundary of the unit ball.

One of the uses for clamped splines is to construct deformations. Let $\{s_i\}_1^n$ be a set of sites inside the unit ball, which are to be mapped to a second set of sites $\{t_i\}_1^n$ also inside the unit ball. Construct d clamped splines $\Psi_j(s)$ taking the $\{s_i\}_1^n$ to each component $\{t_i[j]\}_1^n$ in turn, $j = 1, \ldots, d$. The clamped splines can be constructed by methods similar to those of Chapter 7, though there is no null space here. Write $\Psi(s) = [\Psi_1(s), \ldots, \Psi_d(s)]^T$ to be the vector-valued mapping. Then the mapping

$$\Phi(s) = \Psi(s)$$

is a mapping from $\mathbb{R}^d$ to $\mathbb{R}^d$, which takes the old sites $\{s_i\}_1^n$ to the new sites $\{t_i\}_1^n$ and which holds the boundary of the unit ball fixed. For more details, see, e.g., Davies et al. (2008).

With $A = A(s,t)$ given by (8.11), the first few clamped splines take the form

$$\begin{aligned}
G_{1,1}(s,t) &\propto |s-t|(A-1) = 1 - st - |s-t|,\\
G_{2,1}(s,t) &\propto |s-t|^3\left(\frac{1}{3}A^3 - A + \frac{2}{3}\right)\\
&= \frac{1}{3}(1-st)^3 - (s-t)^2(1-st) + \frac{2}{3}|s-t|^3,\\
G_{2,2}(s,t) &\propto -|s-t|^2\left\{\frac{1}{2}(A^2-1) - \log A\right\},\\
G_{2,3}(s,t) &\propto |s-t|\left\{A + \frac{1}{A} - 2\right\},\\
G_{3,3}(s,t) &\propto |s-t|^3\left\{A^3 - 6A - \frac{3}{A} + 8\right\}.
\end{aligned}$$

The first two lines correspond to the linear and cubic spline in one dimension, the next line corresponds to the thin-plate spline in two dimensions, and the final line corresponds to what is known as the tri-harmonic spline in three dimensions.

8.8 Gaussian Markov Random Field Approximations

Lindgren et al. (2011) make the point that when modeling data on a continuous domain D, a continuously indexed covariance model should be used. However, for numerical work, a discretization of the domain is needed. Further, numerical computations are more computationally efficient if the discretized model is a Gaussian Markov random field (GMRF); i.e. the inverse covariance matrix is sparse. They propose using finite element approximations for a stochastic partial differential equation (SPDE) to achieve the approximation. This section sketches their main ideas.

(a) On $\mathbb{R}^d$ consider the SPDE

$$D_{\kappa,\alpha} = (\kappa^2 - \Delta)^{\alpha/2} X(t) = \varepsilon(t), \tag{8.14}$$

where $\Delta = \sum_{\ell=1}^{d} \partial^2/\partial t[\ell]^2$ is the Laplacian operator, $\kappa > 0$, $\alpha > 0$, and where $\{\varepsilon(t)\}$ is white noise, a generalized random field. This equation was studied in Section 4.5.2. It was noted that provided α is an even integer and $\alpha > d/2$, this equation has a stationary solution given by a random field with a Matérn covariance function of index $\nu = \alpha - d/2$. By working in the spectral domain, Eq. (8.14) and its solution make sense for all $\alpha > 0$. However, as explained next, the main interest is when α is an integer, both odd or even.

(b) Let $\alpha = 2$ for the moment. A discrete version of (8.14) on $\mathbb{Z}^d$ generates the first-order basic SAR model (Example 4.1). This SAR model can also be interpreted as a second-order basic CAR model (Example 4.5). More generally, if α is an even integer, i.e. $\alpha/2 = k$, is an integer, then the discrete analogue of (8.14) is a kth-order SAR model, which can be identified with a $2k$th-order CAR model. However, it is possible to say a bit more. If α is an odd integer, it is possible to give a SAR model of order $\alpha/2$ an interpretation by identifying it with a CAR model of order α.

(c) For a finite set of sites on rectangular lattice, the methods of Chapter 6 can be used to construct an approximate inverse covariance matrix. For irregularly spaced sites it is necessary to develop more bespoke methods. One way in the two-dimensional case is to divide D into a set of nonintersecting triangles, where any two triangles meet in at most a common edge or corner. Then finite element methods can be used to construct a discrete approximation to (8.14), typically with reflecting boundary conditions. Somewhat surprisingly, the finite element method gives a tractable GMRF approximation whenever α is an integer (odd or even), not just when α is an even integer. The reason seems related to the relationship between SAR and CAR models of orders $\alpha/2$ and α, respectively.

(d) The finite element method can be extended to $\kappa = 0$ and $\nu = 0$ by allowing intrinsic and generalized random fields, respectively. It can also be extended to random fields indexed by sites on a Riemannian manifold.

8.9 Designing a Monitoring Network

Throughout the book, it has been assumed that data values are available at a fixed set of known sites and that data values are not available elsewhere.

However, in many applications, new sites may be added to an existing network and the positioning of the new sites can be chosen to "optimize" the additional information in a suitable sense. A typical example is in environmental science where the sites represent monitoring stations. For simplicity, issues such as geographical factors, costs, and policy considerations are not taken into account in the discussion here.

Thus, suppose there are already monitoring stations at n sites $t_1, \ldots, t_n$ and suppose the data are assumed to follow a stationary or intrinsic model where the covariance parameters are known. One new site, at a location t, say, is to be added. The question is how to choose the new location. One intuitively natural approach is to choose the new site to improve prediction as much as possible. That is, find the location with maximum kriging variance (based on the existing sites). In order for the problem to be well-defined, it is necessary to constrain the new site to lie within a compact region. Otherwise, the optimal location will typically be as far as possible from the existing sites. Therefore, for the discussion here, the new site is constrained to lie within the convex hull of the existing sites. The use of this criterion was introduced by McBratney et al. (1981) and further developed by Cressie et al. (1990) and Mardia and Goodall (1993). There is a substantial literature on the topic of site selection; for further details, see e.g., Smith (2001) and Zidek and Zimmerman (2010).

In general, the optimization problem must be tackled numerically. But in the following elementary example, there is a simple analytic answer.

Example 8.1 Consider $n = 2$ sites in $d = 1$ dimension at $t_1 = -1$, $t_2 = 1$. Suppose that the random field is stationary with unknown mean and with the exponential covariance function $\sigma(h) = \exp\{-|h|/\varphi\}$, where $\varphi > 0$ is the range parameter. From part (a) of Exercise 7.3, it follows that the kriging variance is

$$\sigma_K^2(t) = 1 - \frac{\alpha_1^2 + \alpha_2^2 - 2\rho\alpha_1\alpha_2}{1-\rho^2} + \frac{[1+\rho-(\alpha_1+\alpha_2)]^2}{2(1+\rho)}, \tag{8.15}$$

where $\alpha_j = \alpha_j(t) = \sigma(t_j - t)$, $j = 1, 2$, and $\rho = \sigma(t_1 - t_2)$, i.e. $\alpha_1 = \exp\{(-1-t)/\varphi\}$, $\alpha_2 = \exp\{(-1+t)/\varphi\}$ for $t \in [-1, 1]$, and $\rho = \exp(-2/\varphi)$. After a bit of simplification,

$$\sigma_K^2(t) = 1 - \frac{A(t)}{1-\rho^2} + \frac{B^2(t)}{2(1+\rho)},$$

where $A(t) = 2\rho\{\cosh(2t/\varphi) - \rho\}$ and $B(t) = 1 + \rho - 2e^{-1/\varphi}\cosh(t/\varphi)$. It can be checked that $B(t) \geq 0$ for $t \in [-1, 1]$ and that $\cosh(t/\varphi)$ and $\cosh(2t/\varphi)$ are minimized when $t = 0$. Hence, $-A(t)$ and $B^2(t)$ are maximized when $t = 0$ and, therefore, so is $\sigma_K^2(t)$. That is, $t = 0$, the midpoint of the interval $[-1, 1]$ should be the site of the new monitoring station, as might be expected.

Some extensions of this example are investigated in Exercises 8.7–8.8. □

As well as the problem of adding a site, there is sometimes the problem of deleting one of the existing sites. In this case, it is natural to choose the site that minimizes the kriging variance. This modified problem is not discussed further here; see Cressie et al. (1990) and Mardia and Goodall (1993).

A related problem involves choosing a new site to improve the estimation of any unknown parameters in the spatial model. In this case, a natural criterion to maximize is a summary measure of the information matrix, e.g. the determinant, as a function of the new site. In this case, the new site is chosen to improve estimation rather than prediction.

Exercises

8.1 Let $\boldsymbol{Y} \sim N_n(\boldsymbol{\mu}, \Sigma)$ follow a multivariate normal distribution and set $X_i = \exp(Y_i)$, $i = 1, \ldots, n$. The purpose of this exercise is to find the moments of $\boldsymbol{X}$. They are most easily calculated using the moment generating function for $\boldsymbol{Y}$

$$M(\boldsymbol{u}) = E\left\{\exp\left(\boldsymbol{u}^T\boldsymbol{Y}\right)\right\} = \exp\left(\boldsymbol{u}^T\boldsymbol{\mu} + \frac{1}{2}\boldsymbol{u}^T\Sigma\boldsymbol{u}\right)$$

as a function of $\boldsymbol{u} = (u_1, \ldots, u_n)^T$.

Let $\boldsymbol{e}_i$ denote an n-vector with a one in the ith place and zeros elsewhere. Show that $E(X_i) = M(\boldsymbol{e}_i)$ and $\boldsymbol{E}(X_iX_j) = M(\boldsymbol{e}_i + \boldsymbol{e}_j)$, $i,j = 1, \ldots, n$, and hence verify the moments in (8.1)–(8.2).

8.2 Let $\boldsymbol{S} \sim N_n(\boldsymbol{\mu}, \Sigma)$ be a multivariate normal latent "signal," and, given $\boldsymbol{S}$, consider independent Poisson distributed observations $Z_i|\boldsymbol{S} = \boldsymbol{s} \sim P(\lambda_i)$, where $\log \lambda_i = s_i$, $i = 1, \ldots, n$. Show that the first two moments of the observations and signal are given by

$$E(Z_i) = \exp\left(\mu_i + \frac{1}{2}\sigma_{ii}\right) = g_i, \text{ say,}$$
$$E(Z_i^2) = g_i^2 \exp\left(\sigma_{ii}\right) + g_i, \quad \text{var}(Z_i) = g_i^2\left\{\exp\left(\sigma_{ii}\right) - 1\right\} + g_i,$$
$$E(Z_iZ_j) = g_ig_j \exp\left(\sigma_{ij}\right), \quad \text{cov}(Z_i, Z_j) = g_ig_j\left\{\exp\left(\sigma_{ij}\right) - 1\right\}, \quad i \neq j,$$
$$E(S_i) = \mu_i, \quad \text{var}(S_i) = \sigma_{ii},$$
$$E(S_iZ_j) = \left(\mu_i + \sigma_{ij}\right)g_i, \quad \text{cov}(S_i, Z_j) = \sigma_{ij}g_j,$$

where $i,j = 1, \ldots, n$. Note that the formulas for $E(Z_iZ_j)$ and $\text{cov}(Z_i, Z_j)$ are valid only for $i \neq j$; the formula for $\text{var}(z_i)$ includes a nugget-like term g_i not present for the covariances. On the other hand, the formulas for $E(S_iZ_j)$ and $\text{cov}(S_i, Z_j)$ are valid for all $i,j = 1, \ldots, n$.

Hint: Using the tower law (Exercise 4.1), write $E(Z_i) = E[E(Z_i|S_i)]$ where the inner expectation is over the conditional distribution of Z_i given S_i and the outer expectation is over the marginal distribution of S_i. Proceed similarly with the other two expectations. Use the results from Exercise 8.1 and Section 8.2 for the moments of the log-normal distribution.

8.3 In the same setting as Exercise 8.2, suppose it is desired to predict a new signal S_0 given observations $\boldsymbol{Z} = [Z_1, \ldots, Z_n]^T$. Here, it is assumed that $[S_0, \boldsymbol{S}^T]^T$ are jointly multivariate normal with $E(S_0) = \mu_0$, $\text{var}(S_0) = \sigma_{00}$ and $\text{cov}(S_0, S_i) = \sigma_{0i}, i = 1, \ldots, n$. The best linear predictor takes the form

$$\hat{S}_0 = \mu_0 + \text{cov}(S_0, \boldsymbol{Z})^T \, \text{var}(\boldsymbol{Z})^{-1}(\boldsymbol{Z} - E(\boldsymbol{Z}))$$

with prediction variance

$$\sigma_{00} - \text{cov}(S_0, \boldsymbol{Z})^T \, \text{var}(\boldsymbol{Z})^{-1} \, \text{cov}(\boldsymbol{Z}, S_0).$$

Show that $\text{cov}(S_0, \boldsymbol{Z})$ and $\text{var}(\boldsymbol{Z})$ have elements

$$\text{cov}(S_0, Z_i) = \sigma_{0i} g_i,$$

$$\text{cov}(z_i, z_j) = \begin{cases} g_i g_j \left\{\exp\left(\sigma_{ij}\right) - 1\right\}, & i \neq j, \\ g_i^2 \left\{\exp\left(\sigma_{ii}\right) - 1\right\} + g_i, & i = j, \end{cases}$$

for $i \neq j = 1, \ldots, n$ and where $g_i = \exp(\mu_i + \frac{1}{2}\sigma_{ii})$.

8.4 Suppose the signal $\boldsymbol{S}$ in Exercise 8.2 comes from a stationary Gaussian process with mean μ and covariance function $\sigma(h)$, observed at sites $t_1, \ldots, t_n$, with $\sigma(0) = \sigma^2$. Show that the elements of the observation vector $\boldsymbol{Z}$ have a constant mean and covariances, which can be expressed in terms of a new covariance function $\psi(h)$ and a nugget effect as

$$\text{cov}(\boldsymbol{Z}_i, \boldsymbol{Z}_j) = c_1 \psi(t_i - t_j) + c_2 I[i = j],$$

where $\psi(h) = \exp(\sigma(h)) - 1$ and where the indicator function $I[i = j]$ is 1 if $i = j$ and 0 otherwise. What are the values of c_1 and c_2? What happens if σ^2 is small so that $\exp(\sigma(h)) - 1 \approx \sigma(h)$ for all h?

8.5 (Mardia and Goodall, 1993). Using the notation from Section 8.5, assume the tensor product model (8.4) for processes $X_1(t), \ldots, X_q(t)$. Given data on each process at sites, $t_1, \ldots, t_n$, confirm that the covariance matrix of the data can be written in the form $\Sigma^{**} = A \otimes \Sigma$. Show that in the formula for the GLS estimate of $\boldsymbol{\mu}^*$ and for the kriging predictor $\boldsymbol{x}^\circ$ that the matrix A cancels out, and these quantities are the same as if GLS estimation and prediction were carried out on each variable separately.

8.6 Consider a tensor product covariance model for a zero-mean bivariate stationary spatial process on $\mathbb{R}^1$ where

$$\operatorname{cov}(X_u(t), X_v(t+h)) = c_{uv} \exp(-0.5|h|), \quad u, v = 1, 2, \quad h \in \mathbb{R}^1,$$

and

$$C = \begin{bmatrix} 1 & b \\ b & 1 \end{bmatrix}, \quad b = 0.7.$$

Suppose the following observations are available, $x_1(0)$, $x_1(1)$, $x_2(0)$, and that we wish to predict the value of $x_2(1)$. Show that the joint covariance function of $X_1(0)$, $X_1(1)$, $X_2(0)$, $X_2(1)$ is

$$\Sigma = \begin{bmatrix} 1 & a & b & ab \\ a & 1 & ab & b \\ b & ab & 1 & a \\ ab & b & a & 1 \end{bmatrix}, \quad a = e^{-0.5}, \ b = 0.7.$$

Show that the predictor of $x_2(1)$ using just the value $x_2(0)$ from the x_2-process is given by $\hat{x}_2(1) = ax_2(0)$ with prediction variance $1 - a^2 = 0.632$.
Show that the predictor of $x_2(1)$ using $x_2(0)$ together with the values $x_1(0)$, $x_1(1)$ from the x_1-process is given by $\hat{x}_2(1) = -0.425x_1(0) + 0.700x_1(1) + 0.607x_2(0)$ with prediction variance 0.322. That is, the prediction variance drops by nearly half by including the x_1-data in the prediction process.

8.7 Suppose the mean of the random field in Example 8.1 is assumed known. Show that the kriging variance takes the same form as in (8.15), but without the final term. Deduce that the kriging variance is still maximized at $t = 0$, so that the optimal location of the new site is unchanged.

8.8 This exercise generalizes Example 8.1 in two ways. First, the stationary covariance function $\sigma(h)$ is replaced by a limiting intrinsic covariance function. If the range parameter φ tends to ∞ and $\sigma(h)$ is multiplied by φ, then a limiting IRF-0 random field is obtained with an intrinsic covariance function given by the linear scheme $\sigma_I(h) = -|h|$. Second, the number of sites is increased from $n = 2$ sites to $n \geq 2$ sites, located at $t_1 < \cdots < t_n$. It was shown in Eq. (7.6) that the kriging variance is given by

$$\sigma_K^2(t) = \begin{cases} 2(t_{i+1} - t)(t - t_i), & t_i \leq t \leq t_{i+1}, \\ 2(t_1 - t), & t < t_1, \\ 2(t - t_n), & t > t_n. \end{cases}$$

Thus, $\sigma_K^2(t)$ is quadratic on each interval between successive points. Show that on the interval $t_i \le t \le t_{i+1}$, the kriging variance is maximized at $t = (t_i + t_{i+1})/2$ with kriging variance $\sigma_K^2(t) = (t_{i+1} - t_i)^2/2$. Hence, deduce that if the intervals have different lengths, the new monitoring station should go at the midpoint of the longest interval.

Appendix A

Mathematical Background

Spatial analysis uses range of ideas across mathematics. Many of the ideas used in the book are collected here for easy reference. Different spaces of functions and sequences are summarized in Sections A.1–A.2. Tools from matrix algebra are given in Section A.3. The Fourier transform and related concepts useful for circulant and lattice processes are given in Sections A.4–A.11. The final two sections A.12 and A.13 are different in character, They deal with the theory behind maximum likelihood estimation, especially as it relates to Gaussian random fields

A.1 Domains for Sequences and Functions

Several classes of sequences and functions are of interest in this book. This appendix gathers some key facts and properties about these classes and describes their Fourier transforms (FTs). In general, a function is written in the form $f(u)$ as u varies in a continuous domain, and a sequence is written in the form f_k as k varies in a discrete domain. Four important domains are as follows:

(a) *The d-dimensional Euclidean space* $\mathbb{R}^d$. A typical element is a vector of real numbers, written $u = (u[1], \ldots, u[d])$.
(b) *The d-dimensional integer lattice,* $\mathbb{Z}^d$. A typical element is a vector of integers, written $k = (k[1], \ldots, k[d])$.
(c) *The d-dimensional continuous torus* S_1^d. A typical element is a vector of angles, written $u = (u[1], \ldots, u[d])$. To describe the continuous torus, start with dimension $d = 1$ and recall the circle S_1 denotes the set of angles, i.e. the set of real numbers mod 2π. For $u, v \in \mathbb{R}$, write

$$u = v \bmod 2\pi \tag{A.1}$$

if $u - v$ is an integer multiple of 2π. Thus, the two real numbers u and $u + 2\pi$ represent the same angle. It is often convenient to represent an angle as a real

Spatial Analysis, First Edition. John T. Kent and Kanti V. Mardia.

number lying in a specific interval of length 2π, e.g. $[0,2\pi)$ or $[-\pi,\pi)$, with opposite ends identified with one another.

For $d > 1$, the continuous torus is a direct product of circles. A block version of the modulo operation can also be used in this setting. For real vectors $u, v \in \mathbb{R}^d$, use the notation

$$u = v \text{ Mod } 2\pi \tag{A.2}$$

to mean

$$u[\ell] = v[\ell] \bmod 2\pi, \ \ell = 1, \dots, d, \tag{A.3}$$

so that u and v represent the same vector of angles in S_1^d.

(d) *The d-dimensional lattice torus.*

$$\mathbb{Z}_N^d = \{k \in \mathbb{Z}^d : 0 \le k[\ell] \le n - 1, \ \ell = 1, \dots, d\}, \tag{A.4}$$

where N denotes a multi-index of orders $N = (n[1], \dots, n[d])$. A typical element of $\mathbb{Z}_N^d$ is a vector of cyclic integers, written $k = (k[1], \dots, k[d])$, $0 \le k[\ell] \le n[\ell] - 1$, $\ell = 1, \dots, d$.

To describe the lattice torus in more detail, start with dimension $d = 1$. Given a "period" $n \ge 2$, the discrete circle $\mathcal{Z}_n^1 = \mathcal{Z}_n$ denotes the group of *cyclic integers* mod n. Thus, the two integers k and $k + n$ represent the same cyclic integer, i.e. $k = (k + n) \bmod n$. It is often convenient to represent a cyclic integer as an integer lying in a specific set of n consecutive integers, e.g. $0, \dots, n - 1$ or $1, \dots, n$, with the points 0 and n identified with one another.

For $d > 1$, the lattice torus is a direct product of discrete circles. The modulo operation can also be extended to this setting. Given a multi-index of orders $N = (n[1], \dots, n[d])$ and integer vectors $j, k \in \mathbb{Z}^d$, use the notation $j = k \text{ Mod } N$ to mean

$$j[\ell] = k[\ell] \bmod n[\ell], \ \ell = 1, \dots, d, \tag{A.5}$$

so that j and k represent the same vector of cyclic integers.

Thus, a domain can be continuous or discrete, and can be unbounded or periodic. Table A.1 sets out the choices as a two-way table.

Table A.1 Types of domain.

	Unbounded	Periodic
Continuous	$\mathbb{R}^d$	S_1^d
Discrete	$\mathbb{Z}^d$	$\mathbb{Z}_N^d$

A.2 Classes of Sequences and Functions

For each domain, it is useful to define several classes of sequences or functions, respectively, satisfying various types of regularity condition.

A.2.1 Functions on the Domain $\mathbb{R}^d$

- $L_1(\mathbb{R}^d) = \{f(u),\ u \in \mathbb{R}^d : \int |f(u)| < \infty\}$, the space of *integrable* functions.
- $L_2(\mathbb{R}^d) = \{f(u),\ u \in \mathbb{R}^d : \int |f(u)|^2 < \infty\}$, the space of *square-integrable* functions.
- $C_b(\mathbb{R}^d) = \{f(u),\ u \in \mathbb{R}^d : f(u)$ is a bounded continuous function$\}$.
- $C_0(\mathbb{R}^d) = \{f(u),\ u \in \mathbb{R}^d : f(u)$ is continuous and $f(u) \to 0$ as $|u| \to \infty\}$.
- $B(\mathbb{R}^d) = \{f(u),\ u \in \mathbb{R}^d : f(u)$ is a bounded measurable function $\}$.
- $\mathcal{S}(\mathbb{R}^d) = \{f(u),\ u \in \mathbb{R}^d : f(u)$ is infinitely differentiable and $f(u)$ and all of its partial derivatives of all orders are rapidly vanishing at infinity$\}$.
- $\mathcal{K}(\mathbb{R}^d) = \{f(u),\ u \in \mathbb{R}^d : f(u)$ is infinitely differentiable and has compact support$\}$.

A function $f(u)$ is said to be *rapidly vanishing at infinity* if it tends to 0 faster than any power of u, that is, if

$$|u|^n f(u) \to 0 \text{ as } |u| \to \infty$$

for any $n \geq 0$, where $|u|^2 = u[1]^2 + \cdots + u[d]^2$.

In addition to this list of function spaces, add a subscript b, c or 0 on a class of functions to restrict the functions to be bounded, to have compact support, or to vanish at infinity, respectively, where relevant. For example, $L_{1,b}(\mathbb{R}^d)$ denotes the L_1 functions that are bounded and $L_{2,c}(\mathbb{R}^d)$ denotes the L_2 functions that have compact support. The spaces $C_b(\mathbb{R}^d)$ and $C_0(\mathbb{R}^d)$ are important enough to be given entries in the above list. Note the set inclusions

$$C_c(\mathbb{R}^d) \subset C_0(\mathbb{R}^d) \subset C_b(\mathbb{R}^d)$$

since a function with compact support vanishes for large $|u|$, and since a continuous function that tends to 0 for large u must be bounded. Also note that $\mathcal{K}(\mathbb{R}^d) = \mathcal{S}_c(\mathbb{R}^d)$.

Neither $L_1(\mathbb{R}^d)$ nor $L_2(\mathbb{R}^d)$ is a subset of the other. The best that can be said is

$$L_{1,b}(\mathbb{R}^d) \subset L_{2,b}(\mathbb{R}^d) \quad \text{and} \quad L_{2,c}(\mathbb{R}^d) \subset L_{1,c}(\mathbb{R}^d).$$

A.2.2 Sequences on the Domain $\mathbb{Z}^d$

- $L_1(\mathbb{Z}^d) = \{f_k,\ k \in \mathbb{Z}^d : \sum |f_k| < \infty\}$, the space of *summable* sequences.
- $L_2(\mathbb{Z}^d) = \{f_k,\ k \in \mathbb{Z}^d : \sum f_k^2 < \infty\}$, the space of *square summable* sequences.

- $B(\mathbb{Z}^d) = \{f_k,\ k \in \mathbb{Z}^d : \max |f_k| < \infty\}$, the space of *bounded* sequences.
- $B_0(\mathbb{Z}^d) = \{f_k,\ k \in \mathbb{Z}^d : f_k \to 0 \text{ as } |k| \to \infty\}$.

Note the set inclusions

$$L_1(\mathbb{Z}^d) \subset L_2(\mathbb{Z}^d) \subset B_0(\mathbb{Z}^d) \subset B(\mathbb{Z}^d).$$

A.2.3 Classes of Functions on the Domain S_1^d

- $L_1(S_1^d) = \{f(u),\ u \in S_1^d : \int |f(u)| < \infty\}$, the space of *integrable* functions.
- $L_2(S_1^d) = \{f(u),\ u \in S_1^d : \int |f(u)|^2 < \infty\}$, the space of *square-integrable* functions.
- $C(S_1^d) = \{f(u),\ u \in S_1^d : f(u) \text{ is a (bounded) continuous function}\}$.
- $B(S_1^d) = \{f(u),\ u \in S_1^d : f(u) \text{ is a bounded measurable function}\}$.

Note that since S_1^d is compact, a continuous function is automatically bounded. Further, the L_2 functions are a subset of the L_1 functions; hence

$$C(S_1^d) \subset B(S_1^d) \subset L_2(S_1^d) \subset L_1(S_1^d). \tag{A.6}$$

A.2.4 Classes of Sequences on the Domain $\mathbb{Z}_N^d$, Where $N = (n[1], \ldots, n[d])$

In this setting, there is no need to impose regularity conditions. A sequence of real numbers indexed by $\mathbb{Z}_N^d$ is just a finite set of numbers.

A.3 Matrix Algebra

A.3.1 The Spectral Decomposition Theorem

A fundamental theorem in matrix algebra says that an $n \times n$ symmetric matrix A (i.e. $A^T = A$) can be decomposed as

$$A = \Gamma \Lambda \Gamma^T, \tag{A.7}$$

where Γ is an $n \times n$ orthogonal matrix (so $\Gamma^{-1} = \Gamma^T$) whose columns are eigenvectors of A and where the vector of eigenvalues $\lambda = (\lambda_1, \ldots, \lambda_n)^T$ has been stored as a diagonal $n \times n$ matrix $\Lambda = \text{diag}(\lambda)$. In particular, (A.7) implies $A\Gamma = \Gamma\Lambda\Gamma^T\Gamma = \Gamma\Lambda$, which can be written columnwise as

$$A\gamma_{(k)} = \lambda_k \gamma_{(k)}, \quad k = 1, \ldots, n,$$

where $\gamma_{(k)}$ is the kth column of Γ, thus confirming that A takes the eigenvector $\gamma_{(k)}$ to a multiple of itself, where the multiple is the eigenvalue λ_k. A symmetric matrix A is called *positive definite (p.d.)* if all the eigenvalues are positive, $\lambda_k > 0,\ k = 1, \ldots, n$. Similarly, A is called *positive semidefinite (p.s.d.)* if all the eigenvalues are nonnegative, $\lambda_k \geq 0,\ k = 1, \ldots, n$.

If A is symmetric and positive definite, then the *symmetric positive definite matrix square root* of A is defined by

$$A^{1/2} = \Gamma\Lambda^{1/2}\Gamma^T, \tag{A.8}$$

where $\Lambda^{1/2} = \mathrm{diag}(\lambda_k^{1/2})$ is a diagonal matrix containing the square roots of the eigenvalues. It can be easily checked that $A^{1/2}A^{1/2} = A$, and that the inverse of $A^{1/2}$ is

$$A^{-1/2} = \Gamma\Lambda^{-1/2}\Gamma^T.$$

There is also a related decomposition for complex matrices. An $n \times n$ Hermitian matrix A ($A^* = A$) can be decomposed as

$$A = \Gamma\Lambda\Gamma^*, \tag{A.9}$$

where Γ is now a complex-valued unitary matrix ($\Gamma^{-1} = \Gamma^*$) whose columns are eigenvectors of A, and $\Lambda = \mathrm{diag}(\lambda_k)$ is a diagonal matrix of real-valued eigenvalues. Here, $A^* = \bar{A}^T$ denotes the complex conjugate of the transpose of the matrix A.

A.3.2 Moore–Penrose Generalized Inverse

Let A be a symmetric matrix with spectral decomposition (A.7). When some of the eigenvalues are 0, the inverse of A does not exist. However, it is possible to define a restricted sort of inverse by taking the reciprocals of the nonzero eigenvalues and leaving the zero eigenvalues unchanged. Suppose p of the eigenvalues are nonzero. Partition the diagonal eigenvalue matrix as

$$\Lambda = \begin{bmatrix} \Lambda_1 & 0 \\ 0 & \Lambda_2 \end{bmatrix},$$

where $\Lambda_1 = \mathrm{diag}(\lambda_1, \ldots, \lambda_p)$ contains the nonzero eigenvalues and $\Lambda_2 = 0$ contains the zero eigenvalues. Similarly, partition the eigenvector matrix $\Gamma = [\Gamma_1\ \Gamma_2]$. Then the spectral decomposition can be expressed in *reduced form*

$$A = \Gamma_1\Lambda_1\Gamma_1^T = \sum_{j=1}^{p} \lambda_j \boldsymbol{\gamma}_{(j)}\boldsymbol{\gamma}_{(j)}^T. \tag{A.10}$$

The Moore–Penrose generalized inverse of A is defined by

$$A^- = \Gamma_1\Lambda_1^{-1}\Gamma_1^T = \sum_{j=1}^{p} \lambda_j^{-1} \boldsymbol{\gamma}_{(j)}\boldsymbol{\gamma}_{(j)}^T. \tag{A.11}$$

It is straightforward to check that $AA^-A = A$ and $A^-AA^- = A^-$.

If A can be represented in partitioned form as

$$A = \begin{bmatrix} 0 & 0 \\ 0 & A_{22} \end{bmatrix}, \tag{A.12}$$

where A_{22} is nonsingular, then the Moore–Penrose generalized inverse takes the simple form

$$A^- = \begin{bmatrix} 0 & 0 \\ 0 & A_{22}^{-1} \end{bmatrix}. \tag{A.13}$$

A.3.3 Orthogonal Projection Matrices

Let F be an $n \times p$ matrix, where $p \leq n$, and write

$$F = [\boldsymbol{f}_{(1)} \quad \cdots \quad \boldsymbol{f}_{(p)}]$$

in terms of its columns. Suppose F has full rank, which in this setting means that the columns of F are linearly independent of one another; that is, any nontrivial linear combination of the columns cannot vanish. More specifically, if $\boldsymbol{\alpha}$ is a nonzero p-vector of coefficients, then

$$F\boldsymbol{\alpha} = \sum_{j=1}^{p} \alpha_j \boldsymbol{f}_{(j)} \neq \mathbf{0}.$$

Set $B = F^T F$, a $p \times p$ matrix. Then B must be a positive definite since if $\boldsymbol{\alpha} \neq \mathbf{0}$, then $\boldsymbol{\alpha}^T B \boldsymbol{\alpha} = \boldsymbol{\beta}^T \boldsymbol{\beta} = \sum_{j=1}^{p} \beta_j^2 > 0$, where $\boldsymbol{\beta} = F\boldsymbol{\alpha}$.

Define two matrices $G = FB^{-1/2}$ and $P_F = FA^{-1}F^T = GG^T$. Then, F and G have the same column space and P_F is a symmetric matrix. Further, it is straightforward to verify the following properties:

(a) $G^T G = I_p$ so the columns of G are orthonormal.
(b) $P_F^2 = P_F$, i.e. P_F is *idempotent.*
(c) $P_F F = F$, so P_F leaves the columns of F unchanged.
(d) If $\boldsymbol{x}$ is an n-vector orthogonal to F, i.e. $F^T \boldsymbol{x} = \mathbf{0}$, then $P_F \boldsymbol{x} = \mathbf{0}$.

Since the columns of G are orthonormal, the definition of P_F is actually a reduced spectral decomposition, $P_F = GG^T = G\Lambda G^T$, where $\Lambda = I_p$ as in (A.10). In particular, the eigenvalues of P_F are 1 (with multiplicity p) and 0 (with multiplicity $n - p$). The eigenvectors corresponding to the eigenvalue 1 are given by the columns of G, or equivalently by the columns of F.

Properties (c) and (d) mean that P_F can be described as an *orthogonal projection matrix* on to the column space of F.

A.3.4 Partitioned Matrices

Let $A(n \times n)$ be an invertible symmetric matrix and suppose that it and its inverse have been partitioned compatibly as

$$A = \begin{bmatrix} A_{11} & A_{12} \\ A_{21} & A_{22} \end{bmatrix}, \quad A^{-1} = \begin{bmatrix} A^{11} & A^{12} \\ A^{21} & A^{22} \end{bmatrix},$$

where the diagonal blocks are square matrices. Then, provided the relevant inverses exist, the blocks of A^{-1} can be found from the blocks of A by

$$\begin{aligned} A^{11} &= (A_{11} - A_{12}A_{22}^{-1}A_{21})^{-1}, \\ A^{22} &= (A_{22} - A_{21}A_{11}^{-1}A_{12})^{-1}, \\ A^{12} &= (A^{21})^T = -A_{11}^{-1}A_{12}A^{22} = -A^{11}A_{12}A_{22}^{-1}. \end{aligned} \tag{A.14}$$

This result is easily checked by multiplying out AA^{-1} and confirming that the result is the identity matrix. It is often convenient to use the shorthand notation

$$A_{11.2} = A_{11} - A_{12}A_{22}^{-1}A_{21}, \quad A_{22.1} = A_{22} - A_{21}A_{11}^{-1}A_{12}. \tag{A.15}$$

In particular, using this notation, it is straightforward to show that the determinant of A can be written as

$$|A| = |A_{11}|\ |A_{22.1}| = |A_{22}|\ |A_{11.2}|, \tag{A.16}$$

by noting that

$$BAB^T = \begin{bmatrix} A_{11} & 0 \\ 0 & A_{22.1} \end{bmatrix}, \text{ where } B = \begin{bmatrix} I & 0 \\ -A_{21}A_{11}^{-1} & I \end{bmatrix}$$

and recalling that a lower triangular matrix with ones along the diagonal has determinant 1.

The representation (A.14) is helpful in describing the multivariate normal distribution. Let a random vector $\boldsymbol{x}$ follow a multivariate normal distribution with mean vector $\boldsymbol{\mu}$ and covariance matrix Σ. Partition $\boldsymbol{x}$ into two blocks

$$\boldsymbol{x} = \begin{bmatrix} \boldsymbol{x}_1 \\ \boldsymbol{x}_2 \end{bmatrix}$$

of dimensions p_1 and p_2, and similarly partition $\boldsymbol{\mu}$ and Σ. Then, the following results hold:

(a) The marginal distribution of $\boldsymbol{x}_1$ is multivariate normal,

$$\boldsymbol{x}_1 \sim N_{p_1}\left(\boldsymbol{\mu}_1, \Sigma_{11}\right). \tag{A.17}$$

(b) The conditional distribution of $\boldsymbol{x}_1$ given $\boldsymbol{x}_2 = \boldsymbol{x}_2^0$ is multivariate normal,

$$\boldsymbol{x}_1 | \boldsymbol{x}_2 = \boldsymbol{x}_2^0 \sim N_{p_1}\left(\boldsymbol{\mu}_1 + \Sigma_{12}\Sigma_{22}^{-1}\left(\boldsymbol{x}_2^0 - \boldsymbol{\mu}_2\right), \Sigma_{11.2}\right). \tag{A.18}$$

A.3.5 Schur Product

Let $A = (a_{ij})$ and $B = (b_{ij})$ be two $n \times n$ symmetric matrices. The *elementwise product* of the two matrices, $C = (c_{ij})$ with elements $c_{ij} = a_{ij}b_{ij}$ is also called the *Schur product* of A and B and written $C = A\#B$. Then, the following properties hold:

If A and B are positive semidefinite, then C is positive semidefinite. Similarly, if A and B are positive definite, then C is positive definite.

To prove this result, write A using the spectral decomposition theorem in Section A.3.1,

$$A = \Gamma\Lambda\Gamma^T = \sum_{k=1}^{n} \lambda_k \boldsymbol{\gamma}_{(k)}\boldsymbol{\gamma}_{(k)}^T,$$

so that $C = \sum \lambda_k\{(\boldsymbol{\gamma}_{(k)}\boldsymbol{\gamma}_{(k)}^T)\#B\}$. Let $\mathbf{d}$ be a coefficient vector and consider the quadratic form

$$\mathbf{d}^T C\mathbf{d} = \sum \lambda_k \mathbf{d}^T\{(\boldsymbol{\gamma}_{(k)}\boldsymbol{\gamma}_{(k)}^T)\#B\}\mathbf{d} = \sum \lambda_k \mathbf{g}_{(k)}^T B\mathbf{g}_{(k)},$$

where $\mathbf{g}_{(k)}$ has elements $g_{jk} = \gamma_{jk}d_j$, $j = 1, \ldots, n$. Since Γ is an orthogonal matrix, its rows are unit vectors. Hence, for every row j, there is at least one choice of k for which $\gamma_{jk} \neq 0$. If $\mathbf{d} \neq \mathbf{0}$, there is at least one index j such that $d_j \neq 0$, and so for this value of j there is at least one choice of k for which $g_{jk} = \gamma_{jk}d_j \neq 0$. That is, $\mathbf{g}_{(k)} \neq \mathbf{0}$ for at least one index k.

If A and B are p.s.d., then $\lambda_k \geq 0$ and $\mathbf{g}_{(k)}^T B\mathbf{g}_{(k)} \geq 0$ for all k. Hence, $\mathbf{d}^T C\mathbf{d} \geq 0$ and so C is p.s.d. If A and B are p.d., then $\lambda_k > 0$ for all k and $\mathbf{g}_{(k)}^T B\mathbf{g}_{(k)} > 0$ for at least one k. Hence, $\mathbf{d}^T C\mathbf{d} > 0$ and so C is p.d.

A related result states that if $\sigma_1(s,t)$ and $\sigma_2(s,t)$ are positive (semi)definite functions of the sites $s, t \in \mathbb{R}^d$, then the product $\sigma_1(s,t)\sigma_2(s,t)$ is also positive (semi)definite.

A.3.6 Woodbury Formula for a Matrix Inverse

This identity describes how the inverse of a matrix changes if the matrix is altered. It is known by a variety of names, including the Sherman–Morrison–Woodbury formula, or just the Woodbury formula (see, e.g. Mardia et al., 1979, p. 458, equation (A.2.4f)). Let B be an $n \times n$ matrix, which can be written as

$$B = A + UCV^T, \tag{A.19}$$

where $A(n \times n)$, $U(n \times k)$, $V(n \times k)$, and $C(k \times k)$ are compatibly dimensioned. Then, the inverse of B can be expanded as

$$B^{-1} = A^{-1} - A^{-1}UG^{-1}V^TA^{-1}, \quad G = C^{-1} + V^TA^{-1}U, \tag{A.20}$$

assuming A, C, and G are invertible. The proof is straightforward. Just substitute (A.19) and (A.20) for B and B^{-1}, and check that $BB^{-1} = I$.

The formula is most useful when $k \ll n$ because, once A^{-1} has been found, it is only necessary to compute the inverse of a $k \times k$ matrix to find the inverse of B.

A.3.7 Quadratic Forms

Consider minimizing the quadratic form

$$Q = Q(\boldsymbol{\beta}) = (\boldsymbol{x} - F\boldsymbol{\beta})^T A(\boldsymbol{x} - F\boldsymbol{\beta}),$$

over a parameter vector $\boldsymbol{\beta} \in \mathbb{R}^q$, where $\boldsymbol{x} \in \mathbb{R}^n$ is a data vector, $A(n \times n)$ is a positive definite matrix, and where $F(n \times q)$ is a coefficient matrix. Further, suppose F can be partitioned as

$$F = \begin{bmatrix} F_1 \\ 0 \end{bmatrix},$$

where $F_1(q \times q)$ is nonsingular, and similarly partition

$$\boldsymbol{x} = \begin{bmatrix} \boldsymbol{x}_1 \\ \boldsymbol{x}_2 \end{bmatrix}, \quad A = \begin{bmatrix} A_{11} & A_{12} \\ A_{21} & A_{22} \end{bmatrix}.$$

Differentiating Q with respect to $\boldsymbol{\beta}$, setting the derivative to $\mathbf{0}$, and solving for $\boldsymbol{\beta}$ yields

$$\hat{\boldsymbol{\beta}} = (F_1^T A_{11} F_1)^{-1}(F_1^T A_{11}\boldsymbol{x}_1 + F_1^T A_{12}\boldsymbol{x}_2) = F_1^{-1}\boldsymbol{x}_1 + F_1^{-1}A_{11}^{-1}A_{12}\boldsymbol{x}_2,$$

so that

$$F\hat{\boldsymbol{\beta}} = \begin{bmatrix} \boldsymbol{x}_1 + A_{11}^{-1}A_{12}\boldsymbol{x}_2 \\ \mathbf{0} \end{bmatrix}.$$

Substituting $\hat{\boldsymbol{\beta}}$ into Q and simplifying yields the minimized quadratic form

$$Q(\hat{\boldsymbol{\beta}}) = \boldsymbol{x}_2^T A_{22.1}\boldsymbol{x}_2, \tag{A.21}$$

where $A_{22.1}$ is given in (A.15). If $A = \Sigma^{-1}$ is the inverse of a covariance matrix Σ, then from (A.14) $A_{22.1} = \Sigma_{22}^{-1}$ can also be written as the inverse of the lower-right block of Σ.

A version of this result is also available when $F(n \times q)$ cannot be partitioned, but still has full column rank q. In this case, let $G(n \times (n - q))$ be a column orthonormal matrix, which is orthogonal to F, so that

$$G^T G = I_{n-q}, \quad G^T F = 0.$$

Then writing $\Sigma_{GG} = G^T \Sigma G$ and $\boldsymbol{x}_G = G^T\boldsymbol{x}$, the minimized quadratic form is given by

$$Q_{\min} = Q(\hat{\boldsymbol{\beta}}) = \boldsymbol{x}_G^T \Sigma_{GG}^{-1}\boldsymbol{x}_G. \tag{A.22}$$

A.3.8 Toeplitz and Circulant Matrices

Spatial statistics often involves data $\{x_k,\ k \in D\}$ on a rectangular set of sites

$$D = \{k \in \mathbb{Z}^d : 1 \leq k[\ell] \leq n[\ell],\ \ell = 1, \ldots, d\}. \tag{A.23}$$

For practical work, it is helpful to list the sites in *lexicographic order*, say, where the final index varies most quickly. This ordering was introduced in Section 4.8.1. If $d = 2$ and the sites are listed as $(k[1], k[2]),\ k[1] = 1, \dots, n[1],\ k[2] = 1, \dots, n[2]$, then the lexicographic order is

$$(1,1), \dots, (1, n[2]),\ (2,1), \dots, (2, n[2]),\ \dots (n[1], 1), \dots, (n[1], n[2]).$$

Thus, the data can be represented by a vector $\boldsymbol{x}$ with elements x_k, where the sites $k = (k[1], \dots, k[d])$ are listed in lexicographic order.

The same ordering can be used to define matrices indexed by the sites in D. Suppose a $|D| \times |D|$ matrix A has elements a_{jk} where the rows $j = (j[1], \dots, j[d])$ and columns $k = (k[1], \dots, k[d])$ are listed in lexicographic order.

This $|D| \times |D|$ matrix A is called *Toeplitz* if its elements can be written

$$a_{jk} = a_{j-k}, \quad j, k \in D,$$

so that a_{jk} just depends on the difference $j - k$ between the sites. Sometimes, the name *block Toeplitz* is used for dimensions $d > 1$. Toeplitz matrices arise naturally as covariance matrices for data following a stationary random field model.

Similarly, a $|D| \times |D|$ matrix A is called *circulant* if its elements can be written

$$a_{jk} = a_{j-k} \text{ Mod } N, \quad j, k \in D,$$

so that a_{jk} just depends on the difference $j - k$ mod N between the sites. Sometimes, the name *block circulant* is used for dimensions $d > 1$. Circulant matrices play an important role in the discrete Fourier transform (DFT); see Section A.7.3.

A.3.9 Tensor Product Matrices

Consider two matrices $A = (a_{j[1],k[1]})$ $(n[1] \times n[1])$ and $B = (b_{j[2],k[2]})$ $(n[2] \times n[2])$. The *tensor* or *Kronecker* product takes the form

$$A \otimes B = \begin{bmatrix} a_{11}B & \cdots & a_{1,n[1]}B \\ \vdots & \ddots & \vdots \\ a_{n[1],1}B & \cdots & a_{n[1],n[1]}B \end{bmatrix}.$$

Then, for a D-dimensional vector $\boldsymbol{x}$,

$$\boldsymbol{x}^T (A \otimes B) \boldsymbol{x} = \sum_{j[1],k[1]=1}^{n[1]} \sum_{j[2],k[2]=1}^{n[2]} x_{j[1],\,j[2]}\ a_{j[1],k[1]}\ b_{j[2],k[2]}\ x_{k[1],k[2]}.$$

This construction for $d = 2$ carries over naturally to higher dimensions.

A.3.10 The Spectral Decomposition and Tensor Products

The spectral decomposition carries over naturally to tensor products. Here are the details for a $d = 2$-fold product. If $A^{(1)} = \Gamma^{(1)}\Lambda^{(1)}\Gamma^{(1)T}(n[1] \times n[1])$ and $A^{(2)} = \Gamma^{(2)}\Lambda^{(2)}\Gamma^{(2)T}(n[2] \times n[2])$ are two symmetric matrices with the specified spectral decompositions, then the tensor product has spectral decomposition

$$A^{(1)} \otimes A^{(2)} = (\Gamma^{(1)} \otimes \Gamma^{(2)})(\Lambda^{(1)} \otimes \Lambda^{(2)})(\Gamma^{(1)} \otimes \Gamma^{(2)})^T.$$

That is, the eigenvalues $\lambda^{(1)}_{k[1]}\lambda^{(2)}_{k[2]}$ are products of the individual eigenvalues, with corresponding eigenvectors $\gamma^{(1)}_{(k[1])} \otimes \gamma^{(2)}_{(k[2])}$, $1 \leq k[1] \leq n[1]$, $1 \leq k[2] \leq n[2]$.

A.3.11 Matrix Derivatives

Let $\Sigma = \Sigma(\theta)$ be an $n \times n$ symmetric positive definite matrix whose elements $\sigma_{ij} = \sigma_{ij}(\theta)$ are functions of a scalar parameter θ. Write $\Sigma' = (\sigma'_{ij})$ to be the matrix of derivatives where $\sigma'_{ij} = d\sigma_{ij}/d\theta$.

Then the following results hold.

$$d \log |\Sigma| / d\theta = \mathrm{tr}(\Sigma^{-1}\Sigma'), \tag{A.24}$$

$$d\Sigma^{-1}/d\theta = -\Sigma^{-1}\Sigma'\Sigma^{-1}. \tag{A.25}$$

It is also useful to consider scalar functions of a vector variable. For example, consider the quadratic form

$$f(\boldsymbol{u}) = \boldsymbol{u}^T A \boldsymbol{u}, \tag{A.26}$$

where A is a symmetric $n \times n$ given matrix and $\boldsymbol{u}$ is an n-vector. The derivative of f with respect to $\boldsymbol{u}$ is the column vector

$$\frac{df(\boldsymbol{u})}{d\boldsymbol{u}} = 2A\boldsymbol{u}. \tag{A.27}$$

Similarly, the second derivative

$$\frac{d^2 f(\boldsymbol{u})}{d\boldsymbol{u}d\boldsymbol{u}^T} = 2A \tag{A.28}$$

is an $n \times n$ matrix.

A.4 Fourier Transforms

Here is a list of the Fourier transforms in the four different settings of Section A.2.

$$\text{Setting 1} \quad \tilde{f}(v) = \int_{\mathbb{R}^d} \exp(iv^T u) f(u)\, du, \tag{A.29}$$

$$\text{Setting 2} \quad \tilde{f}(v) = \sum_{k \in \mathbb{Z}^d} \exp(iv^T k) f_k, \tag{A.30}$$

$$\text{Setting 3} \quad \tilde{f}_j = \int_{S_1^d} \exp(ij^T u) f(u)\, du, \tag{A.31}$$

$$\text{Setting 4} \quad \tilde{f}_j = |N|^{-1/2} \sum_{k \in \mathbb{Z}_N^d} \exp\left\{ 2\pi i \sum_{\ell=1}^{d} j[\ell]k[\ell]/n[\ell] \right\} f_k. \tag{A.32}$$

The exponents involve the inner products between two d-dimensional vectors, e.g. $v^T u = \sum_{\ell=1}^{d} v[\ell]u[\ell]$. In Setting 4, $|N|$ is shorthand for $|N| = n[1] \times \cdots \times n[d]$.

Here is a list of the inverse Fourier transforms (IFTs) in the four different settings.

$$\text{Setting 1} \quad f(u) = \frac{1}{(2\pi)^d} \int_{\mathbb{R}^d} \exp(-iv^T u) \tilde{f}(v)\, dv, \tag{A.33}$$

$$\text{Setting 2} \quad f_k = \frac{1}{(2\pi)^d} \int_{S_1^d} \exp(-iv^T k) \tilde{f}(v)\, dv, \tag{A.34}$$

$$\text{Setting 3} \quad f(u) = \frac{1}{(2\pi)^d} \sum_{\mathbb{Z}^d} \exp(-ij^T u) \tilde{f}_j, \tag{A.35}$$

$$\text{Setting 4} \quad f_k = |N|^{-1/2} \sum_{j \in \mathbb{Z}_N^d} \exp\left\{ -2\pi i \sum_{\ell=1}^{d} j[\ell]k[\ell]/n[\ell] \right\} \tilde{f}_j. \tag{A.36}$$

In each setting, the Fourier transform takes a function or sequence on a *primary domain* to a corresponding function or sequence on a *dual domain*. A function on the primary domain is written $f(u)$; a sequence is written f_k. Similarly, a function on the dual domain, i.e. the Fourier domain, is written $\tilde{f}(v)$; a sequence is written $\tilde{f}_j$. For the standard Fourier transform, both domains are $\mathbb{R}^d$. For other settings, the domains are set out in Table A.2.

The equation numbers giving the Fourier transforms and IFTs are listed in Table A.2. In the literature, the definitions of the FT appear in several versions, depending on whether $+i$ or $-i$ is used and where factors of 2π appear. For the versions given here in Settings 1–3, the IFT is effectively the same as the Fourier transform for the dual domain, with two small differences: the substitution of $-i$ for i and the introduction of the scaling factor $1/(2\pi)^d$.

Table A.2 Domains for Fourier transforms and inverse Fourier transforms in various settings.

Setting	Primary domain	Dual domain	Fourier transform	Inverse Fourier transform
1	$u \in \mathbb{R}^d$	$v \in \mathbb{R}^d$	(A.29)	(A.33)
2	$k \in \mathbb{Z}^d$	$v \in S_1^d$	(A.30)	(A.34)
3	$u \in S_1^d$	$j \in \mathbb{Z}^d$	(A.31)	(A.35)
4	$k \in \mathbb{Z}_N^d$	$j \in \mathbb{Z}_N^d$	(A.32)	(A.36)

However, in Setting 4, it is convenient to change the convention. The FT and IFT still differ in the replacement of i by $-i$, but there is no factor of $1/(2\pi)^d$. For further discussion, see Section A.7.

In order for the Fourier transform to exist, it is necessary to impose some regularity conditions on the function or sequence.

(i) The simplest condition to impose is that f lies in $L_1(\mathbb{R}^d)$, $L_1(\mathbb{Z}^d)$, or $L_1(S_1^d)$ for Settings 1,2,3, respectively, of Table A.2. No regularity conditions are needed for Setting 4 since there are only a finite number of terms in the sum.
(ii) It is also possible to extend the definition of the Fourier transform to all square-integrable functions $L_2(\mathbb{R}^d)$ and square summable sequences $L_2(\mathbb{Z}^d)$ for Settings 1 and 2 of Table A.2, respectively. In Setting 3, since $L_2(S_1^d) \subset L_1(S_1^d)$, no extension is needed.
(iii) In the continuous Settings 1 and 3, the definition of the Fourier transform can also be extended by replacing the function $f(u)$ by a finite measure $\mu(du)$, say. In particular, when $\mu(du)$ is a probability measure, the Fourier transform is also known as the the *characteristic function* of the measure. More generally, when $\mu(du)$ is a finite nonnegative measure, the Fourier transform provides a way of generating valid positive definite covariance functions; see Section A.5.

A.5 Properties of the Fourier Transform

For notational simplicity, the properties in this section are described for Setting 1. But the same properties hold in other settings.

(a) *Even functions.* If f is a real-valued even function, i.e. $f(u) = f(-u)$, then $\tilde{f}$ is real-valued. This property is a simple result of the identity

$$e^{iu} + e^{-iu} = 2\cos u.$$

(b) *Nonnegative functions and positive semidefinite Fourier Transforms.* If a real-valued function f is nonnegative ($f(u) \geq 0$ for all u), then its Fourier transform has the property of *positive semidefiniteness*; that is, for all integers $m \geq 1$ and all real or complex coefficients $\alpha_1, \ldots, \alpha_m$ and all values for the dual variables $v_1, \ldots, v_m$, the following quadratic form is nonnegative, $\sum_{r,s=1}^m \alpha_r \overline{\alpha}_s \tilde{f}(v_r - v_s) \geq 0$. This property holds since

$$\begin{aligned}\sum_{r,s=1}^m \alpha_r \overline{\alpha}_s \tilde{f}(v_r - v_s) &= \int_{\mathbb{R}^d} \sum_{r,s=1}^m \alpha_r \overline{\alpha}_s \exp\{i(v_r - v_s)^T u\} f(u)\, du \\ &= \int_{\mathbb{R}^d} \left| \sum_{r=1}^m \alpha_r \exp(i v_r^T u) \right|^2 f(u)\, du \geq 0.\end{aligned}$$

(c) *Reflection about the origin.* Given a complex-valued function $f(u)$, define the *reflected function* by $\check{f}(u) = \overline{f(-u)}$ (in practice, $f(u)$ will typically be real-valued so that the complex conjugate is not needed). Then, the Fourier transform of the reflected function is the complex conjugate of the Fourier transform of the original function,

$$\tilde{\check{f}}(v) = \overline{\tilde{f}(v)}.$$

In particular, if $f(u) = \check{f}(u)$, then $\tilde{f}(v)$ is real-valued.

(d) *Convolution.* Given two functions $f(u)$ and $g(u)$ in L_1, the *convolution* is defined by $f * g = h$, say, where

$$h(u) = \int_{\mathbb{R}^d} f(w)g(u-w)\, dw = \int_{\mathbb{R}^d} f(u-w)g(w)\, dw.$$

The convolution also lies in L_1, with FT given by the product of the individual FTs

$$\tilde{h}(v) = \tilde{f}(v)\tilde{g}(v).$$

In particular, if $g(u) = \check{f}(u)$, then

$$\tilde{h}(v) = \tilde{f}(v)\overline{\tilde{f}(v)} = |\tilde{f}(v)|^2 \geq 0$$

is a nonnegative function of v.

(e) *Differentiation.* Given a function $\varphi(u)$, let

$$\psi(u) = \Delta\varphi(u) = (\partial^2/\partial u[1]^2 + \cdots + \partial^2/\partial u[d]^2)\varphi(u)$$

denote the Laplacian of φ. If $\varphi(u)$ is sufficiently smooth and integrable, then the Fourier transform of ψ is related to the Fourier transform of φ by

$$\begin{aligned}
\tilde{\psi}(v) &= \int \exp(iv^T u)\psi(u)du = \int \exp(iv^T u)\Delta\varphi(u)\, du \\
&= -\int (v[1]^2 + \cdots + v[d]^2)\exp(iv^T u)\varphi(u)\, du \\
&= -|v|^2\tilde{\varphi}(v), \qquad \text{(A.37)}
\end{aligned}$$

using integration by parts twice. A convenient sufficient condition on φ is that it lies in the space $\mathcal{K}$ defined in Section A.2.1.

(f) *Parseval relationship.* Given a function $\varphi(u)$ in the primary domain with a Fourier transform $\tilde{\varphi}(v)$ and a function $\psi(v)$ in the dual domain with a Fourier transform $\tilde{\psi}(u)$, the Parseval relationship states that

$$\int \tilde{\varphi}(v)\psi(v)dv = \int \varphi(u)\tilde{\psi}(u)du = \int\int \varphi(u)\psi(v)\exp\{iv^T u\}\, du\, dv. \quad \text{(A.38)}$$

(g) *Interpretation.* In the application to spatial analysis, a stationary covariance function (a function of spatial lag denoted h), can be represented as the Fourier transform of a spectral density (a function of frequency denoted ω).

The frequency domain generally viewed as the primary domain and the spatial domain as the dual domain. For example, in Setting 1, the covariance function for a continuous stationary spatial process $\{X(t) : t \in \mathbb{R}^d\}$ takes the form

$$\sigma(h) = \int_{\mathbb{R}^d} \exp(ih^T\omega)f(\omega)\, d\omega = \int_{\mathbb{R}^d} \cos(h^T\omega)f(\omega)\, d\omega,\ h \in \mathbb{R}^d, \tag{A.39}$$

where $f(\omega)$ is an even, nonnegative, integrable function.
Similarly, in Setting 2 for a lattice stationary spatial process $\{X_t : t \in \mathbb{Z}^d\}$, the covariance function takes the form

$$\sigma_h = \int_{(-\pi,\pi)^d} \exp(ih^T\omega)f(\omega)\, d\omega = \int_{(-\pi,\pi)^d} \cos(h^T\omega)f(\omega)\, d\omega,\ h \in \mathbb{Z}^d. \tag{A.40}$$

(h) *Lattice approximations for continuous processes*. Consider a stationary spatial process $X(t), t \in \mathbb{R}^d$ with covariance function $\sigma(h)$ and spectral density $f(\omega)$ satisfying (A.39). Consider approximating the continuous process by a stationary discrete process on the δ-lattice $\delta\mathbb{Z}^d$, where $\delta > 0$ is a small resolution parameter. Let $X_\delta(t),\ t = k\delta,\ k \in \mathbb{Z}^d$ denote the approximating process and let $f_\delta(\omega)$ denote its spectral density, with support on $(-\pi/\delta, \pi/\delta)^d$. The approximating covariance function is then

$$\sigma_\delta(h) = \int_{(-\pi/\delta,\pi/\delta)^d} \exp\{2\pi h^T\omega\}f_\delta(\omega)\, d\omega, \quad h = k\delta,\ k \in \mathbb{Z}^d.$$

Here $\delta > 0$ is a small resolution parameter. Provided the approximating spectral densities can be uniformly bounded by an integrable function, i.e.

$$f_\delta(\omega) \le g(\omega) \text{ for all } \omega \in \mathbb{R}^d,$$

for some function g satisfying $\int_{\mathbb{R}^d} g(\omega)\, d\omega < \infty$, the approximating process converges to the original continuous process. Here are two natural ways to construct an approximating spectral density.

(i) *Aliasing*. Define the approximating spectral density by combining aliased frequencies

$$f_\delta(\omega) = \sum_{m\in\mathbb{Z}^d} f(\omega + \delta^{-1}m), \quad \omega \in (-\pi/\delta, \pi/\delta)^d.$$

Under this approximation, the approximating covariance function on the δ-lattice is identical to the original covariance function, restricted to the δ-lattice, $\sigma_\delta(\delta k) = \sigma(\delta k),\ k \in \mathbb{Z}^d$.

(ii) *Truncation*. Define

$$f_\delta(\omega) = f(\omega)I[\omega \in (-\pi/\delta, \pi/\delta)^d].$$

so that the approximating spectral density is the same as the original spectral density on $(-\pi/\delta, \pi/\delta)^d$. In this case $\sigma_\delta(\delta k) \neq \sigma(\delta k)$.

A.6 Generalizations of the Fourier Transform

If the integrability conditions on a function or sequence f are relaxed, then it is still possible to define a Fourier transform, but the Fourier transform is no longer a simple function. These considerations form the basis of the spectral representations for intrinsic and generalized random fields in Chapter 3, where the details are spelled out.

(a) In particular, if f is allowed to be nonintegrable or nonsummable for large u or k, respectively, in Settings 1 and 2, then the Fourier transform can still be defined as a *generalized function*.
(b) On the other hand, nonintegrability of $f(u)$ for u near 0 can be allowed in Settings 1 and 3; the price paid is that the Fourier transform is no longer a single-valued function, but is now an *equivalence class of functions*. A similar extension holds in Settings 2 and 4 if the coefficient f_0 for the constant term in the Fourier transform is dropped from the sum.

A.7 Discrete Fourier Transform and Matrix Algebra

There is a link between the DFT in Setting 4 and the spectral decomposition theorem for real symmetric (or complex Hermitian) matrices.

A.7.1 DFT in $d = 1$ Dimension

To understand the connection, start in dimension $d = 1$ with period $n \geq 2$. Define an $n \times n$ complex-valued matrix $G = G_n^{(\mathrm{DFT,com})}$ with entries

$$g_{jk} = n^{-1/2} \exp\{2\pi i jk/n\}, \quad 0 \leq j, k \leq n-1. \tag{A.41}$$

Then G is a *unitary* matrix, i.e.

$$GG^* = I_n,$$

where * denotes complex conjugate transpose, since GG^* has elements

$$(GG^*)_{j_1 j_2} = \sum_{k=0}^{n-1} g_{j_1 k}\bar{g}_{j_2 k} = (1/n)\sum_{k=0}^{n-1} \exp\{2\pi i k(j_1 - j_2)/n\} = \begin{cases} 1, & j_1 = j_2, \\ 0, & j_1 \neq j_2. \end{cases}$$

To verify this statement, note that if $j_1 = j_2$, then the (j_1, j_2) element reduces to $1/n$ times a sum of ones. If $j_1 \neq j_2$, then the (j_1, j_2) element reduces to a geometric series

$$(1/n)\sum_{k=0}^{n-1} \alpha^k = (1/n)(\alpha^0 - \alpha^n) = 0,$$

where $\alpha = \exp\{2\pi i(j_1 - j_2)/n\}$, since $\alpha^n = \alpha^0 = 1$.

Let $\boldsymbol{f} = [f_k,\ k = 0, \ldots, n-1]^T$ denote a vector of n values indexed by the cyclic integers. Then the DFT is another n-vector $\tilde{\boldsymbol{f}} = [\tilde{f}_j,\ j = 0, \ldots, n-1]^T$ also indexed by the cyclic integers. In matrix form (A.32) becomes

$$\tilde{\boldsymbol{f}} = G\boldsymbol{f}.$$

That is, the jth element of $\tilde{\boldsymbol{f}}$ is given by

$$\tilde{f}_j = n^{-1/2} \sum_{k=0}^{n-1} f_k \exp\{2\pi ijk/n\}, \quad j = 0, \ldots, n-1.$$

A.7.2 Properties of the Unitary Matrix G, $d = 1$

Again let $G = G_n^{(\mathrm{DFT,com})}$ denote the complex unitary matrix used in the DFT.

1. The columns $\mathbf{g}_{(k)}$, say, of G can be split into different types.
   (a) $k = 0$. In this case, $g_{jk} = n^{-1/2}$ is constant and real-valued for all j.
   (b) $1 \le k < n/2$ or $n/2 < k \le n-1$. In this case, columns k and $n-k$ are complementary in the sense that

$$g_{jk} + g_{j,n-k} = (2/n^{1/2}) \cos(2\pi jk/n) \tag{A.42}$$

   is real-valued, and

$$g_{jk} - g_{j,n-k} = (2i/n^{1/2}) \sin(2\pi jk/n) \tag{A.43}$$

   is imaginary.
   (c) If n is even, then there is an additional column for $k = n/2$, with entries

$$g_{jk} = n^{-1/2}(-1)^j.$$

2. If a vector $\boldsymbol{f}$ is real-valued and symmetric, i.e. $f_k = f_{n-k}$ for $0 < k < n/2$, then the Fourier transform is real-valued since $f_k g_{jk} + f_{n-k} g_{j,n-k} = 2f_k \cos\{2\pi jk/n\}$, $0 < k < n/2$ is real-valued, and since g_{j0} and $g_{j,n/2}$ (n even) are real-valued.
3. Although the DFT is most elegantly defined using the unitary matrix $G_n^{(\mathrm{DFT,com})}$ of complex numbers, it can also be constructed using an orthogonal matrix $G_n^{(\mathrm{DFT,rea})}$, say, of real numbers whose columns are constructed as follows, where the rows of $G_n^{(\mathrm{DFT,rea})}$ are indexed by $j = 0, \ldots, n-1$ and the columns are indexed by $k = 0, \ldots, n-1$. To avoid notational overload, write $G_n^{(\mathrm{DFT,rea})} = G^\dagger$ for this section. The columns of $G^\dagger$ are defined as follows:
   (a) For $k = 0$, let $g_{jk}^\dagger = g_{jk} = n^{-1/2}$.
   (b) For $1 \le k < n/2$,

$$g_{jk}^\dagger = \left(\frac{1}{\sqrt{2}}\right)(g_{jk} + g_{j,n-k}) = (2/n)^{1/2} \cos\{2\pi jk/n\}, \tag{A.44}$$

$$g^{\dagger}_{j,n-k} = \left(-\frac{i}{\sqrt{2}}\right)(g_{jk} - g_{j,n-k}) = (2/n)^{1/2}\sin\{2\pi jk/n\}. \tag{A.45}$$

That is, columns k and $n-k$ for $G_n^{(\text{DFT,rea})}$ are proportional to the sum and difference of columns k and $n-k$ for $G_n^{(\text{DFT,com})}$, respectively.

(c) For $k = n/2$, $g^{\dagger}_{jk} = g_{jk} = n^{-1/2}(-1)^j$. This column is present only if n is even.

A.7.3 Circulant Matrices and the DFT, $d = 1$

Fix $n \geq 2$ and let $\boldsymbol{\alpha} = [\alpha_j : j = 0, \ldots, n-1]^T$ be a given vector indexed by the cyclic integers. Define an $n \times n$ matrix $A = \text{circ}(\boldsymbol{\alpha})$ by

$$a_{j_1 j_2} = \alpha_{(j_1 - j_2) \bmod n}, \quad j_1, j_2 = 0, \ldots, n-1.$$

As defined in Section A.3.8, such a matrix is called a *circulant matrix*. All the rows are cyclic permutations of one another.

Circulant matrices are special because a version of the spectral decomposition of Section A.3.1 is still valid, even though A is not necessarily symmetric. Further, the spectral decomposition is closely related to the DFT. In particular, the eigenvectors of A are given by the columns of $G_n^{(\text{DFT,com})}$ in (A.41) and the vector of eigenvalues $\boldsymbol{\lambda} = [\lambda_0, \ldots, \lambda_{n-1}]^T$ is given by $n^{1/2}$ times the inverse DFT of $\boldsymbol{\alpha}$. To verify this claim, note that

$$\begin{aligned}(A\mathbf{g}_{(k)})_h &= \sum_{j=0}^{n-1} a_{hj} g_{jk} \\ &= n^{-1/2} \sum_{j=0}^{n-1} \alpha_{(h-j) \bmod n} \exp(2\pi ijk/n) \\ &= n^{-1/2} \sum_{j'=0}^{n-1} \alpha_{j' \bmod n} \exp\{2\pi ik(h-j')/n\} \\ &= \lambda_k g_{hk},\end{aligned} \tag{A.46}$$

where

$$\lambda_k = \sum_{j=0}^{n-1} e^{-2\pi ijk/n} \alpha_j$$

and $\mathbf{g}_{(k)}$ denotes the kth column of $G_n^{(\text{DFT,com})}$. The third line of the derivation makes the substitution $j' = h - j$ and uses the fact that as j ranges between 0 and $n-1$, j' mod n also ranges between 0 and $n-1$. Further, it is only the value of j' mod n that matters in the exponential term since if $j' = mn$ is a multiple of n, $\exp\{2\pi ikmn/n\} = \exp\{2\pi ikm\} = 1$ factors out of the formula.

In vector form, with $G = G_n^{(\text{DFT,com})}$, (A.46) can be written as $A\mathbf{g}_{(k)} = \lambda_k \mathbf{g}_{(k)}$, $k = 0, \ldots, n-1$, which in matrix form becomes $AG = G\Lambda$. Multiplying both sides by G^{-1} yields

$$A = G\Lambda G^{-1} = G\Lambda G^{*}.$$

The last form is the same as the spectral decomposition in (A.9) and follows since G is a unitary matrix, $GG^* = I$, so that $G^{-1} = G^*$.

In the language of Setting 4 in Section A.4, if we identify the vectors $\boldsymbol{f} = n^{-1/2}\boldsymbol{\lambda}$, $\tilde{\boldsymbol{f}} = \boldsymbol{\alpha}$, then $\boldsymbol{\alpha}$ is the DFT of $n^{-1/2}\boldsymbol{\lambda}$ and $n^{-1/2}\boldsymbol{\lambda}$ is the inverse discrete Fourier transform (IDFT) of $\boldsymbol{\alpha}$.

If $\alpha_j = \alpha_{n-j}$ for all $j = 0, \dots, n-1$ so that A is a symmetric matrix ($a_{j_1 j_2} = a_{j_2 j_1}$, $0 \le j_1, j_2 \le n-1$), then the spectral decomposition can also be written in real coordinates. In particular, the eigenvalues are real

$$\lambda_k = \lambda_{n-k} = \sum_{j=0}^{n-1} \cos\{2\pi jk/n\}\, \alpha_j, \qquad k = 0, \dots, n-1,$$

and the complex eigenvector matrix $G_n^{(\mathrm{DFT,com})}$ can be replaced by the real eigenvector matrix $G_n^{(\mathrm{DFT,rea})}$ defined in (A.44) and (A.45). The representation $A = G_n^{(\mathrm{DFT,rea})}\Lambda\{G_n^{(\mathrm{DFT,rea})}\}^T$ can be viewed as an example of the real spectral decomposition (A.7) in Section A.3.1.

A.7.4 The Case $d > 1$

The preceding discussion has assumed dimension $d = 1$. To deal with higher dimensions, block circulant matrices are needed.

Consider the lattice torus of Section A.1, $\mathbb{Z}_N^d = \prod_{\ell=1}^d \{0, \dots, n[\ell]-1\} \subset \mathbb{Z}^d$, with opposite faces treated as adjacent, where $N = (n[1], \dots, n[d])$. Write $j_1 = j_2 \text{ Mod } N$ if $j_1[\ell] - j_2[\ell]$ is an integer multiple of $n[\ell]$ for each $\ell = 1, \dots, d$. Let $|N| = n[1] \times \cdots \times n[d]$ denote the number of elements of $\mathbb{Z}_N^d$. Starting with a vector $\boldsymbol{\alpha} = \{\alpha_j : j \in \mathbb{Z}_N^d\}$ (with the elements listed in lexicographic order), define a $|N| \times |N|$ matrix $A = \mathrm{Circ}(\boldsymbol{\alpha})$ with entries

$$a_{j_1 j_2} = \alpha_{(j_1 - j_2)\text{ Mod } N}, \quad j_1, j_2 \in \mathbb{Z}_N^d.$$

Such a matrix is called *block-circulant*. Suppose $\alpha_j = \alpha_{N-j \text{ Mod } N}$ for all j, so that A is symmetric, where $j = (j[1], \dots, j[d])$ is a multi-index. Then, the eigenvalues of A are given by

$$\lambda_k = \sum_{j\in\mathbb{Z}_N^d} \exp\left\{-2\pi i \sum_{\ell=1}^d j[\ell]k[\ell]/n[\ell]\right\} \alpha_j, \quad k \in \mathbb{Z}_N^d,$$

$$= \sum_{j\in\mathbb{Z}_N^d} \cos\left\{2\pi \sum_{\ell=1}^d j[\ell]k[\ell]/n[\ell]\right\} \alpha_j;$$

note $\lambda_k = \lambda_{-k \text{ Mod } N}$. The corresponding eigenvector, $\boldsymbol{w}_{(k)}$, say, written in complex notation, has entries

$$w_{j;k} = |N|^{-1/2} \exp\left\{2\pi i \sum_{\ell=1}^d j[\ell]k[\ell]/n[\ell]\right\}.$$

The eigenvectors $\boldsymbol{w}_{(k)}$, listed in lexicographic ordering, can be combined together to form the matrix of eigenvectors. This matrix can be written as the tensor product

$$G_N^{(\mathrm{DFT,com})} = G_{n[1]}^{(\mathrm{DFT,com})} \otimes \cdots \otimes G_{n[d]}^{(\mathrm{DFT,com})}, \tag{A.47}$$

of one-dimensional unitary matrices.

A.7.5 The Periodogram

Let $\{x_j : j \in \mathbb{Z}_N^d\}$ be a collection of observations on the lattice torus. The *periodogram* is defined by the function

$$I(\omega) = \left| \sum_{j \in \mathbb{Z}_N^d} x_j \exp\{2\pi i j \omega\} \right|^2 / |N|, \quad \omega \in \mathbb{R}^d. \tag{A.48}$$

In particular, if $\omega_k = (k[1]/n[1], \ldots, k[d]/n[d])$, $k \in \mathbb{Z}_N^d$, then $I(\omega_k)$ is the same as the squared absolute value of the kth element of $G\boldsymbol{x}$, with $G = G_N^{(\mathrm{DFT,com})}$ given by (A.47). The periodogram was called the biased periodogram in (6.46).

A.8 Discrete Cosine Transform (DCT)

A transform related to the DFT is the discrete cosine transform (DCT); it is useful for data with reflecting boundary conditions. As with the DFT, it is helpful to describe the one-dimensional situation first, and then the higher dimensional case.

In the literature, the DCT comes in several different versions. However, for the purposes of this book, only one version and its inverse are needed.

A.8.1 One-dimensional Case

Define an $n \times n$ matrix $C = C_n$ with entries

$$c_{jk} = \alpha_k \cos\{\pi k (j + 1/2)/n\}, \quad j, k = 0, \ldots, n-1, \tag{A.49}$$

where $\alpha_k = \sqrt{1/n}$ if $k = 0$ and $\alpha_k = \sqrt{2/n}$ if $1 \le k \le n-1$. Using the same arguments as in Section A.7.2, it can be checked that the columns of C are orthonormal, so that C is an orthogonal matrix.

Given a vector $\boldsymbol{f} = [f_0, \ldots, f_{n-1}]^T$, the *DCT* of $\boldsymbol{f}$ is defined by

$$\hat{\boldsymbol{f}} = C\boldsymbol{f}, \tag{A.50}$$

and the *inverse discrete cosine transform (IDCT)* of $\hat{\boldsymbol{f}}$ is defined by

$$\boldsymbol{f} = C^T\hat{\boldsymbol{f}}. \tag{A.51}$$

The IDCT arises in the treatment of reflecting boundary conditions for spatial data; see Sections 4.6.5, 6.3, and 6.5. Mathematical details are given below in Sections A.10–A.11. For $n > 1$, let $\boldsymbol{x} = [x_0, \ldots, x_{n-1}]^T$ denote a set of spatial data in one dimension. Define the "doubled" data set $\boldsymbol{y} = [x_0, \ldots, x_{n-1}, x_{n-1}, \ldots, x_0]^T$ of length $2n$, so that $y_j = y_{2n-1-j} = x_j$, $j = 0, \ldots, n-1$.

From (A.36), the IDFT of $\boldsymbol{y}$ is given by

$$\begin{aligned}
\text{IDFT}(\boldsymbol{y})_k &= (2n)^{-1/2} \sum_{j=0}^{2n-1} \exp\{-2\pi i jk/(2n)\} y_j \\
&= (2n)^{-1/2} \sum_{j=0}^{n-1} \left[\exp\{-2\pi i jk/(2n)\} + \exp\{-2\pi ik(2n-1-j)/(2n)\}\right] x_j \\
&= (2n)^{-1/2} \exp\{\pi ik/(2n)\} \sum_{j=0}^{n-1} [\exp\{-2\pi ik(j+1/2)/(2n)\} \\
&\qquad + \exp\{2\pi ik(j+1/2)/(2n)\}] x_j \\
&= (2/n)^{1/2} \exp\{\pi ik/(2n)\} \sum_{j=0}^{n-1} \cos\{\pi k(j+1/2)/n\} x_j \\
&= \alpha_k^{-1} (2/n)^{1/2} \exp\{\pi ik/(2n)\} \sum c_{jk} x_j.
\end{aligned}$$

Thus up to scaling constants, the IDCT of $\boldsymbol{x}$ is the same as the IDFT of $\boldsymbol{y}$. Following the convention in Section A.5, we use the adjective "inverse" in the name for the transformation in (A.51) since it takes a set of values in the spatial domain to a set of values in the frequency domain.

A.8.2 The Case $d > 1$

In higher dimensions, the DCT and IDCT can be defined using tensor products. For example, if $\{x_j : j \in \mathbb{Z}_N^d\}$ is an array of values on the lattice torus in d dimensions, listed in lexicographic order as $\boldsymbol{x}$, say, then the IDCT on a domain of size $n[1] \times \cdots \times n[d]$ can be defined by

$$\text{IDCT}(\boldsymbol{x}) = \{C_{n[1]}^T \otimes \cdots \otimes C_{n[d]}^T\} \boldsymbol{x},$$

where $C_{n[\ell]}$ denotes the matrix from (A.49) in $n[\ell]$ dimensions.

A.8.3 Indexing for the Discrete Fourier and Cosine Transforms

Start with $d = 1$ dimension. The DFT is a map from the frequency domain ($k \in \mathbb{Z}_n$) to the spatial domain ($j \in \mathbb{Z}_n$). The inverse DFT is a map in the opposite direction.

To represent the mapping explicitly, it is necessary to choose a range of n consecutive integers for j in $\mathbb{Z}_n$. The choice made above is $j, k = 0, \ldots, n-1$. However, in practical settings, an alternative choice is $j, k = 1, \ldots, n$. The effect on $G = G_n^{(\text{DFT,com})}$ and $C = C_n$ is to shift the indexing so that G and C now have elements

$$g_{jk} = \exp\{2\pi i(j-1)(k-1)/n\}, \quad c_{jk} = \alpha_k \cos\{\pi(j-1/2)(k-1)/n\}, \tag{A.52}$$

$j, k = 1, \ldots, n$. It is in this reindexed form that these transforms are used below.
The same changes carry over to $d > 1$ dimensions.

A.9 Periodic Approximations to Sequences

Starting from a sequence $\{\alpha_h, h \in \mathbb{Z}^d\}$, and a collection of periods $N = (n[1], \ldots, n[d])$ all greater than or equal to 2, it is possible to approximate sequence $\{\alpha_h\}$ by a periodic sequence $\{\alpha_h^{(C)}\}$ (C for circulant) as follows:

(a) The one-dimensional case, $d = 1$. Given a period $N = n[1] = n \geq 2$, set

$$\alpha_h^{(C)} = \begin{cases} \alpha_h, & |h| < n/2, \\ \frac{1}{2}(\alpha_h + \alpha_{-h}), & h = n/2. \end{cases}$$

This formula defines $\alpha_h^{(C)}$ for n successive values of h, $-n/2 < h \leq n/2$, whether or not n is even, and this definition can be extended periodically to all h by requiring $\alpha_h^{(C)}$ to depend only on h mod n.
Note that if n is even, there is ambiguity about how to define $\alpha_h^{(C)} = \alpha_{-h}^{(C)}$ at lag $h = n/2$; the ambiguity is resolved here by taking the average of two values. The ambiguity is not present when $\{\alpha_h\}$ is symmetric ($\alpha_h = \alpha_{-h}$ for all h) or when n is odd.

(b) The two-dimensional case, $d = 2$. Given a period vector $N = (n[1], n[2])$, define a periodic approximation $\{\alpha_h^{(C)}\}$ by

$$\alpha_h^{(C)} = \begin{cases} \alpha_h, & |h[1]| < n[1]/2 \text{ and } |h[2]| < n[2]/2, \\ \frac{1}{2}\{\alpha_{(h[1],h[2])} + \alpha_{(-h[1],h[2])}\}, & h[1] = n[1]/2 \text{ and } |h[2]| < n[2]/2, \\ \frac{1}{2}\{\alpha_{(h[1],h[2])} + \alpha_{(h[1],-h[2])}\}, & |h[1]| < n[1]/2 \text{ and } h[2] = n[2]/2, \\ \frac{1}{4}\{\alpha_{(h[1],h[2])} + \alpha_{(-h[1],h[2])} + \alpha_{(h[1],-h[2])} + \alpha_{(-h[1],-h[2])}\}, & \\ & h[1] = n[1]/2 \text{ and } h[2] = n[2]/2. \end{cases}$$

This formula defines $\alpha_h^{(C)}$ for an $n[1] \times n[2]$ block of values for h, $-n[1]/2 < h[1] \leq n[1]/2, -n[2]/2 < h[2] \leq n[2]/2$, whether or not $n[1]$ and $n[2]$ are even, and this definition can be extended periodically to all h by requiring $\alpha_h^{(C)}$ to depend only on h Mod N.

(c) The higher dimensional case, $d > 2$. The same construction can be extended to higher dimensions, at the price of more cumbersome notation.

This construction has been used in Chapter 2 to give a fast and exact method for simulating a stationary Gaussian process on a finite rectangular domain D in $\mathbb{Z}^d$. The method is based on the DFT and merely assumes an explicit form for the covariance function.

A.10 Structured Matrices in $d = 1$ Dimension

This section describes three types of structured matrix that specify neighborhood relationships for a sequence of sites in $d = 1$ dimension. These matrices can form building blocks (Section A.11) for finite approximations to an inverse covariance matrix for a stationary random field in $d = 1$ and higher dimensions.

The three types of matrices to consider are called *Toeplitz, circulant, and folded circulant*, respectively. Table A.3 illustrates the corresponding boundary conditions in one dimension.

In each case, the general matrix can be viewed as a linear combination of elementary building blocks. The building blocks are matrices with zeros everywhere except for ones in a band about the diagonal, and perhaps a few other places. In this section, we describe the form of the building blocks.

Let $s, t = 1, \dots, n$ denote $n \geq 2$ equally spaced "sites" in $d = 1$ dimension, and let $r \in \mathbb{Z}$ denote a lag. First define two $n \times n$ matrices, the *banded matrix* $B^{(n,r)}$ and the *circulant banded matrix* $C^{(n,r)}$, by

$$b_{st} = \begin{cases} 1, & t - s - r = 0, \\ 0, & \text{otherwise}, \end{cases} \tag{A.53}$$

$$c_{st} = \begin{cases} 1, & t - s - r = 0 \bmod n, \\ 0, & \text{otherwise}. \end{cases} \tag{A.54}$$

Next we use the banded matrices to construct elementary neighborhood matrices. Various choices for a neighborhood of order $|r|$ can be made, with the information in each case stored in an $n \times n$ matrix. If sites s and t are separated by a lag $+r$ or $-r$, an entry of $1/2$ is made to the elementary neighborhood matrix. Occasionally, a double entry $2 \times 1/2 = 1$ is made to the matrix if both lag separations hold, e.g. when $r = 0$ and $s = t$. A double entry also occurs in the circulant case when n is even, $r = n/2$ and $t = s + r \bmod n = s - t \bmod n$. All other entries in the neighborhood matrix are 0. In the definitions below, the indices are restricted to the range $|r|, |s| \leq n - 1$ to avoid vacuous matrices. Table A.4 provides some examples of these matrices.

(a) *Elementary Toeplitz neighborhood matrix of order r,*

$$U^{(n,r)} = U^{(n,-r)} = \{B^{(n,r)} + B^{(n,-r)}\}/2. \tag{A.55}$$

Each row and column has either 0, 1, or 2 nonzero entries. If $r = 0$, then $U^{(n,0)} = I_n$, the identity matrix. Otherwise, for $r \neq 0$, each nonzero entry equals $1/2$.

(b) *Elementary circulant neighborhood matrix of order r,*

$$V^{(n,r)} = V^{(n,-r)} = \{C^{(n,r)} + C^{(n,-r)}\}/2. \tag{A.56}$$

In this case, a *periodic* or *circulant* boundary condition is imposed so that sites 1 and n are treated as adjacent to each other. The matrix $V^{(n,r)}$ is similar to $U^{(n,r)}$, but with extra elements in the upper-right and lower- left corners, corresponding to an increased number of neighbors.

The Toeplitz and circulant matrices are related by

$$V^{(n,r)} = U^{(n,r)} + U^{(n,n-r)}.$$

If $r = 0$, then $V^{(n,0)} = I_n$, the identity matrix. The family of circulant matrices is closed under matrix multiplication

$$V^{(n,r)}V^{(n,s)} = \{V^{(n,r+s)} + V^{(n,r-s)}\}/2, \tag{A.57}$$

where $r + s$ and $r - s$ are interpreted mod n. The formula (A.57) is a matrix version of the corresponding identity for complex numbers

$$\left(\frac{e^{ir\omega} + e^{-ir\omega}}{2}\right)\left(\frac{e^{is\omega} + e^{-is\omega}}{2}\right) = \frac{1}{2}\left(\frac{e^{i(r+s)\omega} + e^{-i(r+s)\omega}}{2} + \frac{e^{i(r-s)\omega} + e^{i(s-r)\omega}}{2}\right),$$

for any real number ω.

(c) *Elementary folded circulant matrix of order r*, denoted $W^{(n,r)} = W^{(n,-r)}$. The easiest way to understand this matrix is by using a reflecting boundary. Starting with the original list of sites $1, \ldots, n$, append a copy of the list in reverse order to get the doubled list $1, \ldots, n, n, \ldots, 1$, and use a periodic boundary condition on the doubled list of sites. Let t be a site in the original list, $1 \le t \le n$. Say that another site s is an rth-order folded circulant neighbor, $1 \le r \le n/2$, if either of the two copies of s in the doubled list is an rth-order circulant neighbor. Then divide the number of neighbors by two since each pairing appears twice. Note that site 1 is a first-order neighbor of itself; similarly for site n. Hence the matrix $W^{(5,1)}$ in Table 5.4 has the entry 1 in the diagonal positions $(1, 1)$ and $(5, 5)$. A more explicit formula for the elements of $W^{(n,r)} = (w_{st})$ is given in terms of the elements of the doubled circulant matrix $V^{(2n,r)} = (v_{st})$,

$$w_{st} = \{v_{st} + v_{s't} + v_{st'} + v_{s't'}\}/2, \quad s, t = 1, \ldots, n, \tag{A.58}$$

where $s' = 2n + 1 - s$, $t' = 2n + 1 - t$. Each row and column has one or two nonzero entries and sums to 1. For $r = 0$, the matrix reduces to the identity matrix, $W^{(n,0)} = I_n$.

The folded circulant matrices are important because a circulant quadratic form in the doubled data equals twice a folded circulant quadratic form in the original data. That is, for $1 \le r < n$,

$$\boldsymbol{y}^T V^{(2n,r)} \boldsymbol{y} = 2\boldsymbol{x}^T W^{(n,r)} \boldsymbol{x}. \tag{A.59}$$

Further, the family of folded circulant matrices is closed under matrix multiplication

$$W^{(n,r)}W^{(n,s)} = \{W^{(n,r+s)} + W^{(n,r-s)}\}/2, \quad r, s \ge 0, \tag{A.60}$$

where again $r + s$ and $r - s$ are interpreted mod n, using the same arguments as in (A.57).

Table A.3 illustrates the three boundary conditions and Table A.4 illustrates the corresponding matrices.

A key advantage of the elementary circulant matrices $V^{(n,r)}$ is that they all have the same eigenvectors, given by the columns of $G_n^{(\mathrm{DFT,com})}$ in (A.52) using complex coordinates or by the columns of $G_n^{(\mathrm{DFT,rea})}$ in (A.44) and (A.45) using real coordinates.

Similarly, a key advantage of the elementary folded circulant matrices $W^{(n,r)}$ is that they all have the same eigenvectors, given by the columns of C in (A.52), because they are constructed from a circulant matrix of twice the size (see (A.59)).

A.11 Matrix Approximations for an Inverse Covariance Matrix

Let $\{\sigma_h, h \in \mathbb{Z}^d\}$ be a stationary covariance function with spectral representation

$$\sigma_h = \int_{(-\pi,\pi)^d} \cos(h^T\omega) f(\omega)\, d\omega.$$

Given a dimension vector $N = (n[1], \ldots, n[d])$ of size $|N|$, consider the domain

$$D = \{j \in \mathbb{Z}^d : 1 \le j[\ell] \le n[\ell],\ \ell = 1, \ldots, d\} \tag{A.61}$$

and define an $|N| \times |N|$ symmetric Toeplitz matrix Σ with entries $\sigma_{jk} = \sigma_{j-k}$, $j, k \in D$.

Given data $\{x_t,\ t \in D\}$, the likelihood function involves the quadratic form $\boldsymbol{x}^T\Sigma^{-1}\boldsymbol{x}$ and the normalizing constant $\log|\Sigma|$. These quantities are typically

Table A.3 Three types of boundary condition for a one-dimensional set of data, $x_1, \ldots, x_n$.

Layout	Name
$-x_1, \ldots, x_n-$	Null boundary condition
$\to x_1, \ldots, x_n \to$	Periodic boundary condition
$\to x_1, \ldots, x_n x_n, \ldots, x_1 \to$	Reflecting boundary condition

The hyphen "-" in the first row indicates that x_1 and x_n have no neighbors to the left and right, respectively. The arrows "→" in the second row indicate a periodic boundary condition so that x_1 is adjacent to x_n. The arrows in the third row indicate a periodic boundary condition for the doubled data, so that the left x_1 is adjacent to the right x_1.

Table A.4 Some examples of Toeplitz, circulant, and folded circulant matrices $U^{(n,r)}$, $V^{(n,r)}$, $W^{(n,r)}$, respectively, for $n = 5$ and $r = 1,2,3,4$.

Toeplitz	Circulant	Folded circulant
$U^{(5,1)} = \begin{bmatrix} 0 & 1 & 0 & 0 & 0 \\ 1 & 0 & 1 & 0 & 0 \\ 0 & 1 & 0 & 1 & 0 \\ 0 & 0 & 1 & 0 & 1 \\ 0 & 0 & 0 & 1 & 0 \end{bmatrix}$	$V^{(5,1)} = \begin{bmatrix} 0 & 1 & 0 & 0 & 1 \\ 1 & 0 & 1 & 0 & 0 \\ 0 & 1 & 0 & 1 & 0 \\ 0 & 0 & 1 & 0 & 1 \\ 1 & 0 & 0 & 1 & 0 \end{bmatrix}$	$W^{(5,1)} = \begin{bmatrix} 1 & 1 & 0 & 0 & 0 \\ 1 & 0 & 1 & 0 & 0 \\ 0 & 1 & 0 & 1 & 0 \\ 0 & 0 & 1 & 0 & 1 \\ 0 & 0 & 0 & 1 & 1 \end{bmatrix}$
$U^{(5,2)} = \begin{bmatrix} 0 & 0 & 1 & 0 & 0 \\ 0 & 0 & 0 & 1 & 0 \\ 1 & 0 & 0 & 0 & 1 \\ 0 & 1 & 0 & 0 & 0 \\ 0 & 0 & 1 & 0 & 0 \end{bmatrix}$	$V^{(5,2)} = \begin{bmatrix} 0 & 0 & 1 & 1 & 0 \\ 0 & 0 & 0 & 1 & 1 \\ 1 & 0 & 0 & 0 & 1 \\ 1 & 1 & 0 & 0 & 0 \\ 0 & 1 & 1 & 0 & 0 \end{bmatrix}$	$W^{(5,2)} = \begin{bmatrix} 0 & 1 & 1 & 0 & 0 \\ 1 & 0 & 0 & 1 & 0 \\ 1 & 0 & 0 & 0 & 1 \\ 0 & 1 & 0 & 0 & 1 \\ 0 & 0 & 1 & 1 & 0 \end{bmatrix}$
$U^{(5,3)} = \begin{bmatrix} 0 & 0 & 0 & 1 & 0 \\ 0 & 0 & 0 & 0 & 1 \\ 0 & 0 & 0 & 0 & 0 \\ 1 & 0 & 0 & 0 & 0 \\ 0 & 1 & 0 & 0 & 0 \end{bmatrix}$	$V^{(5,3)} = \begin{bmatrix} 0 & 0 & 1 & 1 & 0 \\ 0 & 0 & 0 & 1 & 1 \\ 1 & 0 & 0 & 0 & 1 \\ 1 & 1 & 0 & 0 & 0 \\ 0 & 1 & 1 & 0 & 0 \end{bmatrix}$	$W^{(5,3)} = \begin{bmatrix} 0 & 0 & 1 & 1 & 0 \\ 0 & 1 & 0 & 0 & 1 \\ 1 & 0 & 0 & 0 & 1 \\ 1 & 0 & 0 & 1 & 0 \\ 0 & 1 & 1 & 0 & 0 \end{bmatrix}$
$U^{(5,4)} = \begin{bmatrix} 0 & 0 & 0 & 0 & 1 \\ 0 & 0 & 0 & 0 & 0 \\ 0 & 0 & 0 & 0 & 0 \\ 0 & 0 & 0 & 0 & 0 \\ 1 & 0 & 0 & 0 & 0 \end{bmatrix}$	$V^{(5,4)} = \begin{bmatrix} 0 & 1 & 0 & 0 & 1 \\ 1 & 0 & 1 & 0 & 0 \\ 0 & 1 & 0 & 1 & 0 \\ 0 & 0 & 1 & 0 & 1 \\ 1 & 0 & 0 & 1 & 0 \end{bmatrix}$	$W^{(5,4)} = \begin{bmatrix} 0 & 0 & 0 & 1 & 1 \\ 0 & 0 & 1 & 0 & 1 \\ 0 & 1 & 0 & 1 & 0 \\ 1 & 0 & 1 & 0 & 0 \\ 1 & 1 & 0 & 0 & 0 \end{bmatrix}.$

time-consuming to compute if D is large. Hence, it is useful to consider approximations for Σ^{-1}. The inverse covariance function, defined in Section A.11.1, provides the basis for various approximations.

A.11.1 The Inverse Covariance Function

For simplicity, assume the spectral density is bounded away from 0 and ∞

$$0 < c_1 \leq f(\omega) \leq c_2 < \infty, \tag{A.62}$$

and define the inverse covariance function by

$$\psi_h = \frac{1}{(2\pi)^{2d}} \int_{(-\pi,\pi)^d} \cos(h^T\omega) \, \{1/f(\omega)\} \, d\omega. \tag{A.63}$$

The scaling factor $(2\pi)^{-2d}$ is chosen so that the two sequences are inverse to each other

$$\sum \sigma_h \psi_{t-h} = \delta_t,$$

where δ_t is the Kronecker δ.

The bounds ensure that $\{\sigma_h\}$ and $\{\psi_h\}$ are square summable sequences. Thus, defining infinite Toeplitz matrices Σ_∞ and Ψ_∞ with entries $\sigma_{jk} = \sigma_{j-k}$, $\psi_{jk} = \psi_{j-k}$, $j, k \in \mathbb{Z}$, it follows that $\Sigma_\infty \Psi_\infty = I_\infty$, the infinite-dimensional identity matrix.

The inverse spectral density is particularly simple in the case of a conditional autoregression (CAR) model. In this case, ψ_h vanishes outside a small neighborhood of the origin. The approximations in this section are well suited to this situation.

A covariance function always has the property of *symmetry*, $\sigma_h = \sigma_{-h}$. In some cases, a covariance function has a stronger property of *full symmetry*, $\sigma_h = \sigma_k$ whenever $|h[\ell]| = |k[\ell]|$, $\ell = 1, \ldots, d$. The inverse covariance function possesses the same symmetry properties as the covariance function.

The representation in (A.63) uses cosines of a linear combinations of angles. Under the assumption of full symmetry, it is possible to rewrite this representation in terms of products of cosines of single angles

$$1/f(\omega) = \sum \psi_h \prod_{\ell=1}^{d} \cos(h[\ell]\omega[\ell]). \tag{A.64}$$

For example, in $d = 2$ dimensions, the coefficients $\psi_{1,1} = \psi_{1,-1}$ are equal under full symmetry and

$$\psi_{1,1} \cos(\omega[1] + \omega[2]) + \psi_{1,-1} \cos(\omega[1] - \omega[2]) = 2\psi_{1,1} \cos(\omega[1]) \cos(\omega[2]).$$

Using this principle repeatedly yields (A.64).

Example A.1 Consider a CAR model in $d = 2$ dimensions, defined with respect to the first-order full neighborhood of the origin. The spectral density is given by $f(\omega) = (2\pi)^{-d}\sigma_\eta^2/\tilde{b}(\omega)$ where

$$\begin{aligned}\tilde{b}(\omega) &= 1 - 2\beta_1 \cos(\omega[1]) - 2\beta_2 \cos(\omega[2]) \\ &\quad - 2\beta_3\{\cos(\omega[1] + \omega[2]) + \cos(\omega[1] - \omega[2])\} \\ &= 1 - 2\beta_1 \cos(\omega[1]) - 2\beta_2 \cos(\omega[2]) - 4\beta_3 \cos(\omega[1]) \cos(\omega[2]).\end{aligned}$$

This model can be viewed as a mild generalization of the first-order basic CAR model in $d = 2$ dimensions of Example 4.4. The horizontal and vertical autoregressive coefficients (β_1 and β_2) are allowed to be different, but coefficients for the diagonal neighbors must be the same (β_3). In Example 4.4, there was an "isotropy assumption" $\beta_1 = \beta_2$ and β_3 was not present.

The coefficients ψ_h are given by the Fourier coefficients of $\tilde{b}(\omega)$ divided by σ_η^2. □

A.11.2 The Toeplitz Approximation to Σ^{-1}

Let Ψ denote the finite $|N| \times |N|$ matrix obtained by restricting Ψ_∞ to the indices in D. Then Ψ is the a Toeplitz approximation to Σ^{-1}. In the finite case, Σ and Ψ are no longer exact inverses of each other, i.e. $\Sigma\Psi \neq I_{|N|}$.

Under the assumption of full symmetry, it is possible to express Ψ in terms of a matrix U built using the one-dimensional elementary Toeplitz matrices of Section A.10; that is,

$$\Psi = U = \sum_{h \in \mathbb{Z}^d} \psi_h \prod_{\ell=1}^{d} U^{(n[\ell], h[\ell])}.$$

The sum only involves a finite number of nonzero terms since $U^{n[\ell],h[\ell]} = 0$ whenever $|h[\ell] \geq n[\ell]$.

Provided that the inverse covariances ψ_h can be computed explicitly, it is straightforward to evaluate the quadratic form. However, there do not seem to be any simplifications available for the computation of $\log|\Psi|$. Hence, we look for more computationally tractable approximations.

A.11.3 The Circulant Approximation to Σ^{-1}

Let $\{\psi_h^{(C)}\}$ denote the periodic approximation of Section A.9 for the inverse covariance function $\{\psi_h\}$, and let $\Psi^{(C)}$ denote the corresponding circulant approximation for Ψ. Since $\Psi^{(C)}$ is circulant, it is straightforward to compute its eigenvalues and hence $\log|\Psi^{(C)}|$ using the DFT.

Under the assumption of full symmetry, it is possible to express $\Psi^{(C)}$ in terms of a matrix V built using the one-dimensional elementary circulant matrices of Section A.10; that is,

$$\Psi^{(C)} = V = \sum_{h \in \mathbb{Z}_N^d} \psi_h^{(C)} \prod_{\ell=1}^{d} V^{(n[\ell], h[\ell])}.$$

A.11.4 The Folded Circulant Approximation to Σ^{-1}

The folded circulant approximation is based on ideas in Besag and Mondal (2005) and Mondal (2018) and requires the assumption of full symmetry defined above (A.64) for the covariance and inverse covariance functions. It is expressed in terms of a matrix W built using the one-dimensional elementary folded circulant matrices of Section A.10; that is,

$$W = \sum \psi_h \prod_{\ell=1}^{d} W^{(n[\ell],|h[\ell]|)}, \tag{A.65}$$

where the sum ranges over lags h such that $|h[\ell]| \le n[\ell] - 1,\ \ell = 1, \dots, d$.

The folded circulant approximation can be best understood through the use of reflecting boundary conditions where the data are "doubled" by being reflected along each coordinate axis. The case $d = 1$ is illustrated in Table A.3. In $d = 2$ dimensions, the data on a rectangular region D can be represented by the matrix

$$\begin{bmatrix} x_{11} & \cdots & x_{1n[2]} \\ \vdots & \ddots & \vdots \\ x_{n[2]1} & \cdots & x_{n[1]n[2]} \end{bmatrix},$$

and the doubled data take the form

$$\begin{bmatrix} x_{11} & \cdots & x_{1n[2]} & x_{1n[2]} & \cdots & x_{11} \\ \vdots & \ddots & \vdots & \vdots & \ddots & \vdots \\ x_{n[1]1} & \cdots & x_{n[1]n[2]} & x_{n[1]n[2]} & \cdots & x_{n[1]1} \\ x_{n[1]1} & \cdots & x_{n[1]n[2]} & x_{n[1]n[2]} & \cdots & x_{n[1]1} \\ \vdots & \ddots & \vdots & \vdots & \ddots & \vdots \\ x_{11} & \cdots & x_{1n[2]} & x_{1n[2]} & \cdots & x_{11} \end{bmatrix}.$$

For formulas involving quadratic forms, the data should be represented as a vector using lexicographic ordering, rather than as a matrix.

A.11.5 Comments on the Approximations

In the following comments about W, the inverse covariance function is assumed fully symmetric; for the comments about $U = \Psi$ and $V = \Psi^{(C)}$, the inverse covariance function is not necessarily assumed to be fully symmetric.

1. Consider an inverse spectral density $g(\omega)$ for a stationary covariance function bounded away from 0 and ∞ as in (A.62), with inverse covariance function $\{\psi_h\}$ in (A.63). Then $g(\omega)$ is positive for all ω so that $\{\psi_h\}$ is a positive definite sequence. It follows that all the eigenvalues of U, V, and W are positive. The reasons are as follows.
   (a) Since Ψ is a finite submatrix of Ψ_∞ and since Ψ_∞ is positive definite, so is $U = \Psi$.
   (b) The eigenvalues of $V = \Psi^{(C)}$ are given by $(2\pi)^d g(\omega_k)$ at the discrete frequencies ω_k with entries $\omega_k[\ell] = 2\pi k[\ell]/n[\ell],\ 0 \le k[\ell] \le n[\ell] - 1,\ \ell = 1, \dots, d$.
   (c) The eigenvalues of W are given by $(2\pi)^d g(\omega_k)$ at the discrete frequencies ω_k with entries $\omega_k[\ell] = \pi k[\ell]/n[\ell],\ 0 \le k[\ell] \le n[\ell] - 1,\ \ell = 1, \dots, d$.

2. In a singular CAR model, the inverse spectral density vanishes at the origin, $g(0) = 0$. In this case, the eigenvalue for the constant eigenfunction vanishes, so V and W are only positive semidefinite, not positive definite. Provided $g(\omega) > 0$ for all $\omega \neq 0$, this is the only singularity that can arise in V and W.
3. An important feature for the circulant and folded circulant approximations is that the eigenvectors arise from circulant matrices. Hence, the DFT can be used to evaluate the log determinant in the likelihood more efficiently than using a direct calculation.

A.11.6 Sparsity

Informally, a matrix A is called *sparse* of most of its elements are 0. In the spatial setting, consider the domain D in (A.61) with dimension vector $N = (n[1], \dots, n[d])$, and let A be an $|N| \times |N|$ matrix with elements a_{st} indexed by sites $s, t \in D$, with components $1 \leq s[\ell], t[\ell] \leq n[\ell]$, $\ell = 1, \dots, d$.

A good example of sparsity arises when A is an approximation to an inverse covariance matrix for a CAR model. Then for some bound $c > 0$, the matrix elements have the property that

$$a_{st} = 0 \text{ if } |s - t| \geq c,$$

with opposite edges of the domain identified with one another in the circulant and folded cases. In particular, if $|N| \to \infty$, the proportion of nonzero elements in each row of A tends to 0.

In general, operations on large matrices are computationally expensive. For example, the computation of the inverse requires $O(|N|^3)$ operations. However, for sparse matrices specialized algorithms can be developed, which are much more efficient. See e.g. Rue and Held (2005) for a discussion of efficient algorithms in the spatial setting.

A.12 Maximum Likelihood Estimation

The previous sections set out the mathematical foundations needed for spatial analysis. This section and Section A.13 set out the statistical foundations for maximum likelihood estimation, especially for spatial analysis.

A.12.1 General Considerations

Let $\boldsymbol{x}$ be an n-dimensional random vector with probability density function $f(\boldsymbol{x}; \boldsymbol{\theta})$ depending on a p-dimensional parameter vector $\boldsymbol{\theta}$. Given a realization of $\boldsymbol{x}$, the maximum likelihood estimator (MLE) is defined to maximize the log-likelihood

$$\log L(\boldsymbol{\theta};\boldsymbol{x}) = \log f(\boldsymbol{x};\boldsymbol{\theta}). \tag{A.66}$$

The log-likelihood is the same as the log probability density function, but regarded as a function of $\boldsymbol{\theta}$, with $\boldsymbol{x}$ held fixed. The MLE is defined by

$$\hat{\boldsymbol{\theta}} = \text{argmax}_{\theta} \log L(\boldsymbol{\theta};\boldsymbol{x}). \tag{A.67}$$

The *score function* is defined as the derivative of the log-likelihood with respect to $\boldsymbol{\theta}$. Typically, the derivative of a function vanishes at the maximum, so the score function vanishes at the MLE

$$\frac{d \log L(\hat{\boldsymbol{\theta}};\boldsymbol{x})}{d\boldsymbol{\theta}} = \mathbf{0}.$$

(An exception can occur for a parameter with an attainable endpoint. For example, for a spatial model with a nugget variance $\tau^2 \geq 0$, the maximum can occasionally be attained on the boundary $\tau^2 = 0$.)

The expected value of the score derivative, after changing sign, is known as the *Fisher information matrix*,

$$\mathcal{I}(\boldsymbol{\theta}) = -E\left\{\frac{\partial^2 \log L}{\partial\boldsymbol{\theta}\partial\boldsymbol{\theta}^T}\right\}. \tag{A.68}$$

It can be estimated in practice by either the *observed Fisher information matrix*

$$\mathcal{I}_{\text{obs}} = -\frac{\partial^2 \log L}{\partial\boldsymbol{\theta}\partial\boldsymbol{\theta}^T}, \tag{A.69}$$

evaluated at $\hat{\boldsymbol{\theta}}$, or by the *expected Fisher information matrix*

$$\mathcal{I}_{\text{exp}} = \mathcal{I}(\hat{\boldsymbol{\theta}}). \tag{A.70}$$

An important question for the MLE is its accuracy. Standard results state that under mild regularity conditions for large n, the information grows at a rate of at least n, i.e. the eigenvalues of $\mathcal{I}(\boldsymbol{\theta})$ are at least as large as $O(n)$. Then the MLE is asymptotically normally distributed and unbiased with variance given by the inverse of the Fisher information matrix

$$\hat{\boldsymbol{\theta}} \sim N_p(\boldsymbol{\theta}, \mathcal{I}(\boldsymbol{\theta})^{-1}). \tag{A.71}$$

In the spatial setting, these regularity conditions are most commonly found in the setting of “outfill asymptotics”; see e.g. Section 5.5.3.

A.12.2 The Multivariate Normal Distribution and the Spatial Linear Model

An important example is given by the multivariate normal distribution, as developed by Mardia and Marshall (1984). To begin with, suppose the mean vanishes, so that

$$\boldsymbol{x} \sim N_n(\mathbf{0}, \Sigma), \tag{A.72}$$

where $\Sigma = \Sigma(\theta)$ is a covariance matrix depending on the parameter vector $\boldsymbol{\theta}$.

The log-likelihood is

$$\begin{aligned} \log L &= -\frac{1}{2}\{\boldsymbol{x}^T\Sigma^{-1}\boldsymbol{x} + \log|\Sigma| + n\log(2\pi)\} \\ &= -\frac{1}{2}\{\mathrm{tr}(\Sigma^{-1}\boldsymbol{x}\boldsymbol{x}^T) + \log|\Sigma| + n\log(2\pi)\}. \end{aligned} \tag{A.73}$$

Let

$$\Sigma_i = \partial\Sigma/\partial\theta_i, \quad \Sigma^i = \partial\Sigma^{-1}/\partial\theta_i, \quad \Sigma_{ij} = \partial^2\Sigma/\partial\theta_i\partial\theta_j, \quad i,j = 1,\ldots,p, \tag{A.74}$$

denote the $n \times n$ matrices of derivatives of Σ and Σ^{-1} with respect to the elements of θ. Standard matrix results in Section A.3.11 give

$$\begin{aligned} \Sigma^i = \partial\Sigma^{-1}/\partial\theta_i &= -\Sigma^{-1}\Sigma_i\Sigma^{-1}, \\ \partial\log|\Sigma|/\partial\theta_i &= \mathrm{tr}\{\Sigma^{-1}\Sigma_i\}. \end{aligned} \tag{A.75}$$

Then the score has elements

$$\partial\log L/\partial\theta_i = -\frac{1}{2}\mathrm{tr}\{-\Sigma^{-1}\Sigma_i\Sigma^{-1}\boldsymbol{x}\boldsymbol{x}^T + \Sigma^{-1}\Sigma_i\},$$

and the score derivative matrix has elements

$$\begin{aligned} \partial^2\log L/\partial\theta_i\partial\theta_j = -\frac{1}{2}\mathrm{tr}\{&\left(\Sigma^{-1}\Sigma_j\Sigma^{-1}\Sigma_i\Sigma^{-1} - \Sigma^{-1}\Sigma_{ij}\Sigma^{-1} + \Sigma^{-1}\Sigma_i\Sigma^{-1}\Sigma_j\Sigma^{-1}\right)\boldsymbol{x}\boldsymbol{x}^T \\ &-\Sigma^{-1}\Sigma_j\Sigma^{-1}\Sigma_i + \Sigma^{-1}\Sigma_{ij}\} \end{aligned} \tag{A.76}$$

for $i,j = 1,\ldots,p$. Taking expectations, $E(\boldsymbol{x}\boldsymbol{x}^T) = \Sigma$, changing the sign and simplifying yields the elements of the Fisher information matrix

$$\begin{aligned} (\mathcal{I})_{ij} &= -E(\partial^2\log L/\partial\theta_i\partial\theta_j) = \frac{1}{2}\mathrm{tr}\{\Sigma^{-1}\Sigma_i\Sigma^{-1}\Sigma_j\} = \frac{1}{2}\mathrm{tr}\{\Sigma\,\Sigma^i\,\Sigma\,\Sigma^j\} \\ &= -\frac{1}{2}\mathrm{tr}\{\Sigma_i\Sigma^j\} = -\frac{1}{2}\mathrm{tr}\{\Sigma_j\Sigma^i\} = -\frac{1}{2}\mathrm{tr}\{\Sigma^i\Sigma_j\} = -\frac{1}{2}\mathrm{tr}\{\Sigma^j\Sigma_i\}. \end{aligned} \tag{A.77}$$

Note the elements of $\mathcal{I}$ can be written in terms of the derivatives of Σ, or the derivatives of Σ^{-1}, or both.

Next, allow a regression or drift term in the model

$$\boldsymbol{x} \sim N_n(F\boldsymbol{\beta}, \Sigma), \tag{A.78}$$

where F is an $n \times q$ matrix of regressor variables. The first column is usually the constant vector $\mathbf{1}_n$ to accommodate an intercept, and if $q = 1$, all elements of $\boldsymbol{x}$ have the same mean. Here $\boldsymbol{\beta}$ $(q \times 1)$ is a vector of regression parameters and $\Sigma = \Sigma(\boldsymbol{\theta})$. In the spatial setting, this model is known as the spatial linear model (Section 5.11). The log-likelihood is

$$\log L(\boldsymbol{\beta}, \boldsymbol{\theta}) = -\frac{1}{2}\{(\boldsymbol{x} - F\boldsymbol{\beta})^T\Sigma^{-1}(\boldsymbol{x} - F\boldsymbol{\beta}) + \log|\Sigma| + n\log(2\pi)\}. \tag{A.79}$$

Write $\boldsymbol{w} = \boldsymbol{x} - F\boldsymbol{\beta}$, so that $\boldsymbol{w}$ is multivariate normal with mean $\mathbf{0}$ and covariance matrix Σ. The first and second derivatives of (A.79) with respect to $\boldsymbol{\beta}$ and $\boldsymbol{\theta}$ are given by

$$\partial \log L/\partial \boldsymbol{\beta} = F^T\Sigma^{-1}\boldsymbol{w}, \tag{A.80}$$

$$\partial \log L/\partial \theta_i = -\frac{1}{2}\text{tr}(\Sigma^{-1}\Sigma_i) + \frac{1}{2}\text{tr}(\boldsymbol{w}^T\Sigma^{-1}\Sigma_i\Sigma^{-1}\boldsymbol{w}), \quad i = 1, \ldots, p, \tag{A.81}$$

$$\partial^2 \log L/\partial \boldsymbol{\beta}\partial \boldsymbol{\beta}^T = -F^T\Sigma^{-1}F, \tag{A.82}$$

$$\partial^2 \log L/\partial \boldsymbol{\beta}\partial \theta_i = -F^T\Sigma^{-1}\Sigma_i\Sigma^{-1}\boldsymbol{w}, \tag{A.83}$$

together with (A.76), with $\boldsymbol{x}$ replaced by $\boldsymbol{w}$, for the second derivative with respect to θ_i and θ_j.

Taking expectations of the second derivatives and changing the sign gives the Fisher information matrix

$$\mathcal{I} = \begin{bmatrix} \mathcal{I}^{(\beta)} & 0 \\ 0 & \mathcal{I}^{(\theta)} \end{bmatrix}, \tag{A.84}$$

where the information matrix for $\boldsymbol{\beta}$ is

$$\mathcal{I}^{(\beta)} = F^T\Sigma^{-1}F, \tag{A.85}$$

and where the elements of $\mathcal{I}^{(\theta)}$ are given by (A.77). Note that $\mathcal{I}$ is block diagonal since the expectation of $\partial^2 \log L/\partial \boldsymbol{\beta}\partial \theta_i$ vanishes.

A.12.3 Change of Variables

Suppose $\boldsymbol{\psi}$ is an alternative parameterization of the model. let $\boldsymbol{\theta} = \boldsymbol{\theta}(\boldsymbol{\psi})$ denote the mapping from $\boldsymbol{\psi}$ to $\boldsymbol{\theta}$ with $p \times p$ Jacobian matrix

$$J = \frac{\partial \boldsymbol{\theta}}{\partial \boldsymbol{\psi}^T}, \tag{A.86}$$

where the elements J are given by $(J)_{ij} = \partial \theta_i / \partial \psi_j$.

Then, the information matrices for $\boldsymbol{\theta}$ and $\boldsymbol{\psi}$, denoted $\mathcal{I}^{(\theta)}$ and $\mathcal{I}^{(\psi)}$, say, are related by

$$\mathcal{I}^{(\psi)} = J^T\mathcal{I}^{(\theta)}J. \tag{A.87}$$

The result follows from the chain rule for differentiating a function of several variables.

A.12.4 Profile Log-likelihood

For the purposes of finding the MLE in a multivariate normal setting, it is helpful to parameterize the covariance parameters as $\boldsymbol{\theta} = (\sigma^2, \boldsymbol{\theta}_c^T)^T$ in terms of an overall scale parameter σ^2 and the remaining correlation parameters $\boldsymbol{\theta}_c$, say. That is,

$\Sigma = \sigma^2 P$ where the correlation matrix $P = (\rho_{ij})$ depends on $\boldsymbol{\theta}_c$. Then the parameters can be split into three parts: (i) the regression parameter $\boldsymbol{\beta}$, (ii) the overall scale parameter σ^2, and (iii) the correlation parameters $\boldsymbol{\theta}_c$.

For a given value of $\boldsymbol{\theta}_c$, the log-likelihood is maximized over the regression parameter by

$$\hat{\boldsymbol{\beta}}(\boldsymbol{\theta}_c) = (F^T\Sigma^{-1}F)^{-1}F^T\Sigma^{-1}\boldsymbol{x} = (F^TP^{-1}F)^{-1}F^TP^{-1}\boldsymbol{x}. \tag{A.88}$$

Substituting (A.88) into the log-likelihood (A.79), and using the simplification for quadratic forms in (A.22), gives the profile log-likelihood for $\boldsymbol{\theta}_c$ and σ^2

$$\log L_{\text{pro}}(\boldsymbol{\theta}_c, \sigma^2) = -\frac{1}{2}\{\sigma^{-2}\boldsymbol{x}_G^T P_{GG}^{-1}\boldsymbol{x}_G + n\log\sigma^2 + \log|P| + n\log(2\pi)\}. \tag{A.89}$$

Here $G(n \times p)$ is a column orthonormal matrix, which is orthogonal to F, so that $G^TG = I_{n-p}$, $G^TF = 0$ and $\Sigma_{GG} = G^T\Sigma G$ and $\boldsymbol{x}_G = G^T\boldsymbol{x}$. Equation (A.89) is maximized over σ^2 by

$$\hat{\sigma}^2 = \boldsymbol{x}_G^T P_{GG}^{-1}\boldsymbol{x}_G/n. \tag{A.90}$$

Substituting (A.90) into (A.89) gives the profile log-likelihood for $\boldsymbol{\theta}_c$

$$\log L_{\text{pro}}(\boldsymbol{\theta}_c) = \frac{1}{2}\{n + n\log(\boldsymbol{x}_G^T P_{GG}^{-1}\boldsymbol{x}_G/n) + \log|P| + n\log(2\pi)\}. \tag{A.91}$$

In general, the profile log-likelihood (A.91) must be maximized numerically over $\boldsymbol{\theta}_c$.

A.12.5 Confidence Intervals

In the general setting of Section A.12.1, let θ_i be one of the elements of $\boldsymbol{\theta}$, and let $\hat{\boldsymbol{\theta}}$ denote the MLE of $\boldsymbol{\theta}$. An approximate 95% confidence interval for θ_i is given by

$$(\hat{\theta}_i - 1.96v_i,\ \hat{\theta}_i + 1.96v_i), \tag{A.92}$$

where $v_i = \sqrt{v_{ii}}$ is the square root of the (i,i)th diagonal element of the inverse Fisher expected information matrix, $V = \mathcal{I}(\hat{\boldsymbol{\theta}})^{-1}$.

An asymptotically equivalent version of the confidence interval can be constructed from the profile log-likelihood, as the interval

$$\{\theta_i : \log L_{\text{profile}}(\theta_i) > \log L_{\text{profile}}(\hat{\theta}_i) - 1.92\}. \tag{A.93}$$

Here $1.92 = 3.84/2$ is half the upper 5% critical value of the χ_1^2 distribution, and is justified by the asymptotic representation

$$\log L_{\text{profile}}(\theta_i) - \log L_{\text{profile}}(\hat{\theta}_i) = -\frac{1}{2}(\theta_i - \hat{\theta}_i)^2/v_{ii}$$

since $\hat{\theta}_i$ is asymptotically normally distributed about θ_i with variance v_{ii}.

A.12.6 Linked Parameterization

In the spatial linear model, sometimes both the mean vector and the covariance matrix depend jointly on a p-dimensional parameter vector $\boldsymbol{\theta}$; see, for example, Smith (2001, pp. 124–250). Thus, suppose

$$\boldsymbol{x} \sim N_n(\boldsymbol{\mu}(\boldsymbol{\theta}), \Sigma(\boldsymbol{\theta})).$$

Write $\boldsymbol{\mu}_i = \partial \boldsymbol{\mu}/\partial \theta_i$ and $\Sigma_i = \partial \Sigma/\partial \theta_i$, $i = 1, \ldots, p$. It can be shown that

$$\partial \log L/\partial \theta_i = (\boldsymbol{x} - \boldsymbol{\mu}_i)^T \Sigma^{-1}(\boldsymbol{x} - \boldsymbol{\mu}) + \frac{1}{2}(\boldsymbol{x} - \boldsymbol{\mu})^T \Sigma^{-1}\Sigma_i\Sigma^{-1}(\boldsymbol{x} - \boldsymbol{\mu}) - \frac{1}{2}\text{tr}\{\Sigma^{-1}\Sigma_i\},$$
$$i = 1, \ldots, p. \tag{A.94}$$

In general, the MLE needs to be found numerically.

Using (A.94), it can be found after some algebra that the information matrix has elements

$$\mathcal{I}_{ij}^{(\theta)} = \frac{1}{2}\{\text{tr}\Sigma^{-1}\Sigma_i\Sigma^{-1}\Sigma_j\} + \boldsymbol{\mu}_i^T\Sigma^{-1}\boldsymbol{\mu}_j. \tag{A.95}$$

Example A.2 Consider a simple example with a $p = 1$-dimensional parameter $\theta = \theta_1$. Suppose $\boldsymbol{\mu} = \theta\mathbf{1}$ and $\Sigma = \theta^2 P$, where P is a fully specified covariance matrix. The log-likelihood takes the form

$$\begin{aligned} -2\log L &= -\frac{1}{2}\{(\boldsymbol{x} - \theta\mathbf{1})^T\theta^{-2}P^{-1}(\boldsymbol{x} - \theta\mathbf{1}) + n\log\theta^2 + n\log(2\pi)\} \\ &= -\frac{1}{2}\{\alpha\varphi^2 - 2\beta\varphi + \mathbf{1}^T P^{-1}\mathbf{1} - n\log\varphi^2 + n\log(2\pi)\}, \end{aligned}$$

where $\alpha = \boldsymbol{x}^T P^{-1}\boldsymbol{x}$, $\beta = \boldsymbol{x}^T P^{-1}\mathbf{1}$, and the parameter θ is replaced by $\varphi = 1/\theta$.

It is easily checked that the log-likelihood tends to $-\infty$ as $|\varphi| \to 0$ and $|\varphi| \to \infty$. Hence, there is at least one local maximum on the intervals $\varphi \in (-\infty, 0)$ and $\varphi \in (0, \infty)$. Differentiating the log-likelihood with respect to φ and setting the derivative to 0 leads to the quadratic equation

$$\alpha\varphi^2 - \beta\varphi - n = 0$$

with the two solutions

$$\varphi_{\pm} = \frac{\beta \pm \sqrt{\beta^2 + 4n\alpha}}{2\alpha}.$$

One solution is negative and one is positive; hence the two solutions maximize the log-likelihood on the intervals $\varphi \in (-\infty, 0)$ and $\varphi \in (0, \infty)$, respectively. After substituting these two choices into the log-likelihood, it can be shown that the choice that maximizes the log-likelihood is given by

$$\hat{\varphi} = \varphi_+ \text{ if } \beta > 0\,, \quad \hat{\varphi} = \varphi_- \text{ if } \beta < 0\,.$$

The information (for θ) is given by

$$\mathcal{I} = (2n + \mathbf{1}^T P^{-1}\mathbf{1})/\theta^2.$$

In particular, if $P = I$ is the identity matrix, then $\mathcal{I} = 3n/\sigma^2$. □

A.12.7 Model Choice

Consider two models M_1 and M_2 where M_1 is nested within M_2. Suppose distributions in M_2 are parameterized by a p_2-dimensional parameter vector $\boldsymbol{\theta}$. Suppose $\boldsymbol{\theta}$ can be partitioned as $\boldsymbol{\theta}^T = (\boldsymbol{\theta}_1^T, \boldsymbol{\theta}_2^T)$ of dimensions p_1 and $p_2 - p_1$ and suppose that $\boldsymbol{\theta}_2 = 0$ under M_1. Given an n-dimensional data vector $\boldsymbol{x}$, let $\log \hat{L}_j$ denote the maximized log-likelihood under model M_j, $j = 1{,}2$.

The standard theory of likelihood ratio tests states that under mild regularity conditions,

$$2(\log \hat{L}_2 - \log \hat{L}_1) \sim \chi^2_{p_2 - p_1} \tag{A.96}$$

asymptotically for large n, where $p_2 - p_1$ is the number of extra parameters under the model M_2. Hence, for a test of size 5%, the simpler model M_1 is preferred unless twice the difference in log-likelihoods exceeds the upper 5% critical value of the relevant χ^2 distribution.

Another way to compare models is through the Akaike information criterion (AIC). Define the AIC for a model M_j by

$$\text{AIC}_j = 2(p_j - \log \hat{L}_j), \quad j = 1, 2 \tag{A.97}$$

i.e. twice the difference between the number of parameters in the model and the maximized log-likelihood. Then the AIC model choice rule says to choose the model with the smallest AIC value. The AIC criterion can be used to compare both nested and nonnested models.

A.13 Bias in Maximum Likelihood Estimation

A.13.1 A General Result

First, a general result is given about bias in maximum likelihood estimation for a large sample size n. Then this result is applied to the spatial linear model.

Suppose that $\log L = \log L(\boldsymbol{\theta})$ is the log-likelihood function with p parameters $\boldsymbol{\theta}$ and is based on n observations. The following notation is used in this section for the expected derivatives of $\log L(\boldsymbol{\theta})$.

$$\ell_{ij} = E\left\{\frac{\partial^2 \log L}{\partial\theta_i \partial\theta_j}\right\}, \quad \ell_{ijk} = E\left\{\frac{\partial^3 \log L}{\partial\theta_i \partial\theta_j \partial\theta_k}\right\},$$

$$\ell_{ij,k} = E\left\{\frac{\partial^2 \log L}{\partial\theta_i \partial\theta_j}\frac{\partial \log L}{\partial\theta_k}\right\} \qquad i,j,k = 1, \ldots, p. \tag{A.98}$$

In addition, let

$$\ell_{ij}^{(k)} = \frac{\partial \ell_{ij}}{\partial\theta_k} = \ell_{ijk} + \ell_{ij,k}. \tag{A.99}$$

In terms of the subscript notation ℓ_{ij}, the Fisher information matrix is given by

$$\mathcal{I}(\boldsymbol{\theta}) = (-\ell_{ij}). \tag{A.100}$$

It is also useful to use the superscript notation ℓ^{ij} to specify the elements of the inverse information matrix

$$\mathcal{I}(\boldsymbol{\theta})^{-1} = (-\ell^{ij}). \tag{A.101}$$

The following bias formula is taken from Cox and Snell (1968).

Theorem A.13.1 (Cox–Snell Theorem) *Consider n independent observations (not necessarily identically distributed) from a statistical model with a p-dimensional parameter vector $\boldsymbol{\theta}$, and suppose the derivatives (A.98) are $O(n)$. Then, under suitable regularity conditions, the bias $b(\theta_s)$ of the sth element θ_s of the MLE of $\boldsymbol{\theta}$ is given by*

$$b(\theta_s) = \sum_{i=1}^{p}\sum_{j=1}^{p}\sum_{k=1}^{p} \ell^{si}\ell^{sj}\left\{\ell_{ij,k} + \frac{1}{2}\ell_{ijk}\right\} + o(n^{-1}), \quad s = 1, \dots, p. \tag{A.102}$$

Cordeiro and Klein (1994) have pointed out that this bias expression still holds even if the observations are dependent, provided that all the terms in Eq. (A.102) are $O(n)$. They have also given the following alternative form for the bias:

$$b(\theta_s) = \sum_{i=1}^{p}\ell^{si}\sum_{j=1}^{p}\sum_{k=1}^{p} \ell^{sj}\left\{\ell_{ij}^{(k)} - \frac{1}{2}\ell_{ijk}\right\} + o(n^{-1}), \quad s = 1, \dots, p. \tag{A.103}$$

In matrix notation, (A.103) becomes

$$b(\boldsymbol{\theta}) = \mathcal{I}(\boldsymbol{\theta})^{-1}\delta(\boldsymbol{\theta}) + o(n^{-1}), \quad \delta(\boldsymbol{\theta})_i = \mathrm{tr}(\mathcal{I}(\boldsymbol{\theta})^{-1}U_i), \quad i = 1, \dots, p, \tag{A.104}$$

where the $p \times p$ matrix U_i has elements $(U_i)_{jk} = \ell_{ij}^{(k)} - \frac{1}{2}\ell_{ijk}, \quad j,k = 1, \dots, p.$

Some simplification takes place when the Fisher information matrix $\mathcal{I}(\boldsymbol{\theta})$ for $\boldsymbol{\theta}$ is block diagonal, such as (A.84) for the spatial linear model. Suppose now that $\boldsymbol{\theta} = (\boldsymbol{\theta}_1^T, \boldsymbol{\theta}_2^T)^T$, where $\boldsymbol{\theta}_1 = (\theta_1, \dots, \theta_{p_1})^T$ and $\boldsymbol{\theta}_2 = (\theta_{p_1+1}, \dots, \theta_{p_1+p_2})^T$. Further, suppose that $\mathcal{I}(\boldsymbol{\theta})$ has the respective block matrix with diagonal blocks $\mathcal{I}^{(\boldsymbol{\theta}_1)}$ and $\mathcal{I}^{(\boldsymbol{\theta}_2)}$. Then

$$b(\theta_s) = \sum_{i=1}^{p_1}\ell^{si}\sum_{j=1}^{p_1}\sum_{k=1}^{p_1} \ell^{sj}\left\{\ell_{ij}^{(k)} - \frac{1}{2}\ell_{ijk}\right\} - \frac{1}{2}\sum_{i=1}^{p_1}\ell^{si}\sum_{j=p_1+1}^{p_1+p_2}\sum_{k=p_1+1}^{p_1+p_2}\ell^{sj}\ell_{ijk} + o(n^{-1}),$$
$$s = 1, \dots, p_1. \tag{A.105}$$

In matrix notation, (A.105) becomes

$$\mathbf{b}(\boldsymbol{\theta}) = \mathcal{I}(\boldsymbol{\theta})^{-1}\delta(\boldsymbol{\theta}) + o(n^{-1}), \quad \delta(\boldsymbol{\theta})_i = \mathrm{tr}\{\mathcal{I}(\boldsymbol{\theta}_1)^{-1}U_i\} + \mathrm{tr}\{\mathcal{I}(\boldsymbol{\theta}_2)^{-1}V_i\},$$
$$i = 1, \dots, p_1, \tag{A.106}$$

where U_i is a $p_1 \times p_1$ matrix and V_i is a $p_2 \times p_2$ matrix with elements $(U_i)_{jk} = \ell_{ij}^{(k)} - \frac{1}{2}\ell_{ijk}, \quad j,k = 1, \dots, p_1,$ and $(V_i)_{jk} = -\frac{1}{2}\ell_{ijk}, \quad j,k = p_1+1, \dots, p_1+p_2.$ Note that the bias in this case is additive.

A.13.2 The Spatial Linear Model

In addition to the notation in (A.74), the following notation is also useful for specifying various derivatives of the log-likelihood for the spatial linear model. Let

$$\begin{aligned} M_i &= \Sigma^{-1}\Sigma_i, \quad M_{ij} = \Sigma^{-1}\Sigma_{ij}, \\ M_{i,j} &= \Sigma^{-1}\Sigma_i\Sigma^{-1}\Sigma_j, \quad M_{i,jk} = \Sigma^{-1}\Sigma_i\Sigma^{-1}\Sigma_{jk}, \\ M_{ij,k} &= \Sigma^{-1}\Sigma_{ij}\Sigma^{-1}\Sigma_k, \quad M_{i,j,k} = \Sigma^{-1}\Sigma_i\Sigma^{-1}\Sigma_j\Sigma^{-1}\Sigma_k, \end{aligned} \tag{A.107}$$

where $i, j, k = 1, \ldots, p$. It can be checked that

$$\begin{aligned} \partial\{\text{tr}(\Sigma^{-1}\Sigma_i)\}/\partial\theta_j &= -\text{tr}(M_{i,j} - M_{ij}), \\ \partial(\Sigma^{-1}\Sigma_i\Sigma^{-1})/\partial\theta_j &= (M_{ij} - M_{i,j} - M_{j,i})\Sigma^{-1}. \end{aligned} \tag{A.108}$$

Use lowercase letters to denote the traces of the matrices in (A.107); in particular, set

$$m_{ij} = \text{tr}(\Sigma^{-1}\Sigma_{ij}), \tag{A.109}$$

$$m_{i,j} = \text{tr}(\Sigma^{-1}\Sigma_i\Sigma^{-1}\Sigma_j) = -\text{tr}(\Sigma_i\Sigma^j), \tag{A.110}$$

$$m_{ij,k} = m_{k,ij} = \text{tr}(\Sigma^{-1}\Sigma_{ij}\Sigma^{-1}\Sigma_k) = -\text{tr}(\Sigma_{ij}\Sigma^k). \tag{A.111}$$

The log-likelihood for the spatial linear model is given by ((A.79). The second derivative of the log-likelihood with respect to θ_i and θ_j, given in (A.76) with $\boldsymbol{x}$ replaced by $\boldsymbol{w} = \boldsymbol{x} - F\boldsymbol{\beta}$, can be written as

$$\partial^2 \log L/\partial\theta_i\partial\theta_j = \frac{1}{2}(m_{i,j} - m_{ij}) + \frac{1}{2}\text{tr}\{(M_{ij} - M_{i,j} - M_{j,i})\Sigma^{-1}\boldsymbol{w}\boldsymbol{w}^T\}. \tag{A.112}$$

Next consider the third-order derivatives of the log-likelihood function. Differentiating (A.82), it follows that

$$\begin{aligned} \partial^3 \log L/\partial\boldsymbol{\beta}\partial\boldsymbol{\beta}^T\partial\beta_i &= 0, \quad i = 1, \ldots, q, \\ \partial^3 \log L/\partial\boldsymbol{\beta}\partial\boldsymbol{\beta}^T\partial\theta_j &= \text{tr}(F^T\Sigma^{-1}\Sigma_i\Sigma^{-1}F), \quad j = 1, \ldots, p, \end{aligned} \tag{A.113}$$

where, as far as possible, derivatives with respect to the q-dimensional vector $\boldsymbol{\beta}$ are represented in terms of vectors and matrices, and derivatives with respect to $\boldsymbol{\theta}$ are given elementwise. Further using (A.112), we get

$$\partial^3 \log L/\partial\boldsymbol{\beta}\partial\theta_i\partial\theta_j = -\frac{1}{2}\text{tr}\{(M_{ij} - M_{i,j} - M_{j,i})\Sigma^{-1}F\}\boldsymbol{w}, \quad i, j = 1, \ldots, p. \tag{A.114}$$

Write

$$\begin{aligned} \partial M_{ij}/\partial\theta_k &= M_{ij}^{(k)} = M_{ijk} - M_{k,ij}, \\ \partial M_{i,j}/\partial\theta_k &= M_{i,j}^{(k)} = M_{ik,j} + M_{i,jk} - M_{i,j,k} - M_{i,k,j}, \quad i, j, k = 1, \ldots, p. \end{aligned} \tag{A.115}$$

Using (A.108) and (A.112) it can be shown that

$$\begin{aligned}\partial^3 \log L/\partial\theta_i\partial\theta_j\partial\theta_k &= \frac{1}{2}\mathrm{tr}\left\{\left(M_{i,j}^{(k)} - M_{ij}^{(k)}\right)\right\}\\ &\quad - \frac{1}{2}\mathrm{tr}\left\{\left(M_{ij}^{(k)} - M_{i,j}^{(k)} - M_{j,i}^{(k)}\right)\Sigma^{-1}\boldsymbol{w}\boldsymbol{w}^T\right\}\\ &\quad + \frac{1}{2}\mathrm{tr}\left\{(M_{ij} - M_{i,j} - M_{j,i})\Sigma^{-1}\Sigma_k\Sigma^{-1}\boldsymbol{w}\boldsymbol{w}^T\right\}.\end{aligned} \tag{A.116}$$

Using (A.115) and taking the expectation of (A.116), we get

$$\begin{aligned}E\{\partial^3 \log L/\partial\theta_i\partial\theta_j\partial\theta_k\} &= \frac{1}{2}\mathrm{tr}(2M_{i,j,k} + 2M_{i,k,j} - M_{ij,k} - M_{ik,j} - M_{jk,i})\\ &= \frac{1}{2}\mathrm{tr}(2m_{i,j,k} + 2m_{i,k,j} - m_{ij,k} - m_{ik,j} - m_{jk,i}).\end{aligned} \tag{A.117}$$

We are now ready to give the bias for the spatial linear model following Mardia (1990) and Watkins (1987) in the setting of outfill asymptotics. Regularity conditions are discussed in Section 5.5.3.

Theorem A.13.2 (Mardia–Watkins Theorem)

(i) The bias in $\boldsymbol{\beta}$ is negligible

$$\mathbf{b}(\boldsymbol{\beta}) = E(\hat{\boldsymbol{\beta}} - \boldsymbol{\beta}) = o(n^{-1}). \tag{A.118}$$

(ii) The bias in $\boldsymbol{\theta}$ can be written as the sum of two terms (the first term is present only when $\boldsymbol{\beta}$ is also to be estimated) given by

$$\mathbf{b}(\boldsymbol{\theta}) = \mathbf{b}_1(\boldsymbol{\theta}) + \mathbf{b}_2(\boldsymbol{\theta}) + o(n^{-1}). \tag{A.119}$$

Here

$$\mathbf{b}_1(\boldsymbol{\theta}) = (\mathcal{I}^{(\theta)})^{-1}\boldsymbol{\delta}_1, \quad \delta_{1i} = \frac{1}{2}\mathrm{tr}\left\{(\mathcal{I}^{(\beta)})^{-1}C_i\right\}, \quad C_i = \partial\mathcal{I}^{(\beta)}/\partial\theta_i = \mathrm{tr}\{F^T\Sigma^i F\}, \tag{A.120}$$

and

$$\begin{aligned}\mathbf{b}_2(\boldsymbol{\theta}) = (\mathcal{I}^{(\theta)})^{-1}\boldsymbol{\delta}_2, \quad \delta_{2i} &= \frac{1}{2}\mathrm{tr}\left\{(\mathcal{I}^{(\theta)})^{-1}D_i\right\},\\ (D_i)_{jk} &= \frac{1}{2}(m_{jk,i} - m_{ij,k} - m_{ik,j}).\end{aligned} \tag{A.121}$$

Here $\boldsymbol{\delta}_1$ and $\boldsymbol{\delta}_2$ are vectors of length p with elements δ_{1i} and δ_{2i}, and D_i is a $p \times p$ matrix. The information matrices $\mathcal{I}^{(\beta)}$ and $\mathcal{I}^{(\theta)}$ are given in (A.85) and (A.77).

Note. The bias $\mathbf{b}(\boldsymbol{\theta})$ is the sum of two terms. The first term $\mathbf{b}_1(\boldsymbol{\theta})$ can be viewed as "external," arising from the estimation of $\boldsymbol{\beta}$ and the second term $\mathbf{b}_2(\boldsymbol{\theta})$ can be viewed as "internal," arising from the estimation of $\boldsymbol{\theta}$.

Proof: We will use form (A.105) of the Cox–Snell theorem since the information matrix of $\boldsymbol{\beta}$ and $\boldsymbol{\theta}$ is the block diagonal matrix (A.84). Further to the information matrix, expression (A.105) needs the additional derivatives

$$\ell_{ij}^{(k)}, \ell_{ijk}, \quad i,j,k = 1, \ldots, p_1; \quad \ell_{ijk}, i = 1, \ldots, p_1, \quad j,k = p_1 + 1, \ldots, p_1 + p_2.$$

(i) For this part of the proof, let $\boldsymbol{\beta}$ and θ correspond to $\boldsymbol{\theta}_1$ and $\boldsymbol{\theta}_2$ with dimensions $q = p_1$ and $p = p_2$, respectively, in the notation of (A.105). Note that since (A.85) does not involve any $\boldsymbol{\beta}$ terms,

$$\ell_{ij}^{(k)} = 0 \quad i,j,k = 1, \ldots, q.$$

Further from (A.113), we have

$$\ell_{ijk} = 0, \quad i,j,k = 1, \ldots, q$$

since there is no term there involving $\boldsymbol{w}$. Finally, using $E(\boldsymbol{w}) = 0$ in (A.114),

$$\ell_{ijk} = 0, \quad i = 1, \ldots, q; \quad j,k = q + 1, \ldots, q + p.$$

Thus, all three terms in (A.105) are zero and (A.118) is proved.

(ii) For this part of the proof, swap the identifications so that $\boldsymbol{\theta}$ and $\boldsymbol{\beta}$ correspond to $\boldsymbol{\theta}_1$ and $\boldsymbol{\theta}_2$ with dimensions $p = p_1$ and $q = p_2$, respectively, in the notation of (A.105). From (A.77), (A.100), and (A.107), it follows that

$$\ell_{ij} = -\frac{1}{2} m_{i,j}, \quad i,j = 1 \ldots, p, \quad \ell_{ij} = -(F^T \Sigma^{-1} F)_{ij}, \quad i,j = p + 1 \ldots, p + q.$$

Note that from (A.115),

$$\ell_{ij}^{(k)} = -\frac{1}{2}(m_{ik,j} + m_{i,jk} - m_{i,j,k} - m_{i,k,j}), \quad i,j,k = 1, \ldots, p.$$

Further from (A.117), we have

$$\ell_{ijk} = \frac{1}{2}(2m_{i,j,k} + 2m_{i,k,j} - m_{ij,k} - m_{ik,j} - m_{jk,i}), \quad i,j,k = 1, \ldots, p.$$

Finally, from (A.113),

$$\ell_{ijk} = (\mathrm{tr}\{F^T \Sigma^{-1} \Sigma_i \Sigma^{-1} F\})_{jk}, \quad i = 1, \ldots, p, \quad j,k = p + 1, \ldots, p + q.$$

Using the last three expressions in (A.105), the proof of (A.119) follows. □

Next, consider some special settings for this theorem. Remember the notational convention (A.74) involving subscripts and superscripts on Σ for derivatives with respect to θ.

(a) *Setting* $p = q = 1$. In this setting, the vector $\boldsymbol{x}$ has a constant mean $\beta = \mu$ and the covariance matrix has a single parameter θ. The blocks of the information matrix take the form

$$\mathcal{I}^{(\beta)} = \mathbf{1}^T \Sigma^{-1} \mathbf{1}, \quad \mathcal{I}^{(\theta)} = \frac{1}{2}\mathrm{tr}\{\Sigma^{-1}\Sigma_1 \Sigma^{-1}\Sigma_1\} = -\frac{1}{2}\mathrm{tr}\{\Sigma_1 \Sigma^1\}. \tag{A.122}$$

Hence from (A.119),

$$b(\theta) = \frac{1}{2}(\mathcal{I}^{(\beta)}\mathcal{I}^{(\theta)})^{-1}\mathbf{1}^{\mathrm{T}}\Sigma^{1}\mathbf{1} + \frac{1}{4}(\mathcal{I}^{(\theta)})^{-2}\mathrm{tr}\{\Sigma_{11}\Sigma^{1}\} + o(n^{-1}). \tag{A.123}$$

As a special case of this setting, suppose $\theta = \sigma^2$ is a scale parameter for Σ, that is, $\Sigma = \sigma^2 P$ where the P is a specified matrix. Then

$$\Sigma_1 = \frac{\partial \Sigma}{\partial \sigma^2} = P, \quad \Sigma_{11} = 0, \quad \Sigma^{-1} = (\sigma^2)^{-1}P^{-1},$$
$$\Sigma^1 = -(\sigma^4)^{-1}P^{-1}, \quad \mathrm{tr}(\Sigma^1\Sigma_1) = -I(\sigma^4)^{-1}.$$

In (A.123), the second term is zero as $\Sigma_{11} = 0$, and the bias becomes

$$b(\sigma^2) = -\frac{\sigma^2}{n}.$$

This result might have been expected by considering the i.i.d. case, but note that the formula here holds for any P. The information matrices (A.122) simplify to

$$\mathcal{I}^{(\beta)} = \frac{\mathbf{1}^{\mathrm{T}}P^{-1}\mathbf{1}}{\sigma^2}, \quad \mathcal{I}^{(\theta)} = -\frac{1}{2}\mathrm{tr}(\Sigma^1\Sigma_1) = \frac{n}{2\sigma^4}.$$

(b) *Setting* $p = 2, q = 1$. Suppose $\beta = \mu$ as in (a) and now let $\boldsymbol{\theta} = (\theta_1, \theta_2)^T$ be two dimensional. Then $\mathcal{I}^{(\beta)} = \mathbf{1}^T\Sigma^{-1}\mathbf{1}, C_1 = \mathbf{1}^T\Sigma^1\mathbf{1}, C_2 = \mathbf{1}^T\Sigma^2\mathbf{1}$, and thus (A.120) becomes

$$\boldsymbol{\delta}_1^T = (\delta_{11}, \delta_{12}), \quad \delta_{11} = \frac{1}{2}\frac{\mathbf{1}^{\mathrm{T}}\Sigma^1\mathbf{1}}{\mathbf{1}^T\Sigma^{-1}\mathbf{1}}, \quad \delta_{12} = \frac{1}{2}\frac{\mathbf{1}^{\mathrm{T}}\Sigma^2\mathbf{1}}{\mathbf{1}^{\mathrm{T}}\Sigma^{-1}\mathbf{1}}. \tag{A.124}$$

The 2×2 information matrix $\mathcal{I}^{(\theta)}$ has elements $(\mathcal{I}^{(\theta)})_{ij} = -\frac{1}{2}\mathrm{tr}(\Sigma_i\Sigma^j)$. For convenience, write the inverse matrix as $(\mathcal{I}^{(\theta)})^{-1} = A$, say, with elements a_{ij}. The first term in (A.120) can be expressed using (A.124) and A

$$b_{11}(\boldsymbol{\beta}) = \delta_{11}a_{11} + \delta_{12}a_{12}, \quad b_{12}(\boldsymbol{\beta}) = \delta_{11}a_{12} + \delta_{12}a_{22}. \tag{A.125}$$

Next, consider the second term in (A.121). It can be shown that the matrices D_1 and D_2 are given by

$$D_1 = -\frac{1}{2}\begin{bmatrix} m_{11,1} & m_{11,2} \\ m_{11,2} & 2m_{12,2} - m_{22,1} \end{bmatrix} \text{ and } D_2 = -\frac{1}{2}\begin{bmatrix} 2m_{12,1} - m_{11,2} & m_{22,1} \\ m_{22,1} & m_{22,2} \end{bmatrix}. \tag{A.126}$$

Then δ_{21}, δ_{22} in (A.121) are given by

$$\begin{aligned} \delta_{21} &= -\frac{1}{4}\{m_{11,1}a_{11} + 2m_{11,2}a_{12} + (2m_{12,2} - m_{22,1})a_{22})\}, \\ \delta_{22} &= -\frac{1}{4}\{(2m_{12,1} - m_{11,2})a_{11} + 2m_{22,1}a_{12} + m_{22,2}a_{22}\}, \end{aligned} \tag{A.127}$$

and so the bias term $\mathbf{b}_2(\boldsymbol{\theta})$ has elements

$$b_{21}(\boldsymbol{\theta}) = \delta_{21}a_{11} + \delta_{22}a_{12}, \quad b_{22}(\boldsymbol{\theta}) = \delta_{21}a_{12} + \delta_{22}a_{22}. \tag{A.128}$$

Adding (A.125) and (A.128) yields the total bias $\mathbf{b}(\boldsymbol{\theta})$ in (A.119).

Table A.5 First and second derivatives of Σ and Σ^{-1} with respect to $\theta_1 = \sigma_\varepsilon^2$ and $\theta_2 = \lambda$.

Matrix A	$[a_e, a_m, a_o]$
Σ	$[1, 1, \lambda]\ \sigma_\varepsilon^2/M$
Σ_1	$[1, 1, \lambda]\ /M$
Σ_{11}	$[0, 0, 0]$
Σ_2	$[2\lambda, 2\lambda, P]\ \sigma_\varepsilon^2/M^2$
Σ_{22}	$[2(1+3\lambda^2), 2(1+3\lambda^2), 2\lambda(3+\lambda^2)]\ \sigma_\varepsilon^2/M^3$
Σ_{12}	$[2\lambda, 2\lambda, P]\ /M^2$
Σ^{-1}	$[1, P, -\lambda]\ /\sigma_\varepsilon^2$
Σ^1	$-[1, P, -\lambda]\ /\sigma_\varepsilon^4$
Σ^2	$[0, 2\lambda, -1]\ /\sigma_\varepsilon^2$

Example A.3 ***AR(1) process***
Consider data at sites $1, \ldots, n$ from an AR(1) process, which is the same as a first-order CAR process (Section 6.3). For the bias calculations here, suppose the process is parameterized by $\boldsymbol{\theta} = (\sigma_\varepsilon^2, \lambda)^T$. The exact covariance matrix and inverse covariance matrix are given by (6.21) and (6.22) in terms of $\sigma^2 = \sigma_\varepsilon^2/(1-\lambda^2)$ and λ.

For the purposes of this section, call an $n \times n$ symmetric matrix A *nearly circulant* if

$$\begin{aligned} a_{11} &= a_{nn} = a_e, \text{ say,} \\ a_{22} &= \cdots a_{n-1,n-1} = a_d, \text{ say, and} \\ a_{12} &= \cdots a_{n-1,n} = a_o, \text{ say,} \end{aligned} \tag{A.129}$$

where the labels e, d, o stand for "end," "diagonal," and "off-diagonal." That is, the elements along the main diagonal equal one another, except possibly the endpoints, which also equal one another, and all the elements on the dominant subdiagonal equal one another. If A and B are nearly circulant, and if at least one of them is tri-diagonal, then

$$\text{tr}(AB) = 2a_e b_e + (n-2)a_d b_d + 2(n-1)a_o b_o \approx n(a_d b_d + 2a_o b_o). \tag{A.130}$$

Table A.5 describes the first two derivatives of the nearly circulant matrices Σ and Σ^{-1} using subscripts and superscripts, respectively, for the derivatives with respect to θ_1 and θ_2.

The elements of the information matrix are given by (A.77) and can be found by substituting the entries from Table A.5 into (A.130), with the result

$$\mathcal{I} \approx n \begin{bmatrix} 1/(2\sigma_\varepsilon^4) & 0 \\ 0 & 1/(1-\lambda^2) \end{bmatrix}, \quad \mathcal{I}^{-1} \approx \frac{1}{n} \begin{bmatrix} 2\sigma_\varepsilon^4 & 0 \\ 0 & 1-\lambda^2 \end{bmatrix}, \tag{A.131}$$

where terms of smaller order than $O(n)$ and $O(n^{-1})$ in $\mathcal{I}$ and $\mathcal{I}^{-1}$, respectively, have been ignored.

Next, consider the two bias terms $\mathbf{b}_1(\boldsymbol{\theta})$ and $\mathbf{b}_2(\boldsymbol{\theta})$ in the Mardia–Watkins theorem. For the first bias term, it is necessary to work out $\mathbf{1}^T\Sigma^{-1}\mathbf{1} = \mathrm{tr}(\Sigma^{-1}\mathbf{1}\mathbf{1}^T)$ and its derivatives with respect to $\boldsymbol{\theta}$. Since $\mathbf{1}\mathbf{1}^T$ is nearly circulant in the sense of (A.129), these quantities can be computed using (A.130)

$$\mathbf{1}^T\Sigma^{-1}\mathbf{1} \approx -n(1-\lambda)^2/\sigma_\varepsilon^2, \quad \mathbf{1}^T\Sigma^{1}\mathbf{1} \approx -n(1-\lambda)^2/\sigma_\varepsilon^4, \quad \mathbf{1}^T\Sigma^{2}\mathbf{1} \approx -2n(1-\lambda)/\sigma_\varepsilon^2. \tag{A.132}$$

Using these values in (A.124) yields $\boldsymbol{\delta}_1 = -\left[1/(2\sigma_\varepsilon^2) \;\; 1/(1-\lambda)\right]^T$.

For the second bias term, it is first necessary to find the quantities $m_{ij,k}$ in (A.110). Using the derivatives in Table A.5 and Eq. (A.130), it follows that $m_{11,1} = m_{11,2} = 0$ and

$$m_{12,1} \approx 0, \quad m_{12,2} \approx \frac{2n}{\sigma_\varepsilon^2(1-\lambda^2)}, \quad m_{22,1} \approx \frac{2n}{\sigma_\varepsilon^2(1-\lambda^2)},$$
$$m_{22,2} \approx \frac{8\lambda n(1-\lambda^2)}{(1-\lambda^2)^2}. \tag{A.133}$$

Using (A.133) and the information matrix (A.131) in (A.127) yields $\boldsymbol{\delta}_2 = -\left[1/(2\sigma_\varepsilon^2) \;\; 2\lambda/(1-\lambda^2)\right]^T$.

Finally, substituting $\boldsymbol{\delta}_1$ and $\boldsymbol{\delta}_2$ in (A.125) and (A.128) yields the bias terms,

$$\mathbf{b}_1(\boldsymbol{\theta}) = -\frac{1}{n}\begin{bmatrix} \sigma_\varepsilon^2 \\ 1+\lambda \end{bmatrix} + o(n^{-1}), \quad \mathbf{b}_2(\boldsymbol{\theta}) = -\frac{1}{n}\begin{bmatrix} \sigma_\varepsilon^2 \\ 2\lambda \end{bmatrix} + o(n^{-1}),$$

$$\mathbf{b}(\boldsymbol{\theta}) = \mathbf{b}_1(\boldsymbol{\theta}) + \mathbf{b}_2(\boldsymbol{\theta}) = -\frac{1}{n}\begin{bmatrix} 2\sigma_\varepsilon^2 \\ 1+3\lambda \end{bmatrix} + o(n^{-1}). \tag{A.134}$$

□

Appendix B

A Brief History of the Spatial Linear Model and the Gaussian Process Approach

> The mere formulation of a problem is far more often essential than its solution, which may be merely a matter of mathematical or experimental skill. To raise new questions, new possibilities, to regard old problems from a new angle requires creative imagination and marks real advances in science. (Einstein and Infeld, 1938, p. 38)

B.1 Introduction

Often, there are two streams in statistical research – one developed by practitioners and other by mainstream statisticians. The development of Geostatistics is a very good example where pioneering work under realistic assumptions came from mining engineers, whereas the links to mainstream work on spatial processes only gradually became explicit. Geostatisticians are particularly interested in prediction, also known as kriging. Two of the pioneering figures in the development of geostatistics were Danie Krige and Georges Matheron (Figure B.1); to see how the word "kriging" was coined by Matheron, see Cressie (1990). There have been excellent historical articles related to this subject including by Agterberg (2004), Baddeley (2001), Cressie (1989, 1990), Cressie and Moores (2021), and Diggle (1997, 2010). Their accounts are exclusively about the relation between on statistical prediction and kriging, for example, as summarized by Diggle (2010):

> For some time, the work of the Fontainebleau School remained relatively unconnected to the mainstream of spatial statistical methodology, and vice versa. Watson (1972) drew attention to the close connections between Fontainebleau-style geostatistical methods and more theoretically oriented work on stochastic process prediction (see, for example, Whittle (1963)).

In addition, statisticians have also focused on inference problems, especially likelihood-based inference. An early contribution was Whittle (1954) for the SAR

Spatial Analysis, First Edition. John T. Kent and Kanti V. Mardia.

Krige

Matheron

Figure B.1 Creators of Kriging: Danie Krige and Georges Matheron.

model. It is now well known that when Matheron created kriging, he followed the least squares approach to inference in his formulation, whereas Mardia (1980) and Mardia and Marshall (1984) used the Fisher's likelihood approach after formulating the problem in terms of the spatial linear model (SLM). Incidentally, Mardia (1980) received some helpful comments in person from some participants from the Fontainebleau School when this paper was presented in 26th International Geology Congress Sciences de la Terre, Paris, 1980.

In particular, the maximum likelihood method was used for estimating the parameters for the SLM, and the predictor (universal kriging) was taken to be the plug-in mean from the fitted Gaussian process. More details are given below.

B.2 Matheron and Watson

While Matheron and his group were developing geostatistics, Geof Watson was keen to draw the attention of statisticians to this new area. Watson wrote a report on this topic (Watson, 1969) followed by a shorter version in print (Watson, 1972). Watson (1984) later spelled out the link between splines and kriging.

Historically, Watson's following observation in Watson (1986) (Reprinted in Mardia (1991, p. xl)) is a key to understanding the development of statistical geostatistics:

> In the mid-1970s the work of Georges Matheron and Jean Serra of the Center for Mathematical Morphology at the Paris School of Mines, attracted my attention. They seemed to be breathing new life into the application of statistics to geology and mining. As a result, I spent a lot of time persuading

> English-speaking geologists and statisticians that this was so, while trying to persuade the Fontainebleau School to integrate their writings with that of the anglophones! Our geologists receive very little mathematical training (indeed they seem less mathematically inclined than any scientific group I know) so French-style 'geostatistics' was just too much for them.

The following comment regarding the "Anglo-Saxon optique" in Watson (1986) (Reprinted in Mardia (1991, p. xli)) was not well received by Matheron.

> While I never persuaded Matheron to adopt the 'Anglo-Saxon optique', I enjoyed his hospitality on many occasions, thereby sampling French family life (at its best, I suspect) which one can never know as a tourist.

When Mardia was preparing the Festschrift volume for Geof, he invited Matheron to write an article for the volume. Matheron did not agree and sent him a letter (reproduced here as Figure B.2 with a translation in B.3) together with a copy of Watson (1986) where the quote had appeared.

B.3 Geostatistics at Leeds 1977–1987

B.3.1 Courses, Publications, Early Dissemination

We were fortunate at Leeds University in late 1970s to have the Department of Mining interested in geostatistics, led by Peter Dowd and Allen Royale. Our own interest was boosted by extensive interactions with them. They were already running their M.Sc in Geostatistics, and we were one of the first departments of statistics to introduce geostatistics into a Statistics M.Sc. In the meantime, spatial statistics was becoming more prominent within the statistics community. Ripley's book (Ripley, 1981) came out in 1981. Mardia (1980), Mardia and Gill (1982), Mardia and Marshall (1982), and Mardia and Marshall (1984) formulated geostatistical models in a more statistical framework, with an emphasis on likelihood-based inference.

One key initiative at Leeds was the development of a long-running annual workshop (which became the Leeds Annual Statistics Research or LASR Workshop) on topical areas of Statistics. Spatial Statistics became a regular theme, and it attracted many leading figures. As early as 1979, the theme was geostatistics, and the speakers included A. Marechal (Centre de Geostatistique, Fountainbleau) from the Matheron group.

Subsequent LASR workshops included short courses from leading speakers, including Richard Martin in 1980, Julian Besag in 1981, Brian Ripley in 1982,

ÉCOLE NATIONALE SUPÉRIEURE
DES
MINES DE PARIS

CENTRE DE GÉOSTATISTIQUE

35, RUE SAINT-HONORÉ
77305 FONTAINEBLEAU (FRANCE)
TÉL. 16 (1) 64.22.48.21
TELEX : MINEFON 694 736 F
TELEFAX : 64 22 39 05

Dr.MARDIA K.V.
Dpt.of Statistics
School of Mathematics
The University,
LEEDS LS2 9JT
England

N/RÉF. : GM/jm n°35/G/90

FONTAINEBLEAU, LE 23 Février 1990

Cher Dr.MARDIA,

Vous trouverez ci-joint photocopie d'un passage de l'article autobiographique "The Craft of Probabilistic Modelling" où G.S.WATSON nous révèle ce qu'il pense de la Géostatistique.

Si je devais faire un commentaire, ce serait sur le thème: "Comment se fait-il que des statisticiens "anglo-saxons" normalement intelligents se révèlent régulièrement incapables de comprendre seulement de quoi il est question en Géostatistique ?

Mais un tel commentaire serait sans doute déplacé dans le livre que vous comptez publier pour le jubilé de G.S.WATSON. Je ne vous enverrai donc aucun article.

Sincèrement,

G.MATHERON

P.J : photocopie de l'article cité.

Figure B.2 Letter from Matheron to Mardia, dated 1990.

Xavier Guyon in 1983, Tata Subba-Rao in 1985, John Haslett in 1986, and Hans Künsch in 1987. A related theme on spatial statistics and image analysis developed in the mid-1980s and speakers included J.S. Durrani, Joseph Kittler, and J.R. Ullman.

G Matheron
École Nationale Supérieure
des Mines de Paris
Centre de Géostatistique
35 Rue Saint-Honoré
77305 Fontainebleau
FRANCE

Fontainebleau, 23 February 1990

Dear Dr. Mardia,

Please find enclosed a photocopy of a passage from the autobiographical article "The Craft of Probabilistic Modelling" where G S Watson reveals to us what he thinks of Geostatistics.

If I had to write a commentary, it would be on the theme: "How is it that normally intelligent 'anglo-saxon' statisticians regularly show themselves to be incapable of understanding even what Geostatistics is about?"

But doubtless such a commentary would be out of place in the book which you are intending to publish for the celebration of G S Watson. Therefore, I shall not be contributing any article.

Yours sincerely,

G Matheron

Enclosed: photocopy of the cited article.

Figure B.3 Translation of the letter from Matheron to Mardia, dated 1990.

B.3.2 Numerical Problems with Maximum Likelihood

At one stage, there was considerable debate about the behavior of maximum likelihood methods for the covariance parameters in a spatial model, starting with Mardia and Marshall (1984). In certain problems, there may be multiple modes of the likelihood or the MLE may be singular. Part of the problem seems to a confounding between short-range and long-range behavior of the covariance function, sometimes leading to multiple modes with different choices of the range parameter. In addition, nondifferentiability of the covariance function (e.g. for the spherical scheme) may be a contributory factor in multimodality (Mardia and Watkins, 1989). We have had no problems fitting the exponential scheme in various numerical examples in Chapter 5.

More recent thinking downplays concerns about multimodality, e.g. Stein (1999, p. 173):

> I do not believe the results in Warnes and Ripley (1987) and Ripley (1988) purporting to show multiple maxima in the likelihood when fitting an exponential autocovariance function.

> Their presence is not a sign of a problem with likelihood methods but rather an entirely correct indication that the data provide essentially no information for distinguishing between parameter values along the log likelihood function (long ridges in the likelihood).

Perhaps the most up-to-date view is given by Zimmerman (2010)

> In many standard statistical problems, a unique ML estimate exists. In the present context, however, there is no guarantee of existence or uniqueness, nor is there even a guarantee that all local maxima of the likelihood function are global maxima. Indeed, Warnes and Ripley (1987) and Handcock (1989) show that the likelihood function corresponding to gridded observations of a stationary Gaussian process with spherical covariance function often has multiple modes, and that these modes may be well separated. Rasmussen and Williams (2006) display a bimodal likelihood surface for a case of a stationary Gaussian process with a Gaussian covariance function, which is observed at seven irregularly spaced locations on a line. However, the results of Handcock (1989) and Mardia and Watkins (1989), plus the experience of this author, suggest that multiple modes are extremely rare in practice for covariance functions within the Matérn class, such as the exponential function, and for datasets of the size typical of most applications. In any case, a reasonable practical strategy for determining whether a local maximum obtained by an iterative algorithm is likely to be the unique global maximum is to repeat the algorithm from several widely dispersed starting values.

For some further details on the historical development of spatial analysis, we refer to Mardia (2007) and the conversations with Mardia in Mukhopadhyay (2015a, pp. 28–29, 36–37) and Mukhopadhyay (2015b, pp. 72–73).

B.4 Frequentist vs. Bayesian Inference

Another question concerns the frequentist approach vs. Bayesian approach; we tend to agree with Cox (2006, p. 197):

> Much of this book has involved an interplay between broadly frequentist discussion and a Bayesian approach, the latter usually involving a wider notion of the idea of probability. In many, but by no means all, situations numerically similar answers can be obtained from the two routes.

Cox (2006, p. 196) also argues:

> For the last 15 years or so, i.e. since about 1990, interest has focused instead on applications, especially encouraged by the availability of software for

> Markov chain Monte Carlo calculations, in particular on models of broadly hierarchical type. Many, but not all, of these applications make no essential use of the more controversial ideas on personalistic probability and many can be regarded as having at least approximately a frequentist justification.

At the same time, there is a strong synergy between the two approaches. The frequentist approach often focuses on the development of tractable models where some analytic understanding is available. On the other hand, a Bayesian treatment can lead to more realistic models and can help to account more carefully for uncertainty, especially when there are a large number of parameters.

Here are some general thoughts on the value of statistical modeling, taken from Speed (2007).

George Box: "All models are wrong, some models are useful."

Basil Rennie: "Every model embodies a half-truth, and as one of our wiser politicians once remarked, half-truths are like half-bricks, they are better because they carry further."

We finish with a quote from Mardia and Gilks (2005) related to holistic statistics:

> Through our brief account, we have identified three themes. First, statistics should be viewed in the broadest possible way for scientific explanation or prediction of any phenomenon. Second, the future of statistics lies in a holistic approach to interdisciplinary research. Third, a change of attitude is required by statisticians — a paradigm shift — for the subject to go forward.

References and Author Index

Abend, K., Hartley, T.J., and Kanal, L.N. (1965). Classification of binary random patterns. *IEEE Transactions on Information Theory* IT-11: 538–544. (p. 151)

Abramowitz, M. and Stegun, I.A. (1964). *Handbook of Mathematical Functions*. New York: Dover. (pp. 40, 44, 71, and 195)

Adler, R.J. (1981). *The Geometry of Random Fields*. New York: Wiley. (pp. xxii, 50, 74, 82, and 99)

Adler, R.J. and Taylor, J. (2007). *Random Fields and Geometry*. New York: Springer. (p. 74)

Agterberg, F.P. (2004). Georges Matheron — Founder of Spatial Statistics. *Earth Sciences History* 23: 325–334. (p. 347)

Anselin, L. (1988). *Spatial Econometrics: Methods and Models*. Dordrecht: Kluwer Academic Publishers.

Araujo, A. and Giné, E. (1980). *The Central Limit Theorem for Banach Valued Random Variables*. New York: Wiley. (pp. 64 and 65)

Baddeley, A. (2001). Georges Matheron (1930–2000). *Bulletin of the International Statistical Institute, 53rd Session Proceedings, Tome LIX, Book 1*, pp. 529–532. (p. 347)

Banerjee, S. and Gelfand, A.E. (2010). Multivariate spatial process models. In: *Handbook of Spatial Statistics* (ed. A.E. Gelfand, P.J. Diggle, M. Fuentes, and P. Guttorp), 495–515. Boca Raton, FL: CRC Press. (p. 288)

Banerjee, S., Carlin, B.P., and Gelfand, A.E. (eds.) (2015). *Hierarchical Modeling and Analysis for Spatial Data*. 2e. New York: Chapman and Hall/CRC. (pp. xxiii, 47, 287, and 292)

Beran, J. (1994). *Statistics for Long-Memory Processes*. New York: Chapman and Hall. (p. 58)

Berger, J.O., De Oliveira, V., and Sansó, B. (2001). Objective Bayesian analysis of spatially correlated data. *Journal of the American Statistical Association* 96 (456): 1361–1374. (p. 287)

Berlinet, A. and Thomas-Agnan, C. (2004). *Reproducing Kernel Hilbert Spaces in Probability and Statistics*. Boston, MA: Kluwer. (pp. xxii, 269 and 275)

Spatial Analysis, First Edition. John T. Kent and Kanti V. Mardia.

Besag, J.E. (1972). Nearest-neighbour systems and the auto-logistic model for binary data. *Journal of the Royal Statistical Society, Series B* 34: 538–544. (pp. 127 and 151)

Besag, J.E. (1974). Spatial interaction and the statistical analysis of lattice systems (with discussion). *Journal of the Royal Statistical Society, Series B* 36: 192–236. (pp. xxii, 127, 145, and 147)

Besag, J. (1975). Statistical analysis of non-lattice data. *Journal of the Royal Statistical Society. Series D (The Statistician)* 24: 179–195. (p. 127)

Besag, J. (1981). On a system of two-dimensional recurrence equations. *Journal of the Royal Statistical Society, Series B* 43: 302–309. (p. 127)

Besag, J. and Kooperberg, C. (1995). On conditional and intrinsic autoregressions. *Biometrika* 82: 733–746. (p. 127)

Besag, J. and Mondal, D. (2005). First order intrinsic autoregressions and the de Wijs process. *Biometrika* 92: 909–920. (pp. 127, 214, and 330)

Besag, J.E. and Moran, P.A.P. (1975). On the estimation and testing of spatial interaction in Gaussian lattice processes. *Biometrika* 62: 555–562. (pp. 127 and 219)

Boggio, T. (1905). Sur la funzioni di Green d'ordine *m*. *Rendiconti — Circolo Matematico di Palermo* 20: 97–135. (p. 294)

Bookstein, F.L. (1989). Principal warps: thin-plate splines and the decomposition of deformations. *IEEE Transactions on Pattern Analysis and Machine Intelligence* PAMI-11: 567–585. (p. xxiii and 276)

Bookstein, F.L. (1992). *Morphometric Tools for Landmark Data: Geometry and Biology*. Cambridge University Press. (p. 276)

Brook, D. (1964). On the distinction between the conditional probability and the joint probability approaches in the specification of nearest-neighbour systems. *Biometrika* 51: 481–483. (pp. 142 and 155)

Caragea, P.C. and Smith, R.L. (2007). Asymptotic properties of computationally efficient alternative estimators for a class of multivariate normal models. *Journal of Multivariate Analysis* 98: 1417–1440. (p. 181)

Chilés, J.-P. and Delfiner, P. (2012). *Geostatistics: Modeling Spatial Uncertainty*, 2e. Hoboken, NJ: Wiley. (pp. xxiii, 61, 69, and 168)

Christakos, G. (1992). *Random Field Models in Earth Sciences*. San Diego, CA: Academic Press. (p. 61)

Cliff, A.D. and Ord, J.K. (1981). *Spatial Processes — Models and Applications*. London: Pion. (p. xxii)

Clifford, P. (1990). Markov random fields in statistics. In: *Disorder in Physical Systems. A Volume in Honour of John M. Hammersley* (ed. G.R. Grimmett and D.J.A. Welsh), 19–32. Oxford: Clarendon Press. (p. 145)

Cordeiro, G.M. and Klein, R. (1994). Bias correction in ARMA models. *Statistics and Probability Letters* 19: 169–176. (p. 339)

Cox, D.R. (2006). *Principles of Statistical Inference*. Cambridge University Press. (p. 352)

Cox, D.R. and Miller, H.D. (1965). *The Theory of Stochastic Processes*. London: Methuen. (p. 46)

Cox, D.R. and Snell, E.J. (1968). A general definition of residuals. *Journal of the Royal Statistical Society, Series B* 30: 248–275. (p. 339)

Cramér, H. and Leadbetter, M.R. (1967). *Stationary and Related Stochastic Processes*. New York: Wiley. (p. 35)

Cressie, N. (1989). Geostatistics. *American Statistician* 43: 197–202. (p. 347)
Cressie, N. (1990). The origins of kriging. *Mathematical Geology* 22: 239–252. (pp. 231 and 347)
Cressie, N.A.C. (1993). *Statistics for Spatial Data*, revised edition. New York: Wiley. (pp. xxii, xxiii, 161, 168, 212, and 292)
Cressie, N. and Burden, S. (2015). Evaluation of diagnostics for hierarchical spatial statistical models. In: *Geometry Driven Statistics* (ed. I.L. Dryden and J.T. Kent), 241–259. Chichester: Wiley. (p. 168)
Cressie, N. and Lahiri, S.N. (1993). The asymptotic distribution of REML estimators. *Journal of Multivariate Analysis* 45: 217–233. (p. 190)
Cressie, N. and Moores, M.T. (2021). Spatial statistics, arXiv. (p. 347)
Cressie, N. and Wikle, C.K. (2011). *Statistics for Spatio-Temporal Data*. Hoboken, NJ: Wiley. (pp. 291 and 292)
Cressie, N.A., Gotway, C., and Grondona, M. (1990). Spatial prediction from networks. *Chemometrics and Intelligent Laboratory Systems* 7: 251–271. (p. 297)
Davies, S. and Hall, P. (1999). Fractal analysis of surface roughness by using spatial data (with discussion). *Journal of the Royal Statistical Society, Series B* 61: 3–37. (p. 195)
Davies, R., Twining, C., and Taylor, C. (2008). *Statistical Models of Shape*. London: Springer. (p. 295)
Davis, J.C. (1973). *Statistics and Data Analysis in Geology*. New York: Wiley. (pp. 3, 5, 166, and 176)
de Wijs, H.J. (1951). Statistics of ore distributions, Part 1. Frequency distribution of assay values. *Geologie en Mijnbouw* 13: 365–375. (p. 99)
Diggle, P.J. (1997). Spatial and longitudinal data analysis: two histories with a common future? In: *Modelling Longitudinal and Spatially Correlated Data* (ed. T.G. Gregoire, D.R. Brillinger, P.J. Diggle et al.), 387–402. New York: Springer-Verlag. (p. 347)
Diggle, P.J. (2010). Historical introduction. In: *Handbook of Spatial Statistics* (ed. A.E. Gelfand, P.J. Diggle, M. Fuentes, and P. Guttorp), 3–14. Boca Raton, FL: CRC Press. (p. 347)
Diggle, P.J. and Giorgi, E. (2019). *Model-based Geostatistics for Global Public Health: Methods and Applications*. London: Chapman and Hall/CRC Press. (pp. xxii, xxiii and 288)
Diggle, P.J. and Ribeiro, P.J. Jr. (2007). *Model-Based Geostatistics*. New York: Springer. (pp. xxii, xxiii, 47, 176, 180, and 286)
Diggle, P.J., Tawn, J.A., and Moyeed, R.A. (1998). Model based geostatistics (with discussion). *Applied Statistics* 47: 299–350. (p. 286)
Dowd, P.A. (1982). Lognormal kriging — the general case. *Mathematical Geology* 14: 475–499. (p. 285)
Dryden, I.L. and Mardia, K.V. (2016). *Statistical Shape Analysis, with Applications in R*, 2e. Chichester: Wiley. (p. xxiii and 276)
Einstein, A. and Infeld, L. (1938). *The Evolution of Physics*. Cambridge University Press. (p. 347)

Emery, X., Arroyo, D., and Porcu, E. (2016). An improved spectral turning-bands algorithm for simulating stationary vector Gaussian random fields. *Stochastic Environmental Research and Risk Assessment* 30: 1863–1873. (p. 61)
Erdelyi, A. (ed.) (1954). *Tables of Integral Transforms*, vol. II. New York: McGraw-Hill. (pp. 40 and 68)
Feller, W. (1966). *An Introduction to Probability Theory and Its Applications*, vol. II. New York: Wiley. (pp. 36, 37, 39, 41, 45, 53, 81, and 82)
Feller, W. (1968). *An Introduction to Probability Theory and Its Applications*, vol. I, 3e. New York: Wiley. (pp. 92 and 107)
Fernique, X. (1978). Continuité et théorème centrale limite pour les transformées de Fourier des mesures aléatoires du second ordre. *Zeitschrift für Wahrscheinlichkeitstheorie un verwandte Gebiete* 42: 57–66. (p. 64)
Finkenstadt, B., Held, L., and Isham, V. (eds.) (2007). *Statistical Methods for Spatial-Temporal Systems*. Boca Raton, FL: Chapman and Hall/CRC. (p. 292)
Fonanella, L. Ippoliti, L. and Mardia, K.V. (2005). Exploring Spatio-Temporal Variability by Eigen-Decomposition Techniques. Proceeding of Italian Statistical Society, CLEUP scarl, Padova, 85–96. (p. 291)
Fraser, G. (1957). A problem in the analysis of geophysical data. *Geophysics* 22: 309–344. (p. 20)
Fuller, W.A. (1996). *Introduction to Statistical Time Series*, 2e. New York: Wiley. (pp. 115 and 227)
Gaetan, C. and Guyon, X. (2010). *Spatial Statistics and Modeling*. New York: Springer. (p. 212)
Gelfand, A.E. and Ghosh, S. (2013). Hierarchical modeling. In: *Bayesian Theory and Applications* (ed. P. Damien, P. Dellaportas, N.G. Polson, and D.A. Stephens), pp. 33–49 Oxford: Oxford University Press,. (pp. 262, 286, and 287)
Gel'fand, I.M. and Shilov, G.E. (1964). *Generalized Functions: Properties and Operations*, vol. 1. New York: Academic Press. (pp. 74 and 87)
Gel'fand, I.M. and Shilov, G.E. (1968). *Generalized Functions: Spaces of Fundamental and Generalized Functions*, vol. 2. New York: Academic Press. (pp. 74 and 87)
Gel'fand, I.M. and Vilenkin, N.Y. (1964). *Generalized Functions: Applications of Harmonic Analysis*, vol. 4. New York: Academic Press. (pp. 74, 89, 92, and 94)
Gelfand, A.E., Diggle, P.J., Fuentes, M., and Guttorp, P. (eds.) (2010). *Handbook of Spatial Statistics*. Boca Raton, FL: CRC Press. (pp. xxiii, 287 and 292)
Georgii, H.O. (1988). *Gibbs Measures and Phase Transitions, de Gruyter Studies in Mathematics* 9. Berlin: Walter de Gruyter. (pp. 132 and 148)
Gneiting, T. (2002). Nonseparable, stationary covariance functions for space-time data. *Journal of the American Statistical Association* 97: 590–600. (pp. 291 and 292)
Gneiting, T. and Guttorp, P. (2010). Continuous parameter spatio-temporal processes. In: *Handbook of Spatial Statistics* (ed. A.E. Gelfand, P.J. Diggle, M. Fuentes, and P. Guttorp), 427–436. Boca Raton, FL: CRC Press. (p. 291)
Gneiting, T., Genton, M.G., and Guttorp, P. (2007). Geostatistical space-time models, stationarity, separability and full symmetry. In: *Statistical Methods for*

Spatial-Temporal Systems (ed. B. Finkenstadt, L. Held, and V. Isham), 151–175. Boca Raton, FL: Chapman and Hall/CRC. (pp. 292 and 293)

Golub, G.H. and Van Loan, C.F. (1989). *Matrix Computations*, 2e. Baltimore, MD: Johns Hopkins University Press. (p. 61)

Goodall, C. and Mardia, K.V. (1994). Challenges in Multivariate Spatio-temporal Modelling. Invited Session "Spatial-Temporal Model in Environmental Statistics" Proceedings XVII International Biometric Conference, *Hamilton, Ontario, Canada*, Vol. 1, 1–17. (p. 292)

Gradshteyn, I.S. and Ryzhik, I.M. (1980). *Table of Integrals, Series, and Products*, 2e. London: Academic Press. (pp. 71, 110, 111, 224, and 228)

Green, P.J. and Silverman, B.W. (1994). *Nonparametric Regression and Generalized Linear Models: A Roughness Penalty Approach*. London: Chapman and Hall. (p. 269)

Grenander, U. and Miller, M.I. (2007). *Pattern Theory: From Representation to Inference*. Oxford: Oxford University Press. (p. xxiii)

Gu, C. (2002). *Smoothing Spline ANOVA Models*. New York: Springer-Verlag. (p. 269)

Guttorp, P. and Gneiting, T. (2006). Studies in the history of probability and statistics XLIX on the matérn correlation. *Biometrika* 93 989–995. (p. 45)

Guyon, X. (1982). Parameter estimation for a stationary process on a *d*-dimensional lattice. *Biometrika* 69: 95–105. (pp. 213 and 222)

Handcock, M.S. (1989). Inference for spatial Gaussian random fields when the objective is prediction. PhD thesis. Department of Statistics, University of Chicago. (p. 352)

Handcock, M.S. and Stein, M.L. (1993). A Bayesian analysis of kriging. *Technometrics* 35: 403–410. (pp. 262 and 287)

Harville, D.A. (1977). Maximum likelihood approaches to variance component estimation and to related problems. *Journal of the American Statistical Association* 72: 320–340. (p. 179)

Hastie, T., Tibshirani, R., and Friedman, J. (2009). *The Elements of Statistical Learning: Data Mining, Inference, and Prediction*, 2e. New York: Springer-Verlag. (p. xxii)

Helson, H. and Lowdenslager, D. (1958). Prediction theory and Fourier series in several variables. *Acta Mathematica* 99: 165–202. (p. 137)

Huang, C., Yao, Y., Cressie, N., and Hsing, T. (2009). Multivariate intrinsic random functions for cokriging. *Mathematical Geosciences* 41: 887–904. (pp. 288 and 290)

Isaaks, E.H. and Srivastava, R.M. (1989). *An Introduction to Applied Geostatistics*. Oxford University Press. (p. 15)

Istas, J. and Lang, G. (1997). Quadratic variations and estimation of the local Hölder index of a gaussian process. *Annales de l'institut Henri Poincaré (B) Probabilités et Statistiques* 33: 407–436. (p. 194)

Joshi, S.C. and Miller, M.I. (2000. Landmark matching via large deformation diffeomorphisms. *IEEE Transactions on Image Processing* 9: 1357–1370. (p. 276)

Journel, A.G. and Huijbregts, C.J. (1978). *Mining Geostatistics*. London: Academic Press. (pp. 288 and 290)

Kalaitzis, A.A. and Lawrence, N.D. (2011). A simple approach to ranking differentially expressed gene expression time courses through Gaussian process regression. *BMC Bioinformatics* 12: 180. (p. 266)

Kanevski, M. and Maignan, M. (2004). *Analysis and Modelling of Spatial Environmental Data*. Lausanne: EPFL Press. (p. xxii)

Kaufman, C. and Shaby, B. (2013). The role of the range parameter for estimation and prediction in geostatistics. *Biometrika* 100: 473–484. (p. 193)

Kent, J.T. (1989). Continuity properties for random fields. *Annals of Probability* 17: 1432–1440. (pp. 51 and 194)

Kent, J.T. and Mardia, K.V. (1988). Spatial classification using fuzzy membership models. *IEEE Transactions on Pattern Analysis and Machine Intelligence* PAMI-10: 659–671. (p. 132)

Kent, J.T. and Mardia, K.V. (1994). The link between kriging and thin plate splines. In: *Probability, Statistics and Optimization: A Tribute to Peter Whittle* (ed. F.P. Kelly), 325–339. Chichester: Wiley. (pp. 241 and 292)

Kent, J.T. and Mardia, K.V. (1996). Spectral and circulant approximations to the likelihood for stationary Gaussian randomfields. *Journal of Statistical Inference and Planning* 50: 379–394. (p. 213)

Kent, J.T. and Mardia, K.V. (2002). Modelling strategies for spatial-temporal data. In: *Spatial Cluster Modelling* (ed. A.B. Lawson and D.G.T. Denison), 213–226. Chapman and Hall/CRC. (p. 291)

Kent, J.T. and Wood, A.T.A. (1995). Estimating the Fractal Dimension of a Locally Self-Similar Gaussian Process Using Increments. *Statistics Research Report SSR 034-95*. Australian National University. (p. 195)

Kent, J.T. and Wood, A.T.A. (1997). Estimating the fractal dimension of a locally self-similar Gaussian process using increments. *Journal of the Royal Statistical Society, Series B* 59: 679–699. (p. 194)

Kent, J.T., Mardia, K.V., Morris, R.J., and Aykroyd, R.G. (2001). Functional models of growth for landmark data. In: *Proceedings in Functional and Spatial Data Analysis* (ed. K.V. Mardia and R.G. Aykroyd), 109–115. Leeds University Press. (p. 291)

Kent, J.T., Mohammadzadeh, M., and Mosammam, A.M. (2011). The dimple in Gneiting's spatial-temporal covariance model. *Biometrika* 98: 489–494. (pp. 291 and 293)

Kingman, J.F.C. (1963). Random walks with spherical symmetry. *Acta Mathematica* 109: 11–53. (p. 37)

Kitandis, P.K. (1983). Statistical estimation of polynomial generalized covariance functions and hydrologic applications. *Water Resources Research* 19: 909–921. (p. 189)

Kitandis, P.K. (1986). Parameter uncertainty in estimation of spatial functions: Bayesian analysis. *Water Resources Research* 22: 499–507. (p. 262)

Kitandis, P.K. (1991). Orthonormal residuals in geostatistics: model criticism and parameter estimation. *Mathematical Geology* 23: 741–758. (p. 182)

Kotz, S. and Nadarajah, S. (2004). *Multivariate t Distributions and their applications*. Cambridge: Cambridge University Press. (p. 110)

Krige, D.G. (1951). A statistical approach to some basic mine valuation problems on the Witwatersrand. *Journal of the Chemical, Metallurgical and Mining Society of South Africa* 52: 119–139. (p. xxii)

Krige, D.J. (1976). Some basic considerations in the application of geostatistics to the valuation of ore in South African gold mines. *Journal of the South African Institute of Mining and Metallurgy* 76: 383–391. (p. 28)

Künsch, H. (1987). Intrinsic autoregressions and related models on the two-dimensional lattice. *Biometrika* 74: 517–524. (pp. 131 and 213)

Laslett, G.M., McBratney, A.B., Pahl, P.J., and Hutchinson, M.F. (1987). Comparison of several spatial prediction methods for soil PH. *Journal of Soil Science* 38: 325–341. (p. 21)

Lawson, A.B. and Denison, D.G.T. (eds.) (2002). *Spatial Cluster Modelling*. Chapman and Hall/CRC. (p. xxii and 292)

Le, N.D. and Zidek, J.V. (1992). Interpolation with uncertain spatial covariances: a Bayesian alternative to kriging. *Journal of Multivariate Analysis* 43: 351–374. (p. 262)

Lindgren, F., Rue, H., and Lindström, J. (2011). An explicit link between Gaussian fields and Gaussian Markov random fields: the stochastic partial differential equation approach (with discussion). *Journal of the Royal Statistical Society, Series B* 73: 423–498. (pp. 127, 284, and 295)

Ma, C. (2003). Families of spatio-temporal stationary covariance models. *Journal of Statistical Planning and Inference* 116: 489–501. (p. 293)

Mandelbrot, B.B. (1982). *The Fractal Geometry of Nature*. New York: W.H. Freeman and Co. (pp. xxii, 24 and 50)

Mandelbrot, B.B. and van Ness, J.W. (1968). *Fractional Brownian motion, fractional noises and applications* 10: 422–437. (p. 194)

Mantoglue, A. and Wilson, J.L. (1982). The turning bands method for simulation of random fields using line generation by a spectral method. *Water Resources Research* 18: 1379–1394. (p. 65)

Mardia, K.V. (1980). Some statistical inference problems in Kriging II: theory. *Proceedings 26th International Geology Congress Sciences de la Terre: Advances in Automatic Processing and Mathematical Models in Geology, Series "Informatique Geologie"*, Volume 15, Paris, pp. 113–131. (pp. 160, 348, and 349)

Mardia, K.V. (1988). Multi-dimensional multivariate Gaussian Markov random fields with application to image processing. *Journal of Multivariate Analysis* 24: 265–284. (pp. 127 and 148)

Mardia, K.V. (1990). Maximum likelihood estimation for spatial models. In: *Proceedings Spatial Statistics: Past, Present and Future, Institute of Mathematical Geology* (ed. D.A. Griffith), 203–225. Michigan Document Service. (pp. 160, 173, 176 and 341)

Mardia, K.V. (ed.) (1991). *The Art of Statistical Science: A Tribute to G.S. Watson*. Wiley. (pp. 348 and 349)

Mardia, K.V. (ed.) (1994). *Statistics and Images*, Volume 2. Oxford: Carfax. (p. xxiii)

Mardia, K.V. (2007). Should geostatistics be model-based? *Proceedings of the IAMG 2007 Conference: Geomathematics and GIS Analysis of Resources, Environment and Hazards*, Beijing, China, pp. 4–9. (p. 352)

Mardia, K.V. (2011). Discussion to an explicit link between Gaussian fields and Gaussian Markov random fields: the stochastic partial differential equation approach by F. Lindgren, H. Rue and J. Lindstrom. *Journal of the Royal Statistical Society, Series B* 73: 481–482. (p. 181)

Mardia, K.V. and Gilks, W. (2005). Meeting the statistical needs of 21st-century science. *Significance* 2: 162–165. (p. 353)
Mardia, K.V. and Gill, C.A. (1982). Some statistical inference problems in kriging I: numerical applications. *Journal of the Geological Society, Computer Applications in Geology* 14 56–72. (pp. 160 and 349)
Mardia, K.V. and Goodall, C. (1993). Spatio-temporal analyses of multivariate environmental monitoring data. In: *Multivariate Environmental Statistics* (ed. G.P. Patil and C.R. Rao), 347–386. Amsterdam: Elsevier. (pp. 285, 288, 290, 292, 297, 298 and 299)
Mardia, K.V. and Jupp, P.E. (2000). *Directional Statistics*. Chichester: Wiley. (pp. 57 and 67)
Mardia, K.V. and Kanji, G.K. (eds) (1993). *Statistics and Images*, Volume 1. Oxford: Carfax. (p. xxiii)
Mardia, K.V. and Marshall, R.J. (1982). Maximum Likelihood Fitting of a Spatial Covariance in the Linear Model. *Department of Statistics Research Report*. University of Leeds. (pp. 333 and 349)
Mardia, K.V. and Marshall, R.J. (1984). Maximum likelihood estimation of models for residual covariance in spatial regression. *Biometrika* 71: 135–146. (pp. 160, 173, 188, 190, 197, 333, 348, 349, and 351)
Mardia, K.V. and Pardo-Iguzquiza, E. (2006). Numerical Results on Vecchia (1988) Approximated MLE for Large Spatial Data Sets. *Department of Statistics Research Report 2*. University of Leeds. (p. 7)
Mardia, K.V. and Watkins, A.J. (1989). On multimodality of the likelihood in the spatial linear model. *Biometrika* 76: 289–295. (pp. 45, 70, 160, 172, 351, and 352)
Mardia, K.V., Kent, J.T., and Bibby, J.M. (1979). *Multivariate Analysis*. London: Academic Press. (pp. 38, 43, 46, 170, and 310)
Mardia, K.V., Kent, J.T., Hughes, G., and Taylor, C.C. (2010). Maximum likelihood estimation using composite likelihoods for closed exponential families. *Biometrika* 96: 975–982. (p. 218)
Mardia, K.V., Kent, J.T., and Walder, A.N. (1991). Statistical shape models in image analysis. In: *Computing Science and Statistics: Proceedings of the 23rd Symposium on the Interface* (ed. E.M. Keramidas), 550–555. Fairfax Station, VA: Interface Foundation of North America. (p. 241)
Mardia, K.V., Kent, J.T., Little, J., and Goodall, C.R. (1996). Kriging and splines with derivative information. *Biometrika* 83: 207–221. (p. 259)
Mardia, K.V., Goodall, C., Redfern, E.J., and Alonso, F.J. (1998). The Kriged Kalman filter (with discussion). *Test* 7: 217–252. (p. 291)
Mardia, K.V., Angulo, J.M., and Goitía, A. (2006a). Synthesis of image deformation strategies. *Image and Vision Computing* 24: 1–12. (p. 276)
Mardia, K.V., Bookstein, F.L., Kent, J.T., and Meyer, C.R. (2006b). Intrinsic random fields and image deformations. *Journal of Mathematical Imaging and Vision* 26: 59–71. (p. 265)
Marechal, A. and Serra, J. (1970). Random kriging. In: *Geostatistics, A Colloquium* (ed. D.F. Merriam), 91–112. New York: Plenum Press. (pp. 6, 8, and 166)

Marshall, R.J. and Mardia, K.V. (1985). Minimum norm quadratic estimation of components of spatial covariance. *Mathematical Geology* 17: 517 525. (pp. 192 and 197)
Martin, R. (1979). A subclass of lattice processes applied to a problem in planar sampling. *Biometrika* 66: 209–217. (pp. 70, 140, 161, and 162)
Matérn, B. (1960). Spatial Variation. *Technical Report*. Stockholm: Statens Skogsforskningsinstitut. (pp. xxii, 43 and 69)
Matérn, B. (1986). *Spatial Variation*. Berlin: Springer-Verlag. (pp. xxii and 43)
Matheron, G. (1962). Precision of exploring a stratified formation by boreholes by boreholes with rigid spacing — application to a bauxite deposit. In: *International Symposium on Mining Research*, vol. 1 (ed. G.B. Clark), 407–423. Oxford: Pergamon Press. (p. 27)
Matheron, G. (1963). Principles of geostatistics. *Economic Geology* 58: 1246–1266. (p. xxii)
Matheron, G. (1965). *Les Variables Régionalisées et leur Estimation. Une Application de la Théorie des Fonctions Aléatoires aux Sciences de la Nature*. Paris: Masson. (p. 45)
Matheron, G. (1971). *The Theory of Regionalized Variables, and its Applications*, Vol. Fasc. No. 5. Fontainebleau: Cahiers du Centre de Morphologie Mathématique. (p. 77)
Matheron, G. (1973). The intrinsic random functions and their applications. *Advances in Applied Probability* 5: 439–468. (p. 65)
Matheron, G. (1975). *Random Sets and Integral Geometry*. New York: Wiley. (p. 80)
McBratney, A.B., Webster, R., and Burgess, T.M. (1981). The design of optimal sampling schemes for local estimation and mapping of of regionalized variables—I: theory and method. *Computers & Geosciences* 7: 331–334. (p. 297)
Mercer, W.B. and Hall, A.D. (1911). The experimental error of field trials. *Journal of Agricultural Science* 4: 123–456. (pp. 24 and 161)
Mondal, D. (2018). On edge correction of conditional and intrinsic autoregressions. *Biometrika* 105: 447–454. (pp. 214 and 330)
Mosamam, A. and Kent, J.T. (2010). Semi-reproducing kernel Hilbert spaces, splines and increment kriging. *Journal of Nonparametric Statistics* 22: 711–722. (p. 275)
Mukhopadhyay, N. (2015a). A conversation with Kanti Mardia. In: *Geometry Driven Statistics* (ed. I.L. Dryden and J.T. Kent), 3–58. Chichester: Wiley. (p. 352)
Mukhopadhyay, N. (2015b). A conversation with Kanti Mardia: Part 2. In: *Geometry Driven Statistics* (ed. I.L. Dryden and J.T. Kent), 59–85. Chichester: Wiley. (p. 352)
Pardo-Igúzquiza, E. and Dowd, P.A. (1997). AMLE3D: A computer program for the statistical inference of covariance parameters by approximate maximum likelihood estimation. *Computers & Geosciences* 7: 793–905. (p. 181)
Pardo-Igúzquiza, E., Mardia, K.V., and Chica-Olmo, M. (2008). MLMATERN: A computer program for maximum likelihood inference with the spatial Matérn covariance model. *Computers & Geosciences* 35: 1139–1150. (p. xxiv)
Rao, C.R. (1973). *Linear Statistical Inference and its Applications*, 2e. New York: Wiley. (p. 192)

Rasmussen, C.E. and Williams, C.K.I. (2006). *Gaussian Processes for Machine Learning*. Cambridge, MA: The MIT Press. (pp. xxii, xxiii, 47, 266, 267, and 352)

Rathbun, S. (1998). Spatial modelling in irregularly shaped regions: kriging estuaries. *Environmetrics* 9: 109–129. (p. 233)

Ribeiro, P.J. Jr. and Diggle, P.J. (2001). geoR: A package for geostatistical analysis. *R-News* 1(2): 15–18. (pp. xxiii, xxiv, 15 and 16)

Ripley, B.D. (1981). *Spatial Statistics*. New York: Wiley. (p. 349)

Ripley, B.D. (1988). *Statistical Inference for Spatial Processes*. New York: Wiley. (pp. 176 and 351)

Rue, H. (2001). Fast sampling of Gaussian Markov random fields. *Journal of the Royal Statistical Society, Series B* 63: 325–338. (p. 61)

Rue, H. and Held, L. (2005). *Gaussian Markov Random Fields*. Boca Raton, FL: Chapman and Hall/CRC. (pp. 134 and 332)

Rue, H. and Held, L. (2010). Discrete spatial varation. In: A. E. Gelfand, P. J. Diggle, M. Fuentes, and P. Guttorp (eds). *Handbook of Spatial Statistics*. Boca Raton: CRC Press, pp. 171–200. (p. 194)

Sahu, S.K. (2022). *Bayesian Modeling of Spatio-Temporal Data with R*. New York: Chapman and Hall/CRC. (p. 292)

Sahu, S.K. and Mardia, K.V. (2005). A Bayesian kriged Kalman model for short-term forecasting of air pollution levels. *Applied Statistics* 54: 223–244. (p. 291)

Sahu, S.K., Jona Lasini, G., Orasi, A. and Mardia, K.V. (2005). A Comparison of Spatio-Temporal Bayesian Models for Reconstruction of Rainfall Fields in a Cloud Seeding Experiment. *Journal of Mathematics and Statistics* 1: 273–281. (p. 291)

Sambasivan, R., Das, S., and Sahu, S.K. (2020). A Bayesian perspective of statistical machine learning for big data. *Computational Statistics* 35: 893–930. (p. 266)

Schabenberger, O. and Gotway, C.A. (2005). *Statistical Methods for Spatial Data Analysis*. Boca Raton, FL: Chapman and Hall/CRC. (p. xxii and 262)

Schoenberg, I.J. (1938). Metric spaces and completely monotone functions. *Annals of Mathematics* 39: 811–841. (p. 37)

Schölkopf, B. and Smola, A. (2002). *Learning with Kernels: Support Vector Machines, Regularization, Optimization, and Beyond*. Cambridge: MIT Press. (pp. 266 and 275)

Sherman, M. (2011). *Spatial Statistics and Spatio-Temporal Data*. Chichester: Wiley. (p. 47)

Smith, H.F. (1938). An empirical law describing heterogeneity in the yields of agricultural crops. *Journal of Agricultural Science* 28: 1–23. (p. 60)

Smith, R.L. (2001). Environmental statistics. version 5. unpublished. https://rls.sites.oasis.unc.edu/postscript/rs/envnotes.pdf (accessed 7 September 2021). (pp. 297 and 337)

Sonka, M., Hlavac, V., and Boyle, R. (2013). *Image Processing, Analysis and Machine Vision*, 4e. Springer. (p. xxiii)

Speed, T. (2007). Terence's stuff: model skeptics. *IMS Bulletin* 36: 11. (p. 353)

Srivastava, A. and Klassen, E.P. (2016). *Functional and Shape Data Analysis*. Springer. (p. 276)

Steel, M.F.J. and Fuentes, M. (2010). Non-Gaussian and nonparametric models for continuous spatial data. In: *Handbook of Spatial Statistics* (ed. A.E. Gelfand, P.J. Diggle, M. Fuentes, and P. Guttorp), 149–167. Boca Raton, FL: CRC Press. (p. 287)

Stein, M.L. (1999). *Interpolation of Spatial Data: Some Theory for Kriging*. New York: Springer. (pp. 168, 193, and 351)

Stein, M.L. (2005). Statistical methods for regular monitoring data. *Journal of the Royal Statistical Society, Series B* 67: 667–687. (pp. 292 and 294)

Stein, M.L., Chi, Z., and Welty, L.J. (2004). Approximating likelihoods for large spatial data sets. *Journal of the Royal Statistical Society, Series B* 66: 275–296. (pp. 181 and 182)

Taylor, G.I. (1938). The spectrum of turbulence. *Proceedings of the Royal Society of London, Series A* 164: 476–490. (p. 293)

Upton, G.J.G. and Fingleton, B. (1985). *Spatial Data Analysis by Example: Point Pattern and Quantitative Data*, vol. 1. Chichester: Wiley. (p. xxii)

Upton, G.J.G. and Fingleton, B. (1989). *Spatial Data Analysis by Example: Categorical and Directional Data*, vol. 2. Chichester: Wiley. (p. xxii)

van Lieshout, M.-C. (2019). *Theory of Spatial Statistics: A Concise Introduction*. Boca Raton, FL: Chapman and Hall/CRC. (p. xxiii)

Vecchia, A.V. (1988). Estimation and model identification for continuous spatial processes. *Journal of the Royal Statistical Society, Series B* 50: 297–312. (p. 180)

Wackernagel, H. (2003). *Multivariate Geostatistics*, 3e. Berlin: Springer-Verlag. (p. 288)

Wahba, G. (1990). *Spline Models for Observational Data*. Philadelphia, PA: Society for Industrial and Applied Mathematics. (pp. xxii, 245, 269, 275, and 292)

Warnes, J.J. and Ripley, B.D. (1987). Problems with likelihood estimation of covariance functions of spatial Gaussian processes. *Biometrika* 74: 640–642. (pp. 172, 176, 351, and 352)

Watkins, A.J. (1987). Some aspects of statistical inference in the spatial linear model. PhD thesis. University of Leeds. (pp. 341)

Watkins, A.J. and Mardia, K.V. (1992). Maximum likelihood estimation and prediction mean square error in the spatial linear model. *Journal of Applied Statistics* 19: 49–59. (p. 263)

Watson, G.S. (1969). Trend Surface Analysis and Spatial Correlation. *Technical report*. Department of Statistics, John Hopkins University. https://apps.dtic.mil/sti/pdfs/AD0699163.pdf (accessed 7 September 2021). (p. 348)

Watson, G.S. (1972). Trend surface analysis and spatial correlation. *Geological Society of America Special Paper* 146: 39–46. (pp. 347 and 348)

Watson, G.S. (1984). Smoothing and interpolating by kriging and with splines. *Mathematical Geology* 16: 601–615. (p. xxii and 348)

Watson, G.S. (1986). A boy from the bush. In: *The Craft of Statistical Modelling* (ed. J.M. Gani), 43–60. New York: Springer. Also reprinted in: Mardia, K.V. (ed.) *The Art of Statistical Science*, xxix–lii. New York: Wiley. (pp. 348 and 349)

Webster, R. and Oliver, M.A. (2001). *Geostatistics for Environmental Scientists*. Chichester: Wiley. (p. xxii)

Wendland, H. (2005). *Scattered Data Approximation*. Cambridge: Cambridge University Press. (p. 47)

Whittle, P. (1954). On stationary processes in the plane. *Biometrika* 41: 434–449. (pp. xxii, 44, 212, 213, and 347)

Whittle, P. (1956). On the variation of yields variance with plot size. *Biometrika* 43: 337–343. (pp. 44, 59, and 60)

Whittle, P. (1962). Topographic correlation, power-law covariance functions, and diffusion. *Biometrika* 49: 305–314. (pp. 60 and 127)

Whittle, P. (1963). *Prediction and Regulation*, 2e. Minneapolis, MN: University of Minnesota. (p. 347)

Whittle, P. (1986). *Systems in Stochastic Equilibrium*. Chichester: Wiley. (pp. 60, 127, 212, and 291)

Wikle, C.K. and Cressie, N. (1999). A dimension-reduced approach to space-time Kalman filtering. *Biometrika* 86: 815–829. (p. 291)

Wikle, C.K., Zammit-Mangion, A., and Cressie, N. (2019). Spatio-Temporal Statistics with R. Chapman & Hall/CRC, Boca Raton, FL. (p. 292)

Wilson, A.G. (2000). *Complex Spatial Systems: The Modelling Foundations of Urban and Regional Analysis*. London: Routledge. (p. xxii)

Wood, A.T.A. and Chan, G. (1994). Simulation of stationary processes in $[0, 1]^d$. *Journal of Computational and Graphical Statistics* 3: 409–432. (p. 66)

Zhang, H. (2004). Inconsistent estimation and asymptotically equal interpolations in model-based geostatistics. *Journal of the American Statistical Association* 99: 250–261. (pp. 174 and 193)

Zhang, H. and Zimmerman, D.L. (2005). Towards reconciling two asymptotic frameworks in spatial statistics. *Biometrika* 92: 921–936. (p. 193)

Zidek, J.L. and Zimmerman, D.L. (2010). Monitoring network design. In: *Handbook of Spatial Statistics* (ed. A.E. Gelfand, P.J. Diggle, M. Fuentes, and P. Guttorp), 131–148. Boca Raton, FL: CRC Press. (p. 297)

Zimmerman, D.L. (2010). Likelihood-based methods. In: *Handbook of Spatial Statistics* (ed. A.E. Gelfand, P.J. Diggle, M. Fuentes, and P. Guttorp), 45–56. Boca Raton, FL: CRC Press. (pp. 172 and 352)

Zimmerman, D.L. and Stein, M. (2010). Classical geostatistical methods. In: *Handbook of Spatial Statistics* (ed. A.E. Gelfand, P.J. Diggle, M. Fuentes, and P. Guttorp), 29–44. Boca Raton, FL: CRC Press. (p. 167)

Index